Modern Methods
in the Analysis
and Structural Elucidation
of Mycotoxins

Modern Methods in the Analysis and Structural Elucidation of Mycotoxins

Edited by

Richard J. Cole

U. S. Department of Agriculture
Agricultural Research Service
National Peanut Research Laboratory
Dawson, Georgia

1986

ACADEMIC PRESS, INC.

Harcourt Brace Jovanovich, Publishers
Orlando   San Diego   New York   Austin
Boston   London   Sydney   Tokyo   Toronto

ACADEMIC PRESS, INC.
Orlando, Florida 32887

United Kingdom Edition published by
ACADEMIC PRESS INC. (LONDON) LTD.
24–28 Oval Road, London NW1 7DX

Library of Congress Cataloging in Publication Data

Modern methods in the analysis and structural elucidation of mycotoxins.

Includes index.
1. Mycotoxins—Analysis.   2. Molecular structure.
I. Cole, Richard J.
QP632.M9M63   1986   615.9'52925   86-8022
ISBN 0—12—179515—2 (alk. paper)

PRINTED IN THE UNITED STATES OF AMERICA

86 87 88 89   9 8 7 6 5 4 3 2 1

Contents

3   Chemical Survey Methods for Mycotoxins

Odette L. Shotwell

4   Application of Ultraviolet and Infrared Spectroscopy in the Analysis and Structural Elucidation of Mycotoxins

Charles P. Gorst-Allman

5   Application of Nuclear Magnetic Resonance to Structural Elucidation of Mycotoxins

Richard H. Cox

6   X-Ray Crystallography as a Tool for Identifying New Mycotoxins

James P. Springer

7   Application of Biosynthetic Techniques in the Structural Studies of Mycotoxins

Pieter S. Steyn and Robert Vleggaar

8   Immunoassays for Mycotoxins

F. S. Chu

13 Mass Spectra of Selected Trichothecenes

C. J. Mirocha, S. V. Pathre, R. J. Pawlosky, and D. W. Hewetson

14 Mass Spectrometry–Mass Spectrometry as a Tool for Mycotoxin Analysis

R. D. Plattner

15 Taxonomic Approaches to Mycotoxin Identification (Taxonomic Indication of Mycotoxin Content in Foods)

Jens C. Frisvad

Preface

Mycotoxins are toxic secondary metabolites of fungi that represent numerous and diverse chemical classes. Fungi from important genera, including but by no means exclusive to *Penicillium, Aspergillus,* and *Fusarium,* are capable of producing these toxins in food and feed commodities during all stages of production. Food and feed products can become contaminated prior to, during, or after harvest or during the manufacturing process. The latter can occur during the production of various fermentation products such as cheese, sausage, or other fermented foods. Much remains to be learned about the quality of these toxins occurring in our food and feed supplies. Knowledge of the presence or absence of mycotoxins in our food and feed supplies requires the combined technology of numerous disciplines. New mycotoxins must be isolated and chemically identified; analytical methodology, including sampling methods, must be developed to permit surveillance of those commodities most likely to be contaminated.

This volume presents current methods of analysis and structural elucidation of mycotoxins by recognized experts in the various disciplines. The philosophy in the various chapters of the book is to present each method initially in theoretical terms and then to review the method as it specifically applies to the analysis and/or structural elucidation of mycotoxins. The first chapter deals with screening methods for mycotoxins and toxigenic fungi. Chapters 2 and 3 are concerned with sampling and survey methods. Chapters 4 through 7 deal mostly with methods for structural elucidation, although the more modern methods, such as NMR and X-ray crystallography, have largely supplanted the older methods, such as infrared and ultraviolet spectroscopy. Ultraviolet spectroscopy can still provide valuable information concerning UV chromophores; however, most recent applications of UV are in detection systems such as for high performance liquid chromatography. Chapter 7 applies a novel approach by using biosynthetic techniques to aid in the structural elucidation of mycotoxins. Chapters 8 through 14 are primarily concerned with the analytical methods for mycotoxin analyses. Perhaps the newest approaches for analysis of mycotoxins are the enzyme-linked immunosorbent assay system (ELISA) and mass spectrometry–mass spectrometry. Finally, the last chapter approaches the mycotoxin analytical problem in taxonomic or ecological terms; that is, the knowledge of what fungi frequently occur on a particular substrate along with knowledge of what mycotoxin(s) can be produced by these fungi.

I wish to express my sincere appreciation to all the contributors for their excellent contributions and their cooperation and patience in producing this treatise.

Richard J. Cole

Modern Methods
in the Analysis
and Structural Elucidation
of Mycotoxins

1

Biological Screening Methods for Mycotoxins and Toxigenic Fungi

RICHARD J. COLE[1], **HORACE G. CUTLER**[2],
AND JOE W. DORNER[1]

[1]*U.S. Department of Agriculture, Agricultural Research Service*
National Peanut Research Laboratory
Dawson, Georgia 31742
[2]*U.S. Department of Agriculture, Agricultural Research Service*
Richard B. Russell Research Center
Athens, Georgia 30613

I. INTRODUCTION

Historically, the use of bioassays for detecting and isolating mycotoxins and toxigenic fungi has been quite successful. In the isolation and identification of

1

MODERN METHODS IN THE ANALYSIS
AND STRUCTURAL ELUCIDATION
OF MYCOTOXINS

the aflatoxins, which marked the birth of modern mycotoxicology, British scientists relied almost exclusively on a duckling bioassay to evaluate the degree of toxicity of Brazilian peanut meal and to monitor purification of the unknown toxic component(s). Since that time, numerous mycotoxins have been isolated and identified, and bioassays, using various organisms, have played a vital role in a vast majority of these cases. Today, since the advent of modern analytical methodology, chemical assays are generally preferred over biological assays in screening food, feedstuffs, and fungi for the presence of known toxins. Chemical assays are faster, more specific, and more reproducible than bioassays, but they are limited to known mycotoxins where authentic standards and analytical methodology are available. Obviously, a chemical assay cannot be implemented until a toxin is chemically characterized and appropriate methodology is developed. Therefore, chemical and biological screening methods complement each other. Hence, the greatest value of a biological screening method is generally in the initial detection and isolation of unknown mycotoxins. This, then, provides the basis for the development of subsequent chemical assays.

Before using a bioassay as a routine procedure for detecting mycotoxins, several factors should be considered. The bioassay should be capable of responding to a diverse range of mycotoxins. The sensitivity and repeatability of the bioassay must be acceptable, the cost of conducting the bioassay should be relatively low in terms of resources required, and results should be obtained in a short time (preferably less than 24 hr). The bioassay should be unambiguous. That is, the occurrence and frequency of false positives must be minimal to none.

Bioassays can be classified on the basis of the type of organisms used. These may be vertebrate animals, invertebrate animals, higher plants, and microorganisms. Obviously, vertebrate animal assays are the most dependable type for the detection of vertebrate toxins and they include those that use intact animals (domestic and laboratory animals) as well as cytological assays, skin tests, and chick embryos.

Bioassays may be highly specific or general in nature. The desirability of one type or the other depends on the particular situation. If a feed or food source is causing a disease with a particular pathological manifestation (i.e., bile duct proliferation in ducklings), a specific assay would be desirable. However, if one is screening for toxigenic fungi, a nonspecific assay may be more desirable.

Previous reviews related to screening for mycotoxins have been published (Cole, 1978; Watson and Lindsay, 1982; Patterson, 1983; Cole, 1984). It is the purpose of this chapter to highlight the past use of biological screening methods for mycotoxins and toxigenic fungi and to discuss the advantages, disadvantages, applications, and limitations of the various available bioassays as they relate to mycotoxins.

II. ANIMAL ASSAYS

A. Vertebrates

1. Intact Organisms

Most recognized mycotoxins have been initially detected and isolated with the benefit of a wide array of biological assays; however, the most successful assay systems used vertebrate animals including mice, ducklings, and chickens (Table I). In the classical studies of 1960, British scientists studying the turkey "X" syndrome noted that Peking ducklings were very sensitive to the effects of the toxic Brazilian groundnut meal (Sargeant *et al.*, 1961). They further noted that ducklings affected by this meal exhibited characteristic histological lesions in their livers, characterized by extensive and rapid proliferation of cords of cells radiating from portal tracts (bile duct proliferation). They found that orally dosing methanol extracts of this peanut meal to ducklings made it feasible to test poultry feed samples fairly quickly (3–4 days) to get an approximation of the toxic level of aflatoxins present in these meals. Even though the intoxication primarily involved turkeys, it was recognized that ducklings were more desirable as primary subjects for the assay system. This represented one of the first mycotoxin bioassays to be reported, which was utilized effectively to isolate and identify mycotoxins causing a natural intoxication (excluding ergotism). The biological activity in ducklings was subsequently correlated with the fluorescent properties of the aflatoxins, which became the basis for most present-day analytical techniques for the aflatoxins.

South African scientists have also utilized ducklings dosed *ad libitum* as a bioassay to detect new mycotoxins representing a variety of chemical groups. Two of the more significant mycotoxins detected with this assay were ochratoxin A (Van der Merwe *et al.*, 1965) and α-cyclopiazonic acid (Holzapfel, 1968) (Table I). Another similar vertebrate animal bioassay that has been used in the detection and isolation of a number of different mycotoxins utilized day-old chickens fed *ad libitum* or dosed orally via crop intubation (Kirksey and Cole, 1974) (Table I). The latter has been particularly useful for the initial detection of tremorgenic mycotoxins such as paxilline (Cole *et al.*, 1974b), verruculogen (Cole *et al.*, 1972), and the paspalitrems (Cole *et al.*, 1977). Other classes of mycotoxins detected with this bioassay were moniliformin (Cole *et al.*, 1973), viridiol group metabolites (Cole *et al.*, 1975b), and certain trichothecenes such as neosolaniol monoacetate (Lansden *et al.*, 1978) and 3'-hydroxy-HT-2 toxin (Cole *et al.*, 1981a).

Laboratory animals, particularly rats and mice, have been used successfully for the detection and isolation of several mycotoxins including several trem-

TABLE I

**Primary Bioassay System Used in the Original Detection
and Isolation of Various Mycotoxins**

Bioassay type/organism	Mycotoxin	Reference
Vertebrate animals		
Ducklings	Aflatoxins	Sargeant *et al.* (1961)
	Ochratoxin A	Van der Merwe *et al.* (1965)
	Cyclopiazonic acid	Holzapfel (1968)
Day-old chicks	Verruculogen	Cole *et al.* (1972)
	Moniliformin	Cole *et al.* (1973)
	Paxilline	Cole *et al.* (1974b)
	Verruculotoxin	Cole *et al.* (1975a)
	Neosolaniol monoacetate	Lansden *et al.* (1978)
Mice	Aflatrem	Wilson and Wilson (1964)
	Penitrem A	Wilson *et al.* (1968)
	Zearalenone	Stobb *et al.* (1962)
	Cytochalasin H	Patwardhan *et al.* (1974)
	Territrems	Ling (1976)
	Rubratoxins	Wilson and Wilson (1962)
		Townsend *et al.* (1966)
Skin test	T-2 toxin	Bamburg *et al.* (1968)
Cell cultures	Cytochalasins A, B, C, D	Rothweiler and Tamm (1966)
		Aldridge *et al.* (1967)
		Carter (1967)
Invertebrate animals		
Brine shrimp	Satratoxins	Eppley and Bailey (1973)
Plants		
Wheat coleoptile	Chaetoglobosin K	Cutler *et al.* (1980b)
Bean, corn	Malformin A	Curtis (1958)
Pea, lettuce, other plants	Diacetoxyscirpenol	Brian *et al.* (1961)
Microbiological specimens		
Bacteria		
Staphylococcus aureus	Patulin	Birkinshaw *et al.* (1943)
Bacillus subtilis, S. aureus	Chetomin	Geiger *et al.* (1944)
Fungi		
Rhizoctonia solani	Gliotoxin	Weindling and Emerson (1936)
Botrytis allii and others	Verrucarins	Brian and McGowan (1946)
B. allii	Viridin	Brian and McGowan (1945)
B. allii	Trichothecin	Freeman and Morrison (1948)

orgenic mycotoxins (Glinsukon et al., 1979; Wilson and Wilson, 1964) (Table I). Mice are small, commercially available, relatively inexpensive and, therefore, provide an excellent bioassay for the detection and isolation of mycotoxins.

A vertebrate bioassay that is highly specific for the aflatoxins (similar to bile duct proliferation in ducklings) is the production of liver carcinoma in rainbow trout fed aflatoxin-contaminated rations (Sinnhuber et al., 1977). The trout bioassay is probably one of the most sensitive assays for the aflatoxins; however, it is impractical as a screening method because it takes a minimum of 4 months to complete an assay and it is technically difficult.

The only vertebrate bioassay adopted by the Association of Official Analytical Chemists (AOAC) is the chick embryo assay (Platt et al., 1962; Verrett et al., 1964) in which test solutions are injected into the air cell of fertile chicken eggs. A measure of toxicity is mortality to the developing embryo in excess of control background mortality. The method is simple, inexpensive, and sensitive; however, it is relatively nonspecific, has the disadvantage of bypassing the normal ingestion, digestion, and absorption processes, and it is somewhat prone to false positives.

The skin test is a vertebrate bioassay that is specific for certain trichothecenes. It involves removing the hair from the back of an experimental animal such as a rabbit, rat, or guinea pig, and applying the toxin to the backskin of the animal in an appropriate solvent. This test does not detect those trichothecenes that are relatively nontoxic via oral administration. Sensitivities for this bioassay have been reported to levels of 0.05 to 0.10 μg using T-2 toxin (Wei et al., 1972). Rabbits were more sensitive to topical administration of T-2 toxin than rats or guinea pigs (Bamburg et al., 1968).

Another trichothecene bioassay involves the rejection or acceptance of drinking water by mice (Burmeister et al., 1980). Test materials are dissolved in 95% ethanol, added to distilled water, and presented to mice preconditioned on ethanol–water solution. The volume of water inbibed is compared with controls. A positive test occurs when the presence of trichothecenes causes refusal of water.

A specific vertebrate bioassay for zearalenone is based on the hyperestrogenic activity of this mycotoxin. Swine are the most sensitive of the large domestic animals, although other domestic animals affected are dairy cattle, lambs, chickens, and turkeys (Mirocha et al., 1977). Experimental animals affected by zearalenone are rats, mice, guinea pigs, and monkeys. The rat uterotopic bioassay provides a reasonably practical and specific bioassay for zearalenone if the necessary analytical capability is lacking (Mirocha et al., 1977). This involves feeding suspect feed directly to rats or applying chemical extracts of the feed to the shaved skin of rats. After a test period of 5 days the rats are sacrificed, the uterus excised, weighed, and compared to control animals. Topical application is more sensitive than oral administration in the rat.

Other vertebrate animals used in mycotoxin bioassays included fish and Japanese quail (De Waart *et al.,* 1972; Abdei and McKinley, 1968).

A major consideration in using intact animals as a bioassay is the method of administration of the toxin. Oral routes are the most dependable and valid, since they represent the most natural route. Alternative routes of administration when using intact domestic or laboratory animals are injection types such as IP, IV, or subcutaneous routes. Although these methods do have some advantages, they are not recommended as a primary screening tool for mycotoxins or mycotoxigenic fungi, since they bypass the processes that occur during ingestion, digestion, and absorption. The latter conditions can affect the toxicity or lack of toxicity of a metabolite by activation, deactivation, and selective or nonselective absorption of the metabolite. Most toxins are more active via injection when compared to oral administration; therefore, these treatments are more prone to false positives than false negatives.

Since mycotoxins as a group have quite varied solubility ranging from water (moniliformin) to lipophilic solvents (aflatoxins), another consideration in the use of intact vertebrate animals as a bioassay is the vehicle used for administration of the toxin. Ideally, administration by feeding *ad libitum,* by capsule, or in an inert carrier, such as water, would produce the least undesirable effects. However, because of solubility problems many toxins cannot be readily formulated in inert carriers and, therefore, possible adverse carrier effects must be considered in evaluating an assay method. Carriers themselves may be toxic, or predispose the animal to the effects of the toxin, or dramatically affect the absorption process.

2. Tissue and Organ Cultures

Various vertebrate tissue and organ cultures have been used as bioassay systems for detecting mycotoxins. These include tracheal organ cultures from day-old chicks (Nair *et al.,* 1970; Cardeilhac *et al.,* 1972), calf kidney cells (Juhasz and Greczi, 1964), embryonic lung cells (Legator and Withrow, 1964), human Chang liver cells (Gabliks *et al.,* 1965), mouse leukemia cells (Nakano *et al.,* 1973), rabbit reticulocytes (Ueno *et al.,* 1973), catfish cell cultures (Vosdingh and Neff, 1974), HeLa cells (Tatsuno *et al.,* 1973), human karyoblast cells (Eppley, 1975), rat kidney, liver, and muscle cells (Umeda, 1977), human fibroblasts (San and Stich, 1975), and FM3A cells (Umeda, 1977).

Cytological assays involve exposure of various concentrations of mycotoxins or test materials to various cell cultures for various periods of time. Cytotoxicity can be categorized as general toxicity or hereditable toxicity. Cytological, morphological, and biochemical methods are used to evaluate general toxicity, morphological changes, and effects on macromolecular synthesis or enzyme regulation (Umeda, 1977). Induction of DNA strand breakages, impairment of unscheduled DNA synthesis, mutagenicity, transformability, and chromosomal aberrations are used as measures of hereditable toxicity.

Advantages of these bioassays are that they are relatively available, inexpensive, and sensitive; however, they may suffer from a lack of mycotoxin specificity. Also, they are generally less desirable as a primary screening system, since they bypass normal routes of absorption, that is, the digestive system. Cell culture represents a totally different environment for absorption of toxin into cells compared to the digestive system of animals and, in the case of intact animals, the toxin may be transported to an activation site (liver) prior to affecting particular organ systems within the animal.

B. Invertebrates

Brown *et al.* (1968) described a bioassay for aflatoxin B_1 using brine shrimp (*Artemia salina* L.). They demonstrated that 0.5 µg/ml aflatoxin B_1 resulted in mortality of over 60% of the shrimp in artificial seawater. In a subsequent study, the larvae were also sensitive to ochratoxin A and to an acetone extract of *Fusarium tricinctum* (Brown, 1969).

Harwig and Scott (1971) described the use of brine shrimp as a screening system for toxigenic fungi and some known mycotoxins. In these studies brine shrimp showed a wide range of sensitivities to different mycotoxins. It was concluded that most of the known mycotoxins tested could have been discovered with the brine shrimp bioassay. However, it was further noted that a number of filtrates and extracts were found to possess toxicity to shrimp that could not be explained by the presence of known mycotoxins.

The brine shrimp bioassay was used successfully as a monitor in the isolation of five naturally occurring trichothecene mycotoxins produced by *Stachybotrys atra* (Eppley and Bailey, 1973). The sensitivity range of the bioassay for 10 naturally occurring trichothecenes was 0.04–0.4 µg/ml, with roridin H, verrucarin A, and T-2 toxin being the most toxic (Eppley, 1974). Several factors such as temperature, time, and age of the shrimp increased the sensitivity of the brine shrimp to the toxin. The system has the advantage that brine shrimp eggs are commercially available and live shrimp can be obtained from eggs within 1 to 2 days.

The major disadvantage of the brine shrimp bioassay is that some naturally occurring fatty acids possess toxicity toward the shrimp comparable to that of several known mycotoxins (Curtis *et al.,* 1974; Prior, 1979). Chemical and biological testing of diagnostic specimens indicated that the brine shrimp bioassay identified 2 of 4 chemically positive samples and 59 of 135 chemically negative samples (Prior, 1979); thus, unidentified larvicidal compounds present in normal feedstuffs gave a high percentage of false positive bioassay results when compared to chemical analyses for these mycotoxins.

A method using the ciliated protozoan *Colpidium campylum* has been described for the detection of toxic fungal metabolites (Dive *et al.*, 1978). Known mycotoxins were added to cultures at a concentration range of 0.1 to 10 μg/ml. Toxicity of mycotoxins to cultures was evaluated by means of generation number after 43 hr at 20°C. The study showed that this bioassay was a convenient system for testing some mycotoxins. However, aflatoxin B_1, ochratoxin B, and sterigmatocystin were not detected.

Aflatoxin B_1 was toxic to groups of brown planaria (*Dugesia tigrina*) exposed to solutions containing 3.125 μg/ml toxin (LD_{50} in 2 days) (Llewellyn, 1973). Greater toxicity was observed when cut portions of planaria were used; the head section was more sensitive than tail sections. Even when low expense and limited care required to use *D. tigrina* were considered, the sensitivity of the species to aflatoxin B_1 warrants that the organism should be given limited consideration as a bioassay organism for aflatoxin B_1.

In a bioassay using the mollusk *Bankia setacea,* aflatoxin B_1 inhibited cell cleavage in fertilized eggs without preventing fertilization or nuclear division (Townsley and Lee, 1967). This resulted in multinucleate eggs whereas controls multiplied to multicellular larvae. The bioassay required minimum technique and training and required 2–4 hr to complete when observing egg division and 18 hr for swimming larvae. The sensitivity to concentrations of 0.05 μg/ml aflatoxin B_1 were observed; however, the concentration required could be reduced below 0.05 μg when the reaction was observed with the aid of a low-magnification (70×) microscope.

Hayes and Wyatt (1970) demonstrated growth inhibition of *Volvox aureus* and *Tetrahymena pyriformis* by pure rubratoxin B at concentrations of less than 50 μg/ml. *Tetrahymena pyriformis* strain w was also tested as a bioassay organism for rubratoxins A and B by Wyatt and Townsend (1974). Cultures of *T. pyriformis* were incubated with concentrations of 0.05 to 150 μg/ml of rubratoxins A and B. Results showed an approximate linear range of inhibition from 2.6 to 25 μg/ml for rubratoxin A and a range of 8.5 to 30 μg/ml for rubratoxin B, with an ED_{50} of 8.75 μg/ml for A and 20.5 μg/ml for B. This assay system, using pure toxins, had the advantage of being quantitative, rapid, and repeatable. However, because of the influence of contaminating compounds from biological extracts, further investigation of the system is needed before it is recommended for use.

Hayes *et al.* (1976) evaluated *T. pyriformis* ASM as a bioassay organism for purified citrinin. Citrinin had an inhibitory effect on growth of *T. pyriformis* as measured by optical density of cell culture and cell count. It also caused a marked shift to smaller cell size at concentrations of 25 to 100 μg/ml. The lower limit of detection for citrinin was between 1 and 5 μg/ml.

Another study utilizing *T. pyriformis* as an assay organism for aflatoxin B_1, ochratoxin A, and rubratoxin B showed that rubratoxin B was the most inhibitory mycotoxin tested against growth of this protozoan (Hayes *et al.*, 1974). *Tetrahymena pyriformis* was sensitive to 1 to 5 µg/ml of rubratoxin B. Aflatoxin B_1 and ochratoxin A were considerably less toxic at comparable concentrations.

Bijl *et al.* (1981) studied six different assay methods using *T. pyriformis* strain GL, *Colpidium campylum, Artemia salina,* and *Daphnia magna* strains for the detection of diacetoxyscirpenol and trichothecin. They concluded that the ciliated protozoa were more sensitive to these toxins than crustaceans. *Artemia salina* was preferred if simplicity was a major consideration. However, if sensitivity was preferred, synchronous *Tetrahymena pyriformis* was recommended.

Amoebae of the *Limax* type were exposed to aflatoxin B_1 at concentrations from 0.2 to 50.0 µg/ml. Cysts that formed in the presence of the toxin degenerated shortly after formation followed by lysis of the cells (Bauer and Leistner, 1969). However, there were no detectable harmful effects of the toxin on vegetative amoebae cells, and generation time was only slightly reduced at concentrations up to 2 µg/ml toxin. The amoebae test was proposed as a potentially useful bioassay for aflatoxin B_1, since the test was technically not difficult, was rapid, and showed specificity for the cyst stage.

Insects are a group of invertebrates that are relatively complex, available, and highly dependent on chemicals for growth, development, and behavior. Therefore, insects should provide an excellent bioassay system for the detection of mycotoxins. Matsumura and Knight (1967) evaluated the effects of aflatoxin on development processes of certain insect species. Insects used were larvae and adults of the mosquito (*Aedes aegypti*) and adults of the housefly (*Musca domestica*) and the fruit fly (*Drosophila melanogaster*). In all cases, the number of eggs produced and the percentages of eggs hatched were reduced. It was not possible to determine the exact number of insects that completed the process of copulation and oviposition. The percentage of hatchable eggs produced was definitely affected in all cases. The concept of mycotoxins as a promising source of stable chemosterilants was noted.

Subsequent studies examined the effects of nine representative mycotoxins on silkworm larvae (Murakoshi *et al.*, 1973). The second and fourth instar were fed *ad libitum* with an artificial diet containing different concentrations of purified mycotoxins for 10 days. The larvae (mostly fourth instar) were highly susceptible to aflatoxins B_1, G_1, and ochratoxin A. They were relatively susceptible to sterigmatocystin, diacetoxyscirpenol, and β-nitropropionic acid and were less susceptible to kojic acid, patulin, and penicillic acid. Based on the results of this study it was recommended that further investigation be conducted to evaluate the use of silkworm as agents for a mycotoxin bioassay.

Moore and Davis (1983) described a sensitive bioassay utilizing the bertha armyworm (*Mamestra configurata*), which is closely related to *Heliothis virescens*. The studies used dietary T-2 toxin to evaluate the bioassay. They concluded that this method may be useful either in testing contaminated feed samples or in determining toxicity of crude extracts containing mixtures of mycotoxins without further purification.

A similar study by McMillian *et al.* (1980), using three lepidopterous insects, was concerned with the effects of dietary aflatoxin B_1 on growth and development. Similar conclusions were presented; however, the aflatoxin B_1 was added to the surface of the insect diet and as a result the relative amounts of mycotoxin ingested could not be determined with any degree of accuracy.

C. Immunoassays

Immunoassays represent a highly specific, sensitive assay for the detection and quantification of mycotoxins and have been developed for various mycotoxins. A detailed discussion of immunoassays appears in Chapter 8 and therefore will not be discussed here.

III. PLANT ASSAYS

Bioassays using higher plants to detect mycotoxins either in the pure state or in extracts of fungally contaminated commodities have had limited application. But there are individual cases in which the results have been excellent and the need for a convenient plant bioassay to detect fungal toxins exists. Plant bioassays are also important in the study of structure–activity relationships of biologically active compounds and in the evaluation of mycotoxins and other metabolites for their potential use in the regulation of plant growth. Although the use of plant bioassays in these latter areas has been the subject of many reports, the scope of this chapter will be limited to the use of plant assays in the screening of fungi for mycotoxins.

Plant bioassays offer an alternative to vertebrate and invertebrate assays because of the relative ease of preparation and low cost. It is desirable in any study designed to detect unknown toxins that the ultimate bioassay be the target organism of the causal agent (except, of course, in the case of humans!). However, plant assays may offer an economical alternative before final testing in vertebrate systems. An example is the use of the Ames test to detect potential mutagens (Ames *et al.*, 1973). Furthermore, the final evaluation as to degree of vertebrate toxicity may be carried out at an industrial laboratory, or in conjunction with a veterinary college or diagnostic laboratory on a collaborative basis. This especially applies to situations involving the isolation and identification of new fungal toxins where the bioassay is used as a primary monitoring system to detect the biological activity of crude extracts, partially purified fractions, and, ultimately, the pure toxin.

A. Wheat Coleoptile

During the past decade evidence has accumulated to show that the etiolated wheat coleoptile bioassay is a useful tool for detecting mycotoxins. The assay is based on a combination of the coleoptile straight-growth test (Hancock *et al.*, 1964) and a method developed for the oat first internode assay (Nitsch and Nitsch, 1956). Specifically, what seeds (*Triticum aestivum* L., cv. Wakeland) (Cutler, 1984) are sown on moist sand in plastic trays and covered with a fine layer of sand. Aluminum foil is used to seal the trays, which are placed in the dark at 22° ± 1°C for 4 days. Following germination and growth, the etiolated seedlings are removed from the sand, and the roots and caryopses are cut away from the shoots. All coleoptile preparations are done under a green safelight at 540 nm. The apex of each shoot is placed in a Van der Wiej guillotine, and the first 2 mm is removed with a razor blade and discarded. The next 4-mm section is cut for use in the bioassay. Generally, 10 sections are placed in a test tube containing 2 ml phosphate–citrate buffer at pH 5.6 plus 2% sucrose (Nitsch and Nitsch, 1956), along with the material to be tested. The test tubes containing the test material and coleoptile sections are placed in a roller-tube apparatus that rotates at 0.25 rpm, for approximately 18 hr. Controls containing only buffer and sucrose are included. At completion of the assay, the sections are placed on a glass sheet, put into a photographic enlarger, and the images (magnified 3×), are measured, recorded (Cutler and Vlitos, 1962), and the data statistically analyzed (Kurtz *et al.*, 1965). The metabolites are tested in logarithmic concentrations from 10^{-3} to 10^{-6} M, although with certain highly active metabolites the range may be increased to 10^{-7} and 10^{-8} M. The effects produced by toxins in the bioassay are a measure of inhibition of coleoptile growth (elongation). Although the other phases of plant growth (i.e., cell division, cell maturation) may be affected, the elongation phase is measured for effects on plant growth.

While few mycotoxins have been initially detected using the coleoptile assay, many toxins have been tested in the assay and found to be active. Chaetoglobosin K, however, is one example of a mycotoxin that was first detected using the wheat coleoptile bioassay (Cutler *et al.*, 1980b). The toxin, a member of the cytochalasin group of mycotoxins, was very inhibitory to the growth of wheat coleoptiles at concentrations ranging from 10^{-3} to 10^{-7} M (Table II). The vertebrate toxicity of chaetoglobosin K was subsequently confirmed using day-old chicks and found to have an LD_{50} of between 25 and 62.5 mg/kg (Cutler *et al.*, 1980b).

Other cytochalasins that have been tested in the coleoptile assay include cytochalasin H (Wells *et al.*, 1976), deacetylcytochalasin H (Cole *et al.*, 1981c), epoxycytochalasin H (Cole *et al.*, 1982), epoxydeacetylcytochalasin H (Cole *et al.*, 1982), and cytochalasin B (Wells *et al.*, 1981).

Cytochalasin H (also named kodocytochalasin-1) was originally isolated using

TABLE II

Inhibition of Wheat Coleoptiles (*Triticum aestivum* L., cv. Wakeland), Relative to Controls, Obtained with Various Mycotoxins[a]

Compound	Inhibition (%) based on molar concentration					Reference
	10^{-3}	10^{-4}	10^{-5}	10^{-6}	10^{-7}	
Trichothecenes						
Trichodermin	91	80	15	0	0	Cutler and Le Files (1978)
Neosolaniol monoacetate	87	69	45	16	—	Lansden *et al.* (1978)
3′-Hydroxy-HT-2 toxin	100	81	19	0	—	Cole *et al.* (1981a)
15-Acetoxy-T-2 tetraol	60	0	0	0	—	Cole *et al.* (1981a)
3′-Hydroxy-T-2 triol	96	54	37	0	—	Cole *et al.* (1981a)
HT-2 Toxin	100	62	14	0	—	Cole *et al.* (1981a)
T-2 Toxin	100	100	50	39	—	Cole *et al.* (1981a)
Roridin A	100	81	40	0	—	Cutler and Jarvis (1985)
Isororidin E	100	100	60	0	—	Cutler and Jarvis (1985)
Baccharinol B4	91	82	60	19	0	Cutler and Jarvis (1985)
Verrucarin A	100	97	80	41	37	Cutler and Jarvis (1985)
Verrucarin J	100	100	82	61	43	Cutler and Jarvis (1985)
Trichoverrin B	100	81	62	40	16	Cutler and Jarvis (1985)
Cytochalasins						
Cytochalasin A	44	15	0	0	—	Cox *et al.* (1983)
Cytochalasin B	65	62	0	0	—	Wells *et al.* (1981)
Cytochalasin E	16	0	0	0	—	Cox *et al.* (1983)
Epoxycytochalasin H	81	63	46	0	—	Cole *et al.* (1982)
Deacetylepoxycytochalasin H	100	81	33	0	—	Cole *et al.* (1982)
Cytochalasin H	84	80	76	10	—	Wells *et al.* (1976)
Deacetylcytochalasin H	100	82	40	16	—	Cole *et al.* (1981c)

a mouse bioassay and shown to have an LD_{100} of 2.5 mg/kg (Patwardhan *et al.*, 1974). The toxin was subsequently isolated from cultures of *Phomopsis* sp. with the day-old chick bioassay and then assayed with wheat coleoptiles (Wells *et al.*, 1976). Coleoptiles were significantly inhibited ($P < 0.01$) relative to controls at concentrations ranging from 10^{-3} to 10^{-6} M (Table II). In addition, all coleoptile segments were distorted and curved, which is a unique effect of the cytochalasins tested on wheat coleoptiles. The effects of other cytochalasins on coleoptile growth are reported in Table II.

The most active mycotoxins tested in the wheat coleoptile bioassay were the 12,13-epoxytrichothecenes. Trichodermin, 4β-acetoxy-12,13-epoxy-Δ⁹-trichothecene, a fungal metabolite from *Trichoderma viride,* has been shown to be relatively nontoxic to mice. The LD_{50} was 1 g/kg orally and 0.5–1 g/kg by subcutaneous injection (Bamburg and Strong, 1971). The compound has a very

TABLE II

(*Continued*)

Compound	Inhibition (%) based on molar concentration					Reference
	10^{-3}	10^{-4}	10^{-5}	10^{-6}	10^{-7}	
7-Acetoxycytochalasin H	100	80	17	0	—	Cox *et al.* (1983)
2,7-Diacetoxycytochalasin H	51	39	0	0	—	Cox *et al.* (1983)
2,7,18-Triacetoxycytochalasin H	0	0	0	0	—	Cox *et al.* (1983)
Chaetoglobosins						
Chaetoglobosin K	100	84	62	43	14	Cutler *et al.* (1980b)
Other compounds						
Citreoviridin	100	21	0	0	—	Cole *et al.* (1981b)
Oosporein	100	44	14	0	—	Cole *et al.* (1974a)
Moniliformin	(57)	$(24)^b$			—	Cole *et al.* (1973)
Wentilactone A	81	81	48	0	—	Dorner *et al.* (1980)
Wentilactone B	100	81	0	0	—	Dorner *et al.* (1980)
Toxin A[c]	100	0	0	0	—	Cole *et al.* (1976)
Toxin B[c]	0	0	0	0	—	Cole *et al.* (1976)
Aflatoxin B_1	43	0	0	0	—	H. G. Cutler (unpublished)
Aflatoxin G_1	41	0	0	0	—	H. G. Cutler (unpublished)
Aflatoxin G_2	37	0	0	0	—	H. G. Cutler (unpublished)
Desmethoxyviridiol	100	100	96	58	—	Cole *et al.* (1975b)
Kotanin	0	0	0	0	—	Cutler *et al.* (1979)
Diplodiol	0	0	0	0	—	Cutler *et al.* (1980a)
Austin	0	0	0	0	—	Chexal *et al.* (1976)

[a] All whole numbers shown represent significant inhibition ($P < 0.01$).
[b] Tested at 200 and 20 ppm, respectively.
[c] Isolated from *Penicillium islandicum* as isomers ($C_{16}H_{15}O_2N_2Cl$).

limited effect on bacterial growth but inhibits the growth of several fungi, especially *Candida albicans* (Godtfredsen and Vangedal, 1965). Wheat coleoptiles were significantly inhibited 91, 80, and 50% at 10^{-3}, 10^{-4}, and 10^{-5} *M*, respectively ($P < 0.01$) (Cutler and Le Files, 1978).

In a recent study the macrocyclic trichothecenes, including roridin A, isororidin E, baccharinol, verrucarins A and J, and trichoverrin B, were tested in the coleoptile assay (Cutler and Jarvis, 1985). Roridin A and isororidin E significantly inhibited coleoptiles at 10^{-3}, 10^{-4}, and 10^{-5} *M*. In addition, verrucarins A and J and trichoverrin B inhibited coleoptiles down to 10^{-7} *M* (Table II). Of all the growth inhibitory or phytotoxic substances tested in the coleoptile bioassay, these three latter compounds rank among the most potent. Other trichothecenes and their activity in the coleoptile assay are included in Table II.

Many other chemically diverse mycotoxins have been tested in the wheat coleoptile bioassay. Aflatoxins B_1, G_1, and G_2 inhibited growth 43, 31, and 37%, respectively, at $10^{-3} M$ (P. G. Vincent, 1969 personal communication; J. W. Dorner, 1983 personal communication). Because the inhibitory or phytotoxic response is only elicited by relatively high concentrations of these particular mycotoxins, the utility of the bioassay would be limited as a means to evaluate these three toxins in an "unknown" situation, for example, where the presence of these toxins is suspected but physical and biological vertebrate assays are not available.

The inhibitory effects of other mycotoxins on wheat coleoptiles are presented in Table II. Kotanin (Büchi et al., 1971, 1977), diplodiol (Cutler et al., 1980a), and austin (Chexal et al., 1976) are included as three examples of mycotoxins that did not elicit a response in the coleoptile assay. There are also compounds, such as orlandin (Cutler et al., 1979), that were inhibitory to coleoptile growth but subsequently shown to be nontoxic in a vertebrate assay.

All these examples point out the attractiveness and the limitations of the wheat coleoptile assay as a primary screening method for mycotoxins. In the case of the 12,13-epoxytrichothecenes, it can be concluded that any of these metabolites could have been detected in, and isolated from, a crude extract using wheat coleoptiles as the primary bioassay system. Furthermore, the generally high specific activity of the trichothecenes in this assay makes them readily detectable. Assuming an average molecular weight of 350 for this class of compounds and an average detectability at $10^{-6} M$, the amount of metabolite needed to give a response is 700 ng (the amount contained in 2 ml of test solution). But there is no rule stating that less than 2 ml cannot be used, so that in critical assays it is possible to use 1 ml or 350 ng.

There are certain parameters to be considered during testing with the wheat coleoptile assay. One is the possibility for obtaining false positives when using coleoptiles as a primary screening method. Obviously in these cases, corroboration of toxicity must be established in a vertebrate animal assay. In addition, all mycotoxins do not inhibit coleoptile growth. However, because the assay is simple, relatively nonspecific, repeatable, inexpensive, carried out in a short time, and generally very sensitive, it certainly has a place in biological screening for mycotoxins and warrants further consideration.

B. Intact Plants

As in the examples of mycotoxin detection with the wheat coleoptile bioassay, the utility of intact plants as a primary screening method for mycotoxins has been limited. Obviously, this type of assay is most often used in the detection and isolation of plant growth regulators. However, since many mycotoxins also affect plant growth, an intact plant assay could be used as a mycotoxin-screening assay with the awareness of the need for confirming positive results in a vertebrate animal assay.

Intact plant bioassays may be used independently or included to complement results obtained with the wheat coleoptile bioassay (Cutler and Le Files, 1978). However, they have two major disadvantages in being used as a primary screening assay in lieu of the coleoptile assay. First, larger quantities of metabolite and more elaborate facilities are required. Second, the responses elicited are more diverse and may range from necrosis and chlorosis to stunting or mild malformations of plant parts. These effects may occur in concert or individually. In addition, some genera may be susceptible to toxins while others may give no visible responses (see below).

Among the first mycotoxins to be detected and isolated using an intact plant assay was malformin A. This toxin and related compounds are produced by isolates of *Aspergillus niger, Aspergillus ficum, Aspergillus awamori,* and *Aspergillus phoenicis.* Malformin A was initially isolated as a plant growth regulator using assays consisting of bean plants (*Phaseolus vulgaris*) and corn seedlings (*Zea mays*). Effects observed in these assays were malformations of the stems and petioles of bean plants and curvature of the roots of corn (Curtis, 1958). Many years later, malformin A was shown to have an LD_{50} in male mice of 3.1 mg/kg (IP) (Yoshizawa *et al.,* 1975).

Diacetoxyscirpenol is a member of the 12,13-epoxytrichothecene group of mycotoxins that is acutely toxic to mice and produces dermal toxicity in the rabbit, mouse, and guinea pig (Cole and Cox, 1981). But it was originally studied as a phytotoxin by Brian *et al.* (1961). Their work surveyed the action of diacetoxyscirpenol against several plants. Concentrations of 2.73×10^{-5} M inhibited stem growth and scorched leaves of two pea varieties (*Pisum sativum* L., cv. Meteor and cv. Improved Pilot). The same effects were observed in two lettuce varieties (*Lactuca sativa* L., cv. May King and cv. All-the-year-round) and in winter tares. However, there was stimulation of cress root growth at concentrations ranging from 1.37×10^{-6} to 2.73×10^{-8} M. Surprisingly, beet root, carrot, mustard, and wheat were unaffected by diacetoxyscirpenol at 2.73×10^{-5} $M,$ indicating a phytotoxic selectivity. Diacetoxyscirpenol was also inhibitory to 5-mm pea stem sections (*Pisum sativum* L., cv. Meteor) cut from the upper internodes of seedlings and floated in phosphate buffer (pH 6.1) containing 1% sucrose and 10 μg/ml indole-3-acetic acid.

While few other mycotoxins have been initially detected using intact plant assays, many mycotoxins have been subjected to assays using such plants as bean (*Phaseolus vulgaris* L., cv. Black Valentine), corn (*Zea mays* L., cv. Norfolk Market white), and tobacco (*Nicotiana tabacum* L., cv. Hicks). Results of many of these studies have been detailed in a review (Cutler, 1984).

The higher plant bioassays have most certainly not been pursued to their limits as detectors of mycotoxins, and in most cases, these have been used as secondary, not primary, systems for assaying toxins. In other areas of research the plant growth regulators have been shown to elicit different responses depending on the chosen cultivar within a genus and species (H. G. Cutler, unpublished observations; Dashek and Llewellyn, 1983). That is, the results are genetically dependent. Similarly, the types of tissue used in bioassays have been limited and there are acute questions that need to be resolved if plant assays are to be developed specifically for mycotoxin studies. Among these are the following. Is the right genotype being used for the assay? Is the assay exclusively genotypically dependent for the mycotoxin being assayed? Is the most sensitive tissue being utilized (root, shoot, meristem, parenchyma)? Does light or temperature affect the response of the assay? Can certain cofactors enhance the toxic response? Can highly specific tissue cultures be developed for mycotoxin studies?

The role of plant bioassays for detecting mycotoxins is still undeveloped. Given the proper investigations, the future for these assays seems promising.

C. Pollen

The germination and growth of pollen tubes in tobacco pollen (*Nicotiana sylvestris*) has been used as a semiquantitative and confirmatory test for T-2 toxin and diacetoxyscirpenol (Siriwardana and Lafont, 1978). As little as 10 ng/ml of T-2 toxin inhibited pollen germination by 24%, and 20 ng/ml of diacetoxyscirpenol inhibited germination 26%. In both cases, 200 ng/ml of toxin was needed to inhibit pollen germination 100%. The same bioassay was very sensitive to a mixture of verrucarins A and B, although the exact amounts necessary to induce inhibition were not determined. By comparison, 0.01 of the amount necessary to produce an LD_{50} in chicken embryos caused inhibition of pollen germination. The assay is relatively simple and therefore does not require extensive training to perform. Tobacco plants are easy to grow, and flower throughout the year providing a consistent source of pollen. But so far the application of the bioassay has had only limited success with the 12,13-epoxytrichothecenes. Aflatoxins, ochratoxins, and sterigmatocystin have not given a consistent inhibitory response.

Another bioassay used lily pollen (*Lilium longiflorum* L., cv. Ace) for the detection of mycotoxins (Dashek *et al.*, 1982). The pollen was collected from plants and stored at 4°C for 4 weeks. Aflatoxin G_1 was incorporated into 10-ml quantities of Dickinson's medium at rates of 0.234, 0.468, and on up to 15.000 μg/ml, and 20 mg of pollen was added to each tube. Treatments were incubated at 24° ± 2°C for 2, 4, and 8 hr, then 0.1 ml was removed and fixed in test tubes containing 40% aqueous formaldehyde. Microscopic examination of the pollen was carried out, and both germination counts and pollen tube lengths were recorded.

Insofar as lily pollen germination was concerned, the results changed as a function of time. At 2 hr, 0.468 μg/ml elicited a 3.32-fold increase, but at 0.937, 1.875, 3.750, 7.500, and 15.000 μg/ml there was 20, 13, 61, 68, and 63% inhibition, respectively. At 4 hr after treatment with aflatoxin G_1 at 0.234, 0.937, and 1.875 μg/ml, there was increased germination 1.68, 1.65, and 1.74 times greater than controls. The rate at 7.500 μg/ml was the same as the controls and at 13.200 μg/ml there was 14% inhibition. At 8 hr following initiation of treatments, all the concentrations up to 7.500 μg/ml had increased germination and, more noticeably, the 0.480 μg/ml treatment was twice that of the controls. But at 7.500 and 15.000 μg/ml, inhibition was 13 and 30% compared to controls.

From these results it is possible that a specific assay for aflatoxin G_1 using logarithmic concentrations of the toxin and plotting the percentage germination of lilly pollen could be devised. Detailed studies would have to be initiated to determine the least state of purity of aflatoxin G_1 that could be detected in the assay and to determine what effect the impurities would have on the results. However, the response on pollen tube elongation was not consistent, and it is very doubtful that an assay based on pollen tube elongation could be developed for this plant species.

IV. MICROBIOLOGICAL ASSAYS

A. Bacteria and Fungi

A variety of bioassays have been used to demonstrate the inhibitory effects of mycotoxins on the growth of microorganisms. Initial surveys were carried out on 329 microorganisms by Burmeister and Hesseltine (1966) to determine which were sensitive to mycotoxins, and these included 30 genera of bacteria, 34 genera of fungi, 4 genera of algae, and 1 protozoan. These were tested with a crude extract from *Aspergillus flavus* that contained the following amounts of aflatoxins per 5 μg: 1.19 μg of B_1, 0.32 μg of B_2, 0.34 μg of G_1, and 0.045 μg of G_2. The assay medium was specifically tailored for each organism. *Bacillus megaterium* NRRL B-1368 and a strain of *Bacillus brevis* were the most sensitive to the crude mixture, but neither the fungi nor the protozoan were inhibited.

The effects of aflatoxin B_1 on *B. megaterium* NRRL B-1368 have been incorporated into a rapid confirmatory test for the AOAC using the impregnated-disk assay (Clements, 1968a,b). Among other features, which include growth inhibition, the method states that cells obtained from the margins of the inhibitory zones were, upon staining, aberrant in form and were similar to the elongated cells noted in *Escherichia coli* cultures by Wragg *et al.* (1967). Full details of the experimental procedure were given, with the caveat that while the disk assay with *B. megaterium* was considerably faster than the chick embryo assay (Buckelew *et al.*, 1972), it was not as sensitive. Aflatoxin B_1 (1.0 µg) induced inhibition within 6 to 7 hr following treatment, and very obvious clear zones were visible in 15 to 18 hr. The ochratoxins were detected in the bioassay, but the sensitivity was low.

There was evidence to suggest that certain substances gave false positives in bacterial assays. While it was shown that certain mycotoxins inhibited the growth of *B. megaterium,* other compounds, which may be toxic substances, but which have not yet been found being produced by fungi or bacteria, also inhibited growth (Buckelew *et al.*, 1972). For example, pentachlorophenol, a known inhibitor of fungal growth, was equally as inhibitory to *B. megaterium* as ochratoxin B at 1 mg/ml and 0.1 mg/ml (20 µg/disk) in dimethyl sulfoxide (DMSO). Dicoumarol was as toxic as aflatoxin B_1; both *d*- and *l*-usnic acid were more toxic than aflatoxin B_1; polyporenic acid was almost as active as aflatoxin B_2, ochratoxin A (polyporenic acid C) was inactive, and 18β-glycyrrhetic acid was less active than aflatoxins B_1 and B_2.

Buckelew *et al.* (1972) reported that most of the toxins that had a conjugated carbonyl system (e.g., the α,β-unsaturated lactones, pyrones, and quinones) were inhibitory to *B. megaterium.* Exceptions to this general rule were rubratoxin B, santonin, and kojic acid, which have carbonyls conjugated to double bonds. 4-Hydroxycoumarins with extensive side chains and aromatic substitutions at the C-3 position were active. Compounds known to be uncouplers of oxidative phosphorylation were also inhibitory.

Aflatoxin B_1 was tested on a number of *Bacillus* spp. including *B. cereus, B. anthracoides, B. megaterium, B. mycoides, B. subtilis, Brevibacterium* sp., *Pseudomonas aeruginosa, Streptococcus lactis,* and *E. coli* (Eka, 1972). *Bacillus mycoides* and *Brevibacterium* sp. were sensitive to aflatoxin B_1 at rates of 0.05 to 0.45 mg/ml in paper disk assays. At rates of 0.1 mg/ml aflatoxin B_1 retarded the uptake of oxygen by *Bacillus mycoides,* but the effects were nonlinear with time. The same concentration inhibited cell division, and giant cells formed, indicating the possible binding of aflatoxin B_1 to DNA and concomitant inhibition of DNA polymerase (Clifford and Rees, 1966; Harley *et al.*, 1969).

A turbidimetric assay using *B. megaterium* to evaluate the presence of aflatoxins B_1, B_2, G_1, and G_2 was devised by Viitasalo and Gyllenberg (1968). The amount of aflatoxin that was detected turbidimetrically was relatively high compared to other methods. The quantities of aflatoxin necessary to produce no growth were 100 µg/ml for B_1, 250 µg/ml for B_2, 200 µg/ml for G_1, and 250 µg/ml for G_2.

Assays using *Bacillus cereus mycoides* LSU appeared to have overcome the limitation in detecting the ochratoxins as seen with the *B. megaterium* assay system (Broce *et al.*, 1970). However, it should be noted that some of the pigments produced by *Aspergillus ochraceous* were toxic to the test organism, and accordingly the system appeared to be limited to detecting pure samples of ochratoxins A and B.

The most extensive testing of an array of mycotoxins with a bacterial species has been carried out challenging *Bacillus thuringiensis* with known mycotoxins (Boutibonnes and Auffray, 1977; Boutibonnes, 1979, 1980; Boutibonnes *et al.*, 1983). Examination of the morphological changes and comparison with those obtained for penicillin G (a bacteriostatic or lethal effect) and mitomycin C (a genotoxic agent that produced cell abnormalities including cell volume enlargement) was included in the study.

Nearly all of 47 mycotoxins tested in the *B. thuringiensis* bioassay induced some effect and represented toxins from diverse chemical groups. These included aflatoxins B_1, B_2, G_1, G_2, aflatoxicol, citrinin, citreomontanin, citreoviridin, destruxin, diacetoxyscirpenol, diplodiol, diplosporin, fumitoxin A, fusarenon, fusicoccin, gliotoxin, isochroman, kojic acid, luteoskyrin, moniliformin, nigeron, nivalenol, ochratoxin A, patulin, paxilline, penicillic acid, peptide B, peptide BB 384, PR toxin, roseotoxin B, roridin A, rubratoxin, rugulosin, sterigmatocystin, T-2 toxin, T-2 tetraol, HT-2 toxin, toxin X, verrucarin A, verruculogen, versicolorin A, versiconal acetate, versiconol acetate, viomellin, xanthomegnin, zearalenol, and zearalenone. Of these, aflatoxicol, diplosporin (pH 7.0 and 8.0), nivalenol, peptide B (pH 8.0), T-2 tetraol, toxin X (pH 7.0 and 8.0), verrucarin A (pH 7.0 and 8.0), verruculogen, versicolorin A (pH 8.0), and versiconol acetate did not produce an inhibitory zone. Those compounds which were inactive at the higher pH values were active at pH 6.0. The bacterial length was significantly increased at pH 7.0 (>10 μm relative to controls of 2.7 ± 0.4 μm) with aflatoxins B_1, B_2, G_1, G_2, ochratoxin A, penicillic acid, roseotoxin B, sterigmatocystin, versiconol acetate, zearalenol, and zearalenone. The internal standard mitomycin increased bacterial cell length, but penicillin G did not.

It is evident from the above data that antibiotics may produce similar responses to mycotoxins in bacterial assays and, therefore, confusion may result in attempting to assay extracts of fungi or bacteria in which one, or the other, or both may be present.

Bacillus subtilis spores have been used as a sensitive bioassay for patulin (Reiss, 1975a). Their germination was inhibited by patulin but not by aflatoxins B_1 and G_1, rubratoxin B, or diacetoxyscirpenol. The assay comes in a kit and is therefore simple to use. Test plates, supplied in aluminum foil pouches, are designed for testing the presence of antibiotics in milk and cream (Bactiastrip S.A.). Patulin produced a 32-mm corrected inhibition zone with 100 μg/disk, a 20-mm zone with 10 μg/disk, and a 4-mm zone with 1 μg/disk. With the exception of aflatoxin B_1 (5-mm zone with 100 μg/disk), the other toxins tested had no inhibitory effect.

A disk bioassay using *B. subtilis* as the target organism has been devised to test for penicillic acid (Olivigni and Bullerman, 1978). Tryptone–yeast extract–sucrose agar is the substrate for the organism. The minimum concentration of penicillic acid that gave an inhibitory response was 1 μg/disk.

A novel variation of the disk assay has been reported and refined (Reiss, 1975b). Impregnated spore strips which contain approximately 10^5 spores of *Bacillus stearothermophilus* are available in sealed packets (Oxoid Deutschland). Tablets, which contained 10 g casein–peptone, 5 g glucose, and 0.04 g bromocresol purple per liter (Oxoid, CM 74), were dissolved in screw-cap vials in 5 ml of water, then autoclaved and stored in a refrigerator until used. Mycotoxins were dissolved in aqueous DMSO (0.3%). Spore strips were added to the solutions, using sterilized tweezers so that they became impregnated with the toxin. The process was allowed to continue for 15 min at 22°C to ensure complete impregnation, and controls were prepared identically with aqueous DMSO. The impregnated spore strips were transferred to the nutrient tubes, incubated in the dark at 55°C, and observed at 30-min intervals. If the color changed from purple to yellow, spores were *not* inhibited or killed by the toxin. In the report (Reiss, 1975b), controls had changed to yellow after 16.5 hr, but treatments with aflatoxin B_1, patulin, rubratoxin B, and diacetoxyscirpenol at 0.001, 0.01, 0.1, 10, and 100 μg/ml failed to change the color after a 60-hr incubation with the exception of 0.001 μg/ml of patulin and diacetoxyscirpenol. The method is quite sensitive to small amounts of toxin, but the range of toxins to be assayed needs to be extended.

Kopp and Rehm (1979) developed a plate diffusion assay for quantitative determination of roquefortine with *B. stearothermophilus*. Assay concentrations of 25 μg/ml could be measured with this system.

Fungi have been used to detect toxins in disk assays (Burmeister and Hesseltine, 1970). Two mycotoxins that normally require extensive preparation for detection and require sophisticated bioassays or expensive analytical equipment are T-2 toxin [4β,15-diacetoxy-8α-(3-methylbutyryloxy)-12,13-epoxytrichothec-9-en-3-ol] and butenolide (4-acetamido-4-hydroxy-2-butenoic acid γ-lactone), both *Fusarium* toxins. A number of bacteria and fungi growing on YM agar were challenged with both metabolites at various concentrations on filter paper disks. Of the 54 bacterial and 11 fungal genera, species, or strains tested, 100 μg of butenolide inhibited *Spirillum serpens*, *Vibrio tyrogenus*, and *Xanthomonas campestris*. While T-2 toxin inhibited none of the 54 bacterial organisms, 6 of 11 fungi tested were inhibited. Concentrations of 4 μg of T-2 toxin per disk inhibited *Rhodotorula rubra* and *Penicillium digitatum*.

From these and other reports it seems that a bioassay based on the use of a single bacterial or fungal genus or species for detecting a broad spectrum of mycotoxins is unlikely. Certainly, one or two bacterial strains may be used to detect specific mycotoxins, and this may serve a very limited use for those researchers surveying food or livestock rations for known mycotoxins. But for the researcher who must evaluate dozens of different mycotoxins, some of which may occur together, chemical and electronic analytical procedures are superior. The most successful bioassay to date was that using *B. thuringiensis*.

While the use of bacterial assays may appear to be very limited at the moment, it is quite possible that through the use of suitable genetic engineering techniques a susceptible clone may be generated that is highly sensitive to individual or a range of unrelated mycotoxins. The development of such an organism depends on our understanding of the mode of action of mycotoxins in microorganisms and our ability to transfer the correct characteristics for sensitivity induction.

B. Bioluminescence

A method in the developmental stage which holds promise as a diagnostic tool for determining the presence of mycotoxins is based on the properties of bacterial bioluminescence (Yates and Porter, 1982, 1984). It is based on an earlier system for detecting toxins in aquatic environments (e.g., benzene, mercury(II), formaldehyde, copper, nitrate, and other industrial wastes) (Bulich and Isenberg, 1981). The method appears to have several advantages over some of the procedures already discussed. It is simple to perform, relatively inexpensive, and relies more on instrumentation and less on human manipulations that may introduce error.

The technique consists of reconstituting freeze-dried bacteria, *Photobacterium phosphoreum*, with ultrapure water, and addition of a suitable diluent, such as 2% sodium chloride. DMSO and methanol may be used to dissolve the mycotoxins to be tested and may be present in concentrations of less than 8 μg/ml. The toxicity of the mycotoxins tested was determined at 5, 10, 15, and 20 min, although the times can be extended or limited. An EC_{50} was the concentration of toxin that reduced bioluminescence by 50%. Freshly reconstituted bacteria gave slightly different values compared to aged bacteria, and bioassay values obtained to give an EC_{50} were, for fresh and aged organisms, respectively: patulin 7.53 (6.17), PR toxin 7.79 (9.49), penicillic acid 15.95 (14.67), citrinin 27.74 (30.70), zearalenone 14.37 (11.59), ochratoxin A 18.49 (18.53), aflatoxin B 21.97 (24.79), and rubratoxin B 31.79 (26.68) μg/ml (Yates and Porter, 1982).

In more recent experiments (Yates and Porter, 1984), aflatoxin B_1, patulin, penicillic acid, and zearalenone were assayed against *P. phosphoreum* using time intervals of 5, 10, 15, and 20 min. It was determined that the amount of toxin necessary to induce an EC_{50} in *P. phosphoreum* decreased as time increased, with the exception of zearalenone. The pH had a significant effect on bioluminescence, and the toxicity of patulin increased with increasing pH; zearalenone and penicillic acid decreased with increasing pH, and aflatoxin B_1 had a maximum toxicity at pH 7.0. Temperature tended to have less influence than pH. The maximum effect obtained for these mycotoxins was at 20°C, or greater, with the exception of zearalenone, which had an optimum of 15°C. The results of these experiments showed that critical parameters are needed to evaluate thoroughly the toxic effects elicited by mycotoxins in a bioluminescence bioassay. Patulin required pH 8.0 and 30°C; penicillic acid, pH 6.5 and 25–30°C; zearalenone, pH 6.0 and 10°–15°C; aflatoxin B_1, pH 7.0–7.5 and 20°–30°C.

As more insight is gained into the environmental factors that influence the bioassay and the modes of action by which the various mycotoxins influence the bioluminescent properties of *P. phosphoreum*, so the utuility of the assay as a method for detecting pure toxins, toxins in mixtures, and toxins in contaminated foodstuffs will become apparent. But it must be emphasized that the bioassay is still in the developmental stage and may require further refinement as a diagnostic tool.

V. CONCLUSION

Many factors must be considered in selecting a bioassay for the detection and isolation of mycotoxins. These factors include economics, availability, validity, specificity, sensitivity, and rapidity. Vertebrate assays using intact animals are more direct and, therefore, they are more valid; however, they may suffer in other areas such as speed and economics.

Since mycotoxins generally affect basic biological processes common to all life forms there is some logic in using biological, nonvertebrate assay systems to detect and purify mycotoxins. However, the use of invertebrate animals, such as insects and brine shrimp larvae, for the detection of vertebrate toxins is an indirect method and could result in erroneous or confusing results concerning the vertebrate toxicity of a fungal metabolite or the toxigenicity of a fungal isolate. The brine shrimp bioassay is particularly susceptible to a high proportion of false positives.

The use of higher plants, particularly the coleoptile assay, as a bioassay for mycotoxins is an attractive alternative to vertebrate assays because of the relative ease of preparation and low cost. While it has been used successfully to detect and isolate mycotoxins, final testing in a vertebrate system is necessary before concluding that a compound so isolated is, indeed, a true mycotoxin.

The same is true of microbiological assays, which have the advantage of being economical, readily available, rapid, and relatively sensitive for the detection of some mycotoxins. However, they suffer from a lack of specificity and detect only those mycotoxins that also have antibiotic activity. Mycotoxins that act principally on the nervous system of animals (i.e., tremorgens) are particularly discriminated against with any nonvertebrate assay system or with a nonintact vertebrate animal system.

A more recent consideration in favor of the increased use of nonvertebrate assay systems or vertebrate tissue culture assay systems in lieu of intact vertebrate animals is the movement promoted by animal welfare groups concerned with misuse of animals in laboratory experiments or in use of animals unnecessarily. In the quest for vertebrate toxins a vertebrate assay system must be employed at some point to establish the toxic nature of the metabolite. However, nonvertebrate or tissue culture assays could serve as the initial screening bioassay.

REFERENCES

Abdei, Z. H., and McKinley, W. P. (1968). Zebra fish eggs and larvae as aflatoxin bioassay test organisms. *J. Assoc. Off. Anal. Chem.* **51**, 902–905.

Aldridge, D. C., Armstrong, J. J., Speake, R. N., and Turner, W. B. (1967). The cytochalasins, a new class of biologically active mould metabolites. *Chem. Comm.* pp. 26–27.

Ames, B. N., Durston, W. E., Yamasaki, E., and Lee, F. D. (1973). Carcinogens are mutagens: a simple test system combining liver homogenates for activation and bacteria for detection. *Proc. Natl. Acad. Sci. U.S.A.* **70**, 2281–2285.

Bamburg, J. R., and Strong, F. M. (1971). 12-13 Epoxytrichothecenes. *In* "Microbial Toxins" (S. Kadis, A. Ciegler, and S. Ajl, eds.), Vol. 7, pp. 207–292. Academic Press, New York.

Bamburg, J. R., Riggs, N. V., and Strong, F. M. (1968). The structures of toxins from two strains of *Fusarium tricinctum. Tetrahedron* **24**, 3329–3336.

Bauer, L., and Leistner, L. (1969). Die Wirkung von Aflatoxin B₁ auf Amoben. *Arch. Hyg.* **153**, 397–402.

Bijl, J., Dive, D., and Van Peteghem, C. (1981). Comparison of some bioassay methods for mycotoxin studies. *Environ. Pollut., Ser. A* **26**, 173–182.

Birkinshaw, J. H., Bracken, A., Michael, S. E., and Raistrick, H. (1943). Patulin in the common cold. II. Biochemistry and chemistry. *Lancet* No. 245, 625–630.

Boutibonnes, P. (1979). Mise en évidence de l'activité antibactérienne de quelques mycotoxines par utilisation de *Bacillus thuringiensis* (Berliner). *Mycopathologia* **67**, 45–50.

Boutibonnes, P. (1980). Antibacterial activity of some mycotoxins. *IRCS Med. Sci. Biochem. Pharmacol.* **8**, 850–851.

Boutibonnes, P., and Auffray, Y. (1977). Propriétés antibactériennes de l'aflatoxine B₁: ses effets cytotoxiques chez *Bacillus thuringiensis* (Berliner). *Ann. Nutr. Aliment.* **32**, 831–840.

Boutibonnes, P., Malherbe, C., Auffray, Y., Kogbo, W., and Marais, C. (1983). Mycotoxin sensitivity of *Bacillus thuringiensis. IRCS Med. Sci. Biochem. Pharmacol.* **11**, 430–431.

Brian, P. W., and McGowan, J. C. (1945). Viridin: a highly fungistatic substance produced by *Trichoderma viride. Nature (London)* **156**, 144–145.

Brian, P. W., and McGowan, J. C. (1946). Biologically active metabolic products of the mould *Metarrhizium glutinosum* S. Pope. *Nature (London)* **157**, 334.

Brian, P. W., Dawkins, A. W., Grove, J. F., Hemming, H. G., Lowe, G., and Norris, G. L. F. (1961). Phytotoxic compounds produced by *Fusarium equiseti. J. Exp. Bot.* **12**, 1–12.

Broce, D., Grodner, R. M., Killebrew, R. L., and Bonner, F. L. (1970). Ochratoxins A and B confirmation by microbiological assay with *Bacillus cereus mycoides. J. Assoc. Off. Anal. Chem.* **53**, 616–619.

Brown, R. F. (1969). The effect of some mycotoxins on the brine shrimp, *Artemia salina. J. Am. Oil Chem. Soc.* **46**, 119.

Brown, R. F., Wildman, J. D., and Eppley, R. M. (1968). Temperature-dose relationships with aflatoxin on the brine shrimp, *Artemia salina*. *J. Assoc. Off. Anal. Chem.* **51**, 905–906.

Büchi, G., Klaubert, D. H., Shank, R. C., Weinreb, S. M., and Wogan, G. N. (1971). Structure and synthesis of kotanin and desmethylkotanin, metabolites of *Aspergillus glaucus*. *J. Org. Chem.* **36**, 1143–1147.

Büchi, G., Luk, K. C., Kobbe, B., and Townsend, J. M. (1977). Four new mycotoxins of *Aspergillus clavatus* related to *tryptoquivaline*. *J. Org. Chem.* **42**, 244–249.

Buckelew, A. R., Chakravarti, A., Burge, W. R., Thomas, V. M., and Ikawa, M. (1972). Effect of mycotoxins and coumarins on the growth of *Bacillus megaterium* from spores. *J. Agric. Food Chem.* **20**, 431–433.

Bulich, A. A., and Isenberg, D. L. (1981). Use of the luminescent bacterial system for the rapid assessment of aquatic toxicity. *Instrum. Soc. Am. Trans.* **20**, 29–33.

Burmeister, H. R., and Hesseltine, C. W. (1966). Survey of the sensitivity of microorganisms to aflatoxin. *Appl. Microbiol.* **14**, 403–404.

Burmeister, H. R., and Hesseltine, C. W. (1970). Biological assays for two mycotoxins produced by *Fusarium tricinctum*. *Appl. Microbiol.* **20**, 437–440.

Burmeister, H. R., Vesonder, R. F., and Kwolek, W. F. (1980). Mouse bioassay for *Fusarium* metabolites: rejection or acceptance when dissolved in drinking water. *Appl. Environ. Microbiol.* **39**, 957–961.

Cardeilhac, P. T., Nair, K. P. C., and Colwell, W. M. (1972). Tracheal organ cultures for the bioassay of nanogram quantities of mycotoxins. *J. Assoc. Off. Anal. Chem.* **55**, 1120–1121.

Carter, S. B. (1967). Effects of cytochalasins on mammalian cells. *Nature (London)* **213**, 261–264.

Chexal, K. K., Springer, J. P., Clardy, J., Cole, R. J., Kirksey, J. W., Dorner, J. W., Cutler, H. G., and Strawter, B. J. (1976). Austin, a novel polyisoprenoid mycotoxin from *Aspergillus ustus*. *J. Am. Chem. Soc.* **97**, 6748–6750.

Clements, N. L. (1968a). Note on a microbiological assay for aflatoxin B_1: A rapid confirmatory test by effects on growth of *Bacillus megaterium*. *J. Assoc. Off. Anal. Chem.* **51**, 611–612.

Clements, N. L. (1968b). Rapid confirmatory test for aflatoxin B_1, using *Bacillus megaterium*. *J. Assoc. Off. Anal. Chem.* **51**, 1192–1194.

Clifford, J. I., and Rees, K. R. (1966). Aflatoxin: A site of action in the rat liver cell. *Nature (London)* **209**, 312–313.

Cole, R. J. (1978). Basic techniques in studying mycotoxins: isolation of mycotoxins. *J. Food Prot.* **41**, 138–140.

Cole, R. J. (1984). Screening for mycotoxins and toxin-producing fungi. *In* "Mycotoxins—Production, Isolation, Separation and Purification" (V. Betina, ed.), pp. 45–58. Elsevier, Amsterdam.

Cole, R. J., and Cox, R. H. (1981). "Handbook of Toxic Fungal Metabolites." Academic Press, New York.

Cole, R. J., Kirksey, J. W., Moore, J. H., Blankenship, B. R., Diener, U. O., and David, N. D. (1972). Tremorgenic toxin from *Penicillium verruculosum*. *Appl. Microbiol.* **24**, 248–257.

Cole, R. J., Kirksey, J. W., Cutler, H. G., Doupnik, B. L., and Peckham, J. C. (1973). Toxin from *Fusarium moniliforme*: effects on plants and animals. *Science* **179**, 1324–1326.

Cole, R. J., Kirksey, J. W., Cutler, H. G., and Davis, E. E. (1974a). Toxic effects of oosporein from *Chaetomium trilaterale*. *J. Agric. Food Chem.* **22**, 517–519.

Cole, R. J., Kirksey, J. W., and Wells, J. M. (1974b). A new tremorgenic metabolite from *Penicillium paxilli*. *Can. J. Microbiol.* **20**, 1159–1162.

Cole, R. J., Kirksey, J. W., and Morgan-Jones, G. (1975a). Verruculotoxin, a new mycotoxin from *Penicillium verruculosum*. *Toxicol. Appl. Pharmacol.* **31**, 465–468.

Cole, R. J., Kirksey, J. W., Springer, J. P., Clardy, J., Cutler, H. G., and Garren, K. H. (1975b). Desmethoxyviridiol, a new toxin from *Nodulisporium hinnuleum*. *Phytochemistry* **14**, 1429–1432.

Cole, R. J., Kirksey, J. W., Cutler, H. G., Wilson, D. M., and Morgan-Jones, G. (1976). Two toxic indole alkaloids from *Penicillium islandicum*. *Can. J. Microbiol.* **22**, 741–744.

Cole, R. J., Dorner, J. W., Lansden, J. A., Cox, R. H., Pape, C., Cunfer, B., Nicholson, S. S., and Bedell, D. M. (1977). Paspalum staggers: Isolation and identification of tremorgenic metabolites from sclerotia of *Claviceps paspali*. *J. Agric. Food Chem.* **25**, 1197–2011.

Cole, R. J., Dorner, J. W., Cox, R. H., Cunfer, B., Cutler, H. G., and Stuart, B. (1981a). The isolation and identification of several trichothecene mycotoxins from *Fusarium heterosporum*. *J. Nat. Prod.* **44**, 324–330.

Cole, R. J., Corner, J. W., Cox, R. H., Hill, R. A., Cutler, H. G., and Wells, J. M. (1981b). Isolation of citreoviridin from *Penicillium charlesii* cultures and molded pecan fragments. *Appl. Environ. Microbiol.* **42**, 677–681.

Cole, R. J., Wells, J. M., Cox, R. H., and Cutler, H. G. (1981c). Isolation and biological properties of deacetylcytochalasin H from *Phomopsis* sp. *J. Agric. Food Chem.* **29**, 205–206.

Cole, R. J., Wilson, D. M., Harper, J. L., Cox, R. H., Cochran, T. W., Cutler, H. G., and Bell, D. K. (1982). Isolation and identification of two new [11] cytochalasins from *Phomopsis sojae*. *J. Agric. Food Chem.* **30**, 301–304.

Cox, R. H., Cutler, H. G., Hurd, R. E., and Cole, R. J. (1983). Proton and carbon-13 nuclear magnetic resonance studies of the conformation of cytochalasin H derivatives and plant growth regulating effects of cytochalasins. *J. Agric. Food Chem.* **31**, 405–408.

Curtis, R. F., Coxon, D. T., and Levett, G. (1974). Toxicity of fatty acids in assays for mycotoxins using the brine shrimp (*Artemia salina*). *Food Cosmet. Toxicol.* **12**, 233–235.

Curtis, R. W. (1958). Root curvatures induced by culture filtrates of *Aspergillus niger*. *Science* **128**, 661–662.

Cutler, H. G. (1984). Biologically active natural products from fungi: templates for tomorrow's pesticides. *In* "Bioregulators: Chemistry and Uses" (R. L. Ory and F. R. Rittig, eds.), ACS Symposium Series, No. 257, pp. 153–170. Am. Chem. Soc., Washington, D.C.

Cutler, H. G., and Jarvis, B. B. (1985). Preliminary observations of the effects of macrocyclic trichothecenes on plant growth. *Environ. Exp. Bot.* **15**, 115–118.

Cutler, H. G., and Le Files, J. H. (1978). Trichodermin: effects on plants. *Plant Cell Physiol.* **19**, 177–182.

Cutler, H. G., and Vlitos, A. J. (1962). The natural auxins of the sugar cane. II. Acidic, basic, and neutral growth substances in roots and shoots from twelve days after germination. A vegetative bid to maturity. *Physiol. Plant.* **15**, 27–42.

Cutler, H. G., Crumley, F. G., Cox, R. H., Hernandez, O., Cole, R. J., and Dorner, J. W. (1979). Orlandin: a nontoxic fungal metabolite with plant growth inhibiting properties. *J. Agric. Food Chem.* **27**, 592–595.

Cutler, H. G., Crumley, F. G., Cox, R. H., Cole, R. J., Dorner, J. W., Latterell, F. M., and Rossi, A. E. (1980a). Diplodiol: a new toxin from *Diplodia macrospora*. *J. Agric. Food Chem.* **28**, 135–138.

Cutler, H. G., Crumley, F. G., Cox, R. H., Cole, R. J., Dorner, J. W., Springer, J. P., Latterell, F. M., Thean, J. E., and Rossi, A. E. (1980b). Chaetoglobosin K: a new plant growth inhibitor and toxin from *Diplodia macrospora*. *J. Agric. Food Chem.* **28**, 139–142.

Dashek, W. V., and Llewellyn, G. C. (1983). Mode of action of the hepatocarcinogens, aflatoxins in plant systems: a review. *Mycopathologia* **81**, 83–94.

Dashek, W. V., Boggs, B. W., Gamber, R. B., Hrycyk, M., Schell, S. L., and Llewellyn, G. C. (1982). Evaluating the potency of the mycotoxin aflatoxin G_1 by lily pollen germination and tube elongation bioassays. *Dev. Ind. Microbiol.* **23**, 273–278.

De Waart, J., Van Aken, F., and Pouw, H. (1972). Detection of orally toxic microbial metabolites in foods with bioassay systems. *Zentralbl. Bakteriol., Parasitenkd., Infektionskr. Hyg., Abt. 1: Orig., Reihe A* **222**, 96–114.

Dive, D., Moreau, S., and Cacan, M. (1978). Use of a ciliate protozoan for fungal toxins studies. *Bull. Environ. Contam. Toxicol.* **19**, 489–495.

Dorner, J. W., Cole, R. J., Springer, J. P., Cox, R. H., Cutler, H. G., and Wicklow, D. T. (1980). Isolation and identification of two new biologically active norditerpene dilactones from *Aspergillus wentii*. *Phytochemistry* **19**, 1157–1161.

Eka, O. U. (1972). Effect of aflatoxins on microorganisms. *Z. Allg. Mikrobiol.* **12**, 593–595.

Eppley, R. M. (1974). Sensitivity of brine shrimp (*Artemia salina*) to trichothecenes. *J. Assoc. Off. Anal. Chem.* **57**, 618–620.

Eppley, R. M. (1975). Methods for the detection of trichothecenes. *J. Assoc. Off. Anal. Chem.* **58**, 906–908.

Eppley, R. M., and Bailey W. J. (1973). 12,13-Epoxy-Δ^9-trichothecenes as the probable mycotoxins responsible for stachybotryotoxicosis. *Science* **181**, 758–760.

Freeman, G. G., and Morrison, R. I. (1948). Trichothecin: an antifungal metabolic product of *Trichothecium roseum* Link. *Nature (London)* **162**, 30.

Gabliks, J., Schaeffer, W., Friedman, L., and Wogan, G. (1965). Effect of aflatoxin B_1 on cell cultures. *J. Bacteriol.* **90**, 720–723.

Geiger, W. B., Conn, J. E., and Waksman, S. W. (1944). Chaetomin a new antibiotic substance produced by *Chaetomium cochliodes*. II. Isolation and concentration. *J. Bacteriol.* **48**, 531–536.

Glinsukon, T., Romruen, K., and Visutasunthorn, C. (1979). Preliminary report on toxigenic fungal isolates of *Aspergillus niger* in market foods and foodstuffs. *Experientia* **35**, 522–523.

Godtfredsen, W. O., and Vangedal, S. (1965). Trichodermin, a new sesquiterpene antibiotic. *Acta Chem. Scand.* **19**, 1088–1102.

Hancock, C. R., Barlow, H. W., and Lacey, H. J. (1964). The East Malling coleoptile straight-growth test method. *J. Exp. Bot.* **15**, 166–176.

Harley, E. H., Rees, H. R., and Cohen, A. (1969). A comparative study of the effect of aflatoxin B_1 and actinomycin on HeLa cells. *Biochem. J.* **114**, 289–298.

Harwig, J., and Scott, P. M. (1971). Brine shrimp (*Artemia salina* L.) larvae as a screening system for fungal toxins. *Appl. Microbiol.* **21**, 1011–1016.

Hayes, A. W., and Wyatt, E. P. (1970). Survey of the sensitivity of microorganisms to rubratoxin B. *Appl. Microbiol.* **20**, 164–165.

Hayes, A. W., Melton, R., and Smith, S. S. (1974). Effect of aflatoxin B_1, ochratoxin and rubratoxin B on a protozoan, *Tetrahymena pyriformis* HSM. *Bull. Environ. Contam. Toxicol.* **11**, 321–325.

Hayes, A. W., Birkhead, H. E., and Hall, E. A. (1976). Antiprotozoal activity in citrinin. *Bull. Environ. Contam. Toxicol.* **15**, 429–436.

Holzapfel, C. W. (1968). The siolation and structure of cyclopiazonic acid, a toxic metabolite of *Penicillium cyclopium* Westling. *Tetrahedron* **24**, 2101–2119.

Juhasz, S., and Greczi, E. (1964). Extracts of mould-infected groundnut samples in tissue culture. *Nature (London)* **203**, 861–862.

Kirksey, J. W., and Cole, R. J. (1974). Screening for toxin-producing fungi. *Mycopathol. Mycol. Appl.* **54**, 291–296.

Kopp, G., and Rehm, H.-J. (1979). A biological assay for quantitative determination of roquefortine. *Z. Lebesm.-Unters. Forsch.* **169**, 90–91.

Kurtz, T. E., Link, R. F., Tukey, J. W., and Wallace, D. L. (1965). Short-cut multiple comparisons for balanced single and double classifications. Part 1, Results. *Technometrics* **7**, 95–161.

Lansden, J. A., Cole, R. J., Dorner, J. W., Cox, R. H., Cutler, H. G., and Clark, J. D. (1978). A new trichothecene mycotoxin isolated from *Fusarium tricinctum*. *J. Agric. Food Chem.* **26**, 246–249.

Legator, M. S., and Withrow, A. (1964). Aflatoxin: effect on mitotic division in cultured embryonic lung cells. *J. Assoc. Off. Anal. Chem.* **47**, 1007–1009.

Ling, K. H. (1976). Study on mycotoxins contaminated in food in Taiwan, tremor inducing compounds from *Aspergillus terreus*. *Proc. Natl. Sci. Counc., Repub. China, Part 2* No. 9, 121–130.

Llewellyn, G. C. (1973). Evaluation of the response of *Dugesia tigrina* to aflatoxin B_1. *J. Assoc. Off. Anal. Chem.* **56**, 1119–1122.

McMillian, W. W., Wilson, D. M., Widstrom, N. W., and Perkins, W. D. (1980). Effects of aflatoxins B_1 and G_1 on three insect pests of maize. *J. Econ. Entomol.* **73**, 26–28.

Matsumura, F., and Knight, S. G. (1967). Toxicity and chemosterilizing activity of aflatoxin against insects. *J. Econ. Entomol.* **60**, 871–872.

Mirocha, C. J., Patre, S. V., and Christensen, C. M. (1977). Zearalenone. *In* "Mycotoxins in Human and Animal Health" (J. V. Rodricks, C. W. Hesseltine, and M. A. Mehlman, eds.), pp. 345–364. Pathotox, Park Forest South, Illinois.

Moore, K. C., and Davis, G. R. F. (1983). Bertha armyworm (*Mamestra configurata*), a sensitive bioassay organism for mycotoxin research. *J. Invertebr. Pathol.* **42**, 413–414.

Murakoshi, S., Ohotomo, T., and Kurata, H. (1973). Toxic effects of various mycotoxins on silkworm (*Bombyx mori* L.) larvae in *ad libitum* feeding test. *Shokuhin Eiseigaku Zasshi* **14**, 65–68.

Nair, K. P. C., Colwell, W. M., Edds, G. T., and Cardeilhac, P. T. (1970). Use of tracheal organ cultures for bioassay of aflatoxin. *J. Assoc. Off. Anal. Chem.* **53**, 1258–1263.

Nakano, N., Kunimoto, T., and Aibara, K. (1973). Studies on chemical and biological assay of fusarenone-x and diacetoxyscirpenol in cereal grains. *Shokuhin Eiseigaku Zasshi* **14**, 56–64.

Nitsch, J. P., and Nitsch, C. (1956). Studies on the growth of coleoptile and first internode sections. A new, sensitive, straight-growth test for auxins. *Plant Physiol.* **31**, 94–111.

Olivigni, F. J., and Bullerman, L. B. (1978). A microbiological assay for penicillic acid. *J. Food Prot.* **41**, 432–434.

Patterson, D. S. P. (1983). Screening procedures for mycotoxins. *Proc. Int. Symp. Mycotoxins* pp. 167–175.

Patwardhan, S. A., Pandey, R. C., Dev, S., and Pendse, G. S. (1974). Toxic cytochalasins of *Phomopsis paspalli*, a pathogen of kodo millet. *Phytochemistry* **13**, 1985–1988.

Platt, B. S., Stewart, R. J. C., and Gupta, S. R. (1962). The chick embryo as a test organism for toxic substances in food. *Proc. Nutr. Soc.* **21**, 30–31.

Prior, M. G. (1979). Evaluation of brine shrimp (*Artemia salina*) larvae as a bioassay for mycotoxins in animal feedstuffs. *Can. J. Comp. Med.* **43**, 352–355.

Reiss, J. (1975a). *Bacillus subtilis*: a sensitive bioassay for patulin. *Bull. Environ. Contam. Toxicol.* **13**, 689–691.

Reiss, J. (1975b). Mycotoxin bioassays using *Bacillus stearothermophilus*. *J. Assoc. Off. Anal. Chem.* **58**, 624–625.

Rothweiler, W., and Tamm, C. (1966). Isolation and structure of phomin. *Experientia* **22**, 750–752.

San, R. H. C., and Stich, H. F. (1975). DNA repair synthesis of cultured human cells as a rapid bioassay for chemical carcinogens. *Int. J. Cancer* **16**, 284–291.

Sargeant, K., O'Kelly, J., Carnaghan, R. B. A., and Allcroft, R. (1961). The assay of a toxic principle in certain groundnut meals. *Vet. Rec.* **73**, 1219–1223.

Sinnhuber, R. O., Wales, J. H., Hendricks, J. D., Putnam, G. B., Nixon, J. E., and Pawlowski, N. E. (1977). Trout bioassay of mycotoxins. *In* "Mycotoxins in Human and Animal Health" (J. V. Rodricks, C. W. Hesseltine, and M. A. Mehlman, eds.), pp. 731–744. Pathotox, Park Forest South, Illinois.

Siriwardana, T. M. G., and Lafont, P. (1978). New sensitive biological assay for 12,13-epoxytrichothecenes. *Appl. Environ. Microbiol.* **35**, 206–207.

Stobb, M., Baldwin, R. S., Tuite, J., Andrews, F. N., and Gillette, K. G. (1962). Isolation of an anobolic uterotrophic compound from corn infected with *Gibberella zeae*. *Nature (London)* **196**, 1318.

Tatsuno, T., Obtsubo, K., and Saito, M. (1973). Chemical and biological detection of 12,13-epoxytrichothecenes isolated from *Fusarium* species. *Pure Appl. Chem.* **35**, 309–313.

Townsend, R. J., Moss, M. O., and Peck, H. M. (1966). Isolation and characterization of hepatoxins from *Penicillium rubrum*. *J. Pharm. Pharmacol.* **18**, 471–473.

Townsley, P. M., and Lee, E. G. H. (1967). Response of fertilized eggs of the mollusk *Bankia setacea* to aflatoxin. *J. Assoc. Off. Anal. Chem.* **50**, 361–363.

Ueno, U., Sato, N., Ishii, K., Sakai, K., Tsunoda, H., and Enomoto, M. (1973). Biological and chemical detection of trichothecene mycotoxins of *Fusarium* species. *Appl. Microbiol.* **25**, 699–704.

Umeda, M. (1977). Cytotoxicity of mycotoxins. *In* "Mycotoxins in Human and Animal Health" (J. V. Rodricks, C. W. Hesseltine, and M. A. Mehlman, eds.), pp. 713–720. Pathotox, Park Forest South, Illinois.

Van der Merwe, K. J., Steyn, P. S., and Fourie, L. (1965). Mycotoxins. Part II. The constitution of ochratoxins A, B, and C, metabolites of *Aspergillus ochraceus* Wilh. *J. Chem. Soc.* pp. 7083–7088.

Verrett, M. J., Marliac, J. P., and McLaughlin, J., Jr. (1964). Use of the chicken embryo in the assay of aflatoxin toxicity. *J. Assoc. Off. Anal. Chem.* **47**, 1003–1006.

Viitasalo, L., and Gyllenberg, H. G. (1968). Toxicity of aflatoxins to *Bacillus megaterium*. *Lebensm.-Wiss. Technol.* **12**, 113–114.

Vosdingh, R. A., and Neff, M. J. C. (1974). Bioassay of aflatoxins by catfish cell cultures. *Toxicology* **2**, 107–112.

Watson, D. H., and Lindsay, D. G. (1982). A critical review of biological methods for the detection of fungal toxins in foods and foodstuffs. *J. Sci. Food Agric.* **33**, 59–67.

Wei, R.-D., Smalley, E. B., and Strong, F. M. (1972). Improved skin test for detection of T-2 toxin. *Appl. Microbiol.* **23**, 1029–1030.

Weindling, R., and Emerson, O. (1936). The isolation of a toxic substance from the culture filtrate of *Trichoderma*. *Phytopathology* **26**, 1068–1070.

Wells, J. M., Cutler, H. G., and Cole, R. J. (1976). Toxicity and plant growth regulator effects of cytochalasin H isolated from *Phomopsis* sp. *Can. J. Microbiol.* **22**, 1137–1143.

Wells, J. M., Cole, R. J., Cutler, H. G., and Spalding, D. H. (1981). *Curvularia lunata*, a new source of cytochalasin B. *Appl. Environ. Microbiol.* **41**, 967–971.

Wilson, B. J., and Wilson, C. H. (1962). Extraction and preliminary characterization of a hepatoxic substance from cultures of *Penicillium rubrum*. *J. Bacteriol.* **84**, 283–290.

Wilson, B. J., and Wilson, C. H. (1964). Toxin from *Aspergillus flavus*: production on food materials of a substance causing tremors in mice. *Science* **144**, 177–178.

Wilson, B. J., Wilson, C. H., and Hayes, A. W. (1968). Tremorgenic toxins from *Penicillium cyclopium* grown on food materials. *Nature (London)* **220**, 77–78.

Wragg, J. G., Ross, V. C., and Legator, M. S. (1967). Effect of aflatoxin B_1 on the deoxyribonucleic acid polymerase of *Escherichia coli*. *Proc. Soc. Exp. Biol. Med.* **125**, 1052–1055.

Wyatt, T. D., and Townsend, R. J. (1974). The bioassay of rubratoxins A and B using Tetrahymena pyriformis strain w. *J. Gen. Microbiol.* **80**, 85–92.

Yates, I. E., and Porter, J. K. (1982). Bacterial bioluminescence as a bioassay for mycotoxins. *Appl. Environ. Microbiol.* **44**, 1072–1075.

Yates, I. E., and Porter, J. K. (1984). Temperature and pH affect the toxicological potential of mycotoxins in the bacterial bioluminescence assay. *In* "Toxicity Screening Procedures Using Bacterial Systems" (D. Liu and B. J. Dutka, eds.), pp. 77–88. Dekker, New York.

Yoshizawa, T., Tsuchiya, Y., Morooka, N., and Sawada, Y. (1975). Malformin A, as a mammalian toxicant from *Aspergillus niger*. *Agric. Biol. Chem.* **39**, 1325–1326.

2

Sampling and Sample Preparation Methods for Mycotoxin Analysis

J. W. DICKENS AND T. B. WHITAKER

Agricultural Research Service
U.S. Department of Agriculture
North Carolina State University
Raleigh, North Carolina 27695-7625

I. INTRODUCTION*

The objectives of sampling for mycotoxin analysis may be to check for the presence of the mycotoxin in a given lot of a commodity, to determine the incidence of the mycotoxin among different lots of a given commodity, to

---

*The views expressed here are those of the authors and do not necessarily reflect the policies of the United States Department of Agriculture. The use of trade names in this publication does not imply endorsement by the United States Department of Agriculture of the products named, nor criticism of similar ones not mentioned.

MODERN METHODS IN THE ANALYSIS
AND STRUCTURAL ELUCIDATION
OF MYCOTOXINS

estimate the average concentration of the mycotoxin in a given lot, or to determine the distribution of a population of lots according to their mycotoxin concentration. The method of sampling should be designed for the intended objective. In order to check for the presence of a mycotoxin or to determine the incidence of a mycotoxin among different lots, the sampling procedure should be biased toward the inclusion of items which are more likely to contain the mycotoxin. For example, kernels with visible mold growth or other defects probably contain more mycotoxin than sound-appearing kernels; so the sampling procedure should be selective or biased to ensure that the analytical sample consists of items with defects or characteristics associated with the presence of the mycotoxin. On the other hand, if the objective is to estimate the average mycotoxin concentration within a lot or to determine the distribution of a population of lots according to their mycotoxin concentrations, the sampling procedure must be unbiased, requiring that every item within the sample be collected in a completely random manner without regard to the physical characteristics of the items or their locations within the population of items that make up the lot. Because the objectives of most mycotoxin analyses require unbiased sampling procedures, this chapter will be confined to this type of sampling.

General recommendations for sampling and subsampling products for chemical analyses have been published by Kratochvil and Taylor (1981). They discuss statistical considerations related to sampling and subsampling of different types of products. A thorough review of sampling and subsampling procedures for mycotoxin analyses has been published by Dickens and Whitaker (1982). They provide a list of various types of equipment used for sample preparation and sources of supply.

The sources of error associated with mycotoxin testing are illustrated in Fig. 1. Sampling errors, subsampling errors, and analytical errors affect the precision of quantitative mycotoxin analyses. Traditional methods of sampling and subsampling of agricultural commodities are usually not adequate for mycotoxin analysis, because the mycotoxin is usually concentrated in a very small percentage of the kernels, grains, or other items that comprise the lot. As will be discussed later, the mycotoxin concentration of individual items is usually so variable that a sample composed of a large number of items is required to achieve reasonable agreement between the sample concentration and the lot concentration.

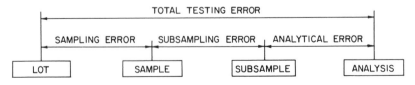

Fig. 1.   Types of error associated with mycotoxin testing.

Most of the published material on sampling and subsampling for mycotoxin analyses concerns aflatoxin. A review of the literature suggests that sampling and subsampling for aflatoxin analyses is more difficult than for other known mycotoxins in agricultural products and that sampling and subsampling procedures recommended for aflatoxins should be adequate for other mycotoxins. Although procedures for sampling and subsampling will be discussed for mycotoxins in general, specific data and recommendations will be confined to aflatoxin.

II. DISTRIBUTION OF MYCOTOXINS WITHIN THE PRODUCT

Extremely high concentrations of aflatoxin are often found at the sites where aflatoxin-producing molds have invaded a commodity. Because most of the kernel is often invaded by the mold, small items such as cottonseed, corn and peanut kernels generally contain higher concentrations of aflatoxin than do larger items. Also, a toxicogenic mold may produce more mycotoxin in some products than in other products with different physical and chemical properties. Cucullu *et al.* (1966) reported aflatoxin concentrations as high as 1,100,000 μg/kg (parts per billion, ppb) in a peanut kernel; Cucullu *et al.* (1977) reported 5,750,000 μg/kg in a cottonseed; and Shotwell *et al.* (1974) reported more than 400,000 μg/kg in a corn kernel. The extremely heterogeneous distribution of aflatoxin among these small kernels is demonstrated by the fact that these commodities would contain 20 μg/kg aflatoxin when only 0.0018% of the peanut kernels, 0.0004% of the cottonseed, and 0.0050% of the corn kernels contain the above aflatoxin concentrations.

Only a small proportion of larger items, such as Brazil nuts, may be invaded by the aflatoxin-producing mold. According to Stoloff *et al.* (1969), aflatoxin concentrations as high as 25,000 μg/kg have been found in individual Brazil nuts. Therefore, 0.0800% of the nuts would have to contain this concentration before the lot would contain 20 μg/kg aflatoxin. Very small proportions of large items such as rounds of cheese may be invaded by the aflatoxin-producing mold. The average aflatoxin concentration in these items may be low compared to the concentration in small items, but the aflatoxin concentration in the molded portions of small or large items may be the same. When sampling large items the sampling procedures must be adjusted according to whether the objective of the analysis is to estimate the average mycotoxin concentration in the entire population of large items, to estimate the average concentration in individual large items or to estimate the average concentration in the molded portions of the large items.

Sampling and subsampling procedures must be designed according to the mode of mycotoxin contamination. If loaves of bread were produced from a blended lot of mycotoxin-contaminated flour, the mycotoxin would be rather uniformly distributed throughout all of the loaves produced from the flour and sampling would be relatively simple. However, if mycotoxin were produced in the loaves after they were formed, the mycotoxin would be concentrated at the sites invaded by the toxicogenic mold and sampling would be much more difficult.

Liquid products, such as milk or beer, are more easily sampled than solid products. The Association of Official Analytical Chemists (AOAC, 1980) specifies that liquids with suspended particles or mixtures of liquids with different densities should be thoroughly stirred before samples are taken. Products that have been made into pastes or powders are more easily sampled than products with larger particles, but mixing and blending of the entire lot is necessary before the sample is taken. Since it is not possible to blend an entire lot of this material, it is necessary to take subsamples from individual batches of thoroughly blended material. The subsamples from each batch may be analyzed separately or they may be combined into a sample representing the entire lot.

Meat and eggs and their products may contain mycotoxins. Some of the products may contain mycotoxins that came from mycotoxin-contaminated feed ingested by the animal, and the mycotoxin may be rather uniformly distributed throughout the organ or product. Cured meats and sausages may contain mycotoxins produced by toxicogenic molds that grew on the products during processing and storage. In this case the mycotoxin may be concentrated in certain portions of the product. Proper sampling and subsampling of these products depends on the type of product and the method of mycotoxin contamination. Lotzsch (1978) has published a summary of techniques for sample preparation and chemical analyses of meat, meat products, eggs, and egg products for aflatoxin.

As indicated by the discussion in this section, the types of products sampled for mycotoxin analysis, the methods of mycotoxin contamination, the ranges in mycotoxin concentrations among individual items within the products, and the types of information desired about the mycotoxin contamination of the products are quite diverse. Consequently, specialized sampling and subsampling procedures must be developed for each product. Although the rationale for good sampling and subsampling procedures is applicable to most products, the remainder of this chapter will emphasize sampling and subsampling of particulate products (such as corn, peanuts, and cottonseed) for aflatoxin analyses.

III. SOURCES OF ERROR IN MYCOTOXIN TESTS

The mycotoxin concentration in a lot of particulate material such as shelled corn, shelled peanuts, or cottonseed is estimated by analyzing a sample taken from the lot. Because it is not feasible to solvent-extract mycotoxins from large samples, it is necessary to comminute the sample and take a small subsample of the comminuted material for solvent extraction and mycotoxin analysis.

The distribution of replicated mycotoxin analyses about their mean is an important consideration in the design and evaluation of mycotoxin testing programs. The distribution of analyses may be skewed or symmetrical depending on the test conditions. Both types of distributions are illustrated in Fig. 2. The distribution of mycotoxin analyses is symmetrical when an equal number of analyses are above and below the mean of all the analyses (the median equals the mean). The distribution is positively skewed when more than half of the replicated analyses are less than the mean of all the analyses (the median is less than the mean) and negatively skewed when more than half of the analyses are more than the mean (the median is more than the mean). When small samples are used, the distribution of sample concentrations about the true aflatoxin concentration in a lot is positively skewed, so that more than half of the sample concentrations are less than the lot concentration (Whitaker and Dickens, 1979a). Consequently, there is more than a 50% probability that the aflatoxin concentration in a small sample is less than the aflatoxin concentration in the lot. As sample size increases, the skewness of the distribution of sample concentrations will decrease and, according to the central limit theorem, will approach a normal distribution (Remington and Schork, 1970). As indicated below, for a given sample size, the skewness also decreases as lot concentration increases.

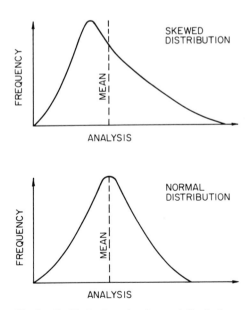

Fig. 2.   Positively skewed and normal distributions.

TABLE I

Distribution of Aflatoxin Analyses for Ten 5.4-kg Samples from Each of Six Lots of Shelled Peanuts[a,b]

Lot number	Aflatoxin analysis (ppb)										Average
1	0	0	0	0	8	8	15	16	16	125	19
2	0	0	0	0	0	0	0	8	22	198	22
3	0	0	0	0	0	0	0	9	12	285	31
4	5	12	56	66	70	92	98	132	141	164	84
5	18	50	53	72	82	108	112	127	182	191	100
6	29	37	41	71	95	117	168	174	183	197	111

[a] After Whitaker et al. (1972).
[b] Analyses and lots are ordered according to aflatoxin concentration.

Whitaker et al. (1972) measured the aflatoxin concentration in ten 5.4-kg samples of shelled peanuts taken from each of 29 lots. Each 5.4-kg sample of peanut kernels was comminuted in a subsampling mill (Dickens and Satterwhite, 1969), and a 250-g subsample of the comminuted material was analyzed for aflatoxin by method II of the AOAC (1980). Table I presents data for the three lots with the lowest average aflatoxin concentration and the three lots with the highest average aflatoxin concentration. Ninety percent of the analyses were below the lot mean for the three lots with the lowest aflatoxin concentrations, but the distribution of the 10 analyses was not skewed for the three lots with the highest aflatoxin concentrations. Although the data are too limited for firm conclusions, they illustrate the positive skewness of the distribution of aflatoxin analyses when 5.4-kg samples of shelled peanuts are used, and the effect of aflatoxin concentration on skewness.

The variance (S^2) of a given number (n) of mycotoxin analyses (X_1, X_2) . . . , X_i) may be computed from the following equation.

$$S^2 = \frac{\sum\limits_{i=1}^{n} X_i^2 - (\sum\limits_{i=1}^{n} X_i)^2/n}{n-1} \tag{1}$$

The square root of the variance is the standard deviation. For a normal distribution of a large number of analyses with a mean of $\overline{X}$, ~68, 95 and 99% of the analyses will be between $\overline{X} \pm S$, $\overline{X} \pm (1.96)S$, and $\overline{X} \pm (2.58)S$, respectively. The variance among analyses is a measure of the total error caused by sampling, subsampling, and analysis.

Several comprehensive reviews have been published on procedures for aflatoxin analyses of agricultural commodities (Nesheim, 1979; Schuller et al., 1976). The following discussion of sampling, subsampling, and analysis of corn for aflatoxin is presented as an example of the statistical considerations related to mycotoxin tests for particulate material.

The official first-action method for corn specified by the AOAC (1980) does not designate sample size, but it requires that the entire sample of shelled corn be ground to pass a No. 14 sieve, and that a 1- to 2-kg subsample of this material be ground to pass a No. 20 sieve. A 50-g subsample of the finely ground material is then analyzed by the CB method (so called because it was developed and tested by the Contaminants Branch of the Food and Drug Administration). The following equations for the variance (error) terms related to this test procedure have been developed (Whitaker et al., 1979).

$$V = S + C + F + Q \tag{2}$$

$$S = 3.9539 \ P/W_s \tag{3}$$

$$C = 0.1196 \ P/W_c \tag{4}$$

$$F = 0.0125 \ P/W_f \tag{5}$$

$$Q = 0.0699 \ P^2/N_q \tag{6}$$

where V, S, C, F, and Q are, respectively, the variances of the total error of the test, error in sampling the shelled corn, error in subsampling the coarsely ground material, error in subsampling the finely ground material, and error in quantification of aflatoxin in the chloroform extract from the finely ground subsample. W_s is the mass of the sample in kilograms, W_c is the mass of the coarsely ground subsample in kilograms, W_f is the mass of the finely ground subsample in kilograms, N_q is the number of times the aflatoxin in the solvent extract is quantified on a separate thin-layer chromatography (TLC) plate, and P is the aflatoxin concentration (micrograms per kilogram) in the lot.

T. B. Whitaker (unpublished observations) has determined that the distribution of aflatoxin analyses about the true lot concentration is practically normal for shelled corn when the lot concentration is 20 µg/kg or more and the sample weight is 4.54 kg or more. Therefore, the standard deviations computed from the variance in Eq. (2) may be used to compute confidence limits for the lot mean when lot concentration and sample weight meet these criteria.

Equations (3)–(5) are based on laboratory studies when sampling and subsampling procedures were designed to eliminate bias. Only the inherent heterogeneity of aflatoxin concentration among the kernels and among the comminuted particles causes the indicated variances. When errors are introduced because of sampling and subsampling procedures, or when the particles in the comminuted samples are larger than in the laboratory study, the variances will be greater than those estimated by the equations.

Equation (6) reflects the variance among replicated measurements of the aflatoxin concentration in the chloroform extract from the finely ground corn according to the CB method. A densitometer was used to quantify the aflatoxin in spots on the TLC plates (Dickens *et al.*, 1980). A study by Whitaker and Dickens (1981) suggests that even when a densitometer is used, most of the error indicated by Eq. (6) is due to errors in quantification of aflatoxin on TLC plates. High-performance liquid chromatography or other analytical procedures may reduce this error, but comparable data on these analytical procedures are not presently available. Equations similar to Eqs. (3)–(6) have been developed to estimate sampling, subsampling, and analytical errors associated with shelled peanuts (Whitaker *et al.*, 1974) and cottonseed (Whitaker *et al.*, 1976).

IV. ERROR REDUCTION

Equations (3)–(5) specify mass rather than the number of kernels or particles in the sample or the subsamples, because mass is directly correlated with the number of kernels or particles for a given material and because mass is a more convenient measurement than number. However, the number of kernels or particles in a sample or subsample is the important criterion. Therefore, for a given mass of subsample, the amount of subsampling error is reduced by grinding the sample more finely and thus increasing the number of particles in the subsample.

Some other methods to increase the precision of aflatoxin tests are to increase sample size, to increase the size of the subsample used for aflatoxin analysis, and to increase the number of analyses. Different costs are associated with each method, and careful study is required to determine the testing program that will provide the most precision for a given cost. The optimum balance in sample size, degree of comminution, subsample size, and number of analyses will vary according to the cost of the sample to be comminuted, the cost of sample comminution and subsampling, the cost of analysis, and other factors. In general, the costs of properly designed aflatoxin testing programs will increase as precision increases.

Equations (2)–(6) may be used to determine if the expected errors in aflatoxin determinations are acceptable and to determine the most efficient method to reduce error when necessary. For example, the expected total error associated with testing a lot of shelled corn when $P = 20$, $W_s = 4.54$, $W_c = 1$, $W_f = 0.05$, and $N_q = 1$ is 52.77 according to Eqs. (2)–(6).

$$V = 17.42 + 2.39 + 5.00 + 27.95 = 52.77 \tag{7}$$

The standard deviation is 7.26 and the coefficient of variation is 36.3%. The major error components in Eq. (7) are related to S (17.42) and Q (27.95). Doubling sample size would halve S, and two independent measurements of aflatoxin in the solvent extract by TLC would halve Q. Although quite small compared to S and Q, the errors related to subsampling can be reduced by finer comminution of the sample or by taking larger subsamples. Choice of the method employed to reduce the expected error should depend on cost as well as other factors.

V. SAMPLING PROCEDURES

Both sample size and sampling procedure affect the error of a sample of particulate material taken for mycotoxin analysis. When the mycotoxin-contaminated particles are uniformly distributed throughout the lot, as illustrated in Fig. 3A, the sampling procedure is relatively unimportant, since a sample taken from any position within the lot would contain about the same number of contaminated particles as samples taken from any other position. In this case, sample size is the most important factor. As indicated by Eq. (3), when a sample is taken from a thoroughly blended population of particles, sampling error is inversely proportional to sample weight. When the mycotoxin-contaminated particles are not uniformly distributed throughout the lot, as illustrated in Fig. 3B, the sampling error is dependent on both sampling method and sample size. The sampling method must ensure that particles from all positions within the lot have an equal probability of being included in the sample.

In some cases the concentration of mycotoxin in individual particles of the product may be correlated with certain physical characteristics of the particles. For example, Huff (1980) has reported that insect-damaged and low-density corn kernels contain higher concentrations of aflatoxin than do other kernels in an aflatoxin-contaminated lot of corn. Consequently, the sampling procedure should provide equal probability that all particles in the lot have an equal probability of being included in the sample irrespective of their location within the lot or their physical characteristics.

Samples may be taken from shelled corn during handling, during storage, and at other points in the production, marketing, and processing system. Each type of

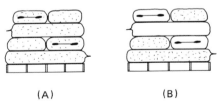

(A)   (B)

Fig. 3. An illustration of uniform (A) and nonuniform (B) distributions of mycotoxin within lots of particulate material. The dots represent mycotoxin-contaminated particles.

sampling presents a different situation regarding the distribution of the aflatoxin-contaminated kernels within the lot sampled and accessibility of the entire lot. When feasible, samples should be taken after the lot has been reduced to a smaller particulate size. For example, it is better to sample shelled corn than ear corn, and it is better to sample ground corn than shelled corn.

When a lot of corn has been recently blended by harvesting, loading and unloading, turning, or other operations, the sampling procedure is less important than when the lot has not been recently blended, because mold growth and mycotoxin production may have occurred in spots. For example, moisture condensation or leaks during storage may cause a portion of the lot to mold and to contain high concentrations of aflatoxin (Shotwell *et al.*, 1975). In similar situations it is impossible to predict where to probe the contents of a storage bin in order to obtain a sample with the same aflatoxin concentration as the concentration in the entire lot.

A. Stream Sampling

The best sampling method is to take small portions from a moving stream at periodic intervals and to combine these portions into a sample. Cross-cut samplers are commercially available that automatically cut through the stream at predetermined intervals. When an automatic cross-cut sampler is not available, a person may be assigned to pass a cup through the stream at periodic intervals and thus collect a sample. The stream should be sampled frequently, but the amount taken at each interval may be small to avoid accumulating too large a sample. The stream must be sampled throughout the time the lot is moved.

A schematic drawing of an automatic cross-cut sampler for grain is shown in Fig. 4. The diverter is shown in three different positions as it is moved through the stream of grain in a direction perpendicular to the direction of grain flow. As the diverter moves through the stream a portion of the stream is diverted to a sample container. The opening in the diverter is long enough to span the entire stream and is wide enough to accept the largest items in the stream. In order to prevent bias against collection of long or large items in the stream, it is generally recommended that the width of the diverter opening be at least two times the longest dimension of any of the items in the stream. The diverter stops in positions 1 and 3. The frequency of movement through the stream is controlled by a timer, and the velocity of diverter movement through the stream is controlled by adjustment of the diverter drive mechanism (usually a two-way air cylinder or a reversible electric gear-motor).

The size of sample collected from the lot by the automatic cross-cut sampler is determined by the quantity of grain in the lot, the frequency of diverter movement through the stream, the velocity of diverter movement through the stream,

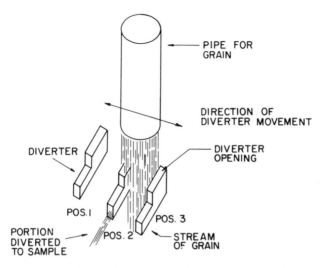

PIPE FOR
GRAIN

DIRECTION OF
DIVERTER MOVEMENT

DIVERTER

DIVERTER
OPENING

POS. I

POS. 3

PORTION
DIVERTED
TO SAMPLE

POS. 2

STREAM
OF GRAIN

Fig. 4. Schematic drawing of an automatic cross-cut sampler.

and the width of the diverter opening. The mathematical relationship of these parameters is shown by the following equation.

$$F = DL/SV \qquad (8)$$

where F is the time between diverter movements through the stream in seconds, D is the width of the diverter opening in centimeters, L is the lot weight in kilograms, S is the total sample weight in kilograms, and V is the velocity of diverter movement through the stream of grain in centimeters per second. For example, Eq. (8) may be used to compute the required frequency of diverter movement to obtain a 10-kg sample from a 25,000-kg lot of shelled corn when the width of the diverter opening is 2 cm and the diverter velocity is 75 cm/sec.

$$F = (2 \text{ cm}) (25,000 \text{ kg})/(10 \text{ kg}) (75 \text{ cm/sec}) = 66.7 \text{ sec} \qquad (9)$$

Therefore, the timer should be set to move the diverter through the stream of corn every 66.7 sec in order to obtain a 10-kg sample during the time 25,000 kg of corn passes through the sampler. Note that the required frequency is independent of the rate of flow for the stream of corn.

B. Probe Sampling

For a variety of reasons, it is often desirable to take samples from material held in bins, sacks, rail cars, barges, trucks, and other containers. As mentioned previously, sampling static populations of grain and other particulate material is

Fig. 5. Probe sampling a truckload of grain.

usually not satisfactory unless the material has recently been thoroughly blended. Samples are generally taken from static populations by using a probe or trier that will take a core of material from top to bottom of the container for bulk material. When the material is in sacks, small probes are often used to penetrate the sacks and take a small portion of material. When sampling bulk material, probes are usually taken at evenly spaced intervals over the surface of the material. A probe sampling operation for a truckload of grain in shown in Fig. 5.

When sacks or other small containers are sampled, the proportion of containers sampled may vary from one-quarter of the containers in small lots to the square root of the number of containers in large lots. Recommended methods for taking probe samples have been published by Bauwin and Ryan (1982).

C. Sampling Equipment

Bauwin and Ryan (1982) have published a comprehensive review of procedures employed for sampling, inspection, and grading of grain. This review includes descriptions of equipment for stream sampling and for probe sampling of grain. Hurbugh and Bern (1983) describe three different types of pneumatic probes which facilitate probe sampling of grain. They discuss the nonuniform distribution of foreign material and broken kernels in corn and soybeans that have been loaded onto trucks at the farm, and they recommend probing patterns to be employed for this material.

VI. SUBSAMPLING PROCEDURES

Because it is not feasible to solvent-extract aflatoxin from the entire sample, it is necessary to take a subsample for aflatoxin analysis. As previously stated, subsampling error may be reduced by increasing the number of particles in the subsample. The number of particles may be increased by comminuting the sample into smaller particles or by increasing the size of the subsample. The subsampling procedures should ensure that each particle in the comminuted sample has an equal chance of being included in the subsample.

One subsampling procedure involves comminuting the sample into a thoroughly blended paste or powder from which a subsample is taken. The HVCM (Hobart Vertical Cutter Mixer, Hobart Corporation, Troy, Ohio 45347) is widely used for simultaneously comminuting and blending samples for mycotoxin analyses (Francis, 1979; Stoloff et al., 1969). The machine will comminute and blend an 11-kg sample of in-shell Brazil nuts in about 3 min. The hard shells of Brazil nuts and pistachio nuts aid the comminuting and blending process for these commodities. It is usually necessary to add an equal weight of oyster shells to shelled nuts in order to achieve satisfactory comminution and blending. The subsample weight should be adjusted in order to account for the weight of nutshells or oyster shells in the sample.

A subsampling mill has been developed by Dickens and Satterwhite (1969) to comminute and subsample simultaneously large samples of granular material. The subsampling mill is shown in Fig. 6. During operation, the cylindrical screen is fastened to a circular platform beneath the blades. As the sample is poured into the chamber thus formed, it is comminuted by the rotating blades. The particles swirl within the chamber until they pass through the screen. A subsample enters the spout formed by two vanes that radiate from the screen surface to the outer shell of the mill. The remainder falls through the opening between the screen and the outer shell. Ratio of subsample to sample weight is determined by the percentage of the total screen area that is between the vanes that form the subsample spout. Comminuted particle size is determined by the openings in the screen. Openings of approximately 3.2 mm are required to prevent the screen from clogging when peanuts are comminuted.

Because the screen openings must be large enough to prevent clogging when oilseeds are comminuted in the subsampling mill, it is necessary to use a larger

Fig. 6. Subsampling mill for granular material.

subsample than would be required if the sample were ground to a paste. For example, a 1100-g subsample of comminuted peanuts from the subsampling mill is recommended compared to the 50-g subsample generally used for finely ground peanut butter. Studies with peanut kernels made radioactive by neutron activation have shown that the subsampling error associated with the subsampling mill is no greater than the error associated with the theoretically optimum procedure of subdividing the thoroughly blended comminuted material with a riffle-type divider (Dickens *et al.*, 1979).

Some agricultural commodities with low oil concentrations may be comminuted into a fine powder or flour without forming a paste. Some of the materials may be comminuted into a fine flour by using a fine screen on the subsampling mill. Fine comminution of corn, which has hard endosperm, is difficult to achieve with the subsampling mill or other devices. A two-stage comminution process has been recommended by the AOAC (1980). The entire sample is comminuted to pass a No. 14 sieve, and a 1- to 2-kg subsample of this coarse material is comminuted to pass a No. 20 sieve with a device such as the Wiley Mill (Arthur H. Thomas Company, Third & Vine St., Philadelphia, Pennsylvania 19105). A 50-g subsample of this finely ground material is then analyzed for aflatoxin. Some laboratories use hammer mills to comminute the entire sample so it will pass a No. 20 sieve, and a 50-g subsample is taken from the material after it is thoroughly blended.

Dust is a major problem when comminuting materials to fine powders. Loss of dust from the sample may cause a subsampling error, since a disproportionate amount of mycotoxin might be in the sample component that is converted to dust. Also, ingestion of dust from the comminution process may cause a health problem for people.

Another subsampling procedure is to comminute sample particles while they are suspended in a liquid to form a slurry. The liquid prevents some materials such as oilseeds from forming a paste when they are comminuted and facilitates thorough blending and more uniform comminution. Some liquids may cause the particles to swell or soften so that finer comminution is possible. If the liquid is a solvent for the mycotoxin, some of the mycotoxin might be extracted so that it is more uniformly distributed throughout the slurry. It is important to have a known amount of sample material in the analytical subsample of the slurry. Usually a weighed proportion of the slurry is taken for the subsample.

A water slurry procedure for peanuts has been developed by Whitaker and Dickens (1980). This procedure has been approved as a modification of method II of the AOAC (1980) and is used for most samples of raw peanuts analyzed for aflatoxin in the United States. This slurry procedure is used to comminute further a 1100-g subsample of comminuted peanuts from the subsampling mill. Approximately 83 g of finely comminuted peanut material suspended in a water slurry is extracted with solvents rather than extracting the entire 1100-g subsample from the subsampling mill. This procedure reduces the cost of solvents and reduces the problems associated with disposal of large quantities of solvents. Another slurry grinding technique which uses heptane has been developed by Stoloff and Dantzman (1972).

Other pieces of equipment such as hammer mills, grinders, food choppers, twin-shell blenders, planetary mixers, and riffle-type dividers may be employed to comminute and subsample samples of products for mycotoxin analyses. The HVCM and the subsampling mill are probably the pieces of equipment most frequently used in the United States for large samples of material.

VII. PRESERVATION OF SAMPLES

Controlling the conditions of sample storage to prevent mold growth and mycotoxin elaboration from the time of sampling until the samples are analyzed is extremely important. The rate of mycotoxin production can be very rapid when sample conditions are conducive to growth of a toxicogenic mold. Recommendations for prevention of mycotoxins in samples have been published by Davis *et al.* (1980).

Quite often, mycotoxin samples are taken from products located a considerable distance from the analytical laboratory. Consequently there is a time delay before the samples can be analyzed, and during shipment to the laboratory the samples may be exposed to temperatures that are favorable for mold growth. If the samples are dry they may be sealed in plastic bags to prevent them from absorbing moisture. If the samples are not completely dry, plastic bags should be avoided, since condensation may occur within the bags and promote mold growth on a portion of the sample.

When the samples contain moisture levels that are conducive to mold growth it is necessary either to dry the samples or to place them in cold storage. The average moisture content of the product may be misleading, because there may be a wide range in the moisture content of individual kernels or particles in the lot and because the moisture measurement may be inaccurate. It may be possible to treat samples with chemicals such as acetic acid or propionic acid to prevent or retard mold growth, but the potential health hazard to personnel and the effects of the chemical on the proposed mycotoxin analysis would have to be considered.

When samples are placed in temporary cold storage, the samples should be spread out so they will cool more rapidly than they would if they were piled together. Moisture condensation may form on samples that are taken from cold storage, because their temperature is below the dew point for conditions outside the cold storage. Therefore, samples taken from cold storage should be sealed in vapor-proof containers such as plastic bags, until they reach the outside temperature, or the samples should be processed and analyzed for mycotoxins immediately after removal from cold storage. In cases where there is a backlog of samples it may be desirable to save cold-storage space by comminuting the samples and storing subsamples rather than storing the samples.

VIII. SURVEY STUDIES

Surveys for mycotoxin in agricultural commodities sometimes are based on the misconception that when one sample is taken from each lot the distribution of mycotoxin concentrations for the survey samples closely approximates the distribution for the lots which were sampled. The distribution of sample concentrations determined in a survey probably will be different from the distribution of the true mycotoxin concentrations of the sampled lots. When small samples are used, most of the samples will contain a lower mycotoxin concentration than the sampled lot, and a few samples will contain much higher concentrations than the sampled lot. Expected agreement between the sample distribution and the lot distribution increases with the size of samples used.

It is difficult to make a general recommendation about the sample size (amount of product per sample) that should be used in surveys. Larger samples will reduce error, but the cost of the product, transportation, and other factors may be limiting. Cost factors must be reconciled with acceptable limits of error for the anticipated survey. The effect of sample size on the agreement between the sample distribution and the lot distribution may be determined according to a procedure developed for aflatoxin in peanuts by Whitaker and Dickens (1979a). This same procedure may be used with other commodities to estimate the lot distribution based on the sample distribution. If the purpose of the survey is limited to a determination of the average mycotoxin concentration in all lots sampled, relatively small samples are acceptable. When an accurate estimate of the distribution of lots according to mycotoxin concentration is desired, larger samples are required.

IX. QUALITY CONTROL

In order to control the amount of aflatoxin in finished products, early detection of aflatoxin contamination is important. For most commodities, the production and marketing system is a giant mixer where many different lots are blended together during handling, storage, and processing. One aflatoxin-contaminated lot may contaminate several aflatoxin-free lots when they are blended together. In addition, the value of the commodity increases as it moves through the production and marketing system. Therefore, detection and segregation of aflatoxin-contaminated material at the farm or at the first point of marketing will greatly reduce the cost of aflatoxin control.

In order to identify those lots that require special handling, it is necessary to determine the aflatoxin concentration of every lot. Because of errors associated with sampling, subsampling, and analysis it is extremely difficult to obtain a good estimate of the true aflatoxin concentration of a lot. A test for a good lot may erronously indicate that the lot is bad (false reject); conversely, a test for a bad lot may indicate that the lot is good (false accept). The errors related to aflatoxin tests and to the probability of false accepts and false rejects have been determined for peanuts (Whitaker *et al.*, 1974), corn (Whitaker *et al.*, 1979), and cottonseed (Whitaker *et al.*, 1976). Both false rejects and false accepts may be reduced by increasing the precision of aflatoxin tests.

Most procedures for dealing with aflatoxin-contaminated products are expensive; so the owners of the product need to minimize their risks. Because the costs of a properly designed aflatoxin testing program increase with the precision of the program, it is not economically feasible to employ a testing program that will eliminate all risk for the owner. A reasonable compromise may be to use a program that will maximize the benefit/cost ratio for the owner of the product. The benefit is the estimated savings to the owner for rejecting the bad lot before its value is increased by further processing. The costs include the cost of the testing program, the cost for rejecting lots, and the cost for accepting bad lots for further processing. The relative magnitudes of these savings and costs depend on many factors including the value of the commodity, the intended use of the commodity, the cost of removing aflatoxin contamination, the cost of diverting aflatoxin-contaminated lots to nonfood uses, the cost of aflatoxin tests, the incidence of bad lots, and many other factors. As a result, different types of testing programs may be desirable depending on the circumstances. A description of the aflatoxin control program used in the United States for peanuts has been published by Dickens (1977), and an evaluation of the aflatoxin testing program used for shelled peanuts has been published by Whitaker and Dickens (1979b).

At many points in the marketing and processing system, the benefits and costs can be treated as economic values rather than questions of human health. Evaluation of health risk can be deferred to the last stage in the marketing and processing system. At this stage, sampling error is often reduced because of comminution and blending during processing operations. For example, aflatoxin tests on cornmeal are usually more accurate than tests on shelled corn, and tests on peanut butter are more accurate than tests on shelled peanut kernels.

X. EXPERIMENTATION

The variability of mycotoxin test results should be recognized, and estimates of this variance should be used to calculate confidence limits for the mean of a given number of replicated mycotoxin analyses. This procedure is necessary to design properly experiments for research on mycotoxin control procedures and on the measurement of mycotoxin's effect on animals. Equations (2)–(6) and published statistical procedures (Remington and Schork, 1970) have been used by us to make calculations that illustrate the need for replicated aflatoxin tests on corn used for experimental purposes. (In order to avoid the introduction of additional statistical terminology, the details of the calculations are not presented.) The conclusions presented in the following two examples are based on the assumption that the distribution of the sample means used in the examples are normally distributed about the true lot mean.

Example 1. Situation: In order to conduct animal feeding trials to determine the effects of aflatoxin concentration in feed on animal performance, it is necessary to determine the aflatoxin concentration in a lot of corn that has been coarsely ground to pass a No. 14 sieve. *Conclusion:* If 1-kg samples of the coarsely ground corn were finely ground, subsampled, and analyzed according to the official first-action method of the AOAC (1980), and if the true lot concentration were 20 ppb aflatoxin, the sum of the variance components in Eqs. (4)–(6) would be 35.35. In this case, eight samples must be analyzed in order for their average to fall between 16 and 24 ppb with 95% confidence.

Example 2. Situation: A comparison of the aflatoxin concentrations in two lots of shelled corn is required in order to determine if there was an effect of experimental treatment on aflatoxin production. *Conclusion:* If 4.54-kg samples of shelled corn from each lot were ground and analyzed according to the procedure of the AOAC (1980), and if one lot contained 20 ppb and the other lot contained 30 ppb aflatoxin, the sum of the variance components in Eqs. (3)–(6) would be 52.77 and 100.12, respectively. It would take twenty 4.54-kg samples from each lot to detect the 10-ppb difference in the two lots with a power of 95% (assuming a type I probability of 5%). Observations on the "inevitability of variability" in most experimental measurements are especially applicable to aflatoxin tests where variance due to sampling, subsampling, and analysis is very large (Horwitz, 1976). This variability must be considered in the design of experiments. Otherwise, treatment effects may go undetected and incorrect conclusions may be drawn. When feeding trials are made to determine animal tolerance for aflatoxin, replicated measurements of the aflatoxin concentration in the feed should be made to provide reasonable confidence limits about the estimated aflatoxin concentration. Further research is needed to develop more information about the statistical considerations related to design of aflatoxin feeding tests for animals and experiments on aflatoxin control methods.

REFERENCES

Association of Official Analytical Chemists (1980). "Official Methods of Analysis," 13th Ed., pp. 229–434. Assoc. Off. Anal. Chem., Washington, D.C.

Bauwin, G. R., and Ryan, H. L. (1982). Sampling, inspection and grading of grain. *In* "Storage of Cereal Grains and their Products" (C. M. Christensen, ed.), Monograph Series, 3rd Ed., Vol. 5, pp. 115–157. Am. Assoc. Cereal Chem., St. Paul, Minnesota.

Cucullu, A. F., Lee, L. S., Mayne, R. Y., and Goldblatt, L. A. (1966). Determination of aflatoxins in peanuts and peanut sections. *J. Am. Oil Chem. Soc.* **43,** 89–92.

Cucullu, A. F., Lee, L. S., and Pons, W. A., Jr. (1977). Relationship of physical appearance of individual mold damaged cottonseed to aflatoxin content. *J. Am. Oil Chem. Soc.* **54,** 235A–237A.

Davis, N. D., Dickens, J. W., Freie, R. L., Hamilton, P. B., Shotwell, O. L., and Wyllie, T. D.

(1980). Protocols for surveys, sampling, post-collection handling, and analysis of grain samples involved in mycotoxin problems. *J. Assoc. Off. Anal. Chem.* **63**, 95–102.

Dickens, J. W. (1977). Aflatoxin control program for peanuts. *J. Am. Oil Chem. Soc.* **54**, 225A–228A.

Dickens, J. W., and Satterwhite, J. B. (1969). Subsampling mill for peanut kernels. *Food Technol.* **23**, 90–92.

Dickens, J. W., and Whitaker, T. B. (1982). Sampling and sampling preparation. *In* "Environmental Carcinogens—Selected Methods of Analysis. Vol. 5: Some Mycotoxins" (H. Egan, L. Stoloff, P. Scott, M. Costegnaro, I. K. O'Neill, and H. Bartsch, eds.), pp. 17–30. ARC. Lyon.

Dickens, J. W., Whitaker, T. B., Monroe, R. J., and Weaver, J. N. (1979). Accuracy of subsampling mill for granular materials. *J. Am. Oil Chem. Soc.* **56**, 842–844.

Dickens, J. W., McClure, W. F., and Whitaker, T. B. (1980). Densitometric equipment for rapid quantitation of aflatoxins on thin layer chromatograms. *J. Am. Oil Chem.* **57**, 205–208.

Francis, O. J., Jr. (1979). Sample preparation of some shelled treenuts and peanuts in a vertical cutter-mixer for myucotoxin analysis. *J. Assoc. Off. Anal. Chem.* **62**, 1182–1185.

Horwitz, W. (1976). The inevitability of variability. *J. Am. Oil Chem. Soc.* **59**, 338–342.

Huff, W. E. (1980). A physical method for the segregation of aflatoxin contaminated corn. *Cereal Chem.* **57**, 236–238.

Hurburgh, C. R., and Bern, C. J. (1983). Sampling corn and soybeans. I Probing method. *Trans. ASAE* **26**, 930–934.

Kratochvil, B., and Taylor, J. K. (1981). Sampling for chemical analysis. *Anal. Chem.* **53**, 924A–932A.

Lotzsch, R. (1978). Sampling, preparation and chemical analysis of meat and meat products and of eggs and egg products suspected of containing aflatoxin residues. *Fleischwirtschaft* **58**, 594–598.

Nesheim, S. (1979). Methods of aflatoxin analysis. *NBS Spec. Publ. (U.S.)* No. 519, 355–372.

Remington, R. D., and Schork, M. A. (1970). "Statistics with Application to the Biological and Health Sciences." Prentice-Hall, Englewood Cliffs, New Jersey.

Schuller, P. O., Horwitz, W., and Stoloff, L. (1976). A review of sampling and collaboratively studied methods of analysis for aflatoxin. *J. Assoc. Off. Anal. Chem.* **59**, 1315–1343.

Shotwell, O. L., Goulden, M. L., and Hesseltine, C. W. (1974). Aflatoxin: Distribution in contaminated corn. *Cereal Chem.* **51**, 492–499.

Shotwell, O. L., Goulden, M. L., Botast, R. J., and Hesseltine, C. W. (1975). Mycotoxins in hot spots in grains. I. Aflatoxin and zearalenone occurrence in stored corn. *Cereal Chem.* **52**, 687–697.

Stoloff, L., and Dantzman, J. (1972). Preparation of lot samples of nut meats for mycotoxin assay. *J. Am. Oil Chem. Soc.* **49**, 264–266.

Stoloff, L., Campbell, A. D., Beckwith, A. C., Neshiem, S., Winbush, J. S., Jr., and Fordham, A. M. (1969). Sample preparation for aflatoxin assay. The nature of the problems and approaches to a solution. *J. Am. Oil Chem. Soc.* **46**, 678–684.

Whitaker, T. B., and Dickens, J. W. (1979a). Estimation of the distribution of lots of shelled peanuts according to aflatoxin concentrations. *Peanut Sci.* **6**, 124–126.

Whitaker, T. B., and Dickens, J. W. (1979b). Evaluation of the Peanut Administrative Committee testing program for aflatoxin in shelled peanuts. *Peanut Sci.* **6**, 7–9.

Whitaker, T. B., and Dickens, J. W. (1980). A water slurry method of extracting aflatoxin from peanuts. *J. Am. Oil Chem. Soc.* **57**, 269–272.

Whitaker, T. B., and Dickens, J. W. (1981). Errors in aflatoxin analyses of raw peanuts by thin layer chromatography. *Peanut Sci.* **8**, 89–92.

Whitaker, T. B., Dickens, J. W., and Monroe, R. J. (1972). Comparison of the observed distribution of aflatoxin in shelled peanuts to the negative binomial distribution. *J. Am. Oil Chem. Soc.* **49,** 590–593.

Whitaker, T. B., Dickens, J. W., and Monroe, R. J. (1974). Variability of aflatoxin test results. *J. Am. Oil Chem. Soc.* **51,** 214–218.

Whitaker, T. B., Whitten, M. E., and Monroe, R. J. (1976). Variability associated with testing cottonseed for aflatoxin. *J. Am. Oil Chem. Soc.* **53,** 502–505.

Whitaker, T. B., Dickens, J. W., and Monroe, R. J. (1979). Variability associated with testing corn for aflatoxin. *J. Am. Oil Chem. Soc.* **56,** 789–794.

3

Chemical Survey Methods for Mycotoxins

ODETTE L. SHOTWELL

U.S. Department of Agriculture
Agricultural Research Service
Northern Regional Research Center
Peoria, Illinois 61604

MODERN METHODS IN THE ANALYSIS
AND STRUCTURAL ELUCIDATION
OF MYCOTOXINS

I. INTRODUCTION

Studies to establish the occurrence of a mycotoxin in a given commodity are conducted to determine the geographical extent and levels of the contaimination and to determine the necessity for a surveillance program to prevent mycotoxins from entering the food chain or animal feeds. A study of the occurrence of a mycotoxin can be of a commodity at any stage of production: harvesting, transporting, storing, processing, or in the finished product. A survey can encompass any area—county, state, region, or nation—and cover any time period (Davis *et al.*, 1980). Before a survey is conducted, the purpose should be determined and a sampling protocol must be designed so that the objectives of the study will be attained. Choice of sampling procedures, discussed in Chapter 2, determines the reliability of the results of all surveys (Schuller *et al.*, 1976).

Preferably, analytical methods used should have been validated by collaborative studies and approved by recognized scientific societies, such as the Association of Official Analytical Chemists (AOAC), the American Oil Chemists Society (AOCS), the American Association of Cereal Chemists (AACC), and the International Union of Pure and Applied Chemistry (IUPAC). A collaborative study should include the determination of the repeatability as established by the coefficient of variation within a laboratory, reproducibility as determined by the coefficient of variation between laboratories, the limit of detection, and recoveries (AOAC, 1982; Anonymous, 1984). The "ruggedness" or repeatability of an analytical method should be established before a method is studied collaboratively. If an officially approved method is not available for the analysis of a mycotoxin in a given commodity, persons conducting a survey should, at least, determine recoveries at several levels of the mycotoxin, limit of detection, and repeatability of the method they select for the study.

Methods of analysis of mycotoxins fall into three categories: (1) presumptive tests that identify lots of cereal grains, oilseeds, or other commodities that are probably contaminated, (2) rapid screening tests, designed to establish the presence or absence of a mycotoxin, that are used to determine whether a lot should be accepted or rejected, and (3) quantitative methods that are used to determine levels of a mycotoxin in an agricultural commodity. Presumptive tests are usually based on visual inspection of a cereal grain or oilseed—sometimes under ultraviolet (UV) light. Cereal grains or oilseeds may be inspected for mold growth. Rapid screening tests involve an extraction, and a purification, that may include a distribution between solvents or precipitation of impurities with inorganic salts. Detection can be carried out for some mycotoxins by placing semipurified extracts on minicolumns inspected visually under UV light after development. Ideally, rapid screening methods should be carried out in 10 to 15 min. Methods that determine levels of mycotoxins are often complicated and time-consuming. They include an extraction, one or more purification steps, and measurement of mycotoxins in semipurified extracts by thin-layer chromatography (TLC), high-performance thin-layer chromatography (HPTLC), high-performance liquid chromatography (HPLC), gas chromatography using an electron capture detector (GC-EC) or a flame ionization detector (GC-FID), and gas chromatography–mass spectrometry (GC-MS).

For each mycotoxin the occurrence and programs of surveillance for appropriate commodities, and toxicities are summarized. The presumptive, screening, and quantitative methods that have been used in surveys and/or monitoring programs are described. An outline of the steps in the analysis will be given excluding details available in references, so readers will know techniques and equipment used in a given method. The time required for preparing material for quantitation depends on the number of the operations in the analysis such as filtrations, liquid–liquid partitions, column chromatographies, concentrations, and transfers. After the steps of a method are summarized, recoveries, limits of detection, coefficients of variation, and other results of collaborative studies are described.

II. AFLATOXINS

Aflatoxins are the mycotoxins causing the most concern worldwide because of their carcinogenic properties and occurrence as natural contaminants of a large number of agricultural commodities. Aflatoxins have been found in peanuts, cottonseed, corn, grain sorghum, rice, wheat, soybeans, pecans, walnuts, almonds, sunflower seeds, pistachio nuts, copra, dairy products, and figs. The levels and frequency of the incidence of aflatoxins in only three commodities— peanuts, cottonseed, and corn—in the United States has led to regulatory activity by the Food and Drug Administration (FDA). However, the incidence in pecans, almonds, pistachio nuts, and walnuts has indicated that continued surveillance of these commodities is warranted (Stoloff, 1976a). Aflatoxins are a group of chemically related difuranocoumarin compounds (Fig. 1) produced by species of the *Aspergillus flavus* series.

A. Peanuts

1. Surveillance

Aflatoxins were originally implicated as toxic animal feed contaminants when they were discovered in peanuts. From the first report of the possibility of naturally occurring carcinogens in peanuts and peanut products, there has been unusually close cooperation between government and industry in the initiation of an aflatoxin control program for peanuts grown in the United States and the development of analytical methods to implement the program (Dickens, 1977).

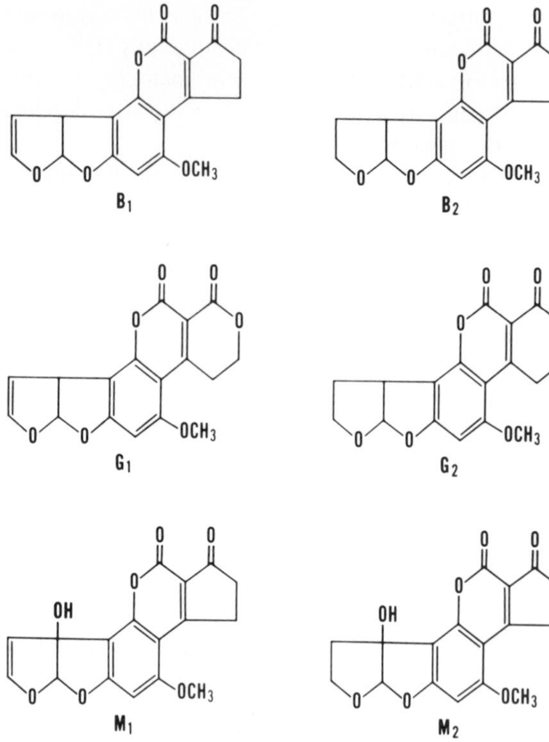

Fig. 1. Structures of aflatoxins.

The aflatoxin control program is administered by the Peanut Administration Committee (PAC) under provisions of a marketing agreement for peanuts (Anonymous, 1983). Members of the committee are growers and shellers from the three peanut production areas (Virginia-Carolina, Southeast, and Southwest).

2. Presumptive Test

The inspection of farmers' stock peanuts for visible *A. flavus* was the first example of a presumptive test for aflatoxins in a commodity (Dickens and Welty, 1969). One of the steps carried out by state-federal inspectors in monitoring peanuts under the requirements of PAC is the examination of peanuts for fungal growth under 20× or 40× magnification with a scanning microscope. If visible *A. flavus* is found during this test, the lot of peanuts is placed in segregation-3 storage to be crushed for oil, which is free of aflatoxin after refining, and the meal is not used for feed or food purposes. Meal produced from segregation-3 peanuts has been used for fertilizer. Lots of peanuts with more than 2% damaged kernels or more than 1% concealed damage caused by rancidity, molds other than *A. flavus,* or decay are placed in segregation-2 storage. Segregation-2 peanuts are crushed for oil and the meal is used for animal feed if chemical assays for aflatoxins do not indicate otherwise. Segregation-1 peanuts with damage less than 2% and no visible *A. flavus* are used for human consumption.

3. Quantitative Determination

All peanuts sold for human consumption must be tested for aflatoxins. After a prescribed sampling procedure, subsamples are prepared for analysis by the best foods (BF) method approved for peanuts by the AOAC (1984) and the AOCS (1973). The BF method was developed and studied collaboratively for approval by Waltking (1970). Steps in the method are as follows.

Extract blended, ground subsample with methanol–water (55 + 45 v/v) and hexane.

Separate phases by centrifugation.

Extract aqueous methanol phase with chloroform.

Evaporate chloroform layer almost to dryness and transfer to vial to be evaporated to dryness under gentle stream of nitrogen.

Dissolve residue in acetonitrile-benzene (2 + 98 v/v) for TLC on silica gel plates to be developed with acetone-chloroform (10 + 90 v/v) or other solvent.

Use any TLC system capable of performance outlined in official methods of AOAC (1984). Measure fluorescing zones on TLC plates by visual comparisons with standards, or densitometrically compare zones of unknowns with standards.

The BF method requires less time, effort, and equipment than other methods proposed for the determination of aflatoxin in peanuts and peanut products (e.g., CB method, see below). Recoveries of total aflatoxin from spiked peanut products were 60–94%, and coefficients of variation were 38–55%. The analysis of naturally contaminated peanut butters (48–49 ng/g) led to results with a coefficient of variation of 24–29%. It is difficult to spike uncontaminated peanut products with standard aflatoxin solutions because of the instability of the toxin to oxidation. In the same collaborative study it was determined that the BF method was equal in accuracy to the CB method as judged by recovery data and precision. The BF method takes one-third to one-fourth the time required for the CB method.

Eppley (1966) reported "a versatile procedure for assay of aflatoxins in peanut products" that was known as the CB method. It is called the CB method because it was developed and tested in the Contaminants Branch of the FDA (the branch no longer exists because of reorganization). It was considered to be a "rapid and sensitive method" (1–5 ng/g). The steps are summarized below.

Extract blended, ground samples with water–chloroform (1 + 10 v/v) in the
presence of Celite filter aid.

Chromatograph filtrate on a silica gel column and wash with hexane and
anhydrous ether.

Elute with methanol–chloroform (3 + 97 v/v).

Concentrate the eluate to near dryness and transfer quantitatively to a vial to
be evaporated to dryness under nitrogen for TLC.

Carry TLC out in the same manner as the BF method.

The CB method was evaluated collaboratively by Epply *et al.* (1968). Analyses of naturally contaminated peanut butter (average 94 ng/g total aflatoxin) and
peanut meal (264 ng/g) gave coefficients of variation of 31 and 37%, respectively. The coefficients of variation of total aflatoxin values for peanut butter
spiked at 30 and 80 ng/g were 37 and 46%, respectively, for all laboratories.
Recoveries of >100% were obtained on the peanut butter spiked at 30 ng/g,
indicating that components in the extract are contributing to the fluorescence of
the aflatoxin. As a result of this study, the method is approved for peanuts by the
AOAC (1984) and has been successfully applied to 50- and 1000-g samples.

The CB and BF methods for determining aflatoxin in peanuts and peanut
products have been compared in several studies. Chang *et al.* (1979) found that
the CB method yielded 60, 112, 35, and 22% higher results for aflatoxins B_1,
B_2, G_1, and G_2, respectively, for the four samples of peanut meals and six
samples of peanut butters analyzed than the BF method. Both reversed-phase
HPLC and TLC were used to measure aflatoxins in extracts. In a 2-year study of
results obtained in peanut meals in the AOCS Smalley Check Sample Series, it
was found that laboratories using the BF method averaged slightly lower results
than those using the CB method (McKinney, 1984). However, the difference
was not significant considering the magnitude of the coefficients of variations for
all analysts. In two international mycotoxin check sample programs (Friesen *et
al.*, 1980; Friesen and Garren, 1982), there were no significant differences
among means for aflatoxin obtained on raw peanut meal with the BF and CB
methods. However, for a finished peanut butter sample (Friesen *et al.*, 1980) and
a raw deoiled peanut meal (Friesen and Garren, 1982), significantly lower results
were obtained by the BF method compared with the CB method.

A number of methods for determining aflatoxins in peanut products by both
normal and reversed-phase HPLC have been proposed. One of the first was
proposed by Pons and Franz (1978), who used 0.1 *N* hydrochloric acid–methanol (1 + 4 v/v) for extraction and purified extracts by partition and a small (2 g)
silica gel column for HPLC. μ-Porasil silica gel (10 μm) was used for the HPLC
column, and UV absorbance or fluorescence was used for detection. DeVries and
Chang (1982) designed a rapid HPLC method for the assay of aflatoxins in corn
and peanuts. The method included a dichloromethane extraction in the presence

of basic cupric carbonate and evaporation to dryness of an aliquot of the dichloromethane solution. The residue was taken up in acetonitrile and the solution was washed with petroleum ether to remove lipid impurities. Aflatoxins were measured by reversed-phase HPLC of water adducts formed by treatment with trifluoroacetic acid. The rapid HPLC method gave results that compared favorably to those obtained by the CB method. The results of a collaborative study of the determination of aflatoxins in peanut butter using both normal and reversed-phase HPLC have been reported (Campbell et al., 1984). Both methods showed improved limits of detection and better within-laboratory precision over current AOAC methods. However, between-laboratory precision is no better, and the reversed-phase method shows evidence of interferences being measured. Neither method was submitted for approval.

The quantitation of aflatoxin in partially purified extracts of peanut butter by normal-phase HPLC and by HPTLC was compared (Tosch et al., 1984). With respect to precision, accuracy, sensitivity, recovery, and linearity of response, HPLC appears to be comparable to HPTLC.

4. Screening Methods

Rapid minicolumn methods have been used to detect aflatoxins in peanuts at the earliest buying points. The first such procedure was based on "millicolumn" chromatography (Holaday, 1968). It was rapid and simple and involved an extraction and development on a small column packed with silica gel by dipping the column in the extract. As the solvent ascended, a blue fluorescent zone, visible under UV light (365 nm) due to aflatoxin, moved up the column. Since then a number of improvements have been made in the minicolumn methods for peanuts and other commodities (Velasco, 1972; Holaday and Lansden, 1975). Minicolumns were developed in a descending mode.

Because of interest of the industry in rapid screening tests for raw peanuts, a collaborative study was conducted on two minicolumn methods, a modified Holaday method and the Holaday–Velasco method (Shotwell and Holaday, 1981). The Holaday method (Holaday and Lansden, 1975) was modified by substituting toluene for benzene in the solvent partition and dichloromethane for chloroform in the minicolumn development to eliminate use of the more hazardous solvents. The neutral alumina in the Holaday column was changed from activity V to activity III to provide a stable column. The Holaday–Velasco method used the extraction and cleanup of the Holaday method and the column designed by Velasco (1972), containing neutral alumina, silica gel, and Florisil (from top to bottom). As a result of the collaborative study in which 15 laboratories participated, the Holaday–Velasco method was recommended for adoption by the AOAC (1984), because there were no false positives and no misses in aflatoxin detection at levels of 13 to 20 ng/g in raw peanuts. The steps are outlined below.

Mix blended, ground peanut subsample with methanol–water (4 + 1 v/v) in a blender.

Filter aliquot of extract into separatory funnel and add salt solution (zinc acetate, sodium chloride, and acetic acid in water).

Shake vigorously and filter aliquot into second separatory funnel.

Extract with toluene and add aliquot of extract to top of minicolumn packed with calcium sulfate, neutral alumina, silica gel, Florisil, calcium sulfate (top to bottom).

Develop minicolumn with dichloromethane using vacuum to hasten development.

If aflatoxin is present a blue fluorescent zone will appear at the top of the Florisil layer.

The Holaday–Velasco method takes 10 min and has a detection limit of 5 to 10 ng/g. It is reliable when applied to a sample representative of a given lot of raw peanuts properly ground and blended to obtain representative subsamples for the test. It is used to accept or reject a lot of peanuts at a given level (20 ng/g) of aflatoxin or to determine the range of the level (e.g., 20–100 ng/g) of aflatoxin in a lot.

B. Cottonseed

1. Surveys and Occurrence

Aflatoxin-contaminated cottonseed was first implicated in an outbreak of hepatoma in rainbow trout. Even before a relationship had been established between aflatoxin contamination of cottonseed and hepatoma in trout, the occurrence of aflatoxin was already being studied by Whitten (1968) in a cooperative effort of the United States Department of Agriculture (USDA) and National Cottonseed Products. Samples were collected during the harvest season from processing plants to be representative of the production. As a result of such early surveys, the California Imperial Valley and southern Arizona were identified as having the most severe aflatoxin problem in cottonseed (Stoloff, 1976a). Aflatoxin also occurs in irrigated areas of the southern Rio Grande valley in Texas, but to a lesser degree. Surveillance data for 1971, 1974, and 1975 collected by the FDA revealed the same regional pattern of aflatoxin contamination of cottonseed, but there were changes in incidences and levels from year to year. In 1967, aflatoxin contamination of cottonseed was particularly severe in the California Imperial Valley. All samples tested contained aflatoxin with a mean aflatoxin B_1 level of 1400 ng/g.

2. Presumptive Test

In 1954, the characteristic bright greenish yellow fluorescence (BGYF) under UV light (365 nm) was first noticed in cotton fibers. It is now the basis of the presumptive test for aflatoxin in both cottonseed and corn (Marsh and Simpson, 1984). BGYF was first associated with the presence of the fungus *A. flavus* in cotton bolls. The term "cat-eye cotton" came into use to designate such fibers. After nine surveys of cotton crops grown between 1957 and 1981, no doubt remained that there was a relationship between BGYF and *A. flavus* boll rot of cotton. Later, BGYF was found to be associated with the presence of aflatoxin in variable, but frequently high levels. A close quantitative parallel between amounts of BGYF fiber and aflatoxin levels has not been established. BGYF is commonly used to identify lots of cottonseed likely to contain aflatoxin. Cotton fibers or seeds are inspected in a darkened room under UV light (365 nm).

Marsh *et al.* (1969) postulated that BGYF is formed through an interaction between kojic acid formed by *A. flavus* and a heat-labile peroxidase from the host. The heat-labile constituent appeared to be necessary for BGYF formation in cotton fiber incubated with *A. flavus,* since a steam-sterilized cotton lock incubated with the fungus did not produce BGYF. The heat-labile constituent was not unique to cotton fibers, because BGYF was also formed when undried corn, milkweed fiber, strawberry fruits, panic grass, and foxtail grass all formed BGYF at the same R_f value as that from cotton fiber when incubated with *A. flavus*. Bothast and Hesseltine (1975) reported that inoculation with *A. flavus* and incubation of wheat, oats, grain sorghum, and barley resulted in the production of BGYF. The fluorescence did not form on the legumes, peanuts, and soybeans, under the same conditions.

3. Screening Methods

Although rapid-screening minicolumn methods have been developed that are applicable to cottonseed and cottonseed products, they have not been used as much as those for corn and peanuts. The first one reported used acetonitrile–water (80 + 20 v/v) as the extracting solvent and partition into benzene to effect a fourfold concentration of aflatoxin (Cucullu *et al.*, 1972). The benzene solution was wicked onto the bottom of a minicolumn packed with acidic alumina and silica gel (bottom to top) and developed ascendingly with chloroform–acetonitrile–2-propanol (93 + 5 + 2 v/v/v). The method took 15 min to complete with a limit of detection of 10 ng/g.

A second screening method reported was applicable to cottonseed meal as well as to corn, peanuts, peanut butter, pistachio nuts, to chicken, pig, and turkey starter rations, and to dairy cattle feed with limits of detection of 5 to 15 ng/g (Romer, 1975). Because of the number of steps, as listed below, required to purify extracts of all of the commodities, the method took longer than most minicolumn procedures.

Extract 50-g sample of a blended, ground commodity in a blender 3 min with acetone–water (85 + 15 v/v) and filter.

To an aliquot of filtrate, add 0.2 N sodium hydroxide and ferric chloride slurry and mix.

Add basic cupric carbonate and mix. Filter.

Extract acidified (0.03% sulfuric acid) filtrate with chloroform and wash chloroform layer with 0.02 N potassium hydroxide.

Transfer aliquot of chloroform extract to minicolumn packed with calcium sulfate, Florisil, silica gel, neutral alumina, and calcium sulfate (bottom to top). Calcium sulfate is used to dehydrate solvents; neutral alumina and silica gel remove impurities from extracts; and Florisil adsorbs aflatoxins.

Develop column with chloroform–acetone (9 + 1 v/v).

Examine column under UV light (365 nm) in a darkened room or in a Chromatoview cabinet. If aflatoxin is present, there will be a blue fluorescent zone at the top of the Florisil layer. Compare with a minicolumn developed with an extract of aflatoxin-free commodity and with one developed with an extract of aflatoxin-free commodity containing a known amount of standard solution of aflatoxin to represent a definite toxin level in the commodity.

The minicolumn method of Romer (1975) was studied collaboratively by Romer and Campbell (1976) and approved by the AOAC (1984) and the AACC (1983) for the listed commodities (mixed feeds, corn, peanut meal, pistachio nuts), including cottonseed.

4. Quantitative Methods

Methods for determining levels of aflatoxins B_1 and B_2 in cottonseed and cottonseed products have been available for surveys, detoxification studies, and regulatory purposes since 1965 (Pons et al., 1980). A collaborative study was made by Pons (1975) on a rapid method for determining aflatoxins in cottonseed products that eliminated a step that required boiling. Where chloroform and benzene had been used in earlier methods, less toxic solvents, dichlormethane and toluene, were substituted. Quantitation was accomplished by TLC, and coefficients of variations were 28, 35, and 35% for B_1, respectively, in meals, meats, and ammoniated meals. Aflatoxins G_1 and G_2 have not been detected in cottonseed. Minor modifications were made in the rapid method for cottonseed products, and aflatoxins were measured by TLC or HPLC to give the method outlined below (Pons et al., 1980).

Extract 50-g samples with acetone–water (85 + 15 v/v) and filter.

Add water and lead acetate solution (200 g lead acetate, hydrated, plus 3 ml acetic acid in 1 liter water) to aliquot of filtrate. Zinc acetate can be used instead.

Let stand to flocculate precipitate, add filter aid, and filter.

Extract filtrate with dichloromethane and drain through layer of Whatman
CF-11 cellulose powder or anhydrous sodium sulfate.

Concentrate dichloromethane for chromatography on silica gel.

Wash column with toluene–acetic acid (9 + 1 v/v) and ether–hexane (3 +
1 v/v).

Elute aflatoxins with dichloromethane–acetone (9 + 1 v/v).

Measure aflatoxins by TLC or by normal-phase HPLC, using either a UV or
fluorescence detector.

In the collaborative study of the above method (Pons *et al.*, 1980) the coefficients of variation for aflatoxin B_1 between laboratories were 33.2, 25.5, and 25.7%, respectively, by TLC, HPLC with a UV detector, and HPLC with a fluorescence detector. The method has been accepted in official first action by the AOAC (1984), and the AOCS (1983). In the Smalley Check Sample Program for the AOCS, the results for cottonseed meal show a within-laboratory precision in the range reported for the collaborative studies (McKinney, 1984). Although the between-laboratory precision has improved since 1980, it is still far from the range observed in collaborative studies.

C. Corn

1. Surveys and Occurrence

The first studies of natural occurrence of aflatoxin in corn of all grades moving in commercial channels were made in 1964, 1965, 1967, 1968, and 1969 (Shotwell, 1977) (Table I). The results of the studies did not appear alarming because of the low incidences (2.1–2.7%) and low levels (3–37 ng/g) of total toxins. Aflatoxin was usually found in samples from the poorest grades. Also, the samples used in these studies originated in the Midwest. The studies summarized in Table I cover an 11-year period and indicate that corn grown in certain regions is more subject to aflatoxin contamination than in others. Surveys of corn grown in the South from 1969 to 1973 showed that 13–32% of the samples analyzed contained 20 or more ng/g aflatoxin. However, isolated outbreaks of aflatoxin also have been reported in the Midwest (e.g., southwestern Iowa, 1975; southern Illinois, 1980; and limited areas of the Midwest, 1983).

Aflatoxin contamination of corn in the Southeast was a serious problem in 1977 and 1980. Economic losses have been experienced by growers, elevator owners, feed manufacturers, and livestock feeders (Nichols, 1983). A large part of the cost of the problem resulted from programs of surveillance of corn for aflatoxin that had to be initiated by state agencies and industry. Variances to the guidelines granted by the FDA over the years to affected areas have necessitated further increases in these programs. In 1978, the FDA advised southeastern states they could feed intrastate corn harvested in 1977 containing up to 100 ng/g

TABLE I

Aflatoxin Incidence in Corn

Year	Agency surveying	Sample origin	Type of samples and source	Number of samples assayed	Percentage samples with indicated level of aflatoxin (ppb)				
					ND[a]	<20	20–49	50–100	>100
1964–1965	NRRC	Corn Belt	Grain inspection AMS	1311	98	2	<0.1	—	—
1965	Wet milling industry	Corn Belt	Corn received by industry	372	97	3	—	—	—
1967	NRRC	Corn Belt	Grain inspection AMS	283	98	1	1	—	—
1968–1969	NRRC	Export cargo	Grain inspection AMS	293	97	2	1	—	—
1969–1970	NRRC	South	Grain inspection AMS	60	65	13	5	8	8
1971	NRRC	Southeast Missouri	Stored ASCS white corn	1283	68	18	7	4	2
1972	FDA	Corn Belt	Elevator and food processing plants	223	98	2	—	—	—
1973	NRRC	South Carolina	Field—freshly harvested	297	49	19	17	8	7
1973	FDA	Corn Belt	Farm and country elevator	169	98	2	—	—	—
1973	FDA	South[b]	Farm and country elevator	146	64	22	7	4	2
1975	NRRC	Iowa	Field—freshly harvested	214	83	15	1	<1	—

[a] ND, Not detected.

[b] Includes Southeast, Appalachia, southeastern Missouri, Kentucky, Tennesee, Oklahoma, Texas, and California.

to mature poultry and swine, and mature non-milk-producing cattle, provided they could prove that such corn would not be used for human consumption (Federal Register, 1978). Blending was also permitted under the same restrictions. Similar exemptions from the 20 ng/g guideline for feed corn and from regulations against blending were allowed for corn harvested in 1980 in North and South Carolina and in Virginia for interstate shipment (Federal Register, 1981). When the aflatoxin problem became more prevalent in corn in 1983, the FDA proposed a cooperative program regarding interstate shipment similar to those in previous years. Any state could participate that had a control program in place that was approved by the FDA (Federal Register, 1983).

Over the years it has become increasingly important to have reliable procedures available for determining levels of aflatoxin in corn that is to be exported or processed into food or feed. Several analytical methods developed for peanuts and cottonseed were applicable to corn with slight modification; others were not. The methods applied to or developed for corn vary in purpose and complexity (Shotwell, 1983).

2. Presumptive Test

The BGYF associated with the presence of *A. flavus* or *Aspergillus parasiticus* was first observed in corn samples naturally contaminated with aflatoxin collected in 1969 to 1970 from the South (Shotwell *et al.*, 1972). Previously BGYF had been seen in corn kernels inoculated in the laboratory with *A. flavus* (Marsh and Simpson, 1984). Analysis of BGYF kernels from 10 corn lots revealed aflatoxin B_1 levels of 284 to 101,000 ng/g (Shotwell *et al.*, 1974). In the same study, kernels with white, blue, or orange fluorescence under UV light (365 nm) contained no aflatoxin. The BGYF or ''black light'' presumptive test for aflatoxin in corn is widely used by the corn industry and by state and federal agencies, including regulatory agencies. The test has been accepted because false negative tests rarely occur. The BGYF test is used to identify lots of corn that should be further analyzed to determine the presence and levels of aflatoxin.

The BGYF test consists of examining a 4.5-kg sample of corn representative of an entire lot under ''black'' or long-wave UV light (365 nm) in a darkened room or chamber. Although a high-intensity light is recommended, lower intensity light may be used in complete darkness. The eyes of the operator must be protected from UV light by appropriate goggles or other means to prevent possible damage from continued exposure. The test is easily adapted for field use. False positive results can be reduced by use of a color reference standard such as Tinopal BHF (Ciba-Geigy Corporation, Greensboro, North Carolina 27409). BGYF has a bright glow, sometimes called a firefly glow, that differentiates it from other fluorescent materials in corn. When a lot of BGYF positive corn is identified, it should be further analyzed for aflatoxin.

It has been recommended that corn be cracked or coarsely ground before critical examination, because BGYF can occur under the seed coat (Shotwell *et al.*, 1974). Sometimes it can be detected as a dull gold color under the seed coat, usually in the germ area, and becomes fully visible when cracked. Most of the corn samples with aflatoxin levels equal to or more than 20 ng/kg have had at least one or two particles or kernels with visible BGYF per kilogram of sample without cracking. A black light viewing apparatus (CPC International, Argo, Illinois) has been described by Barabolak *et al.* (1978), in which corn is discharged in a monolayer onto a vibrating feeder tray moving under UV light (365 nm). The feeder tray and light are enclosed in a cabinet with a viewing port which protects the viewer's eye from UV light. A comparison was made of the BGYF kernels and particles observed in 4.5-kg corn samples in the black light viewer, and the number of BGYF particles counted in the stream from the Straub disk mill at the same samples were coarsely ground (Shotwell and Hesseltine, 1981). It was concluded that the BGYF test can be carried out equally well by using the black light viewer on whole-kernel corn as by inspecting a stream of coarsely ground corn from a mill under UV light (365 nm). Barabolak *et al.* (1978) developed the black light viewer for use in elevators where the creation of dust in a grinding process is highly undesirable. If the airborne dust contains aflatoxin, it presents a potential hazard to workers inhaling the dust. In laboratories where efficient hoods are available, examination of streams of coarsely ground corn from mills is convenient, and the process facilitates the preparation of finely ground samples for aflatoxin analysis.

Attempts have been made to determine aflatoxin levels in corn by the black light test either by the number of BGYF particles per kilogram or by weight percentage of BGYF particles in whole-kernel samples. There was a relationship between number of BGYF particles and kernels and aflatoxin levels in 1971 white corn harvested in southeastern Missouri. The correlation, however, was not high enough to encourage use of the number as the only indication of the aflatoxin content (Shotwell *et al.*, 1975). Barabolak *et al.* (1978) also concluded that BGYF count cannot be used as a quantitative measure of aflatoxin levels in corn. They found that only one-third of the corn lots having more than four BGYF particles per kilogram had more than 20 ng/g aflatoxin. However, BGYF counts can be used to determine whether samples should be further tested for aflatoxin (Shotwell and Hesseltine, 1981).

There have been two studies on the prediction of aflatoxin levels by weighing corn particles exhibiting BGYF in uncracked corn samples. As 1971 white corn was delivered at an elevator in southeastern Missouri in 1972, BGYF material was separated from each sample and weighed. Results obtained by predication equations were so variable that estimates of aflatoxin levels based on weight of BGYF material would be too imprecise for practical use for these samples (Kwolek and Shotwell, 1979). Dickens and Whitaker (1981) studied aflatoxin levels in 1977 and 1978 yellow corn marketed in North Carolina as determined by the CB method, approved by the AOAC and AACC, and by weight of the BGYF particles separated from the same samples. Approximately the same percentage of lots were accepted by both methods, although the same lots were not always accepted by both methods. The authors concluded that more research was required to determine the efficacy of the two methods under a variety of conditions.

3. Screening Methods

Over the years a number of rapid screening methods for aflatoxin in corn have been reported, involving the use of minicolumns, TLC, and a fluorometric iodine method (Shotwell, 1983). Minicolumn screening methods are most widely used and all include the following steps: extraction, purification of the extracts, concentration, and development on a minicolumn for detection. When contamination of corn by aflatoxin reached crisis proportions in the Southeast in 1977 (Nichols, 1983), use of minicolumns to detect the mycotoxin became widespread. A collaborative study of the methods being used (Shannon and Shotwell, 1979) resulted in the adoption of the Holaday–Velasco method, described earlier for peanuts, by the AOAC (1984) and the AACC (1983) in official first action. The limit of detection by experienced analysts was 5 ng/g. Participation of 23 laboratories in the collaborative study is evidence of the interest in minicolumn methods for detecting aflatoxin in corn. The Federal Grain Inspection Service has modified the Holaday–Velasco method for use in their export grading facilities, on demand of the buyers to certify corn as having less than 20 ng/g. Laboratories in large elevators or grain terminals use the minicolumn method for "go–no go" situations. Some states use it to determine whether corn has less than 20 ng/g, 20–100 ng/g, or more than 100 ng/g.

One of the most effective and efficient screening methods is that reported by Dantzman and Stoloff (1972); it can be applied in laboratories that have the capability of performing TLC. A ground, blended corn sample is extracted with chloroform–water by the CB method. The oily residue from the extract is spotted on a TLC plate and first developed with anhydrous ether to move lipid impurities to the solvent front. After drying, the plate is developed in the same direction with chloroform–acetone (90 + 10 v/v) for aflatoxin. Seventeen extracts can be screened with one standard on a TLC plate (20 × 20 cm). When a ground corn sample is extracted by the CB method, two 50-ml aliquots of chloroform can be collected, one of which can be saved for quantitation if aflatoxin appears to be present by the rapid TLC method. The method is approved by the AOAC (1984) and AACC (1983). It was used successfully in 1977 to survey 928 samples collected in the Midwest when reports of aflatoxin contamination threatened the orderly marketing of corn (Shotwell et al., 1980a). Results indicated that aflatoxin in corn was not a problem in 1977 Midwest corn.

4. Quantitative Methods

Methods of determining levels of aflatoxin in corn include TLC, both reversed- and normal-phase HPLC, HPTLC, and the fluorometric iodine method (Shotwell, 1983). The CB method described earlier for peanuts was approved for corn in official first action by the AOAC (1984) and the AACC (1983), after an international collaborative study (Shotwell and Stubblefield, 1972). The International Check Sample Program provided a cornmeal sample free of charge for aflatoxin analysis to 182 laboratories requesting samples (Friesen *et al.,* 1980). Of the 139 laboratories from 34 countries who provided results, 47 used the CB method; the coefficient of variation of results obtained by all methods was 73%. In a second International Check Sample Program, 42 of 142 laboratories reporting results used the CB method with a coefficient of variation of 51% between laboratories (Friesen and Garren, 1982). In a report on results of the AOCS Smalley Check Sample Series of 1980 to 1981 and 1981 to 1982, it was stated that the between-laboratory coefficients of variation for aflatoxin B_1 was around 50% (McKinney, 1984). Most of the laboratories had used the CB method for corn. The disadvantage of the CB method is that large quantities of chloroform, an expensive and a relatively toxic solvent, and ether, a flammable solvent, are used. Dichloromethane can be substituted for chloroform.

Since HPLC equipment has become available, a number of laboratories have quantitated aflatoxins in purified extracts of corn by HPLC. The advantages are that quantitation can be automated and degradation of aflatoxins that sometimes takes place on TLC plates is avoided.

Since 1978, a large number of corn samples have been analyzed by a method that quantitates levels of aflatoxin by normal-phase HPLC in a silica gel-packed flow cell and measured by UV fluorescence (Thean *et al.,* 1980). The steps in the method are summarized below.

> Extract blended, ground subsample with methanol–water (80 + 20 v/v) and filter.
> Treat aliquot of filtrate with 0.61 *M* ammonium sulfate and filter.
> Extract filtrate with chloroform and evaporate chloroform solution to dryness.
> Dissolve residue in chloroform for cleanup on Sep-Pak silica gel cartridge.
> Aflatoxin in residue from eluate of cartridge was measured by normal-phase HPLC.

The method has been applied in surveys of corn and grain sorghum (Wilson *et al.,* 1984; D. M. Wilson, personal communication).

A rapid reversed-phase HPLC method for determining aflatoxin in corn as well as in peanuts has been reported and is being used on corn (DeVries and

Chang, 1982). Comparisons of results obtained by the rapid HPLC method and the CB method indicate there is no significant difference between results.

D. Confirmatory Tests

Analytical results obtained for levels of aflatoxin in agricultural commodities should be supported by proof of identity, so-called confirmatory tests. However, confirmation of identity is not always necessary in cases of commodities known to have high incidences of aflatoxin and for which reliable analytical methods exist. Tests devised for confirmation of identity have been based on toxicological effects, formation of derivatives, color reactions, and mass spectrometry (Nesheim and Brumley, 1981). The method used most often in surveys has been the acid-catalyzed hydration of the vinyl ether group in aflatoxins B_1 and G_1 (Przybylski, 1975; AOAC, 1984; AACC, 1983). Aflatoxins B_1 and G_1 as standards and in unknowns spotted on TLC plates alone and in admixtures are treated with trifluoroacetic acid to form the water adduct. Plates are developed with isopropanol–acetone–chloroform (5 + 10 + 85 v/v/v). Negative-ion chemical ionization mass spectrometry of aflatoxins has been used to confirm the presence of the toxin in corn and peanuts (Brumley et al., 1981). It is feasible for regulatory purposes. The method has been studied collaboratively and has been approved in official first action by the AOAC. More recently, tandem mass spectrometry, also called mass spectrometry–mass spectrometry, has been used to confirm the presence of aflatoxin in crude extracts (Plattner et al., 1984). No matter how elegant or specific the method, confirmatory tests will not correct false positive results that are due to previously contaminated equipment or reagents. All equipment and reagents should be checked out by carrying out all of the steps of the analysis with the equipment and reagents to be used in the analysis. Laboratories that are doing feeding or production studies involving large quantities of aflatoxin as well as studies on the natural occurrence of the toxins are particularly vulnerable to cross-contamination.

E. Aflatoxin M_1

1. Surveys and Occurrence

Aflatoxin M_1 (Fig. 1) is the carcinogenic metabolite that appears in the milk of dairy cattle ingesting aflatoxin in their feed. It causes concern because milk is a major part of the diet of children; the FDA has set a tolerance of 0.5 ng/g aflatoxin M_1. In years when weather conditions are favorable for formation of aflatoxin (e.g., hot, drought), surveillance is increased not only of corn and cottonseed, but of milk for aflatoxin M_1. Contamination of corn with aflatoxin in the Southeast in 1977 resulted in increased incidence and levels of M_1 in milk

(Nichols, 1983). In 1980, aflatoxin contamination of cottonseed in Arizona was a severe problem for dairy farmers.

2. Quantitative Methods

Although a rapid screening minicolumn method was reported for aflatoxin M_1 (Holaday, 1981), it has not been widely used because of difficulties encountered in detecting the mycotoxin on the minicolumn. There are two quantitative TLC methods approved for M_1 in dairy products by the AOAC (1984). The first method (II) approved in official first action by the AOAC (1980) is applicable to milk, cheese, and butter (Stubblefield and Shannon, 1974). It is rather lengthy, as seen by the steps listed below.

> Add appropriate amount of water to dairy product and extract with acetone. Filter.
> Add lead acetate solution to aliquot of filtrate to precipitate impurities that interfere with TLC.
> Add sodium sulfate and diatomaceous earth and filter.
> Wash aliquot of filtrate in a separatory funnel with hexane and extract aflatoxin with chloroform.
> Dry combined chloroform extracts with sodium sulfate.
> Evaporate chloroform solution to dryness and dissolve residue in chloroform (1 ml), benzene (2 ml), and hexane (10 ml) for chromatography on cellulose powder column.
> After solution is placed on cellulose powder column, wash with hexane–benzene (3 + 1 v/v) and then hexane–ether (2 + 1 v/v).
> Elute aflatoxin M_1 (B_1 also, if present) with hexane–chloroform (1 + 1 v/v).
> Evaporate eluate to dryness for TLC.
> Measure aflatoxin M_1 by TLC on silica gel plates developed with isopropanol–acetone–chloroform (5 + 10 + 85 v/v/v).

The second quantitative method for aflatoxin M_1 (I) and a confirmatory test, approved for official first action by the AOAC (1984), has also been adopted as a reference method by IUPAC. The method and a confirmatory test for M_1 (van Egmond and Stubblefield, 1981) were both validated as a result of a joint AOAC–IUPAC collaborative study (Stubblefield et al., 1980). The method is applicable to milk and cheese and is an improvement over the method described before in that it has fewer steps.

> Partition dairy sample, sodium chloride solution, and chloroform in a separatory funnel. Filter chloroform layer in presence of sodium chloride.
> Add filtrate to a silica gel column and wash with toluene–acetic acid (9 + 1 v/v), hexane, and hexane–ether–acetonitrile (5 + 3 + 1 v/v/v).

Elute aflatoxin M_1 with chloroform–acetone (3 + 1 v/v) and evaporate eluate to dryness for TLC.

Measure M_1 in extracts of dairy products by TLC on silica gel plates developed with isopropanol–acetone–chloroform (3 + 10 + 87 v/v/v).

Measure M_1 in purified extracts by two-dimensional TLC on silica gel. Develop the plate in the first direction with water–methanol–ethyl ether (1 + 4 + 95 v/v/v) and in second direction with isopropanol–acetone–water (3 + 10 + 87 v/v/v).

There have not been as many participants in the Smalley series for aflatoxin in milk as one would want for statistical evaluations of analyses (McKinney, 1984). There have only been 8–13 values reported for each of the seven samples in the two series discussed. Evaluations of analyses for aflatoxin in milk reveal large between-laboratory coefficients of variations for all samples. The coefficients of variation for samples analyzed in the 1980–1981 series was 143% and in the 1981–1982 series, 98%. Few analysts, however, used officially approved methods.

3. Confirmatory Test

Treat aflatoxin M_1 as standards and in unknown sample alone and in admixtures with trifluoroacetic acid (TFA)–hexane (1 + 4 v/v) (van Egmond and Stubblefield, 1981).

Remove excess TFA by heating plate and develop with water–acetic acid–methanol–chloroform (0.4 + 1 + 1 + 46 v/v/v/v).

4. High-Performance Liquid Chromatography

A rapid HPLC method has been reported by Ferguson-Foos and Warren (1984). They have used this method with success to monitor fluid milk for aflatoxin M_1 in Florida.

Adsorb aflatoxins M_1 and M_2 from warm diluted (1 → 2) milk sample (20 ml) onto a C_{18} Sep-Pak cartridge.

Wash cartridge with water–acetonitrile (95 + 5 v/v).

Transfer M_1 and M_2 from cartridge to silica gel 60 column with ether.

Elute M_1 and M_2 with methylene chloride–ethanol (95 + 5 v/v).

Evaporate eluate under nitrogen.

Measure aflatoxins M_1 and M_2 by either normal or reversed-phase HPLC.

The rapid HPLC method for aflatoxins M_1 and M_2 in milk was tested in an international collaborative study (Stubblefield and Kwolek, 1986). Recoveries averaged 93.7% for aflatoxin M_1 and 109.8% for aflatoxin M_2. Coefficients of variation averaged 21.4% for M_1 and 35.9% for M_2. The method using reversed-phase HPLC was adopted in official first action by the AOAC. Although the data for normal-phase HPLC were satisfactory, they were insufficient to give a meaningful analysis of variance. Normal-phase HPLC could not be recommended for adoption.

III. ZEARALENONE

A. Surveys and Occurrence

Zearalenone (Fig. 2) is an estrogenic metabolite from *Fusarium* first isolated and characterized by Stob *et al.* (1962). It has been found in cereal grains and feeds associated with feed refusal and hyperestrogenism in swine (Mirocha *et al.*, 1976, 1977). Outbreaks of zearalenone are more prevalent in northern corn-growing areas than in southern areas, although zearalenone occasionally occurs as far south as Georgia (Christensen, 1979). Surveys of corn (Shotwell *et al.*, 1970, 1971; Eppley *et al.*, 1974), wheat (Shotwell *et al.*, 1977), and grain sorghum (Shotwell *et al.*, 1980b) moving in commercial channels detected zearalenone usually in lower levels and in poorest grades. Low levels of zearalenone were found in 10% of the samples collected of 1973 marketable corn stored on farms and in elevators in the Midwest (Stoloff *et al.*, 1976). One-third of the samples of preharvest grain sorghum collected from 64 fields in the Georgia coastal plain in 1980 and 1981 had levels of zearalenone ranging from 2 to 1468 ng/g (McMillian *et al.*, 1983). In the same study, no zearalenone was detected in preharvest grain sorghum collected from 15 fields in 1981 in the Mississippi Delta.

Analytical methods used to determine zearalenone in cereal grains and feeds moving in commercial channels, in preharvest grains, and in feeds implicated in estrogenic problems in farm animals have been reviewed by Bennett and Shotwell (1979).

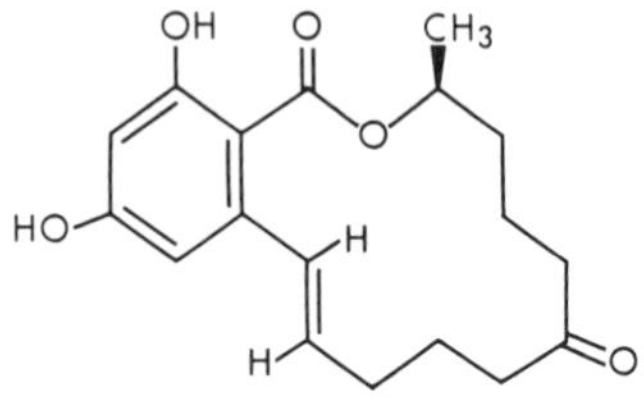

Fig. 2. Structure of trans-zearalenone.

B. Screening Methods

The most widely used screening method for zearalenone has been the one reported by Eppley (1968) for zearalenone, aflatoxin, and ochratoxin. It has been applied to surveys of corn, wheat, soybeans, and grain sorghum. The advantage is that only one extraction and one silica gel chromatography are required for all three mycotoxins, as seen below.

> Extract a blended ground sample of the cereal grain or soybeans with chloroform–water (10 + 1 v/v) in the presence of Celite filter aid, and filter.
>
> Place an aliquot of the filtrate on a column.
>
> Wash column first with hexane and then with benzene.
>
> Elute zearalenone with acetone–benzene (5 + 95 v/v).
>
> Wash column with anhydrous diethyl ether.
>
> Elute aflatoxin with methanol–chloroform (3 + 97 v/v).
>
> Elute ochratoxin with glacial acetic acid–benzene (1 + 9 v/v).
>
> Evaporate eluates for TLC. Benzene or acetonitrile–benzene (2 + 98 v/v) are used to dissolve the three residues for detection and measurement of the mycotoxins on silica gel TLC plates.
>
> The TLC developing solvent for zearalenone is ethanol–chloroform (5 + 95 v/v) or glacial acetic acid–benzene (1 + 9 v/v).
>
> The developing solvent for aflatoxin is acetone–chloroform (1 + 9 v/v).
>
> The developing solvent for ochratoxin is glacial acetic acid–benzene (1 + 9 v/v) or toluene–ethyl acetate–90% formic acid (5 + 4 + 1 v/v/v).

The limits of detection for zearalenone, aflatoxin, and ochratoxin are 200, 5, and 50 ng/g, respectively. Some aflatoxin tends to elute with zearalenone and in the diethyl ether washes; so aflatoxin, if present, has to be quantitated separately by the CB method.

More recently, a rapid TLC method for the determination of zearalenone and zearalenol in grains and animal feeds has been reported (Swanson et al., 1984). It has been applied to corn, wheat, barley, millet, swine rations, and a pelleted feed sample.

> Extract blended ground sample with methanol–water (75 + 25 v/v), and filter.
>
> Precipitate pigments with lead acetate solution from aliquot of filtrate.
>
> Defat filtrate from precipitation with hexane.
>
> Partition mycotoxins into toluene–ethyl acetate (9 + 1 v/v).
>
> Remove solvent by evaporation.
>
> Dissolve residue in acetonitrile–benzene (2 + 98 v/v) for TLC on HPTLC plates with preadsorbent area.

Develop plates with chloroform–ethanol (95 + 5 v/v) or benzene–acetone
(9 + 1 v/v).

Spray plates, after inspection under short-wave UV light (254 nm) for
fluorescence, with Fast Violet B salt solution followed by borate buffer
solution.

Observe presence of pink spots of zearalenone and zearalenol. They turn
violet when sprayed with 30% sulfuric acid in methanol.

Some complex feed samples require a cleanup on a Florisil column before
HPTLC. The limits of detection are 80 and 200 ng/g for zearalenone and
zearalenol. Ten samples can be analyzed in 2 hr.

C. Quantitative Methods

Previously, the only approved method for determining levels of zearalenone in
an agricultural commodity was a modification of the screening method developed by Eppley (1968) for zearalenone, aflatoxin, and ochratoxin. Zearalenone
in the residue from the eluate from silica gel columns can be quantitated by TLC,
either by visual comparisons under short-wavelength UV light (254 nm) or by
densitometric scanning at excitation wavelength of 313 nm and fluorescence
measurement of 443 nm. In a collaborative study (Shotwell *et al.*, 1976), the
coefficient of variation for all samples between laboratories was 40.5%. The
method was approved for corn in official first action by the AOAC (1984) and
the AACC (1983). The limit of detection is 200 ng/g.

A method reported by Mirocha *et al.* (1974) was widely used to assay grains
and feeds implicated in hyperestrogenism in swine as well as for a survey of
preharvest grain sorghum. Although the extraction and purification steps are
lengthy, procedures for quantitation and verification of identity were developed
that are still used worldwide.

Extract blended ground sample with ethyl acetate, decant, and filter two
times.

Concentrate combined filtrates to near dryness and redissolve in chloroform.

Place chloroform solution in a separatory funnel, and layer with 1 N sodium
hydroxide, rotating funnel gently to partition zearalenone into basic
solution.

Separate chloroform layer and treat it again with 1 N sodium hydroxide in a
separatory funnel.

Combine two aqueous layers and wash with chloroform.

Adjust pH of aqueous layer to 9.5 with 2 N phosphoric acid.

Extract aqueous layer (pH 9.5) twice with chloroform and dry with anhydrous sodium sulfate.

Evaporate combined extracts to dryness for either TLC or GC-MS.

Measure zearalenone in the residue on TLC plates by developing with chloroform–ethanol (97 + 3 v/v) and detecting zones under UV light (254 nm) (blue-green fluorescence) or by spraying with 50% sulfuric acid in methanol (yellow to brown color) or potassium ferrocyanide–ferric chloride (blue color).

The other method of measuring zearalenone in the residue from the cleanup prodceure is by GC-MS of dimethoxy, trimethylsilyl ether (TMS), or TMS methyloxime derivatives of zearalenone. The advantage of GC-MS is that minute quantities of a compound can be identified precisely by the fragmentation pattern. The method of identification is known as multiple-ion detection or selected-ion monitoring (Mirocha et al., 1974).

A quantitative HPLC method using fluorescence detection has been reported for zearalenone and α-zearalenol. It has a limit of detection of 10 ng/g (Ware and Thorpe, 1978). The procedure was modified by Bagnaris (1986). The method has fewer steps than others and is reported to have the same sensitivity. It includes a confirmation procedure that involves sequential analysis of the sample and a zearalenone standard, and it compares fluorescence responses at four different excitation wavelengths.

Extract blended ground sample with chloroform and water in the presence of Celite, and filter.

Place aliquot of extract, saturated sodium chloride solution, and 2% sodium hydroxide solution in a separatory funnel and shake to transfer mycotoxins to aqueous alkaline layer.

Wash aqueous layer with chloroform and discard chloroform.

Add citric acid solution to separatory funnel and extract zearalenone and α-zearalenol with dichloromethane solution twice.

Dry combined dichloromethane extracts with anhydrous sodium sulfate and evaporate to dryness.

Dissolve residue in the mobile phase [methanol–acetonitrile–water (1.0 + 1.6 + 1.0 v/v/v)] for HPLC on reversed-phase (C_8 or C_{18}) column.

Confirm identity of zearalenone and α-zearalenol by sequentially injecting unknown into HPLC with fluorescence excitation wavelengths set at 236, 254, 274, and 314 nm. The peak height ratios of the responses were determined at 236/254, 236/274, and 236/313, and compared with the ratios of peak heights obtained in the same way with standards.

A collaborative study on the HPLC determination of α-zearalenol and zearalenone in corn has been carried out (Bennett et al., 1985). The coefficients of variation for spiked samples between laboratories were 25.6 and 33.8% for α-zearalenol and zearalenone, respectively. The coefficients of variation for naturally contaminated samples were 47.0% for α-zearalenol and 37.7% for zearalenone. As a result the method has been adopted by the AOAC in official first action.

D. Confirmatory Tests

When zearalenone is measured in extracts by GC-MS, confirmation of identity is obtained from the fragmentation patterns as mentioned before. The ratios of peak heights of fluorescence responses to different excitation wavelengths are also evidence for identification. Zearalenone can be detected in a crude extract of a gram sample (corn, wheat, rice, and oats) at levels of 0.1 ng/g with no sample cleanup by MS-MS (Plattner and Bennett, 1983). Detection can be accomplished by isobutane chemical ionization–mass spectrometry of zearalenone in extracts. Little fragmentation was observed; the predominant fragment was the protonated molecule at m/z 319. Daughter spectra of m/z 319 were obtained by collisionally activated dissociation that made possible a definitive identification.

IV. TRICHOTHECENES

A. Surveys and Occurrence

Trichothecenes are produced on grains by fungi belonging to the genera *Fusarium, Acremonium, Trichoderma, Trichothecium, Myrothecium, Stachybotrys,* and *Cylindrocarpon.* Studies on the overall occurrence of trichothecenes have been limited. Most reports on naturally occurring trichothecenes have resulted from the analysis of grains and feedstuffs implicated in causing adverse effects in farm animals, such as emesis, feed refusal, lesions, and hemorrhages. Trichothecenes are tetracyclic sesquiterpenes containing a six-membered oxygen-containing ring, an epoxide group in the 12,13-position and an olefinic bond in the 9,10-position (Fig. 3). They have been classified into four groups: A, B, C, and D. Although perhaps 100 trichothecenes and their derivatives have been reported, general information exists on the natural occurrence of only four of these in groups A and B: T-2 toxin, deoxynivalenol (DON) (also known as vomitoxin), diacetoxyscirpenol (DAS), and nivalenol (NIV) (Fig. 3) (National Research Council, 1983). They were found on barley, corn, wheat, grain sorghum, and in animal feeds at levels ranging from 0.02 to 40 μg/g. These analyses were all made in response to health problems in animals.

More extensive studies have been made on the occurrence of DON, T-2 toxin, and NIV in corn and other field crops; but most of the samples examined were selected because of *Fusarium* damage to kernels or because they were associated with infertility or refusal symptoms in swine (Jemmali *et al.,* 1978; Marasas *et al.,* 1979; Vesonder *et al.,* 1979). In 1977, 52 preharvest corn samples were

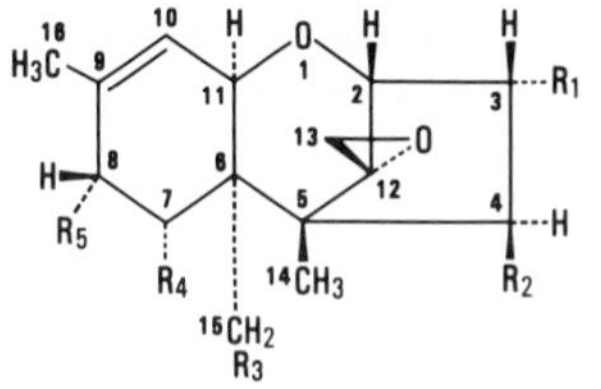

Group A

	R₁	R₂	R₃	R₄	R₅
Diacetoxyscirpenol	OH	OAc	OAc	H	H
T-2 Toxin	OH	OAc	OAc	H	OCOCH₂CH(CH₃)₂

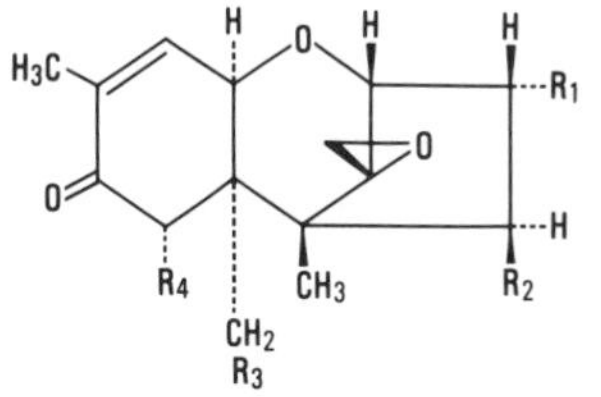

Group B

	R₁	R₂	R₃	R₄
Deoxynivalenol	OH	H	OH	OH
Nivalenol	OH	OH	OH	OH

Fig. 3. Structures of group A and B trichothecenes.

collected in northwestern Ohio and analyzed for DON (Vesonder *et al.*, 1978). DON was found in 24 samples at levels ranging from 0.5 to 10 μg/g. During the harvest of 1980 white winter wheat in Ontario, farmers observed pink discoloration on the kernels which is typical of *Fusarium* mold. Samples of white winter wheat were collected from elevators and boats loading wheat for export for analysis of DON (Trenholm *et al.*, 1981, 1983). DON was present in 58 of the 77 samples at levels ranging from 0.06 to 8.5 μg/g. In a survey of 87 samples of feeding and malting barley in the United Kingdom, DON occurred in 10% of the samples at levels of 0.2 to 0.36 μg/g (Gilbert *et al.*, 1983). The same workers found DON in 10 of 28 imported corn samples at levels of more than 0.1 μg/g. Osborne and Willis (1984) analyzed 199 United Kingdom-grown samples, collected from 1980 to 1982, and 33 imported wheat samples for seven trichothecenes. The only mycotoxin detected in the United Kingdom samples was DON, which occurred in 32 of the 199 samples at levels from 0.02 to 0.40 μg/g. Of the 33 imported samples, 23 had levels of DON from 0.02 to 1.32 μg/g. The positive samples came from eastern Canada and midwestern United States. In 1982, hard red winter wheat grown in certain areas of Kansas and Nebraska had *Fusarium* damage. A survey for DON in hard red winter wheat harvested in these areas was given high priority because of concern about the export market (Shotwell *et al.*, 1985). Of the 161 samples of wheat, 42% contained equal to 1 μg/g DON or less; 25% contained more than 1 μg/g to 2 μg/g; 23% more than 2 to 4 μg/g; and 10% more than 4 μg/g. A study was made of 101 samples of 1982 hard red winter wheat collected in some of the counties of Kansas, predominantly in northern and central Kansas (Seitz, 1983). Samples collected in counties south and west of the area that had scab damage were negative. Fifty samples considered to be an "overview of Kansas wheat moving into commercial channels" were analyzed for DON; only two contained >1 μg/g DON (Koch *et al.*, 1983).

B. Levels of Concern and Tolerances

In October 1982, the United States FDA considered establishing levels of concern for DON in wheat and wheat products, but did not (Anonymous, 1982). For finished wheat products, the level of concern would have been at 1 μg/g; for wheat as it enters the milling process, 2 μg/g; and for wheat and milling by-products intended for use as animal feed ingredients in the diets of swine, beef cattle, and broilers, 4 μg/g. In May 1983, the Canadian government raised its tolerance for DON in wheat (Anonymous, 1983). The Health Protection Branch stated that it would not object to the use of uncleaned soft wheat containing up to 2.0 μg/g DON, and that this level, as in previous years, could be achieved by blending with other soft wheat. The Health Protection Branch estimated that a level of 2.0 μg/g in uncleaned soft wheat would be expected to result in no more than 1.2 μg/g DON in the flour portion of finished products as consumed.

C. Screening Methods

A survey for DON was made by the FDA of 57 wheat samples from four midwestern states where the 1982 winter wheat crop was infected with *Fusarium* (Eppley *et al.*, 1984). The screening method used can be adapted for field use as well as for quantitative determinations. The method is a combination of an extraction with acetonitrile–water (84 + 16 v/v) and purification on a neutral alumina-activated charcoal column, reported by Romer *et al.* (1981), and a TLC method using an aluminum chloride spray solution to give a blue spot, reported by Kamimura *et al.* (1981). The columns are available commercially (Mycolab Co., P.O. Box 321, Chesterfield, Missouri 63017) and can be attached to a filter flask assembly, so vacuum can be applied to hasten development. Trichothecenes other than DON will also react with aluminum chloride, but higher levels are required to give a color.

A field test is being developed based on the reaction of 4-(*p*-nitrobenzylpyridine) (NBP) with the 12,13-epoxide group of trichothecenes (Takitani *et al.*, 1979). A 1% solution of NBP in a chloroform–carbon tetrachloride solution is sprayed on a developed TLC plate, or the plate is dipped in.the NBP solution before use. Plates must be heated at 135° to 150°C for 10 min for the reaction to take place. The blue color becomes visible after spraying with a solution of tetraethylenepentamine or other base (National Research Council, 1983).

D. Quantiative Methods

The method developed by Scott *et al.* (1981) for wheat, corn, barley, and soybeans has been used to determine DON in thousands of Canadian wheat samples. When scab and *Fusarium* damage was a problem in hard and soft red winter wheats in certain areas of the United States, this method was widely used to determine levels and incidences of DON, in spite of the fact it was very time-consuming. Recoveries and detection limits have been established, and it has the advantage that DAS, T-2 toxin, HT-2 toxin, NIV, fusarenon-X, and 3-acetyl-DON can also be detected. Exchange of samples among government agencies, commercial analytical laboratories, and industries monitoring wheat indicate that the method is reliable. The steps in the analysis are outlined below.

Extract the sample with methanol–water (1 + 1 v/v), and centrifuge.

Decant centrifugate and add 30% ammonium sulfate, mix and add diatomaceous earth. Stir 2 min with magnetic stirrer.

Filter and transfer aliquot of filtrate to separatory funnel.

Extract filtrate four times with ethyl acetate.

Dry combined extracts with anhydrous sodium sulfate by stirring 10 min with magnetic stirrer.

Decant through glass wool plug into pear-shaped flask to evaporate to dryness. Wash sodium sulfate with ethyl acetate, and add washes to flask to evaporate to dryness.

Transfer to vial and dry by evaporation before transfer to a silica gel column with dichloromethane.

Wash column with toluene–acetone (95 + 1 v/v), and elute trichothecenes with dichloromethane–methanol (95 + 5 v/v).

Evaporate eluate to dryness *in vacuo* and transfer to vial with dichloromethane. Evaporate to dryness.

To 50 µl heptafluorobutyrate imidazole in a vial, add solution of residue of unknown in toluene–acetonitrile (95 + 5 v/v).

Heat vial 1 hr at 60°C and cool to room temperature.

Stop reaction with 5% aqueous sodium bicarbonate. Dilute 50 µl of reaction mixture with 950 µl hexane for gas chromatography with GC-EC or GC-MS-single-ion monitoring (SIM).

Results obtained by GC-EC or GC-MS(SIM) of DON heptafluorobutyrate are the same. The detection limit for DON in wheat is 0.01 µg/g, and DON can be quantitated down to levels of 0.02 µg/g. Recoveries were 57–86%; coefficients of variation were 10.2–10.0%. The advantage of using GC-MS(SIM) for quantitation of DON is that it also offers a confirmation of identity. The method is applicable for determining DON in barley and corn.

In 1982, when the survey of hard red winter wheat from eastern Kansas and Nebraska and north central Kansas was made by the Northern Regional Research Center, it was urgent that results be available as soon as possible to be used by exporters. The method of Scott et al. (1981) was chosen because it had been thoroughly evaluated and tested. Many other analytical methods for trichothecenes reported in the literature have not been tested extensively for their reliability. The following modifications were made in the Scott method to save time (Bennett et al., 1983).

> Samples were filtered through rapid-flow filter paper rather than centrifuged.
>
> Filtrate and ammonium sulfate solution were stirred with a glass rod rather than a magnetic stirrer.
>
> Three extractions rather than four were made with ethyl acetate.
>
> Combined ethyl acetate extracts were dried by swirling in beaker intermittently with anhydrous sodium sulfate for 3 to 4 min rather than by stirring for 10 min with a magnetic stirrer.
>
> Dried ethyl acetate extracts were evaporated in a round-bottomed flask in vacuo. The residue was transferred quantitatively directly to a silica gel column rather than to a vial to be transferred later to the column.
>
> A hexane wash of the silica gel column was added to remove residual toluene and to reduce dry-down time for the DON eluate.

The modifications did not lower the recoveries of DON or raise the detection limits in wheat, but they increased by a factor of two or three the number of samples that could be analyzed in a day. This modified method, now being used by a number of laboratories, still needs to be validated for DON, DAS, T-2 toxin, HT-2 toxin, and other trichothecenes in a collaborative study.

A method was developed to determine DON in wheat that required only one cleanup step and involved quantitation of the trisheptafluorobutyrate of DON by GC-EC (Ware et al., 1984).

> Extract a 25-g sample of wheat with dichloromethane–ethanol (8 + 1 v/v) and filter.
>
> Evaporate aliquot of filtrate to dryness under nitrogen for silica gel chromatography on a small disposable column.
>
> Dissolve residue in dichloromethane for addition to the column which is centrifuged to remove dichloromethane rapidly.
>
> Wash column with toluene–acetone (8 + 2 v/v) and elute DON with di-

chloromethane–methanol (95 + 5 v/v) removing wash and eluant by
centrifugation.

Concentrate eluant to dryness under nitrogen in a vial.

Add 4-dimethylaminopyridine catalyst to vial and then add heptafluorobutyric acid anhydride.

Warm mixture 20 min at 60°C and cool to room temperature.

Stop reaction with 3% sodium bicarbonate.

Dilute 100 μl of reaction mixture with 900 μl hexane for measurement of
DON by GC-EC.

Recoveries of DON added to wheat averaged 88% with a coefficient of variation of 8.6%. Ware *et al.* (1986) conducted a successful collaborative study on
this method to determine DON in wheat resulting in adoption in official first
action by the AOAC.

As the need arose to analyze both corn and wheat for DON in 1982, personnel
at FDA modified the method, suggested by Romer *et al.* (1981), to decrease the
limit of detection by using an aluminum chloride solution to develop color on
TLC plates (Trucksess *et al.*, 1984). The procedure is summarized below.

Extract blended ground sample with acetonitrile–water (84 + 16 v/v), and
filter.

Wheat: Proceed with column chromatography.

Corn: Transfer aliquot of filtrate to separatory funnel and wash with hexane.
Use hexane-washed filtrate for column chromatography.

Place glass wool and Celite in bottom of a polypropylene column (10 mm
i.d. × 50 mm) with filter disk and reservoir on a filter flask.

Mix charcoal, neutral alumina, and Celite thoroughly, pack in column, and
apply suction.

Apply extract to the column and add more acetonitrile–water (84 + 16 v/v).

Continue aspiration until extract and acetonitrile–water wash are through
the column and the flow stops.

Evaporate the eluate from the column (the impurities are retained on the
column) to dryness.

Dissolve the residue in hot ethyl acetate to transfer to a vial, and evaporate
ethyl acetate for TLC.

For quantitation, incorporate color reagent in prepoured silica gel plates by
dipping in a solution of aluminum chloride hydrate dissolved in water–
ethanol (15 + 85 v/v), and reactivate before developing. Instead, aluminum chloride can be incorporated in the water used to prepare slurry
when TLC plates are prepared.

To quantitate DON, dissolve residue in acetonitrile–chloroform (4 + 1 v/v)
for spotting on TLC plate.

Develop plate with chloroform–acetone–propanol (8 + 1 + 1 v/v/v).

Measure DON densitometrically as done for aflatoxin (AOAC, 1984) or by visual comparisons with standards.

The limit of detection and determination of DON is ~40 ng/g for wheat and 100 ng/g for corn. Recoveries of DON added to wheat and corn at 100 to 1000 ng/g levels ranged from 77 to 93%. Eppley *et al.* (1986) conducted a collaborative study of the method on winter wheat including naturally contaminated samples and wheat spiked with DON (50–1000 ng/g). Recoveries of DON from spiked samples ranged from 78 to 96% with repeatabilities of 30 to 64% and re-producibilities of 33 to 87%. The coefficient of variation for the DON in the wheat with a mean level of 582 ng/g was 34% between laboratories. All of the 15 collaborators detected DON in the sample spiked at 300 ng/g. There were no false positives. The method has been adopted by the AOAC in official first action.

The method developed by Romer *et al.* (1981) for sample cleanup was modified to increase recoveries of DON and to measure the trichothecene by HPLC on a reversed-phase C_{18} column (Chang *et al.*, 1984). The mobile phase was methanol–water (23 + 77 v/v). Results obtained by Chang *et al.* did not differ from those obtained by the time-consuming method of Scott *et al.* (1981).

D. Confirmatory Tests

Trichothecenes have been confirmed by GC-MS and MS-MS (Plattner and Bennett, 1983) as discussed in other chapters of this book.

V. OCHRATOXINS

A. Surveys and Occurrence

Ochratoxins are a group of chemically related mycotoxins produced by species of the genera *Penicillium* and *Aspergillus* and having nephrotoxic properties. Their structures consist of an isocoumarin linked to L-β-phenylalanine (Fig. 4). Ochratoxins A, B, C, and D, and their methyl and ethyl esters have been isolated from cereal grains inoculated with *Penicillium* and *Aspergillus*. Ochratoxin A was first found as a natural contaminant in a survey of corn (Shotwell *et al.*, 1969), and is the predominant ochratoxin. Since then it has been detected in corn, wheat, barley, and coffee beans at levels of 5 to 360 ng/g (Carlton and Krogh, 1979). Ochratoxins have been suspected of being implicated in Balkan endemic nephropathy. Animal feeds suspected of containing ochratoxin because of veterinarian problems in the field have been reported to have levels as high as 27,000 ng/g. Besides being nephrotoxic, ochratoxins are teratogenic. There are conflicting reports as to the carcinogenicity of these mycotoxins. Surveys of

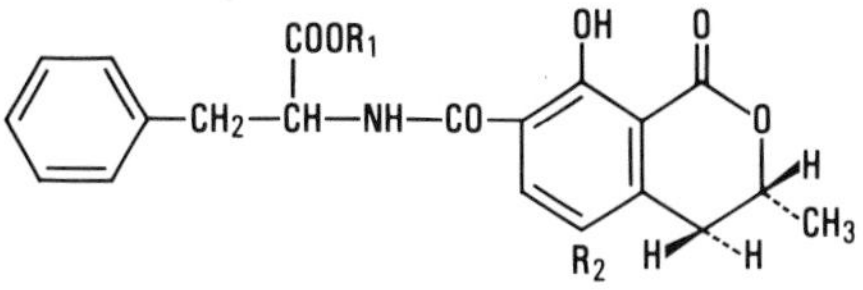

	R₁	R₂
Ochratoxin A	H	Cl
Ochratoxin B	H	H
Ochratoxin C	C₂H₅	Cl

Ochratoxin D is 4-hydroxyochratoxin A

Fig. 4. Structures of ochratoxins.

meat processing plants in Sweden and Denmark revealed residues of ochratoxin A in 25 to 35% of the pigs afflicted with nephropathy (Carlton and Krogh, 1979).

B. Screening Methods

The Eppley method for screening cereal grains for zearalenone, aflatoxin, and ochratoxin has been described in the section on zearalenone (Eppley, 1968). Recoveries of ochratoxin from spiked grain samples are 60%; the limit of detection is 50 ng/g. A simplified procedure using a silica gel minicolumn was developed by Hald and Krogh (1975) to detect ochratoxin A in barley samples; positive samples were quantitated by TLC. Few steps are required.

Extract sample with 0.1 N phosphoric acid–chloroform (1 + 10 v/v), and filter.

Place an aliquot in a separatory funnel and partition into 0.1 N sodium bicarbonate.

Acidify aqueous layer and extract with chloroform.

Evaporate chloroform to dryness and redissolve residue in small volume of chloroform to place on minicolumn.

Develop minicolumn with toluene–ethyl acetate–formic acid (5 + 4 + 1 v/v/v). Ochratoxin appears as a blue-green fluorescent band under UV light (365 nm).

Another minicolumn method was reported by Holaday (1976), in which the same extract is used for detection of aflatoxin and ochratoxin A on two different minicolumns. The method has not been widely used.

C. Quantitative Methods

A method for analysis of ochratoxins A and B and their esters in barley was developed using partition and TLC (Nesheim et al., 1973). It has been approved for ochratoxin A in official first action by the AOAC (1984) after a collaborative study (Nesheim, 1973). The cleanup includes a series of partitions.

> Extract blended ground samples with 0.1 M phosphoric acid–chloroform (1 + 9 v/v).
>
> Mix diatomaceous earth (2 g) thoroughly with 1.25% sodium bicarbonate (1 ml), and pack into column.
>
> Mix aliquot of sample extract with hexane and add to column. Ochratoxins A and B will be adsorbed on column and esters will be eluted.
>
> Add additional chloroform and evaporate combined chloroform eluates for ester analysis.
>
> Elute ochratoxins A and B with formic acid–chloroform (1 + 99 v/v), and evaporate nearly to dryness transferring concentrate to tube with chloroform.
>
> Evaporate chloroform and save residue for TLC.
>
> To separate ochratoxin esters, prepare column as above. Dissolve ester residue in hexane and add to column.
>
> Wash with benzene–hexane (1 + 9 v/v) that has been equilibrated with methanolic sodium bicarbonate solution.
>
> Elute esters with formic acid–benzene–hexane solution, and evaporate to dryness for TLC.
>
> For ochratoxins A and B, develop silica gel plates with acetic acid–benzene (1 + 99 v/v).
>
> For ochratoxin esters, develop silica gel plates with hexane–acetone–acetic acid (180 + 20 + 10 v/v/v).
>
> Measure all ochratoxins on developed plates visually by comparison with standards or densitometrically (excitation wavelength, 310 nm; emission wavelength, 440 nm). Ochratoxins fluoresce blue green on silica gel plates under UV light (365 nm). The color changes to blue when plates are exposed to ammonia fumes or sprayed with sodium bicarbonate solution.

When the method was studied collaboratively in 13 laboratories, the average recovery of ochratoxin A for barley samples spiked at 45 and 90 ng/g was 112% with a coefficient of variation of 27.1% (Nesheim, 1973). Recoveries of ochratoxin B and the esters were unsatisfactory. The method was adopted in official first action by AOAC (1984) as quantitative for ochratoxin A and qualitative for ochratoxin B and the esters. In an international mycotoxin check sample series (Friesen and Garren, 1983), the 27 laboratories using the Nesheim method on animal feed samples made from barley had within-laboratory coefficients of variation of 8.9% and a total coefficient of variation of 65.5%. In the same study, the 17 laboratories that used other methods had coefficients of variation within laboratories and between laboratories of 8.0 and 82.1%, respectively.

D. Confirmatory Tests

Ochratoxins A and B are confirmed by the formation of ethyl esters (Nesheim et al., 1973).

> Dissolve part of residue from chloroform extracts containing ochratoxins A and B, if present, in chloroform.
> Add 14% boron trifluoride in ethanol, boil 5 min, and transfer to a separatory funnel.
> Add water, and extract formed esters with chloroform.
> Evaporate chloroform and conduct TLC of esters with silica gel plates and develop with acetic acid–benzene (1 + 99 v/v).
> Compare with esters of standards of ochratoxins A and B.

It is particularly important to confirm the identity of ochratoxin A in corn. Of the apparently positive corn samples obtained by the Eppley method, only about one-third are found to contain ochratoxin A when extracts are subjected to the confirmatory tests described.

VI. STERIGMATOCYSTIN

A. Surveys and Occurrence

Sterigmatocystin is a toxic metabolite produced by various species of *Aspergillus, Bipolaris,* and *Chaetomium* (Mirocha, 1980). It has been shown to be carcinogenic to rats (Dickens *et al.,* 1966; Purchase and van der Watt, 1970; Ohtsubo *et al.,* 1978). It has been reported as naturally occurring in 1 of 29 samples of heated grains in Canada (Scott *et al.,* 1972), in 12 of 37 samples of moldy rice (Manabe and Tsurata, 1975), and in one subsample (40 half-kernels) of a lot of in-shell pecans (Schroeder and Hein, 1977). Sterigmatocystin has a difuranoxanthone structure (Fig. 5) resembling the difuranocoumarin structure of the aflatoxins (Fig. 1).

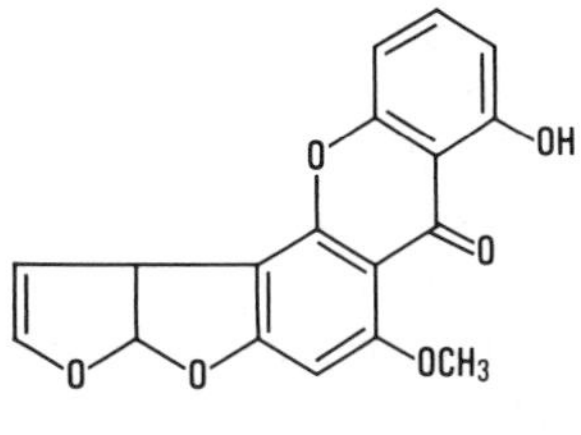

Fig. 5.   Structure of sterigmatocystin.

B. Quantitative Methods

The method approved for barley and wheat by the AOAC (1984) was developed by Stack and Rodricks (1971) for corn, barley, wheat, oats, and rye.

Extract sample with acetonitrile–4% potassium chloride (9 + 1 v/v) on wrist action shaker for 30 min.

Partition filtered extract first against hexane and then against chloroform. Evaporate chloroform solution to dryness to obtain residue for TLC.

Develop silica gel TLC plates with benzene–acetic acid–methanol (90 + 5 + 5 v/v/v) in a lined tank.

Spray developed plate with an aluminum chloride solution (20 g AlCl$_3$·6H$_2$O + 100 ml methanol), and heat 10 min at 80°C. Sterigmatocystin zones fluoresce bright yellow under short-wave light.

Quantitate by visual comparisons with standards.

Confirm the identity of sterigmatocystin by isolating the material with the correct R_f from a preparative TLC plate. Treat a portion of the material with pyridine and acetic anhydride to form an acetate derivative. Treat the remainder of the material with 0.1 N hydrochloric acid to form a derivative that is presumed to be the hemiacetal or water adduct.

Chromatograph the two derivatives of the unknown and of standards on silica gel TLC plates with acetone–chloroform (5 + 95 v/v) in an equilibrated tank.

Examine the developed plate under long-wave UV light (365 nm). The acetate derivative appears as a blue fluorescent zone at R_f about one-half that of sterigmatocystin. The derivative formed by treatment with aqueous hydrochloric acid is seen at R_f about one-fourth that of sterigmatocystin as a yellow fluorescent zone.

If there is too much interference for proper quantitation of sterigmatocystin by TLC, the material can be further purified by silica gel chromatography. The initial water extract is often free of interfering compounds, and sterigmatocystin in such extracts can be quantitated without further purification. Sterigmatocystin fluoresces brick red under UV light (365 nm), but the lowest amount detectable on TLC plates is 500 ng. An aluminum chloride spray is used to enhance fluorescence.

The method developed by Stack and Rodricks (1971) was evaluated in a collaborative study for sterigmatocystin in corn, wheat, barley, and oats (Stack and Rodricks, 1973). As a result of the study the method was adopted by the AOAC (1984) for barley and wheat. Of the 17 collaborators, 7 had difficulty with corn and oat samples because of oily residues which interfered with TLC of sterigmatocystin. Levels of 100 ng/g could be determined easily in wheat and barley. In naturally contaminated barley samples with a mean level of sterigmatocystin of 122 ng/g, the coefficient of variation was 24.6%. In barley, spiked at 100 to 200 ng/g, recoveries were 90%; the coefficient of variation was 12.2%. The coefficient of variation for naturally contaminated wheat with a mean level of 160 ng/g sterigmatocystin was 19.8%. In wheat, spiked at 100 to 200 ng/g, the mean recovery was 100%; the mean coefficient of variation was 21.4%.

The FDA analyzed 457 samples of small grains for sterigmatocystin by the method of Stack and Rodricks (1971). No sterigmatocystin was detected in any of the samples (Stoloff, 1976b). Starting with the official method for barley and wheat, Stack et al. (1976) added purification steps and quantitated sterigmatocystin in wheat and oats by HPLC.

A method was devised for sterigmatocystin that was applied to white and yellow corn, barley, oats, rye, grain sorghum, brown and wild rice, and soybeans (Shannon and Shotwell, 1976). Steps in the method are summarized below.

> Extract ground commodity with acetonitrile–4% potassium chloride (9 + 1 v/v), and concentrate an aliquot to one-fifth volume.
>
> Purify on Florisil column by washing with hexane and eluting with acetone–dichloromethane (5 + 95 v/v).
>
> Measure sterigmatocystin in solution of the residue from the column eluate by the approved AOAC method (1984), except that plates were developed in benzene–ethanol–acetic acid (90 + 5 + 5 or 94 + 3 + 3 v/v/v). At levels of 200 to 500 ng/g, the mean recovery was 84% for all commodities except brown rice. The mean recovery from brown rice was 110% when it was spiked at 200 to 500 ng/g sterigmatocystin, indicating an enhancement of fluorescence on TLC plates by interfering impurities in the extract not removed by the Florisil column. The method was used to analyze 200 corn samples collected in South Carolina; sterigmatocystin was not detected in any sample.

Aspergillus versicolor, a sterigmatocystin producer, was found in cheese warehouses and on cheeses in The Netherlands. The procedure for assaying cheeses (van Egmond et al., 1980) is based on the method of Shannon and Shotwell (1976). A polyamide column cleanup step was added to obtain cleaner extracts. Sterigmatocystin was measured by two-dimensional TLC. The plates were developed in the first direction with chloroform–methanol (100 + 2 v/v); in the second direction with hexane–ethyl ether–acetic acid (75 + 25 + 10 v/v/v). Developed plates were sprayed with an aluminum chloride solution to enhance fluorescence. The identity of sterigmatocystin was confirmed by treatment with benzene–trifluoroacetic acid on the TLC plate to form the water adduct. The method of van Egmond et al. (1980) was used to analyze the upper 1-cm layer of 39 warehouse cheeses (Northolt et al., 1980). Of the 39 cheeses, 9 contained 5–600 ng/g sterigmatocystin.

A one-dimensional TLC method has been developed for determining sterigmatocystin in cheese in which a simplified liquid–liquid partition cleanup is used
(Francis *et al.*, 1985).

> Extract 36 g cheese with acetonitrile–4% potassium chloride (85 + 15 v/v)
> and filter.
> Transfer aliquot of filtrate to separatory funnel and add calcium chloride–
> water (1 + 2 w/v), shake, and allow layers to separate. Discard bottom
> layer.
> Add water and hexane to upper layer and shake. Let layers separate.
> Transfer bottom layer to another separatory funnel and extract twice with
> dichloromethane.
> Pass dichloromethane extracts through column of cupric carbonate–di-
> atomaceous earth (1 + 2 v/v).
> Evaporate purified dichloromethane solution to dryness for TLC.
> Measure sterigmatocystin from dichloromethane solution by TLC on preco-
> ated silica gel plates (Sil G-25 HR) developing with benzene–methanol–
> acetic acid (85 + 10 + 5 v/v/v).
> Spray dried developed plates with a solution of aluminum chloride in
> ethanol.
> Quantitate sterigmatocystin visually by comparisons with standards.

The average recovery of sterigmatocystin in spiked samples was 87% and the
limit of detection was 2 ng/g. A collaborative study for the determination of
sterigmatocystin in cheese by this method is being conducted.

VII. CONCLUSION AND FUTURE DEVELOPMENTS

There are a number of methods available for detection, determination, and
confirmation of identity of mycotoxins. Presumptive tests exist only for aflatoxins in peanuts, corn, and cottonseed. Corn or wheat kernels containing zearalenone and/or trichothecenes may exhibit symptoms of the *Fusarium* infection
responsible for the contamination. However, infected kernels are not always
evident. Minicolumn and TLC screening procedures are routinely used for
aflatoxin in corn and are available for peanuts and cottonseed if needed. Several
agencies are working on the development of rapid screening procedures for
trichothecenes. One of the most promising approaches in the future to rapid
screening may be enzyme-linked immunosorbent assays (ELISA) for mycotoxins. ELISA methods have been reported for aflatoxins including aflatoxin M_1
(El-Nakib *et al.*, 1981; Pestka *et al.*, 1980, 1981a), zearalenone (J. J. Pestka,
personal communication), ochratoxin (Pestka *et al.*, 1981b; Morgan *et al.*,
1983), and T-2 toxin (Pestka *et al.*, 1981c). An ELISA method for mycotoxins

using nylon beads and terasaki plate solid phases has been reported (Pestka and Chu, 1984). Research is being done on development of ELISA assays for other mycotoxins and the production of monoclonal antibodies to be used in the assays. Commercial production of ELISA kits for detection of aflatoxin and zearalenone is being considered. ELISA techniques have been applied to the determination of levels of mycotoxins in agricultural commodities, particularly in corn or peanuts. Although reliable analytical methods for mycotoxins are available and being used by government agencies, academic institutions, and industry, improvements are needed. Two desirable improvements would be minimizing the use of expensive and relatively toxic solvents as well as modifications to avoid labor-intensive methods.

REFERENCES

American Association of Cereal Chemists (1983). "Book of Methods," 8th Ed. AACC, St. Paul, Minnesota.
American Oil Chemists Society (1973). "Official Book of Methods: Tentative Method Ab 6-68: AOCS, Champaign, Illinois.
American Oil Chemists Society (1983). Aflatoxin in cottonseed. In "Official and Tenative Methods of AOCS: Aa 8-83," pp. 1–10. AOCS, Champaign, Illinois.
Anonymous (1982). Food Chem. News Oct. 4, p. 3.
Anonymous (1983). Food Chem. News June 6, p. 34.
Anonymous (1983). "Marketing Agreement for Peanuts," No. 146. Peanut Adm. Comm., Atlanta, Georgia.
Anonymous (1984). Guidelines for interlaboratory collaborative study procedure to validate characteristics of a method of analysis. J. Assoc. Off. Anal. Chem. 67, 433–440.
Association of Official Analytical Chemists (1980). Changes in methods. J. Assoc. Off. Anal. Chem. 63, 391–394.
Association of Official Analytical Chemists (1982). "Handbook for AOAC Members," 5th Ed. AOAC, Arlington, Virginia.
Assocation of Official Analytical Chemists (1984). "Book of Methods," 14th Edition. AOAC, Washington, D.C.
Bagnaris, R. W. (1986). High performance liquid chromatography method for the determination of zearalenone and zearalenol in grains and feeds. J. Assoc. Off. Anal. Chem. (in press).
Barabolak, R., Colburn, C. R., Just, D. E., Kurtz, F. A., and Schleichert, E. A. (1978). Apparatus for rapid inspection of corn for aflatoxin contamination. Cereal Chem. 55, 1065–1067.
Bennett, G. A., and Shotwell, O. L. (1979). Zearalenone in cereal grains. J. Am. Oil Chem. Soc. 56, 812–819.
Bennett, G. A., Stubblefield, R. D., Shannon, G. M., and Shotwell, O. L. (1983). Gas chromatographic determination of DON in wheat. J. Assoc. Off. Anal. Chem. 66, 1478–1480.
Bennett, G. A., Shotwell, O. L., and Kwolek, W. F. (1985). Determination of α-zearalenol and zearalenone in corn by high performance liquid chromatography: Collaborative study. J. Assoc. Off. Anal. Chem. 68, 954–957.
Bothast, R. J., and Hesseltine, C. W. (1975). Bright greenish-yellow fluorescence and aflatoxin in agricultural commodities. Appl. Microbiol. 30, 337–338.
Brumley, W. C., Nesheim, S., Trucksess, M. W., Trucksess, E. W., Dreifus, P. A., Roach, J. A.

G., Andrzejewski, D., Eppley, R. M., Pohland, A. E., Thorpe, C. W., and Sphon, J. A. (1981). Negative ion chemical ionization mass spectrometry of aflatoxins and related mycotoxins. *Anal. Chem.* **53,** 2003–2006.

Campbell, A. D., Francis, O. J., Jr., Beebe, R. A., and Stoloff, L. (1984). Determination of aflatoxins in peanut butter, using two liquid chromatographic methods: Collaborative study. *J. Assoc. Off. Anal. Chem.* **67,** 312–316.

Carlton, W. W., and Krogh, P. (1979). Ochratoxins. *Proc. Conf. Mycotoxins Anim. Feeds Grains Relat. Anim. Health* (W. Shimoda, ed.), pp. 165–287. Natl. Tech. Inf. Serv., Springfield, Virginia.

Chang, H. L., DeVries, J. W., and Hobbs, W. E. (1979). Comparative study of two methods for extraction of aflatoxin from peanut meal and peanut butter. *J. Assoc. Off. Anal. Chem.* **62,** 1281–1284.

Chang, H. L., DeVries, J. W., Larson, P. A., and Patel, H. H. (1984). Rapid determination of deoxynivalenol (Vomitoxin) by liquid chromatography using modified Romer column cleanup. *J. Assoc. Off. Anal. Chem.* **67,** 52–54.

Christensen, C. M. (1979). Zearalenone. *Proc. Conf. Mycotoxins Anim. Feeds Grains Relat. Anim. Health* (W. Shimoda, ed.), pp. 1–75. Natl. Tech. Inf. Serv., Springfield, Virginia.

Cucullu, A. F., Pons, W. A., Jr., and Goldblatt, L. A. (1972). Fast screening method for detection of aflatoxin contamination in cottonseed products. *J. Assoc. Off. Anal. Chem.* **55,** 1114–1119.

Dantzman, J., and Stoloff, L. (1972). Screening method for aflatoxin in corn and various corn products. *J. Assoc. Off. Anal. Chem.* **55,** 139–141.

Davis, N. D., Dickens, J. W., Freie, R. L., Hamilton, P. B., Shotwell, O. L., and Wyllie, T. D. (1980). Protocols for surveys, sampling, post-collection handling, and analysis of grain samples involved in mycotoxin problems. *J. Assoc. Off. Anal. Chem.* **63,** 95–102.

DeVries, J. W., and Chang, H. L. (1982). Comparison of rapid high pressure liquid chromatographic and CB methods for determination of aflatoxins in corn and peanuts. *J. Assoc. Off. Anal. Chem.* **65,** 206–209.

Dickens, F., Jones, H. E. H., and Waynforth, H. B. (1966). Oral, subcutaneous, and intratracheal administration of carcinogenic lactones and related substances: the intratracheal administration of cigarette tar in the rat. *Br. J. Cancer* **20,** 134–144.

Dickens, J. W. (1977). Aflatoxin control program for peanuts. *J. Am. Oil Chem. Soc.* **54,** 225A–228A.

Dickens, J. W., and Welty, R. E. (1969). Detecting farmers' stock peanuts containing aflatoxin by examination for visible growth of *Aspergillus flavus. Mycopathol. Mycol. Appl.* **37,** 65–69.

Dickens, J. W., and Whitaker, T. B. (1981). Bright greenish-yellow fluorescence and aflatoxin in recently harvested yellow corn marketed in North Carolina. *J. Am. Oil Chem. Soc.* **58,** 973A–975A.

El-Nakib, O., Pestka, J. J., and Chu, F. S. (1981). Determination of aflatoxin B_1 in corn, wheat, and peanut butter by enzyme-linked immunosorbant assay and solid phase radioimmuno-assay. *J. Assoc. Off. Anal. Chem.* **64,** 1077–1082.

Eppley, R. M. (1966). A versatile procedure for assay and preparatory separation of aflatoxin from peanut products. *J. Assoc. Off. Anal. Chem.* **49,** 1218–1223.

Eppley, R. M. (1968). Screening method for zearalenone, aflatoxin, and ochratoxin. *J. Assoc. Off. Anal. Chem.* **51,** 74–78.

Eppley, R. M., Stoloff, L., and Campbell, A. D. (1968). Collaborative study of "a versatile procedure for assay of aflatoxins in peanut products," including preparatory separation and confirmation of identity. *J. Assoc. Off. Anal. Chem.* **51,** 67–73.

Eppley, R. M., Stoloff, L., Trucksess, M. W., and Chung, C. W. (1974). Survey of corn for *Fusarium* toxins. *J. Assoc. Off. Anal. Chem.* **57,** 632–635.

Eppley, R. M., Trucksess, M. W., Nesheim, S., Thorpe, C. W., Wood, G. E., and Pohland, A. E.

(1984). Deoxynivalenol in winter wheat: Thin Layer Chromatographic method and survey. *J. Assoc. Off. Anal. Chem.* **67**, 43–45.

Eppley, R. M., Trucksess, M. W., Nesheim, S., Thorpe, C. W., and Pohland, A. E. (1986). Thin layer chromatographic method for determination of deoxynivalenol in wheat: Collaborative study. *J. Assoc. Off. Anal. Chem.* **69**, 37–40.

Federal Register (1978). Aflatoxin-contaminated corn, limited exemption from blending prohibiton. Vol. 43, Apr. 4, pp. 14122–14123.

Federal Register (1981). Aflatoxin-contaminated corn: Limited exemptions from prohibition of interstate shipping and blending. Vol. 46, Jan. 23, pp. 7447–7449.

Federal Register (1983). Aflatoxin-Contaminated Corn: FDA's Policy Regarding Interstate Shipment of Corn Harvested in 1983. Vol. 48 (November 25, 1983), pp. 53175–53176.

Ferguson-Foos, J., and Warren, J. D. (1984). Rapid high performance liquid chromatographic determination of aflatoxin M_1 and M_2 in fluid milk. *J. Assoc. Off. Anal. Chem.* **67**, 1111–1114.

Francis, O. J., Ware, G. M., Carman, A. S., and Kuan, S. S. (1985). Thin layer chromatographic determination of sterigmatocystin in cheese. *J. Assoc. Off. Anal. Chem.* **68**, 643–645.

Friesen, M. D., and Garren, L. (1982). International check sample program: Part I. Report on laboratory performance for determination of aflatoxins B_1, B_2, G_1, and G_2 in raw peanut meal, deoiled peanut meal, and yellow corn meal. *J. Assoc. Off. Anal. Chem.* **65**, 855–863.

Friesen, M. D., and Garren, L. (1983). International mycotoxin check sample survey program. Part III. Report on performance of participating laboratories for determining ochratoxin A in animal feed. *J. Assoc. Off. Anal. Chem.* **66**, 256–259.

Friesen, M. D., Walker, E. A., and Castegnaro, M. (1980). International mycotoxin check sample program. Part I. Report on the performance of participating laboratories. *J. Assoc. Off. Anal. Chem.* **63**, 1057–1066.

Gilbert, J., Shepherd, M. J., and Startin, J. R. (1983). A survey of the occurrence of the trichothecene mycotoxin deoxynivalenol (vomitoxin) in UK grown barley and in imported maize by combined gas chromatography-mass spectrometry. *J. Sci. Food Agric.* **34**, 86–92.

Hald, B., and Krogh, P. (1975). Detection of ochratoxin A in barley, using silica gel minicolumns. *J. Assoc. Off. Anal. Chem.* **58**, 156–158.

Holaday, C. E. (1968). Rapid method for detecting aflatoxins in peanuts. *J. Am. Oil Chem. Soc.* **45**, 680–682.

Holaday, C. E. (1976). A rapid screening method for the aflatoxins and ochratoxin A. *J. Am. Oil Chem. Soc.* **53**, 603–605.

Holaday, C. E. (1981). Rapid screening method for aflatoxin M_1 in milk. *J. Assoc. Off. Anal. Chem.* **64**, 1064–1066.

Holaday, C. E., and Lansden, J. (1975). Rapid screening method for aflatoxin in a number of products. *J. Agric. Food Chem.* **23**, 1134–1136.

Jemmali, M., Ueno, Y., Ishii, K., Frayssinet, C., and Etienne, M. (1978). Natural occurrence of trichothecenes (nivalenol, deoxynivalenol, T_2) and zearalenone in corn. *Experientia* **34**, 1333–1334.

Kamimura, H., Nishijima, M., Yasuda, K., Saito, K., Ibe, A., Nagayama, T., Ushiyama, A., and Naoi, Y. (1981). Simultaneous detection of several *Fusarium* mycotoxins in cereals, grains, and foodstuffs. *J. Assoc. Off. Anal. Chem.* **64**, 1067–1073.

Koch, K. B., Behnke, K. C., Burroughs, R., Seitz, L. M., and Saur, D. B. (1983). Deoxynivalenol levels in 1982 Kansas-grown wheat collected at country elevators. *In* "Scab and Deoxynivalenol in the 1982 Wheat Crop," Report of Progress, No. 440, pp. 5–9. Agric. Exp. Stn., Kansas State Univ., Manhattan.

Kwolek, W. F., and Shotwell, O. L. (1979). Aflatoxin in white corn under loan. V. Aflatoxin prediction from weight percent of bright greenish-yellow fluorescent particles. *Cereal Chem.* **56**, 342–345.

McKinney, J. D. (1984). Analyst performance with aflatoxin methods as determined from AOCS Smalley Check Sample Program: Long-term and short-term views. *J. Assoc. Off. Anal. Chem.* **67**, 25–32.

McMillian, W. W., Wilson, D. M., Mirocha, C. J., and Widstrom, N. W. (1983). Mycotoxin contamination in grain sorghum from fields in Georgia and Mississippi. *Cereal Chem.* **60**, 226–227.

Manabe, M., and Tsurata, O. (1975). Mycological damage of domestic brown rice during storage in warehouse under natural conditions (Part 2). Natural occurrence of sterigmatocystin on rice during long time storage. *Nippon Kingakkai Kaiho* **16**, 399–405.

Marasas, W. F. O., van Rensburg, S. J., and Mirocha, C. J. (1979). Incidence of *Fusarium* species and the mycotoxins, deoxynivalenol and zearalenone, in corn produced in the esophageal areas in Transkic. *J. Agric. Food Chem.* **27**, 1108–1112.

Marsh, P. B., and Simpson, M. E. (1984). Detection of *Aspergillus flavus* and aflatoxins in cotton and corn by ultraviolet fluorescence. *J. Environ. Qual.* **13**, 8–17.

Marsh, P. B., Simpson, M. E., Ferretti, R. J., Merola, G. V., Donoso, J., Craig, G. O., Trucksess, M. W., and Work, P. S. (1969). Mechanism of formation of a fluorescence in cotton fiber associated with aflatoxins in the seeds at harvest. *Agric. Food Chem.* **17**, 468–472.

Mirocha, C. J. (1980). Sterigmatocystin. *Conf. Mycotoxins Anim. Feeds Grains Relat. Anim. Health* pp. 177–208. Natl. Tech. Inf. Serv., Springfield, Virginia.

Mirocha, C. J., Shauerhamer, B., and Pathre, S. V. (1974). Isolation, detection, and quantitation of zearalenone in maize and barley. *J. Assoc. Off. Anal. Chem.* **57**, 1104–1110.

Mirocha, C. J., Pathre, S. V., Shauerhamer, B., and Christensen, C. M. (1976). Natural occurrence of *Fusarium* toxin in feedstuff. *Appl. Environ. Microbiol.* **32**, 553–556.

Mirocha, C. J., Pathre, S. V., and Christensen, C. M. (1977). Zearalenone. *In* "Mycotoxins in Human and Animal Health" (J. V. Rodricks, C. W. Hesseltine, and Mehlman, M. A., eds.), pp. 345–364. Pathotox, Park Forest South, Illinois.

Morgan, M. R. A., McNerney, R., and Chan, H. W. S. (1983). Enzyme-linked immunosorbant assay of ochratoxin A in barley. *J. Assoc. Off. Anal. Chem.* **66**, 1481–1484.

National Research Council (1983). "Protection against Trichothecene Mycotoxins." Natl. Acad. Sci., Washington, D.C.

Nesheim, S. (1973). Analysis of ochratoxins A and B and their esters in barley, using partition and thin layer chromatography. II. Collaborative study. *J. Assoc. Off. Anal. Chem.* **56**, 822–826.

Nesheim, S., and Brumley, W. C. (1981). Confirmation of identity of aflatoxins. *J. Am. Oil Chem. Soc.* **58**, 945A–949A.

Nesheim, S., Hardin, N. F., Francis, O. J., Jr., and Langham, W. S. (1973). Analysis of ochratoxins A and B and their esters in barley using partition and thin layer chromatography. I. Development of the method. *J. Assoc. Off. Anal. Chem.* **56**, 817–821.

Nichols, T. E., Jr. (1983). Economic impact of aflatoxin in corn. *In* "Aflatoxin and *Aspergillus flavus* in Corn" (U. L. Diener, ed.), Southern Cooperative Series Bulletin, No. 279, pp. 67–71. Auburn Univ., Auburn, Alabama.

Northolt, M. D., van Egmond, H. P., Soentoro, P., and Deijill, E. (1980). Fungal growth and the presence of sterigmatocystin in hard cheese. *J. Assoc. Off. Anal. Chem.* **63**, 115–119.

Ohtsubo, K., Saito, M., and Kimura, H. (1978). High incidence of hepatic tumours in rats fed mouldy rice contaminated with *Aspergillus versicolor* containing sterigmatocystin. *Food Cosmet. Toxicol.* **16**, 143–149.

Osborne, B. G., and Willis, K. H. (1984). Studies into the occurrence of some trichothecene mycotoxins in UK home grown wheat and in imported wheat. *J. Sci. Food Agric.* **35**, 579–583.

Pestka, J. J., and Chu, F. S. (1984). Enzyme-linked immunosorbent assay of mycotoxins using nylon bead and terasaki plate solid phases. *J. Food Prot.* **47**, 305–308.

Pestka, J. J., Gaur, P. K., and Chu, F. S. (1980). Quantitation of aflatoxin B_1 and aflatoxin B_1

antibody by an enzyme-linked immunosorbent microassay. *Appl. Environ. Microbiol.* **40**, 1027–1031.

Pestka, J. J., Li, Y., Harder, W. O., and Chu, F. S. (1981a). Comparison of radio-immunoassay and enzyme-linked immunosorbent assay for determining aflatoxin M_1 in milk. *J. Assoc. Off. Anal. Chem.* **64**, 294–301.

Pestka, J. J., Steinart, B. W., and Chu, F. S. (1981b). Enzyme-linked immunosorbent assay for detection of ochratoxin A. *Appl. Environ. Microbiol.* **41**, 1472–1474.

Pestka, J. J., Lee, S. C., Lau, H. P., and Chu, F. S. (1981c). Enzyme-linked immunosorbent assay for T-2 toxin. *J. Am. Oil Chem. Soc.* **58**, 940A–944A.

Plattner, R. D., and Bennett, G. A. (1983). Rapid detection of *Fusarium* mycotoxins in grains by quadrupole mass spectrometry/mass spectrometry. *J. Assoc. Off. Anal. Chem.* **66**, 1470–1477.

Plattner, R. D., Bennett, G. A., and Stubblefield, R. D. (1984). Identification of aflatoxins by quadrupole mass spectrometry-mass spectrometry. *J. Assoc. Off. Anal. Chem.* **67**, 734–738.

Pons, W. A., Jr. (1975). Collaborative study of a rapid method for determining aflatoxins in cottonseed products. *J. Assoc. Off. Anal. Chem.* **58**, 746–753.

Pons, W. A., Jr., and Franz, A. O., Jr. (1978). High pressure liquid chromatographic determination of aflatoxins in peanut products. *J. Assoc. Off. Anal. Chem.* **61**, 793–800.

Pons, W. A., Jr., Lee, L. S., and Stoloff, L. (1980). Revised method for aflatoxins in cottonseed products, and comparison of thin layer and high performance liquid chromatography determinative steps. Collaborative study. *J. Assoc. Off. Anal. Chem.* **63**, 899–906.

Przybylski, W. (1975). Formation of aflatoxin derivatives on thin-layer chromatographic plates. *J. Assoc. Off. Anal. Chem.* **58**, 163–164.

Purchase, I. F. H., and van der Watt, J. J. (1970). Carcinogenicity of sterigmatocystin. *Food Cosmet. Toxicol.* **8**, 289–295.

Romer, T. R. (1975). Screening method for the detection of aflatoxins in mixed feeds and other agricultural commodities with subsequent confirmation and quantitative measurement of aflatoxins in positive samples. *J. Assoc. Off. Anal. Chem.* **58**, 500–506.

Romer, T. R., and Campbell, A. D. (1976). Collaborative study of a screening method for the detection of aflatoxins in mixed feeds, other agricultural products, and foods. *J. Assoc. Off. Anal. Chem.* **59**, 110–117.

Romer, T. R., Greaves, D. E., and Gibson, G. E. (1981). Thin layer chromatography of deoxynivalenol. *Annu. Spring Workshop Assoc. Off. Anal. Chem. 6th, Ottawa, Can.*

Schroeder, H. W., and Hein, H., Jr. (1977). Natural occurrence of sterigmatocystin in in-shell pecans. *Can. J. Microbiol.* **23**, 639–641.

Schuller, P. L., Horwitz, W., and Stoloff, L. (1976). A review of sampling plans and collaboratively studied methods of analysis for aflatoxins. *J. Assoc. Off. Anal. Chem.* **59**, 1315–1343.

Scott, P. M., van Walbeck, W., Kennedy, B., and Anyeti, D. (1972). Mycotoxins (ochratoxin A, citrinin, and sterigmatocystin) and toxigenic fungi in grains and other agricultural products. *J. Agric. Food Chem.* **20**, 1103–1109.

Scott, P. M., Lau, P. Y., and Kanhere, S. R. (1981). Gas chromatography with electron capture and mass spectrometric detection of deoxynivalenol in wheat and other grains. *J. Assoc. Off. Anal. Chem.* **64**, 1364–1371.

Seitz, L. (1983). Deoxynivalenol analyses of Kansas wheat quality profile samples. *In* "Scab and Deoxynivalenol in the 1982 Wheat Crop," Report of Progress, No. 440, pp. 10–12. Agric. Exp. Stn., Kansas State Univ., Manhattan.

Shannon, G. M., and Shotwell, O. L. (1976). Thin layer chromatographic determination of sterigmatocystin in cereal grains and soybeans. *J. Assoc. Off. Anal. Chem.* **59**, 963–965.

Shannon, G. M., and Shotwell, O. L. (1979). Minicolumn detection methods for aflatoxin in yellow corn: collaborative study. *J. Assoc. Off. Anal. Chem.* **62**, 1070–1075.

Shotwell, O. L. (1977). Aflatoxin in corn. *J. Am. Oil Chem. Soc.* **54**, 216A–224A.

Shotwell, O. L. (1983). Aflatoxin detection and determination in corn. *In* "Aflatoxin and *Aspergillus flavus* in Corn" (U. L. Diener, ed.), Southern Cooperative Series, Bulletin No. 279, pp. 38–45. Auburn Univ., Auburn, Alabama.

Shotwell, O. L., and Hesseltine, C. W. (1981). Use of bright greenish yellow fluorescence as a presumptive test for aflatoxin in corn. *Cereal Chem.* **58,** 124–127.

Shotwell, O. L., and Holaday, C. E. (1981). Minicolumn detection methods for aflatoxin in raw peanuts: collaborative study. *J. Assoc. Off. Anal. Chem.* **64,** 674–677.

Shotwell, O. L., and Stubblefield, R. D. (1972). Collaborative study of the determination of aflatoxin in corn and soybeans. *J. Assoc. Off. Anal. Chem.* **55,** 781–788.

Shotwell, O. L., Hesseltine, C. W., and Goulden, M. L. (1969). Note on the natural occurrence of ochratoxin A. *J. Assoc. Off. Anal. Chem.* **52,** 81–83.

Shotwell, O. L., Hesseltine, C. W., Goulden, M. L., and Vandegraft, E. E. (1970). Survey of corn for aflatoxins, zearalenone and ochratoxin. *Cereal Chem.* **47,** 700–707.

Shotwell, O. L., Hesseltine, C. W., Vandergraft, E. E., and Goulden, M. L. (1971). Survey of corn from different regions for aflatoxins, ochratoxins, and zearalenone. *Cereal Sci. Today* **16,** 266–271.

Shotwell, O. L., Goulden, M. L., and Hesseltine, C. W. (1972). Aflatoxin contamination: Association with foreign material and characteristic fluorescence in damaged corn kernels. *Cereal Chem.* **49,** 458–465.

Shotwell, O. L., Goulden, M. L., and Hesseltine, C. W. (1974). Aflatoxin distribution in contaminated corn. *Cereal Chem.* **51,** 492–499.

Shotwell, O. L., Goulden, M. L., Jepson, A. M., Kwolek, W. F., and Hesseltine, C. W. (1975). Aflatoxin occurrence in some white corn under loan. 1971. III. Association with bright greenish-yellow fluorescence in corn. *Cereal Chem.* **52,** 670–677.

Shotwell, O. L. , Goulden, M. L., and Bennett, G. A. (1976). Determination of zearalenone in corn: collaborative study. *J. Assoc. Off. Anal. Chem.* **59,** 666–670.

Shotwell, O. L., Goulden, M. L., Bennett, G. A., Plattner, R. D., and Hesseltine, C. W. (1977). Survey of 1975 wheat and soybeans for aflatoxin, zearalenone, and ochratoxin. *J. Assoc. Off. Anal. Chem.* **60,** 778–783.

Shotwell, O. L., Bennett, G. A., Goulden, M. L., Shannon, G. M., Stubblefield, R. D., and Hesseltine, C. W. (1980a). Survey of 1977 midwest corn at harvest for aflatoxin. *Cereal Foods World* **25,** 12, 14.

Shotwell, O. L., Bennett, G. A., Goulden, M. L., Plattner, R. D., and Hesseltine, C. W. (1980b). Survey for zearalenone, aflatoxin, and ochratoxin in U.S. grain sorghum from 1975 and 1976 crops. *J. Assoc. Off. Anal. Chem.* **63,** 922–926.

Shotwell, O. L., Bennett, G. A., Stubblefield, R. D., Shannon, G. M., Kwolek, W. F., and Plattner, R. D. (1985). Deoxynivalenol in hard red winter wheat: relationship between toxin levels and grading factors. *J. Assoc. Off. Anal. Chem.* **68,** 954–957.

Stack, M. E., and Rodricks, J. V. (1971). Method for analysis and chemical confirmation of sterigmatocystin. *J. Assoc. Off. Anal. Chem.* **54,** 86–90.

Stack, M. E., and Rodricks, J. V. (1973). Collaborative study of the quantitative determination and chemical confirmation of sterigmatocystin in grains. *J. Assoc. Off. Anal. Chem.* **56,** 1123–1125.

Stack, M. E., Nesheim, S., Brown, N. L., and Pohland, A. C. (1976). Determination of sterigmatocystin in corn and oats by gel permeation and high pressure liquid chromatography. *J. Assoc. Off. Anal. Chem.* **59,** 966–970.

Stob, M., Baldwin, R. S., Tuite, J., Andrews, F. N., and Gillette, K. G. (1962). Isolation of an anabolic uterotrophic compound from corn infected with *Gibberella zeae*. *Nature (London)* **196,** 1318–1319.

Stoloff, L. (1976a). Incidence, distribution, and disposition of products containing aflatoxins. *Proc. Am. Phytopathol. Soc.* **3**, 156–172.

Stoloff, L. (1976b). Report on mycotoxins. *J. Assoc. Off. Anal. Chem.* **59**, 317–323.

Stoloff, L., Henry, S., and Francis, O. J., Jr. (1976). Survey for aflatoxin and zearalenone in 1973 crop corn stored on farms and in country elevators. *J. Assoc. Off. Anal. Chem.* **59**, 118–121.

Stubblefield, R. D., and Kwolek, W. F. (1986). Rapid HPLC determination of aflatoxins M_1 and M_2 in artificially contaminated fluid milks: Collaborative study. *J. Assoc. Off. Anal. Chem.* (in press).

Stubblefield, R. D., and Shannon, G. M. (1974). Collaborative study of methods for the determination and chemical confirmation of aflatoxin M_1 in dairy products. *J. Assoc. Off. Anal. Chem.* **57**, 852–857.

Stubblefield, R. D., Van Egmond, H. P., Paulsch, W. E., and Schuller, P. L. (1980). Determination and confirmation of identity of aflatoxin M_1 in dairy products: collaborative study. *J. Assoc. Off. Anal. Chem.* **63**, 907–921.

Swanson, S. P., Corley, R. A., White, D. G., and Buck, W. B. (1984). Rapid thin-layer chromatographic method for determination of zearalenone and zearalenol in grains and animal feeds. *J. Assoc. Off. Anal. Chem.* **67**, 580–582.

Takitani, S., Asabe, Y., Kato, T., Suzuki, M., and Ueno, Y. (1979). Spectrodensitometric determination of trichothecene mycotoxins with 4-(p-nitrobenzyl)pyridine on silica gel thin-layer chromatograms. *J. Chromatogr.* **172**, 335–342.

Thean, J. E., Lorenz, D. R., Wilson, D. M., Rodgers, K., and Gueldner, R. C. (1980). Extraction, cleanup, and quantitative determination of aflatoxins in corn. *J. Assoc. Off. Anal. Chem.* **63**, 631–633.

Tosch, D., Waltking, A. E., and Schlesier J. F. (1984). Comparison of liquid chromatography and high performance thin layer chromatography for determination of aflatoxin in peanut products. *J. Assoc. Off. Anal. Chem.* **67**, 337–339.

Trenholm, H. L., Cochrane, W. P., Cohen, H., Elliot, J. I., Farnworth, E. R., Friend, D. W., Hamilton, R. M. G., Neish, G. A., and Standish, J. F. (1981). Survey of vomitoxin contamination of the 1980 white winter wheat crop in Ontario, Canada. *J. Am. Oil Chem. Soc.* **58**, 992A–994A.

Trenholm, H. L., Cochrane, W. P., Cohen, H., Elliot, J. I., Farnworth, E. R., Friend, D. W., Hamilton, R. M. G., Standish, J. F., and Thompson, B. K. (1983). Survey of vomitoxin contamination of 1980 Ontario white winter wheat crop: results of survey and feed trials. *J. Assoc. Off. Anal. Chem.* **66**, 92–97.

Trucksess, M. W., Nesheim, S., and Eppley, R. M. (1984). Thin layer chromatographic determination of deoxynivalenol in wheat and corn. *J. Assoc. Off. Anal. Chem.* **67**, 40–43.

van Egmond, H. P., and Stubblefield, R. D. (1981). Improved method for confirmation of identity of aflatoxins B_1 and M_1 in dairy products and animal tissue extracts. *J. Assoc. Off. Anal. Chem.* **64**, 152–155.

van Egmond, H. P., Paulsch, W. E., Deijll, E., and Schuller, P. L. (1980). Thin layer chromatographic method for analysis and chemical confirmation of sterigmatocystin in cheese. *J. Assoc. Off. Anal. Chem.* **63**, 110–114.

Velasco, J. (1972). Detection of aflatoxin using small columns of Florisil. *J. Am. Oil Chem. Soc.* **49**, 141–142.

Vesonder, R. F., Ciegler, A., Rogers, R. F., Burbridge, K. A., Bothast, R. J., and Jensen, A. H. (1978). Survey of 1977 crop year preharvest corn for vomitoxin. *Appl. Environ. Microbiol.* **36**, 885–888.

Vesonder, R. F., Ciegler, A., Rohwedder, W. K., and Eppley, R. (1979). Re-examination of 1972 midwest corn for vomitoxin. *Toxicon* **17**, 658–660.

Waltking, A. E. (1970). Collaborative study of three methods for determination of aflatoxin in peanuts and peanut products. *J. Assoc. Off. Anal. Chem.* **53,** 104–113.

Ware, G. M., and Thorpe, C. W. (1978). Determination of zearalenone in corn by high pressure liquid chromatography and fluorescence detection. *J. Assoc. Off. Anal. Chem.* **61,** 1058–1062.

Ware, G. M., Carman, A., Francis, O., and Kuan, S. (1984). Gas chromatographic determination of deoxynivalenol in wheat with electron capture detection. *J. Assoc. Off. Anal. Chem.* **67,** 731–734.

Ware, G. M., Francis, O., Carman, A., and Kuan, S. (1986). Gas chromatographic determination of deoxynivalenol in wheat: Collaborative study. *J. Assoc. Off. Anal. Chem.* (in press).

Whitten, M. E. (1968). Occurrence of aflatoxins in cottonseed and cottonseed products. *Proc. Mycotoxin Res. Semin., Washington, D.C., 1967* pp. 7–9. U.S. Dep. Agric., Washington, D.C.

Wilson, D. M., Sangster, L. T., and Bedell, D. M. (1984). Recognizing the signs of porcine aflatoxicosis. *Vet. Med.* **79,** 974–977.

4

Application of Ultraviolet and Infrared Spectroscopy in the Analysis and Structural Elucidation of Mycotoxins

CHARLES P. GORST-ALLMAN

Council for Scientific and Industrial Research
National Chemical Research Laboratory
Pretoria, 0001, Republic of South Africa

I. INTRODUCTION

Ultraviolet (UV) and infrared (IR) spectroscopy have long been important tools for structural elucidation for organic chemists. For many years the information obtained with respect to the nature and environment of organic functional groups was unavailable by other spectroscopic techniques. However, in recent years, the enormous advances made in the field of nuclear magnetic resonance (NMR) spectroscopy, particularly with respect to very high field NMR spec-

MODERN METHODS IN THE ANALYSIS
AND STRUCTURAL ELUCIDATION
OF MYCOTOXINS

troscopy, and the increasing usage of single-crystal X-ray cyrstallography, have resulted in a decline in the use of UV and IR spectroscopy as a tool for structural determination. Coincidental with this has been the development of more sophisticated chromatographic techniques, many of which rely on UV absorption as the method of choice for the detection of the metabolites in the eluent. IR detectors are also finding increased usage when coupled to either gas chromatographic (GC) or high-performance liquid chromatographic (HPLC) equipment. Consequently a shift in emphasis has occurred. UV and IR spectroscopy are now more prominent as analytical tools and in the characterization of known substances and are being used to a lesser extent for structural elucidation of new mycotoxins.

The accurate recognition and analysis of known toxins in contaminated samples is of the utmost importance in studying mycotoxins. IR spectroscopy, with its highly characteristic spectra, is a powerful method for the identification of known metabolites, especially as numerous standard spectra exist (Cole and Cox, 1981; Pohland *et al.*, 1982). As most mycotoxins contain chromophores which absorb in the UV portion of the electromagnetic spectrum, UV detectors, with their high sensitivity and ease of operation, are by far the most commonly used in liquid chromatographic analysis of these compounds. Because of this, a greater emphasis will be placed in this chapter on the analysis of mycotoxins rather than on structural elucidation. The recognition and quantitation of known toxins is of prime importance to the food and feed industry worldwide.

II. INFRARED SPECTROSCOPY

A. Instrumentation

An in-depth examination of the equipment and experimental techniques of IR spectroscopy is outside the scope of this chapter. However, brief mention will be made of the new possibilities available as a result of the development of Fourier transform infrared (FT-IR) spectroscopy. This new generation of IR spectrometers has several advantages over the traditional grating instrument. These include (i) fast scanning time, (ii) high sensitivity, (iii) large resolving power, (iv) high wavenumber accuracy, (v) large wavenumber range per scan, and (vi) low cost of basic optical equipment. In addition, the data systems available with FT-IR spectrometers make possible fast searching for similar spectra in commercial and user-generated libraries. This has obvious advantages in mycotoxin research, where families of toxins are frequently found. The metabolites in a particular group tend to have similar IR spectra, and, thus, a closely related metabolite may be quickly recognized.

B. Structural Elucidation

For almost a hundred years chemists have relied on IR spectroscopy as a powerful tool in the structural elucidation of organic molecules. The information content of an IR spectrum is very high; each compound possesses its own characteristic IR absorptions. Over the years certain characteristic absorption bands have been associated with particular molecular functional groups, so that absorption by an unknown compound around the characteristic frequencies of a functional group is a strong indication of the presence of that particular moiety in the unknown compound under investigation. Tables of these characteristic frequencies have been compiled (Nakanishi, 1962; Bellamy, 1975) and are of great value in the intepretation of spectra of compounds of unknown structure.

In Table I characteristic IR absorptions of 64 of the most common and important mycotoxins have been arranged in order of decreasing wave number (ranging from 3620 to 740 cm^{-1}). They have been collected by a variety of different methods as indicated in Table I. In many instances only one member of a particular family has been included; however, in these cases the absorptions for that compound are fairly characteristic of the family as a whole. For example, the

TABLE I

Infrared Absorptions of Some Common Mycotoxins

Compound	Method[a]	Reference	Wave number (cm^{-1})
Austocystin D	CHCl$_3$	Steyn and Vleggaar (1974)	3620
Verrucarol	CH$_2$Cl$_2$	Gutzwiller et al. (1964)	3610
Patulin	CHCl$_3$	Scott (1974)	3580
Verrucarol	CH$_2$Cl$_2$	Gutzwiller et al. (1964)	3570
Sporidesmin	Para	Ronaldson et al. (1963)	3558
Aflatoxin Q$_1$	KBr	Masri et al. (1974)	3550
Alternariol	Nujol	Freeman (1966)	3550
Viridicatumtoxin	CHCl$_3$	Kabuto et al. (1976)	3550
Roridin A	KBr	Böhner et al. (1965)	3546
Aflatoxin B$_3$ (parasiticol)	KBr	Cole and Kirksey (1971)	3520
Rubratoxin B	—	Moss et al. (1968)	3520
Cytochalasin B	Nujol	Aldridge et al. (1967)	3510
Secalonic acid D	KBr	Andersen et al. (1977)	3505
Citreoviridin	KBr	Sakabe et al. (1964)	3500
Cytochalasin J	Nujol	Patwardhan et al. (1974)	3500
Nivalenol	KBr	Tatsuno et al. (1969)	3500
Tryptoquivaline	CHCl$_3$	Büchi et al. (1977)	3490
Scirpenetriol	Nujol	Brian et al. (1961)	3480
α-Cyclopiazonic acid	CHCl$_3$	Holzapfel (1968)	3478
Roridin A	KBr	Böhner et al. (1965)	3472
Aflatoxin Q$_1$	KBr	Masri et al. (1974)	3460
Austdiol	KBr	Vleggaar et al. (1974)	3460
Scirpenetriol	Nujol	Brian et al. (1961)	3455

(*continued*)

TABLE I *(Continued)*

Compound	Method[a]	Reference	Wave number (cm^{-1})
Dihydrosterigmatocystin	KBr	Davies *et al.* (1960)	3450
Islanditoxin	Nujol	Marumo (1955)	3450
Sterigmatocystin	KBr	Davies *et al.* (1960)	3450
Cytochalasin K	KBr	Steyn *et al.* (1982)	3440
Chaetoglobosin A	KBr	Natori (1977)	3438
Ochratoxin A	KBr	Van der Merwe *et al.* (1965)	3430
Roquefortine	CHCl$_3$	Scott *et al.* (1976)	3430
Aflatoxin M$_1$	Nujol	Holzapfel *et al.* (1966)	3425
Aflatrem	KBr	Gallagher *et al.* (1980)	3420
Paxilline	KBr	Cole *et al.* (1974)	3420
Rhizonin A	CHCl$_3$	Steyn *et al.* (1983)	3418
Cytochalasin E	CHCl$_3$	Steyn *et al.* (1982)	3410
Scirpenetriol	Nujol	Brian *et al.* (1961)	3405
Emodin	KBr	Wells *et al.* (1975)	3400
HT-2 Toxin	CHCl$_3$	Bamburg and Strong (1969)	3400
Janthitrem E	KBr	de Jesus *et al.* (1984)	3400
Penitrem A	KBr	de Jesus *et al.* (1983)	3400
T-2 Toxin	KBr	Bamburg *et al.* (1968)	3400
Zygosporin D	—	Minato and Matsumoto (1970)	3400
Aflatrem	KBr	Gallagher *et al.* (1980)	3390
4-Hydroxyochratoxin A	CHCl$_3$	Hutchison *et al.* (1971)	3380
Cytochalasin B	Nujol	Aldridge *et al.* (1967)	3380
Roquefortine	CHCl$_3$	Scott *et al.* (1976)	3380
Aurovertin		Baldwin *et al.* (1964)	3378
Austdiol	KBr	Vleggaar *et al.* (1971)	3370
Sporidesmin	Para	Ronaldson *et al.* (1963)	3355
Aspertoxin	KBr	Rodricks *et al.* (1968)	3354
Aflatoxin M$_2$	CHCl$_3$	Holzapfel *et al.* (1966)	3350
Cytochalasin A	Nujol	Aldridge *et al.* (1967)	3350
Viridicatumtoxin	CHCl$_3$	Kabuto *et al.* (1976)	3350
Tentoxin	—	Meyer *et al.* (1971)	3345
Paspalicine	KBr	Leutwiler (1973)	3340
Patulin	CHCl$_3$	Scott (1974)	3340
Phomopsin A	KBr	Culvenor *et al.* (1983)	3340
Rhizonin A	CHCl$_3$	Steyn *et al.* (1983)	3340
Cytochalasin H	Nujol	Patwardhan *et al.* (1974)	3300
Maltoryzine	KBr	Iizuka (1974)	3300
Zearalenone	—	Mirocha *et al.* (1977)	3300
Rhizonin A	CHCl$_3$	Steyn *et al.* (1983)	3295
Islanditoxin	Nujol	Marumo (1955)	3270
Malformin A	KBr	Yoshizawa *et al.* (1975)	3260
Chaetoglobosin A	KBr	Natori (1977)	3259

TABLE I   (*Continued*)

Compound	Method[a]	Reference	Wave number (cm^{-1})
Alternariol	Nujol	Freeman (1966)	3250
Tenuazonic acid	CCl$_4$	Stickings (1959)	3236
Cytochalasin A	Nujol	Aldridge *et al.* (1967)	3210
α-Cyclopiazonic acid	CHCl$_3$	Holzapfel (1968)	3200
Citrinin	KBr	Kovac *et al.* (1961)	3200
Cytochalasin J	Nujol	Patwardhan *et al.* (1974)	3200
Nivalenol	KBr	Tatsuno *et al.* (1969)	3200
Roquefortine	CHCl$_3$	Scott *et al.* (1976)	3190
O-Methylsterigmatocystin	CHBr$_3$	Davies *et al.* (1960)	3124
Aspergillic acid	KBr	Masaki *et al.* (1966)	3120
Tenuazonic acid	CCl$_4$	Stickings (1959)	3098
Phomopsin A	KBr	Culvenor *et al.* (1983)	2988
Cytochalasin E	CHCl$_3$	Steyn *et al.* (1982)	2970
HT-2 Toxin	CHCl$_3$	Bamburg and Strong (1969)	2950
Aspergillic acid	KBr	Masaki *et al.* (1966)	2940
T-2 Toxin	KBr	Bamburg *et al.* (1968)	2940
Penitrem A	KBr	de Jesus *et al.* (1983)	2930
Cytochalasin K	KBr	Steyn *et al.* (1982)	2920
Ipomeanine	CCl$_4$	Boyd *et al.* (1974)	2900
Janthitrem E	KBr	de Jesus *et al.* (1984)	2900
Aspergillic acid	KBr	Masaki *et al.* (1966)	2800
Citrinin	KBr	Kovac *et al.* (1961)	2700
α-Cyclopiazonic acid	CHCl$_3$	Holzapfel (1968)	2600
4-Hydroxyochratoxin A	CHCl$_3$	Hutchison *et al.* (1971)	2500
Aspergillic acid	KBr	Masaki *et al.* (1966)	2040
Rubratoxin B	—	Moss *et al.* (1968)	1860
Rubratoxin B	—	Moss *et al.* (1968)	1820
Rubratoxin B	—	Moss *et al.* (1968)	1790
Tryptoquivaline	CHCl$_3$	Büchi *et al.* (1977)	1786
Patulin	CHCl$_3$	Scott (1974)	1782
Moniliformin	KBr	Rabie *et al.* (1978)	1780
Cytochalasin K	KBr	Steyn *et al.* (1982)	1765
Cytochalasin E	CHCl$_3$	Steyn *et al.* (1982)	1762
Aflatoxin B$_1$	CHCl$_3$	Asao *et al.* (1965)	1760
Aflatoxin B$_2$	CHCl$_3$	Chang *et al.* (1963)	1760
Aflatoxin G$_1$	CHCl$_3$	Asao *et al.* (1965)	1760
Aflatoxin M$_1$	Nujol	Holzapfel *et al.* (1966)	1760
Aflatoxin M$_2$	CHCl$_3$	Holzapfel *et al.* (1966)	1760
Patulin	CHCl$_3$	Scott (1974)	1755
Cytochalasin H	Nujol	Patwardhan *et al.* (1974)	1745
Ochratoxin A	KBr	Van der Merwe *et al.* (1965)	1745
Roridin A	KBr	Böhner *et al.* (1965)	1742

(*continued*)

TABLE I   (*Continued*)

Compound	Method[a]	Reference	Wave number (cm^{-1})
Austalide A	CHCl$_3$	Horak *et al.* (1981)	1740
Aurovertin	—	Baldwin *et al.* (1964)	1739
Aflatoxin B$_3$ (parasiticol)	KBr	Cole and Kirksey (1971)	1735
PR Toxin	KBr	Wei *et al.* (1975)	1735
Secalonic acid D	KBr	Andersen *et al.* (1977)	1735
Tenuazonic acid	CCl$_4$	Stickings (1959)	1735
Ochratoxin B	KBr	Van der Merwe *et al.* (1965)	1730
Trichodermin	KBr	Godtfredsen *et al.* (1964)	1730
Tryptoquivaline	CHCl$_3$	Büchi *et al.* (1977)	1728
Viridicatumtoxin	CHCl$_3$	Kabuto *et al.* (1976)	1725
4-Hydroxyochratoxin A	CHCl$_3$	Hutchison *et al.* (1971)	1723
Cytochalasin E	CHCl$_3$	Steyn *et al.* (1982)	1720
Cytochalasin K	KBr	Steyn *et al.* (1982)	1720
HT-2 Toxin	CHCl$_3$	Bamburg and Strong (1969)	1720
Ipomeanine	CCl$_4$	Boyd *et al.* (1974)	1720
PR Toxin	KBr	Wei *et al.* (1975)	1720
T-2 Toxin	KBr	Bamburg *et al.* (1968)	1720
Cytochalasin B	Nujol	Aldridge *et al.* (1967)	1715
Sporidesmin	Para	Ronaldson *et al.* (1963)	1715
Cytochalasin A	Nujol	Aldridge *et al.* (1967)	1714
Penitrem A	KBr	de Jesus *et al.* (1983)	1710
α-Cyclopiazonic acid	CHCl$_3$	Holzapfel (1968)	1708
Moniliformin	KBr	Rabie *et al.* (1978)	1705
Roridin A	KBr	Böhner *et al.* (1965)	1704
Citreoviridin	KBr	Sakabe *et al.* (1964)	1702
Griseofulvin	—	Grove *et al.* (1952)	1700
Janthitrem E	KBr	de Jesus *et al.* (1984)	1700
Maltoryzine	KBr	Iizuka (1974)	1700
Zygosporin D	—	Minato and Matsumoto (1970)	1700
Aflatoxin G$_1$	CHCl$_3$	Asao *et al.* (1965)	1695
Aurovertin	—	Baldwin *et al.* (1964)	1695
Cytochalasin A	Nujol	Aldridge *et al.* (1967)	1692
Cytochalasin B	Nujol	Aldridge *et al.* (1967)	1692
Aflatoxin M$_1$	Nujol	Holzapfel *et al.* (1966)	1690
Aflatoxin M$_2$	CHCl$_3$	Holzapfel *et al.* (1966)	1690
Cytochalasin H	Nujol	Patwardhan *et al.* (1974)	1690
Cytochalasin J	Nujol	Patwardhan *et al.* (1974)	1690
Ochratoxin A	KBr	Van der Merwe *et al.* (1965)	1690
Chaetoglobosin A	KBr	Natori (1977)	1689
Citreoviridin	KBr	Sakabe *et al.* (1964)	1689
Zearalenone	—	Mirocha *et al.* (1977)	1688
Aflatoxin B$_2$	CHCl$_3$	Chang *et al.* (1963)	1685

TABLE I (*Continued*)

Compound	Method[a]	Reference	Wave number (cm^{-1})
Citrinin	KBr	Kovac *et al.* (1961)	1685
Fumitremorgin B	—	Yamazaki *et al.* (1974)	1685
Roquefortine	CHCl$_3$	Scott *et al.* (1976)	1685
Aflatoxin B$_1$	CHCl$_3$	Asao *et al.* (1965)	1684
Trichodermin	KBr	Godtfredsen *et al.* (1964)	1682
Aflatrem	KBr	Gallagher *et al.* (1980)	1680
Asteltoxin	KBr	Kruger *et al.* (1979)	1680
Ipomeanine	CCl$_4$	Boyd *et al.* (1974)	1680
Moniliformin	KBr	Rabie *et al.* (1978)	1680
Nivalenol	KBr	Tatsuno *et al.* (1969)	1680
Ochratoxin B	KBr	Van der Merwe *et al.* (1965)	1680
PR Toxin	KBr	Wei *et al.* (1975)	1680
4-Hydroxyochratoxin A	CHCl$_3$	Hutchison *et al.* (1971)	1678
Scirpenetriol	Nujol	Brian *et al.* (1961)	1676
Verrucarol	CH$_2$Cl$_2$	Gutzwiller *et al.* (1964)	1675
Tenuazonic acid	CCl$_4$	Stickings (1959)	1674
Tryptoquivaline	CHCl$_3$	Büchi *et al.* (1977)	1672
Austdiol	KBr	Vleggaar *et al.* (1974)	1670
Phomopsin A	KBr	Culvenor *et al.* (1983)	1670
Tentoxin	—	Meyer *et al.* (1971)	1670
Paspalicine	KBr	Leutwiler (1973)	1666
Austocystin D	CHCl$_3$	Steyn and Vleggaar (1974)	1665
Ochratoxin A	KBr	Van der Merwe *et al.* (1965)	1665
Roquefortine	CHCl$_3$	Scott *et al.* (1976)	1665
Sporidesmin	Para	Ronaldson *et al.* (1963)	1664
Cytochalasin E	CHCl$_3$	Steyn *et al.* (1982)	1662
Cytochalasin K	KBr	Steyn *et al.* (1982)	1662
O-Methylsterigmatocystin	CHBr$_3$	Davies *et al.* (1960)	1662
Alternariol	Nujol	Freeman (1966)	1660
Malformin A	KBr	Yoshizawa *et al.* (1975)	1660
Aspertoxin	KBr	Rodricks *et al.* (1968)	1656
4-Hydroxyochratoxin A	CHCl$_3$	Hutchison *et al.* (1971)	1655
Fumitremorgin B	—	Yamazaki *et al.* (1974)	1655
Citreoviridin	KBr	Sakabe *et al.* (1964)	1654
Griseofulvin	—	Grove *et al.* (1952)	1650
Islanditoxin	Nujol	Marumo (1955)	1650
Paxilline	KBr	Cole *et al.* (1974)	1650
Penitrem A	KBr	de Jesus *et al.* (1983)	1650
Rhizonin A	CHCl$_3$	Steyn *et al.* (1983)	1650
Sterigmatocystin	KBr	Davies *et al.* (1960)	1650
Viridicatumtoxin	CHCl$_3$	Kabuto *et al.* (1976)	1650
Dihydrosterigmatocystin	KBr	Davies *et al.* (1960)	1648

(*continued*)

TABLE I   (*Continued*)

Compound	Method[a]	Reference	Wave number (cm^{-1})
Phomopsin A	KBr	Culvenor *et al.* (1983)	1645
Zearalenone	—	Mirocha *et al.* (1977)	1645
O-Methylsterigmatocystin	CHBr$_3$	Davies *et al.* (1960)	1643
Aspergillic acid	KBr	Masaki *et al.* (1966)	1640
Citrinin	KBr	Kovac *et al.* (1961)	1640
Cytochalasin B	Nujol	Aldridge *et al.* (1967)	1638
Roridin A	KBr	Böhner *et al.* (1965)	1637
Austocystin D	CHCl$_3$	Steyn and Vleggaar (1974)	1635
Emodin	KBr	Wells *et al.* (1975)	1635
HT-2 Toxin	CHCl$_3$	Bamburg and Strong (1969)	1635
T-2 Toxin	KBr	Bamburg *et al.* (1968)	1635
Aflatoxin B$_1$	CHCl$_3$	Asao *et al.* (1965)	1632
Aflatoxin G$_1$	CHCl$_3$	Asao *et al.* (1965)	1630
Aspertoxin	KBr	Rodricks *et al.* (1968)	1630
Tentoxin	—	Meyer *et al.* (1971)	1630
Tenuazonic acid	CCl$_4$	Stickings (1959)	1630
Sterigmatocystin	KBr	Davies *et al.* (1960)	1627
Citreoviridin	KBr	Sakabe *et al.* (1964)	1626
Aflatoxin B$_2$	CHCl$_3$	Chang *et al.* (1963)	1625
Emodin	KBr	Wells *et al.* (1975)	1625
Cytochalasin A	Nujol	Aldridge *et al.* (1967)	1623
Dihydrosterigmatocystin	KBr	Davies *et al.* (1960)	1622
Asteltoxin	KBr	Kruger *et al.* (1979)	1620
Austdiol	KBr	Vleggaar *et al.* (1974)	1620
Griseofulvin	—	Grove *et al.* (1952)	1620
PR Toxin	KBr	Wei *et al.* (1975)	1620
Viridicatumtoxin	CHCl$_3$	Kabuto *et al.* (1976)	1620
α-Cyclopiazonic acid	CHCl$_3$	Holzapfel (1968)	1618
Austocystin D	CHCl$_3$	Steyn and Vleggaar (1974)	1615
Chaetoglobosin A	KBr	Natori (1977)	1615
Moniliformin	KBr	Rabie *et al.* (1978)	1615
Paspalicine	KBr	Leutwiler (1973)	1613
Zearalenone	—	Mirocha *et al.* (1977)	1612
Alternariol	Nujol	Freeman (1966)	1610
Austalide A	CHCl$_3$	Horak *et al.* (1981)	1610
Nivalenol	KBr	Tatsuno *et al.* (1969)	1610
Secalonic acid D	KBr	Andersen *et al.* (1977)	1610
Sterigmatocystin	KBr	Davies *et al.* (1960)	1610
Tryptoquivaline	CHCl$_3$	Büchi *et al.* (1977)	1610
Roquefortine	CHCl$_3$	Scott *et al.* (1976)	1608
O-Methylsterigmatocystin	CHBr$_3$	Davies *et al.* (1960)	1603
Aflatoxin B$_2$	CHCl$_3$	Chang *et al.* (1963)	1600

TABLE I (*Continued*)

Compound	Method[a]	Reference	Wave number (cm^{-1})
Austdiol	KBr	Vleggaar et al. (1974)	1600
Maltoryzine	KBr	Iizuda (1974)	1600
Aflatoxin B_1	$CHCl_3$	Asao et al. (1965)	1598
Roridin A	KBr	Böhner et al. (1965)	1597
Aflatoxzin G_1	$CHCl_3$	Asao et al. (1965)	1595
Aspergillic acid	KBr	Masaki et al. (1966)	1585
Secalonic acid D	KBr	Andersen et al. (1977)	1585
Dihydrosterigmatocystin	KBr	Davies et al. (1960)	1582
Griseofulvin	—	Grove et al. (1952)	1580
Zearalenone	—	Mirocha et al. (1977)	1578
Aspertoxin	KBr	Rodricks et al. (1968)	1575
Paspalicine	KBr	Leutwiler (1973)	1573
Aflatoxin B_1	$CHCl_3$	Asao et al. (1965)	1562
Ipomeanine	CCl_4	Boyd et al. (1974)	1560
Paspalicine	KBr	Leutwiler (1973)	1549
Aflatoxin G_1	$CHCl_3$	Asao et al. (1965)	1545
Malformin A	KBr	Yoshizawa et al. (1975)	1540
Aurovertin	—	Baldwin et al. (1964)	1536
Ochratoxin A	KBr	Van der Merwe et al. (1965)	1535
Ochratoxin B	KBr	Van der Merwe et al. (1965)	1535
Asteltoxin	KBr	Kruger et al. (1979)	1530
Islanditoxin	Nujol	Marumo (1955)	1530
Maltoryzine	KBr	Iizuka (1974)	1500
Sterigmatocystin	KBr	Davies et al. (1960)	1482
Citrinin	KBr	Kovac et al. (1961)	1480
Phomopsin A	KBr	Culvenor et al. (1983)	1480
O-Methylsterigmatocystin	$CHBr_3$	Davies et al. (1960)	1473
Tryptoquivaline	$CHCl_3$	Büchi et al. (1977)	1470
Aspertoxin	KBr	Rodricks et al. (1968)	1460
PR Toxin	KBr	Wei et al. (1975)	1460
Citreoviridin	KBr	Sakabe et al. (1964)	1452
Dihydrosterigmatocystin	KBr	Davies et al. (1960)	1450
Secalonic acid D	KBr	Andersen et al. (1977)	1432
O-Methylsterigmatocystin	$CHBr_3$	Davies et al. (1960)	1418
Aurovertin	—	Baldwin et al. (1964)	1410
Dihydrosterigmatocystin	KBr	Davies et al. (1960)	1398
Ipomeanine	CCl_4	Boyd et al. (1974)	1390
Citrinin	KBr	Kovac et al. (1961)	1380
PR Toxin	KBr	Wei et al. (1975)	1380
Verrucarol	CH_2Cl_2	Gutzwiller et al. (1964)	1380
Austalide A	$CHCl_3$	Horak et al. (1981)	1370
Paxilline	KBr	Cole et al. (1974)	1365

(*continued*)

TABLE I (*Continued*)

Compound	Method[a]	Reference	Wave number (cm^{-1})
T-2 Toxin	KBr	Bamburg *et al.* (1968)	1365
Sterigmatocystin	KBr	Davies *et al.* (1960)	1362
Paxilline	KBr	Coles *et al.* (1974)	1355
Verrucarol	CH$_2$Cl$_2$	Gutzwiller *et al.* (1964)	1335
Dihydrosterigmatocystin	KBr	Davies *et al.* (1960)	1275
O-Methylsterigmatocystin	CHBr$_3$	Davies *et al.* (1960)	1267
Austalide A	CHCl$_3$	Horak *et al.* (1981)	1260
Aurovertin	—	Baldwin *et al.* (1964)	1250
Trichodermin	KBr	Godtfredsen *et al.* (1964)	1245
HT-2 Toxin	CHCl$_3$	Bamburg and Strong (1969)	1240
T-2 Toxin	KBr	Bamburg *et al.* (1968)	1240
Cytochalasin H	Nujol	Patwardhan *et al.* (1974)	1235
Trichodermin	KBr	Godtfredsen *et al.* (1964)	1225
Aspertoxin	KBr	Rodricks *et al.* (1968)	1133
Dihydrosterigmatocystin	KBr	Davies *et al.* (1960)	1127
Zygosporin D	—	Minato and Matsumoto (1970)	1127
Moniliformin	KBr	Rabie *et al.* (1978)	1110
O-Methylsterigmatocystin	CHBr$_3$	Davies *et al.* (1960)	1075
Zygosporin D	—	Minato and Matsumoto (1970)	1075
Cytochalasin H	Nujol	Patwardhan *et al.* (1974)	1018
Chaetoglobosin A	KBr	Natori (1977)	983
Cytochalasin J	Nujol	Patwardhan *et al.* (1974)	970
Paxilline	KBr	Cole *et al.* (1974)	740

[a] Refers to method of measurement. KBr, Potassium bromide disk; Nujol, Nujol mull; CCl$_4$, CHCl$_3$, CH$_2$Cl$_2$, solutions in carbon tetrachloride, chloroform, and dichloromethane, respectively; Para, parafin.

first entry in Table I is austocystin D, possessing absorptions at 3620, 1665, 1635, and 1615 cm^{-1}. The latter three absorptions are typical of the five other major austocystins (A, B, C, E, and F), and are associated with the α-pyrone and aromatic functional groups (Steyn and Vleggaar, 1974).

The broad diversity in the characteristic frequencies of various functional groups is clearly indicated by Table I. The hydroxyl absorptions range from 3620 cm^{-1} for the free hydroxyl group in austocystin D (Steyn and Vleggaar, 1974) to 3200 cm^{-1} for cytochalasin J (Patwardhan *et al.*, 1974). The most important hydroxyl absorption is usually the stretching vibration of the OH band, which typically appears around 3300 cm^{-1}, but this is affected by concentration, method of measurement, and temperature (Nakanishi, 1962).

Frequently overlapping the hydroxyl absorption is that of the NH stretching vibration. Examples of this from Table I are that of α-cyclopiazonic acid at 3478

cm^{-1} (Holzapfel, 1968) and paspalicine at 3340 cm^{-1} (Leutwiler, 1973). It should be noted that the frequency changes from 3340 to 3480 cm^{-1} for the absorption in paspalicine when the method of measurement is changed from KBr to $CHCl_3$ (Leutwiler, 1973). The NH band can usually be differentiated from the OH band by the relative sharpness of the former.

Absorptions in the region of 3050 to 2900 cm^{-1} are typical of CH stretching vibrations. These are common to all mycotoxins and give relatively little structural information, so they will not be further discussed.

The OH stretching vibration of carboxylic acids appears as a broad band in the region 3000–2500 cm^{-1}. This is typified by the absorptions of citrinin at 2700 cm^{-1} (Kovac et al., 1961) and 4-hydroxyochratoxin A at 2500 cm^{-1} (Hutchison et al., 1971).

One of the most important groups of absorptions is that of the carbonyl moiety. The frequencies for absorptions of this functional group range from 1860 cm^{-1} to the region around 1630 cm^{-1}. Some examples of the different nature of the carbonyl group, and its consequent different characteristic frequency are apparent from Table I. The α,β-unsaturated anhydride carbonyl functionalities present in the rubratoxins typically absorb at between 1860 and 1790 cm^{-1} (Moss et al., 1968). The α,β-unsaturated lactone carbonyl group present in the five-membered ring of patulin absorbs at 1782 cm^{-1} (Scott, 1974), whereas in the six-membered lactone rings found in the aflatoxins, the characteristic frequency shifts to 1760 cm^{-1} (see, e.g., Asao et al., 1965). Moniliformin with its cyclobutenedione structure shows strong absorption at 1780 cm^{-1} (Rabie et al., 1978).

The carbonyl group in a carboxylic acid typically appears in the region around 1735 cm^{-1}; examples are ochratoxin A at 1745 cm^{-1}, ochratoxin B at 1730 cm^{-1} (Van der Merwe et al., 1965), and secalonic acid D at 1735 cm^{-1} (Stickings, 1959). In esters, the characteristic frequency range is from ~1750 cm^{-1} to 1710 cm^{-1}. Thus the acetate carbonyl groups in cytochalasin H and PR toxin absorb at 1745 cm^{-1} (Patwardhan et al., 1974) and 1735 cm^{-1} (Wei et al., 1975), respectively, whereas in the macrocyclic esters, for example, in cytochalasins E and K the characteristic frequency shifts to 1720 cm^{-1} (Steyn et al., 1982).

Many of the aflatoxins contain α,β-unsaturated ketones. Because of the close structural similarity of this family of compounds, this particular range of characteristic frequencies is quite narrow, from 1695 cm^{-1} for aflatoxin G_1 to 1684 cm^{-1} for aflatoxin B_1 (Asao et al., 1965). A much broader diversity is found for amide carbonyl groups. In cytochalasins H and J, this absorption appears at 1690 cm^{-1} (Patwardhan et al., 1974). This value is also found for ochratoxin A, but for ochratoxin B it shifts to 1680 cm^{-1} (Van der Merwe et al., 1965). Values for the cyclic peptides malformin A, rhizonin, and phomopsin A range from 1670 and 1645 cm^{-1} for phomopsin A (Culvenor et al., 1983) to 1660 cm^{-1} for

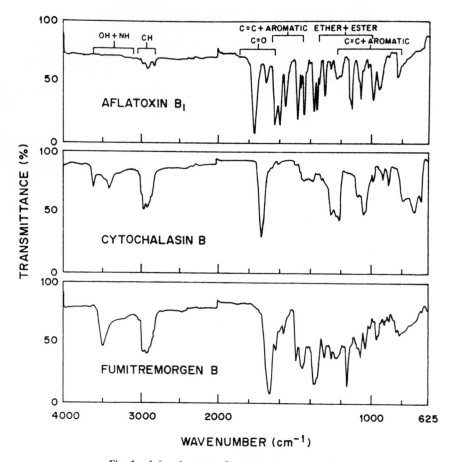

Fig. 1. Infrared spectra of some common mycotoxins.

malformin A (Yoshizawa *et al.*, 1975) and 1650 cm^{-1} for rhizonin A (Steyn *et al.*, 1983). Falling within this broad band of characteristic frequencies, the diketopiperazine carbonyl groups in roquefortine and fumitremorgin B absorb at 1665 cm^{-1} (Scott *et al.*, 1976) and 1655 cm^{-1} (Yamazaki *et al.*, 1974), respectively. Finally, the xanthone moiety in *O*-methylsterigmatocystin gives rise to carbonyl absorption at 1662 cm^{-1} (Davies *et al.*, 1960).

The region between 1650 and 1450 cm^{-1} is characteristic of absorptions due to olefins and aromatic rings. Because practically all mycotoxins possess some form of unsaturation, absorptions in this region are extremely common, though not necessarily very informative. ^{1}H and ^{13}C NMR provide far more useful and detailed information with regard to such structural features. Olefinic absorption

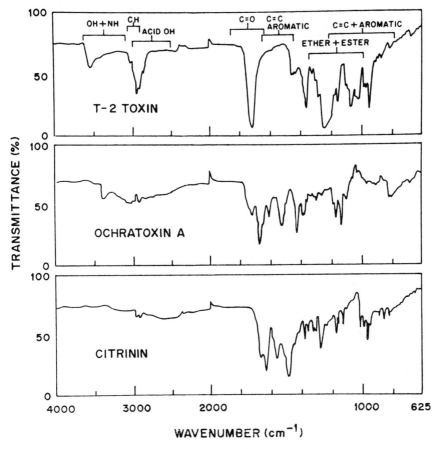

Fig. 1. (*continued*)

of this type is usually accompanied by bands in the region 1000 to 800 cm^{-1}, and aromatic absorption by bands in the regions 1225–950 cm^{-1} and 900–800 cm^{-1}. Additionally, ethers and esters usually absorb in the range 1350–1000 cm^{-1}.

The above provides a brief overview of the more common IR absorptions found for toxic fungal metabolites. The vast diversity of structures found in this field, coupled with the possible ambiguity of the information obtainable from an IR spectrum, means that IR spectroscopy is not particularly useful in analyzing the structures of unknown metabolites. However, the IR spectrum is useful in that it provides an indication of some of the functional moieties present in a molecule, as well as some information regarding their environment (Fig. 1).

Application of a more sensitive probe, such as ^1H and ^{13}C NMR spectroscopy, can frequently supply detail lacking in the IR spectrum.

Table I may also be used as an aid in the structural eludication of an unknown compound, in that comparison of the IR absorptions of the unknown metabolite with those reported for compounds of known structure can lead to the identification of functional groups in the unknown, or even its family or type of compound.

C. Analysis of Known Metabolites

1. Gas Chromatography–Fourier Transform Infrared Spectroscopy (GC-FTIR)

The technique involving the combined use of gas–liquid chromatography and IR spectroscopy has become increasingly more important with the development of Fourier transform IR spectroscopy, with its enhanced sensitivity and decreased scanning times (Erickson, 1979). However, the low volatility of most mycotoxins precludes their purification and analysis by gas chromatography (GC), and it is generally necessary to derivatize these compounds before attempting a gas chromatographic analysis. This inherent disadvantage has restricted the use of gas chromatography in mycotoxin research (Gorst-Allman and Steyn, 1984), and consequently GC-FTIR will not be discussed in this chapter. The review by Erickson referred to above provides details for those interested in the technique.

2. High-Performance Liquid Chromatography–Fourier Transform Infrared Spectroscopy (HPLC-FTIR)

High-performance liquid chromatography (HPLC) is an extremely useful method for the separation and analysis of organic molecules. However, the technique provides no positive identification of a compound, although a comparison of retention time with those of standard samples does indicate possible identities for an unknown. Coupling HPLC equipment to an IR spectrometer allows utilization of the vast amount of accumulated IR data to obtain a method of identifying known compounds with almost 100% accuracy by examination and comparison with reference spectra.

Unlike GC, however, HPLC has several problems inherent in the separation technique which interfere with the analysis. These are associated with the solvents used in the separation which, unlike gas chromatography carrier gases, generally absorb in the most useful IR region (i.e., between 4000 and 400 cm^{-1}). This problem has been approached in two ways (Herres, 1982): the first involves a coupling technique requiring complete removal of the solvent before IR analysis, and the second uses direct coupling to a flow cell device.

The first approach is advantageous in that it completely removes solvent absorptions. However, it is time-consuming and mechanically complex. The second approach is nondestructive and allows collection of chromatographic fractions. A number of factors must be considered in undertaking HPLC using an IR detector if direct coupling to a flow cell is used. The solvent selection influences subsequent choice of window material and cell thickness (Herres, 1982). In the identification of the eluting compounds, the spectral regions made opaque by solvent absorptions must be manipulated so as not to interfere with the important absorption regions of the metabolites themselves. This may be achieved either by solvent selection or by a decrease in path length through the flow cell. Thus when using water or methanol as eluent, a cell thickness of 10 μm is required to achieve full-range spectra, whereas for solvents such as $CHCl_3$, tetrahydrofuran or n-hexane, cell thicknesses of 0.1 to 0.2 mm are used (Herres, 1982). These latter solvents show strong absorptions but are transparent in large spectral "windows" between opaque regions.

Even with the larger cells, however, only a small fraction of the chromatographic peak is present in the detector beam at any one time; thus FTIR, with its high sensitivity, fast measurement time, and high wavenumber accuracy, is mandatory. Moreover, modern data systems which permit up to 10 scans/sec at a resolution of 8 cm^{-1} also permit control of the chromatograph from the spectrometer console, thus allowing stopped-flow techniques for even lower detection limits (Herres, 1982).

It is thus apparent that coupled HPLC-FTIR systems have considerable advantage where the simultaneous elution and identification of known metabolites are required. This would be enhanced by employing a user-generated data bank, in which the spectrum of the eluted material could be immediately compared with that of a reference spectrum as an additional check on purity. The major disadvantage of FTIR spectroscopy as a detection tool is undoubtedly that of cost, in that an IR detector of this capability is likely to cost 20 times as much as a good ultraviolet (UV) or refractive index (RI) detector. The user must decide if the additional information obtained by this method is worth the additional expenditure.

3. Quantitative Analysis

Theoretically, Beer's law may be applied in IR spectroscopy just as it is in UV or visible spectroscopy. However, deviations from Beer's law can be produced by the presence of scattered radiation and the use of radiation which is, of necessity, not monochromatic (Willard et al., 1965). The concentration of the metabolite under investigation may be determined empirically from the spectrum by comparison of the absorption at a particular wavelength with standard curves of known concentration at that same wavelength.

KBr disks can also be used in quantitative analysis. Known amounts of KBr together with known amounts of analyte are used to produce a calibration curve.

The disadvantage of varying disk thicknesses can be overcome by the use of an internal standard such as potassium thiocyanate premixed with KBr. A series of calibration curves of the ratio of the thiocyanate absorption at 2125 cm^{-1} to a chosen absorption of the analyte for different analyte concentrations is then plotted. With the same KBr–KSCN mixture, unknown concentrations of the analyte can then be determined by measuring the ratio of absorptions and using the calibration curves (Willard et al., 1965).

A more detailed theoretical treatment of quantitative IR spectroscopy is given by Brügel (1962).

D. Conclusion

The above description is highly condensed. An immense amount of published material on IR spectroscopy is available, and for more detailed descriptions the reader should refer to the articles cited and the references therein. The use of IR spectroscopy as an aid in structural elucidation is long established and familiar to all chemists. However, it would seem that in the mycotoxin field at least, IR spectroscopy has an increasingly important role to play as an analytical tool. Its high information content—especially in terms of the rapid and unambiguous identification of known toxins—makes it an invaluable aid to the food industry in a world rapidly awakening to the real and serious dangers posed by foodstuffs contaminated with fungal metabolites. Many reguatory laboratories lack highly sophisticated equipment, so it is necessary to resort to routine IR and UV spectrometers.

III. ULTRAVIOLET SPECTROSCOPY

A. Introduction

UV spectorscopy has been utilized for over a hundred years as a method of structural elucidation. The transparency of large sections of many molecules to UV radiation is an inherent limitation in the method and severely restricts the quantity of information to be obtained from a UV spectrum. However, the great sensitivity of UV spectroscopy, especially in the case of strongly absorbing chromophores, makes it an ideal means of detection and quantitation for small amounts of material. Nearly all liquid chromatographic systems presently on the market come with a UV detector as standard equipment.

UV instrumentation, apart from digitalization and microprocessor control, has changed very little since the mid-1960s. As a result it will not be dealt with in this chapter, but many excellent reviews are available to those interested in this field (see, e.g., Willard et al., 1965; Strobel, 1960).

B. Structural Elucidation

Structural elucidation by UV spectroscopy per se is limited to information obtainable about those chromophores which absorb in the segment of the electromagnetic spectrum between 200 and 600 nm. Of these, the most useful chromophores in the mycotoxin field are conjugated allenes, α,β-unsaturated ketones, aromatic and heteroaromatic compounds.

1. Conjugated Allenes

The most important mycotoxins among this group of compounds are citreoviridin, asteltoxin, the aurovertins, and the fusarins. The UV spectra of these metabolites in methanol are similar, with very strong absorption maxima at 367 nm for asteltoxin (ε 32,700) (Kruger et al., 1979) and aurovertin (ε 42,700) (Baldwin et al., 1964), with five conjugated double bonds, and at 383 nm (ε 44,000) for citreoviridin (Sakabe et al., 1977), with six conjugated double bonds. In addition, all these metabolites have absorption maxima associated with the α,β-unsaturated carbonyl moieties between 270 and 295 nm.

As the number of conjugated double bonds decreases, so the position of the absorption maximum associated with this chromosphere moves to lower wavelength. Thus in the case of the roridins, many of which have conjugated diene-carbonyl moieties, the absorption occurs at ~260 nm (ε ~ 20000) (Cole and Cox, 1981). The intensity of the absorption is also decreased.

2. α,β-Unsaturated Carbonyl Moieties (Cole and Cox, 1981)

The absorption due to an α,β-unsaturated carbonyl group normally falls in the range 220–250 nm. Examples of this type of absorption in the mycotoxin field are found in the chaetoglobosins (ε ~ 40,000) (at ±222 nm), in paxilline (230 nm) (ε 41,500), and in PR toxin (249 nm) (ε 15,300).

Increasing the unsaturation causes a shift in the position of the absorption maximum toward higher wavelength. The dienone system found in patulin, for example, absorbs at 276 nm (ε 16,600) and that in baccharin at 259 nm (ε 18,700).

Finally, enediones can appear between 220 and 270 nm, depending on the geometrical environment.

3. Aromatic Compounds (Cole and Cox, 1981)

The shikimate and acetate biosynthetic pathways lead to a plethora of natural products containing aromatic rings. Consequently this particular structural feature is widespread in the mycotoxin field, with many different families possessing substituted benzenoid derivatives. The interpretation of UV spectra of this type of compound is fraught with pitfalls (Scott, 1964). However, certain spectral features are now recognized as being characteristic of particular chromophores. Benzene itself has an absorption maximum at 254 nm which increasing alkyl substitution displaces to longer wavelengths. In the case of phenols and ethers, this band is also displaced to longer wavelengths [e.g., in zearalenone this maximum occurs at 274 nm (ϵ 13,900) and in mycophenolic acid at 304 nm].

Benzenoid derivatives with ketonic, aldehydic, or carboxylic acids or esters in conjugation with the aromatic ring usually display an intense absorption maximum between 230 and 280 nm. In viomellein this band appears at 264 nm (ϵ 20,200) and in maltoryzine at 280 nm (ϵ 1300).

Xanthones, and anthraquinones are a special example of this type of absorption. Sterigmaticoystin shows three strong absorption maxima typical of a 1,6,8-trioxygenated xanthone at 235, 249, and 329 nm (ϵ 24,500, 27,500, and 13,100, respectively). However, the position and pattern of the absorption bands is quite sensitive to changes in hydroxylation pattern (Scott, 1964), as witnessed by comparison with the spectrum of 5-methoxysterigmatocystin, which has maxima at 232, 248, 279, and 331 nm (ϵ 24,100, 26,800, 11,200 and 12,100, respectively).

The spectra of anthraquinones are more complex, showing numerous absorption maxima. Emodin, for example, has five such maxima at 223, 250, 267, 290, and 442 nm (ϵ 35,400, 18,200, 18,600, 21,800, and 12,500, respectively). A partial analysis of the spectra of these type of compounds is possible by regarding them as cross-conjugated acetophenone–benzoquinone systems (Scott, 1964).

4. Heteroaromatic Compounds (Cole and Cox, 1981)

The most important groups of compounds in this category of mycotoxins are the indoles, the pyrones, and the coumarins. The indole nucleus occurs intact and with a variety of modifications in a large number of mycotoxins. Of particular importance are the neurotropic mycotoxins (the penitrems, janthitrems, lolitrems, fumitremorgins, and roquefortines).

Indole itself has three bands with maxima at 220, 262, and 280–290 nm. the positions of these bands may vary, however, according to the substitution pattern, with the central band sometimes being absent (Scott, 1964). Thus in fumitremorgin A they appear at 226, 277, and 296 nm (ϵ 31,700, 5300, and 4900, respectivly), whereas in penitrem A there are only two maxima at 233 (ϵ 3700) and 295 nm (ϵ 11,600). Similarly, paxilline shows only two bands, at 230 (ϵ 41,500) and 281 nm (ϵ 8000). The extended conjugation in the janthitrems shifts the one maximum to $\pm$320 nm.

The indoline chromophore, present in the tryptoquivalines, has an intense absorption at ~250 nm ($\epsilon \sim$ 20,000). This moiety is also present in roquefortine, where the absorption maximum occurs at 240 nm (ϵ 16,200).

Both α- and γ-pyrones have characteristic absorptions, the α-pyrones absorbing at considerably longer wavelength. Thus the α-pyrone moiety in citreoviridin is associated with an intense band at 294 nm (ϵ 27,000), whereas the γ-pyrone in diplosporin has maxima at 252 and 256 nm (ϵ 12,000 and 11,700) (Chalmers *et al.*, 1978). The most important class of mycotoxins containing the coumarin moiety is the aflatoxins. Hydroxylated coumarins have their principal maxima >300 nm (Scott, 1964), which is typified in the aflatoxin series. Those aflatoxins (e.g., B_1) which have the carbonyl group of the cyclopentenone ring in conjugation with the coumarin system display this band at ~360 nm (ϵ ~ 22,000), whereas the aflatoxins with no cyclopentenone moiety (e.g., B_3), have this absorption at ~325 nm (ϵ ~ 9500).

5. *Intensity of Absorptions*

No mention has yet been made of the intensity of the absorption. This is measured in terms of the molar absorption coefficient (ϵ) and is obtained from the Beer–Lambert law $\epsilon = A/cl$ (usually liters mol^{-1} cm^{-1}, or mol^{-1} cm^2 × 10^3), where A is the absorbance measured on the spectrometer, c is the concentration of the absorbing species (mol liter^{-1}), and l the path length of the sample in the light beam (cm).

Absorption bands are usually classified as being of high intensity if $\epsilon > 5000$ liters mol^{-1} cm^{-1}; of medium intensity if $200 < \epsilon < 5000$ liters mol^{-1} cm^{-1}; and of low intensity if $\epsilon < 200$ liters mol^{-1} cm^{-1}. Factors governing the intensities of bands are complicated (Timmons, 1973). In general high-intensity bands arise from transitions involving π-electrons in conjugated systems, and low-intensity bands from transitions involving the promotion of nonbonding electrons from multiply bonded atoms such as oxygen in the carbonyl group. Medium-intensity bands usually arise from transitions involving π-electrons in aromatic rings. Thus a knowledge of the intensity of a band is important in making a correct assignment in structural determination.

6. *Conclusion*

This is only a brief description of some of the more important UV chromophores and their characteristic absorption maxima (see Fig. 2). Several examples of mycotoxins containing these features have been used to illustrate this discussion. Table II gives a listing of the more important groups of mycotoxins and the characteristic absorptions associated with them. A much more detailed listing of the UV spectra of most known mycotoxins is available (Cole and Cox, 1981). Finally, a more complete treatment of the UV chromophores of natural products has been produced by Scott (1964).

C. Analysis of Known Metabolites

The analysis of toxic fungal metabolites using UV spectroscopy can be conveniently divided into qualitative analysis, in which the UV device is used

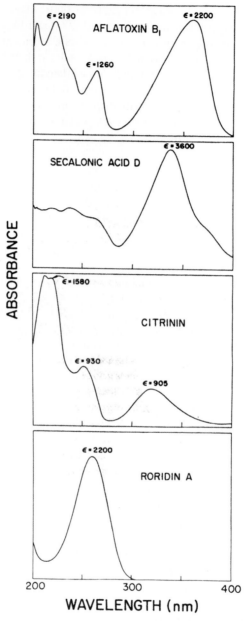

Fig. 2. Ultraviolet spectra of some common mycotoxins (ϵ is expressed in m^2 mol^{-1}).

TABLE II

Characteristic UV Absorptions for Various
Mycotoxin Groups

Mycotoxins	Characteristic absorptions[a] (nm)
Aflatoxins	225, 265, 325, 360, and 425
Sterigmatocystins	210, 235, 250, and 330 (310 for O-methyl derivatives
Ochratoxins	215 and 335 (315 for dechloro derivatives)
8-Ketotrichothecenes	220
Macrocyclic esters of verrucarol	220 and 260 (depending on degree of unsaturation)
Cytochalasins	Absorption depends on chromophores present. Mainly α,β-unsaturated ketones, absorption at 220
Chaetoglobosins	220, 275, 280, and 290
Rubratoxins	250
Fumitremorgins	225, 275, and 295
Penitrems	235 and 295
Paspalitrems	230 and 280
Tryptoquivalines	230, 270, 305, and 320
Sporidesmins	220, 255, and 300
Secalonic acids	250 and 345

[a] Absorptions have been adjusted to nearest 5 nm.

merely as a detector as in an HPLC system, and quantitative analysis, in which the UV spectrum of the metabolite is used to obtain a value for the concentration of the metabolite by application of Beer's law.

1. Ultraviolet-Visible Detectors in High-Performance Liquid Chromatography Systems

The UV-visible detector is probably the most commonly used detector in HPLC systems today. This is the result of both the high incidence of UV-absorbing compounds, and the quality optical systems developed over the years.

These detectors are now capable of monitoring effluents from a column over a wavelength range from 200 to 600 nm. Consequently the analyte can be examined at one of its absorption maxima, greatly enhancing the sensitivity of the technique. Moreover, UV detectors are relatively insensitive to flow rate and back pressure changes, a large advantage over refractive index detectors.

UV detectors presently available allow the detection of about 1 ppb of compounds with molar absorption coefficients of 1 to 2×10^4. In addition, double-beam UV detectors are available which can record the UV spectrum of the analyte using stopped-flow techniques. This has obvious advantages for the characterization and identification of a particular metabolite, although in some instances the UV spectrum is not a reliable diagnostic probe. Numerous mycotoxins have been analyzed with an UV detector on HPLC systems. Examples of these are the ergot alkaloids (Yoshida *et al.*, 1979; Wurst *et al.*, 1979; Szepesi *et al.*, 1980), sporidesmin (Halder *et al.*, 1979), penicillic acid (Chan *et al.*, 1980; Hunt *et al.*, 1978; Engstrom *et al.*, 1977), patulin and zearalenone (Hunt *et al.*, 1978; Engstrom *et al.*, 1977), rubratoxin B, ochratoxin A, and trichothecin (Engstrom *et al.*, 1977), sterigmatocystin (Engstrom *et al.*, 1977; Kingston *et al.*, 1976), citrinin (Phillips *et al.*, 1980), and the aflatoxins (Engstrom *et al.*, 1977; Rao and Anders, 1973; Seitz, 1975; Pons, 1976; Garner, 1975). The use of UV detectors in the HPLC analysis of mycotoxins has also been examined (Gorst-Allman and Steyn, 1984).

2. Ultraviolet Detection in Thin-Layer Chromatography

Thin-layer chromatography (TLC) is by far the most widely used technique in the detection, analysis, and characterization of fungal toxins. One of the most common methods for the detection of noncolored metabolites on a TLC plate is that of examining the plate under UV light. The UV light used is generally of wavelength 254 or 366 nm, and is generated from a source enclosed in a box, to which easy access for the plates is provided. As can be seen from Table II, many of the most common classes of mycotoxins have absorptions around one of these wavelengths, and as the radiation produced by these sources is not monochromatic, visualization of a sizeable proportion of the toxins is possible.

3. Quantitative Analysis

It is possible to effect routine and accurate quantitative analysis of mycotoxins using UV spectroscopy by application of the Beer–Lambert law (i.e., $\epsilon = A/cl$. For a standard path length of 1 cm, this becomes $c = A/\epsilon$ and since ϵ, the molar absorption coefficient at a particular wavelength, is readily available for most mycotoxins (Cole and Cox, 1981), it is possible to relate the concentration in mol liter^{-1} to the absorbance shown on the spectrometer. This method is particularly useful for the quantitation of small amounts of a toxin and to assess the purity of reference standards. This method of analysis is easily extended to mixtures (Jaffe and Orchin, 1962) if the various components are known. The analyte from a liquid chromatography column may also be readily quantified in this way.

Quantitation of mycotoxins by TLC using UV light can be achieved by either (a) evaluation on the plate itself or (b) evaluation after extraction from the plate (Gorst-Allman and Steyn, 1984). In the first procedure, quantitation can be achieved by visually comparing the spot area with the area of reference spots of known concentration. Alternatively, the accuracy and precision of analysis can be improved by using a densitometer, which can determine transmission and reflectance properties on thin solid layers. If the toxin is first extracted from the plate, the Beer–Lambert law may again be used to calculate the concentration.

IV. CONCLUSION

Rather than discussing in detail the role played by IR and UV spectroscopy in the analysis and structural elucidation of mycotoxins, this chapter emphasizes the increasing utilization of these two spectroscopic techniques in the analysis of small quantities of mycotoxins present in contaminated foodstuffs. In the years ahead the increasing need to provide nutrition for rapidly growing populations, especially in Third-World countries, will increase the hazard of using such contaminated samples as a source of food. Thus it may be in this area that IR and UV spectroscopy will assume the most importance.

REFERENCES

Aldridge, D. C., Armstrong, J. J., Speake, R. N., and Turner, R. B. (1967). The structures of cytochalasins A and B. *J. Chem. Soc. C* pp. 1667–1676.

Andersen, R., Büchi, G., Kobbe, B., and Demain, A. L. (1977). Secalonic acids D and F are toxic metabolites of *Aspergillus aculeatus*. *J. Org. Chem.* **42**, 352–353.

Asao, T., Büchi, G., Abdel-Kader, M. M., Chang, S. B., Wick, E. L., and Wogan, G. N. (1965). The structures of aflatoxins B and G₁. *J. Am. Chem. Soc.* **87**, 882–886.

Baldwin, C. L., Weaver, L. C., Brooker, R. M., Jacobsen, T. N., Osborne, C. E., and Nash, H. A. (1964). Biological and chemical properties of aurovertin: a metabolic product of *Calcarisporium arbuscula*. *Lloydia* **27**, 88–95.

Bamburg, J. R., and Strong, F. M. (1969). Mycotoxins of the trichothecane family produced by *Fusarium tricinctum* and *Trichoderma lignorum*. *Phytochemistry* **8**, 2405–2410.

Bamburg, J. R., Riggs, N. V., and Strong, F. M. (1968). The structures of toxins from two strains of *Fusarium tricinctum. Tetrahedron* **24**, 3329–3336.

Bell, R. J. (1972). "Introductory Fourier Transform Spectroscopy." Academic Press, New York.

Bellamy, L. J. (1975). "The Infra-red Spectra of Complex Molecules." Chapman & Hall, London.

Böhner, B., Fetz, E., Härri, E., Sigg, H. P., Stoll, C., and Tamm, C. (1965). Über die isoliering van verrucarin H, verrucarin J, roridin D und roridin E aus *Myrothecium*—Arten. *Helv. Chim. Acta* **48**, 1079–1087.

Boyd, M. R., Burka, L. T., Harris, T. M., and Wilson, B. J. (1974). Lung-toxic furanoterpenoids produced by sweet potatoes (*Ipomoea batatas*) following microbial infection. *Biochim. Biophys. Acta* **377**, 184–195.

Brian, P. W., Dawkins, A. W., Grove, J. F., Hemming, H. G., Lowe, D., and Norris, G. L. F. (1961). Phytotoxic Compounds produced by *Fusarium equisetti. J. Exp. Bot.* **12**, 1–12.

Brügel, W. (1962). "An Introduction to Infrared Spectroscopy." Methuen, London.

Büchi, G., Luk, K. C., Kobbe, B., and Townsend, J. M. (1977). Four new mycotoxins of *Aspergillus clavatus* related to tryptoquivaline. *J. Org. Chem.* **42**, 244–246.

Chalmers, A. A., Gorst-Allman, C. P., Kriek, N. P. J., Marasas, W. F. O., Steyn, P. S., and Vleggaar, R. (1978). Diplosporin, a new mycotoxin from *Diplodia macrospora* Earle. *S. Afr. J. Chem.* **31**, 111–114.

Chan, P. K., Siraj, M. Y., and Hayes, A. W. (1980). High-performance liquid chromatographic analysis of the mycotoxin penicillic acid and its application to biological fluids. *J. Chromatogr.* **194,** 387–398.

Chang, S. B., Abdel-Kader, M. M. A., Wick, E. L., and Wogen, G. N. (1963). Aflatoxin B$_2$: chemical identity and biological activity. *Science* **141,** 1191–1192.

Cole, R. J., and Cox, R. H. (1981). "Handbook of Toxic Fungal Metabolites." Academic Press, New York.

Cole, R. J., and Kirksey, J. W. (1971). Aflatoxin G$_1$ metabolism by *Rhizopus* species. *J. Agric. Food Chem.* **19,** 222–223.

Cole, R. J., Kirksey, J. W., and Wells, J. M. (1974). A new tremorgenic metabolite from *Penicillium paxilli. Can. J. Microbiol.* **20,** 1159–1162.

Culvenor, C. C. J., Cockrum, P. A., Edgar, J. A., Frahn, J. L., Gorst-Allman, C. P., Jones, A. J., Marasas, W. F. O., Murray, K. E., Smith, L. W., Steyn, P. S., Vleggaar, R., and Wessels, P. L. (1983). Structure elucidation of phomopsin A, a novel cyclic hexapeptide mycotoxin produced by *Phomopsis leptostromiformis. J.C.S. Chem. Commun.* pp. 1259–1262.

Davies, J. E., Kirkalday, D., and Roberts, J. C. (1960). Studies in mycological chemistry. Part VII. Sterigmatocystin, a metabolite of *Aspergillus versicolor* (Vuillemin) Tiraboschi. *J. Chem. Soc.* pp. 2169–2178.

de Jesus, A. E., Steyn, P. S., van Heerden, F. R., Vleggaar, R., and Wessels, P. L. (1983). Tremorgenic mycotoxins from *Penicillium crustosum.* Isolation of penitrems A-F and the structure elucidation and absolute configuration of penitrem A. *J.C.S. Perkin I* pp. 1847–1856.

de Jesus, A. E., Steyn, P. S., van Heerden, F. R., and Vleggaar, R. (1984). Structure elucidation of the janthitrems, novel tremorgenic mycotoxins from *Penicillium janthinellum. J.C.S. Perkin I* pp. 697–701.

Engstrom, G. W., Richard, J. L., and Cysewski, S. J. (1977). High-pressure liquid chromatographic method for detection and resolution of rubratoxin, aflatoxin, and other mycotoxins. *J. Agric. Food Chem.* **25,** 833–836.

Erickson, M. D. (1979). Gas chromatography/Fourier Transform infrared spectroscopy applications. *Appl. Spectrosc. Rev.* **15,** 261–325.

Freeman, G. G. (1966). Isolation of alternariol and alternariol monomethyl ether from *Alternaria dauci* (Kühn) Groves and Skolko. *Phytochemistry* **5,** 719–725.

Gallagher, R. T., Clardy, J., and Wilson, B. J. (1980). Aflatrem, a tremorgenic toxin from *Aspergillus flavus. Tetrahedron Lett.* **21,** 239–242.

Garner, R. C. (1975). Aflatoxin separation by high-pressure liquid chromatography. *J. Chromatogr.* **103,** 186–188.

Godtfredsen, W. O., and Vangedal, S. (1964). Trichodermin, a new antibiotic, related to trichothecin. *Proc. Chem. Soc., London* pp. 188–189.

Gorst-Allman, C. P., and Steyn, P. S. (1984). Applications of chromatographic techniques in separation, purification and characterization of mycotoxins. *In* "Mycotoxins: Production, Isolation, Separation and Purification" (V. Betina, ed.). Elsevier, Amsterdam, pp. 59–85.

Grove, J. F., MacMillan, J., Mulholland, T. P. C., and Rogers, M. A. T. (1952). Griseofulvin. Part 1. *J. Chem. Soc.* pp. 3949–3958.

Gutzwiller, J., Mauli, R., Sigg, H. P., and Tamm, C. (1964). Die konstitution van verrucarol und roridin C. *Helv. Chim. Acta* **47,** 2234–2262.

Halder, C. A., Taber, R. A., and Camp, B. J. (1979). High-performance liquid chromatography of the mycotoxin, sporidesmin, from *Pithomyces chartarum* (Berk and Curt) M. B. Ellis. *J. Chromatogr.* **175,** 356–361.

Herres, W. (1982). Normal phase liquid chromatography using flowcell detection. *Bruker FT-IR Appl. Note* No. 16.

Holzapfel, C. W. (1968). The isolation and structure of cyclopiazonic acid, a toxic metabolite of *Penicillium cyclopium* Westling. *Tetrahedron* **24**, 2101–2119.

Holzapfel, C. W., Steyn, P. S., and Purchase, I. F. H. (1966). Isolation and structure of aflatoxins M_1 and M_2. *Tetrahedron Lett.* pp. 2799–2803.

Horak, R. M., Steyn, P. S., van Rooyen, P. H., and Vleggaar, R. (1981). Structures of the austalides A-E, five novel toxic metabolites from *Aspergillus ustus. J.C.S. Chem. Commun.* pp. 1265–1267.

Hunt, D. C., Bourdon, A. T., and Crosby, N. T. (1978). Use of high performance liquid chromatography for the identification and estimation of zearalenone patulin and penicillic acid in food. *J. Sci. Food Agric.* **29**, 239–244.

Hutchison, R. D., Steyn, P. S., and Thompson, D. L. (1971). The isolation and structure of 4-hydroxyochratoxin A and 7-carboxy-3,4-dihydro-8-hydroxy-3-methylisocoumarin from *Penicillium viridicatum. Tetrahedron Lett.* pp. 4033–4036.

Iizuka, H. (1974). Maltoryzine. *In* "Mycotoxins" (I. F. H. Purchase, ed.), pp. 405–418. Elsevier, Amsterdam.

Jaffe, H. H., and Orchin, M. (1962). "Theory and Applications of Ultraviolet Spectroscopy." Wiley, New York.

Kabuto, C., Silverton, J. V., Akiyama, T., Sankawa, U., Hutchison, R. D., Steyn, P. S., and Vleggaar, R. (1976). X-Ray structure of viridicatumtoxin: a new class of mycotoxin from *Penicillium viridicatum* Westling. *J.C.S. Chem. Commun.* pp. 728–729.

Kingston, D. G. I., Chen, P. N., and Vercellotti, J. R. (1976). High-performance liquid chromatography of sterigmatocystin and other metabolites of *Aspergilles versicolor. J. Chromatogr.* **118**, 414–417.

Kovac, S., Nemec, P., Betin, V., and Balan, J. (1961). Chemical structure of citrinin. *Nature (London)* **190**, 1104–1105.

Kruger, G. J., Steyn, P. S., Vleggaar, R., and Rabie, C. J. (1979). X-ray crystal structure of asteltoxin, a novel mycotoxin from *Aspergillus stellatus* Curzi. *J.C.S. Chem. Commun.* pp. 441–442.

Leutwiler, A. (1973). "Die Konstitution des Paspalicins." Juris Druck Verlag, Zurich.

Marumo, S. (1955). Islanditoxin, a toxic metabolite produced by *Penicillium islandicum*. I. *Bull. Agric. Chem. Soc. Jpn.* **19**, 258–261.

Masaki, M., Chigira, Y., and Ohta, M. (1966). Total syntheses of racemic aspergillic acid and neoaspergillic acid. *J. Org. Chem.* **31**, 4143–4146.

Masri, M. S., Haddon, W. F., Lundin, R. E., and Hsieh, D. P. H. (1974). Aflatoxin Q_1. A newly identified major metabolite of aflatoxin B_1 in monkey liver. *J. Agric. Food Chem.* **22**, 512–515.

Meyer, W. L., Templeton, G. E., Grable, C. I., Sigel, C. W., Jones, R., Woodhead, S. H., and Sauer, C. (1971). The structure of tentoxin. *Tetrahedron Lett.* pp. 2357–2360.

Minato, H., and Matsumoto, M. J. (1970). Studies on the metabolites of *Zygosporium masonii*. Part 1. Structure of zygosporin A. *J. Chem. Soc. C* pp. 38–45.

Mirocha, C. J., Pathre, S. V., and Christensen, C. M. (1977). Zearalenone. *In* "Mycotoxins in Human and Animal Health" (J. V. Rodricks, C. W. Hesseltine, and M. A. Mehlman, eds.), pp. 345–364. Pathotox, Park Forest South, Illinois.

Moss, M. O., Robinson, F. V., Wood, A. B., Paisley, H. M., and Fenney, J. (1968). Rubratoxin B, a proposed structure for a bis-anhydride from *Penicillium rubrum* Stoll. *Nature (London)* **220**, 767–770.

Nakanishi, K. (1962). "Infrared Absorption Spectroscopy–Practical." Holden-Day, San Francisco, California and Nankodo, Tokyo.

Natori, S. (1977). Toxic cytochalasins. *In* "Mycotoxins in Human and Animal Health" (J. V.

Rodricks, C. W. Hesseltine, and M. A. Mehlman, eds.), pp. 559–581. Pathotox, Park Forest South, Illinois.

Patwardhan, S. A., Pandey, R. C., Dev, S., and Pendse, G. S. (1974). Toxic cytochalasins of *Phomopsis paspalli,* a pathogen of kodo millet. *Phytochemistry* **13,** 1985–1988.

Phillips, R. D., Hayes, A. W., and Berndt, W. O. (1980). High-performance liquid chromatographic analysis of the mycotoxin citrinin and its application to biological fluids. *J. Chromatogr.* **190,** 419–427.

Pohland, A. E., Schuller, P. L., Steyn, P. S., and van Egmond, H. P. (1982). Physicochemical data for some selected mycotoxins. *Pure Appl. Chem.* **54,** 2219–2284.

Pons, W. A. (1976). Resolution of aflatoxins B_1, B_2, G_1 and G_2 by high-pressure liquid chromatography. *J. Assoc. Off. Anal. Chem.* **59,** 101–105.

Rabie, C. J., Lubben, A., Louw, A. I., Rathbone, E. B., Steyn, P. S., and Vleggaar, R. (1978). Moniliformin, a mycotoxin from *Fusarium fusarioides. J. Agric. Food Chem.* **26,** 375–379.

Rao, G. H. R., and Anders, M. W. (1973). Aflatoxin detection by high-speed liquid chromatography and mass spectrometry. *J. Chromatogr.* **84,** 402–406.

Rodricks, J. V., Lustig, E., Campbell, A. D., Stoloff, L., and Henery-Logan, K. R. (1968). Aspertoxin, a hydroxy derivative of O-methylsterigmatocystin from aflatoxin-producing cultures of *Aspergillus flavus. Tetrahedron Lett.* pp. 2975–2978.

Ronaldson, J. W., Taylor, A., White, E. P., and Abraham, R. J. (1963). Sporidesmins. I. Isolation and characterization of sporidesmin and sporidesmin-B. *J. Chem. Soc.* pp. 3172–3180.

Sakabe, N., Goto, T., and Hirata, Y. (1964). The structure of citreoviridin a toxic compound produced by *P. citreoviride* molded on rice. *Tetrahedron Lett.* pp. 1825–1830.

Sakabe, N., Goto, T., and Hirata, Y. (1977). Structure of citreoviridin, a mycotoxin produced by *Penicillium citreo-viride* molded on rice. *Tetrahedron* **33,** 3077–3081.

Scott, A. I. (1964). "Interpretation of the Ultraviolet Spectra of Natural Products." Pergammon, Oxford.

Scott, P. M. (1974). Patulin. *In* "Mycotoxins" (I. F. H. Purchase, ed.), pp. 383–403. Elsevier, Amsterdam.

Scott, P. M., Merrien, M. A., and Polonsky, J. (1976). Roquefortine and isofumigaclavine A, metabolites from *Penicillium roqueforti. Experientia* **32,** 140–142.

Seitz, L. M. (1975). Comparison of methods for aflatoxin analysis by high pressure liquid chromatography. *J. Chromatogr.* **104,** 81–89.

Steyn, P. S., and Vleggaar, R. (1974). Austocystins. Six Novel Dihydrofuro [3′,2′:4,5] furo [3,2-*b*] xanthenones from *Aspergillus ustus. J.C.S. Perkin I* pp. 2250–2256.

Steyn, P. S., van Heerden, F. R., and Rabie, C. J. (1982). Cytochalasins E and K, toxic metabolites from *Aspergillus clavatus. J.C.S. Perkin I* pp. 541–544.

Steyn, P. S., Tuinman, A. A., Van Heerden, F. R., Van Rooyen, P. H., Wessels, P. L., and Rabie, C. J. (1983). The isolation, structure and absolute configuration of the mycotoxin, rhizonin A, a novel cyclic heptapeptide containing *N*-methyl-3-(3-furyl)alanine, produced by *Rhizopus microsporus. J.C.S. Chem. Commun.* pp. 47–49.

Stickings, C. E. (1959). Metabolites of *Alternaria tenuis* Auct.: the structure of tenuazonic acid. *Biochem. J.* **72,** 332–340.

Strobel, H. A. (1960). "Chemical Instrumentation." Addison-Wesley, Reading, Massachusetts.

Szepesi, G., Gazdag, M., and Terdy, L. (1980). Separation of ergotoxine alkaloids by high-performance liquid chromatography on silica. *J. Chromatogr.* **191,** 101–108.

Tatsuno, T., Fujimoto, Y., and Morita, Y. (1969). Toxicological research on substances from *Fusarium nivale* III. The structure of nivalenol and its monoacetate. *Tetrahedron Lett.* pp. 2823–2826.

Timmons, C. J. (1973). Ultraviolet and visible spectroscopy. *In* "Structure Determination in

Organic Chemistry'' (W. D. Ollis, ed.), MTP International Review of Science, Organic Chemistry Series One, Vol. 1, pp. 63–83. Butterworth, London.

Van der Merwe, K. J., Steyn, P. S., and Fourie, L. (1965). Mycotoxins. Part II. The constitution of ochratoxins A, B and C, metabolites of *Aspergillus ochraceus* Wilh. *J. Chem. Soc.* pp. 7083–7088.

Vleggaar, R., Steyn, P. S., and Nagel, D. W. (1974). Constitution and absolute configuration of austdiol, the main toxic metabolite from *Aspergillus ustus*. *J.C.S. Perkin I* pp. 45–49.

Wei, R. D., Schnoes, H. K., Hart, P. A., and Strong, F. M. (1975). The structure of PR toxin, a mycotoxin from *Penicillium roqueforti*. *Tetrahedron* **31**, 109–114.

Wells, J. M., Cole, R. J., and Kirksey J. W. (1975). Emodin, a toxic metabolite of *Aspergillus wentii* isolated from weevil-damaged chestnuts. *Appl. Microbiol.* **30**, 26–28.

Willard, H. H., Merritt, L. L., and Dean, J. A. (1965). "Instrumental Methods of Analysis." Van Nostrand-Reinhold, New York.

Wurst, M., Flieger, M., and Rehacek, Z. (1979). Analysis of ergot alkaloids by high-performance liquid chromatography. II Cyclol alkaloids (ergopeptines). *J. Chromatogr.* **174**, 401–407.

Yamazaki, M., Sasago, K., and Miyaki, K. (1974). The Structure of Fumitremorgen B (FTB), a Tremorgenic Toxin from *Aspergillus fumigatus* Fres. *J.C.S. Chem. Commun.* pp. 408–409.

Yoshida, A., Yamazaki, S., and Sakai, T. (1979). Etude par chromatographie liquide a haute pression des alcaloides de l'ergot. *J. Chromatogr.* **170**, 399–404.

Yoshizawa, T., Tsuchiya, Y., Morooka, N., and Sawada, V. (1975). Malformin A_1 as a mammalian toxicant from *Aspergillus niger*. *Agric. Biol. Chem.* **39**, 1325–1326.

5

Application of Nuclear Magnetic Resonance to Structural Elucidation of Mycotoxins

RICHARD H. COX

Philip Morris U.S.A.
Research Center
Richmond, Virginia 23261

MODERN METHODS IN THE ANALYSIS
AND STRUCTURAL ELUCIDATION
OF MYCOTOXINS

I. INTRODUCTION

Few spectroscopic techniques have enjoyed the success and provided as much useful information toward solving problems in science as nuclear magnetic resonance (NMR) spectroscopy. After the demonstration of the potential for applications of NMR in chemistry in the early 1950s, developments in both the theoretical and experimental aspects of NMR have been rapidly forthcoming. Advances in electronics have led to low-cost minicomputers and microprocessors which are used in the present-day pulsed Fourier transform NMR (FT-NMR) spectrometers that are now routine instruments in most laboratories.

Two major advances have been made in NMR spectroscopy since 1975. The first has been the development of high-field, stable superconducting magnet-based spectrometers. The higher magnetic fields have improved sensitivity such that NMR spectra may now be obtained in a shorter time on smaller sample sizes. The improvements in chemical shift dispersion associated with high-field spectrometers results in the spectra of complex organic molecules showing more isolated resonances without overlap. This has led to a decrease in the effort required to interpret a spectrum and an increase in the amount of information available from a complex molecule. Furthermore, because the spectrometer is controlled by the data system, the transmitter, decoupler, receiver, and their phases, and the temperature may be programmed such that several spectra may be run without operator intervention. The second development with great potential impact on the analysis of complex NMR spectra is that of multiple-pulse experiments and two-dimensional (2D) NMR. Although first suggested in 1971 (1), it was not until 1975 (2) that the first 2D NMR experiments were reported. Since that time numerous 2D NMR methods have been developed and will continue to be developed in the future.

A normal FT-NMR spectrum is obtained by a sequence in which the nuclei in a sample are excited by a radiofrequency pulse, and the response (free-induction decay, FID) is acquired as a function of an acquisition time. After a suitable delay to allow the nuclear spins to reach equilibrium, this process is repeated a number of times with the FID being added coherently and stored in a computer until a sufficient signal:noise ratio has been obtained. The FID can then be multiplied by a number of possible weighting functions and then Fourier transformed to yield the frequency spectrum.

For purposes of comparison it is convenient to divide the 2D NMR experiment into three time periods: the preparation, evolution, and detection or acquisition periods. During the preparation period the nuclear spins are prepared to some specific state (i.e., equilibrium, decoupled, etc.). In the evolution period defined by the time t_1, the magnetization due to the spins develops according to some prescribed motion that will influence the detected signal. During the detection period, the signal is acquired for a time t_2 similar to that in a normal spectrum.

By varying t_1 in a specified manner, one obtains a series of FIDs, a signal matrix $S(t_1,t_2)$, as a function of the two time variables t_1 and t_2. The first Fourier transformation is carried out with respect to t_2 yielding a series of spectra $S(t_1,F_2)$. If the corresponding data points on each spectrum are followed as a function of t_1, the result is a FID signal $S(t_1)$, which is built up point by point in a series of $S(F_2)$ experiments. A second Fourier transformation is then carried out with respect to t_1, yielding a spectrum $S(F_1,F_2)$ in two frequency dimensions. This is not merely two spectra but a single spectrum in two orthogonal frequency dimensions or a surface in three-dimensional space.

Many types of 2D spectra, both homonuclear and heteronuclear, are possible depending on what is done to the magnetization during the three time periods. Owing to the number of spectra required and the data processing time, the 2D experiments will always require a substantially larger time investment than a conventional NMR experiment. Therefore, care must be exercised to ensure that the information obtained from the 2D NMR experiment justifies the increased time investment. In many cases this is not a critical decision, because the information cannot be obtained easily in any other way.

In most cases a number of parameters including chemical shifts, coupling constants, intensities, relaxation times, and nuclear Overhauser enhancements (NOEs) can be obtained from the interpretation of NMR spectra. The major application of NMR spectroscopy has been in the areas of structural confirmation and structural elucidation. Through the use of empirical relationships that have been established with known compounds, considerable progress can be made toward determining the structure of an unknown compound using only chemical shifts and coupling constants. Examples of the application of NMR to studies of time-dependent phenomena include conformational analysis, rotational isomerism, restricted rotation, and fast chemical exchange. In the area of biological sciences, examples of the application of NMR include structural elucidations, the binding of drugs and other small molecules to macromolecules, and biosynthetic studies. Dramatic progress has also been made in recent years in the area of solid-state NMR.

Because covering all areas of NMR spectroscopy is clearly beyond the scope of this chapter, we cover here the use of high-resolution NMR spectroscopy for structural determinations, with examples from fungal metabolites. As a further limitation, although NMR spectra may be obtained on a host of nuclei, our discussion is limited to proton (^1H NMR) and carbon-13 (^{13}C NMR) NMR spectra with only a brief mention of nitrogen-15 NMR (^{15}N NMR) spectra. Several excellent texts and review articles (3–12) are available, with in-depth discussion of the fundamentals of NMR and 2D NMR (13–15) and specific applications to areas such as conformational analysis and other time-dependent phenomena (16), nuclear Overhauser enhancements (NOEs) (17), and biological applications (18). Excellent review articles relating to specific areas of NMR are available in several continuing series (19–22). A text on the experimental aspects of NMR is available (23). The reader is referred to these texts for a more in-depth discussion of the application of NMR to specific areas.

II. STRATEGIES IN STRUCTURAL DETERMINATION

In determining the structure of complex natural products such as fungal metabolites where the number of structural types isolated to date is seemingly endless, one must make use of every piece of information available. This would include other spectroscopic techniques such as infrared spectroscopy for possible functional groups, ultraviolet spectroscopy for conjugation and functional groups, and mass spectroscopy for molecular weights, for possible formulas, and for various groups from the fragmentation patterns. When taken together with information available from NMR spectroscopy, considerable progress can be made toward the determination of structure of complex molecules. Many compounds have been identified by using spectroscopic techniques, and this has now become an accepted practice.

As mentioned previously, the use of NMR spectra for structural determination depends on the ability to extract the parameters from a complex spectrum and use this data along with data from similar compounds to assign a structure. A general strategy for the complete analysis of complex natural products has been presented by Hall and Sanders (24). This is presented in Fig. 1 along with some additional techniques which have been developed since that time. As indicated in Fig. 1, one uses ^{1}H and ^{13}C NMR and proceeds down the list of experiments until a suitable structure can be proposed. Each of these techniques is discussed in some detail below.

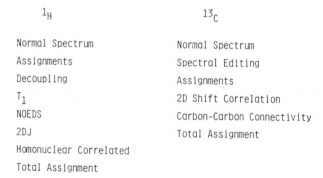

^{1}H

Normal Spectrum
Assignments
Decoupling
T_1
NOEDS
2DJ
Homonuclear Correlated
Total Assignment

^{13}C

Normal Spectrum
Spectral Editing
Assignments
2D Shift Correlation
Carbon-Carbon Connectivity
Total Assignment

Fig. 1. General strategy for determination of structure by NMR and/or assignment of NMR spectra (24).

It should be pointed out that there are limitations to the use of NMR for structural determinations. In many cases, the amount of material initially isolated may be too small to allow one to use all of the techniques available. If the material forms suitable crystals, the structure should be solved by X-ray crystallography and the NMR spectra assigned as far as possible. However, if the material is a liquid or does not form suitable crystals, then one is forced to isolate more material to acquire suitable amounts for the 2D NMR experiments. From 1 to 10 mg of material should be sufficient for most ^{1}H experiments. However, from 50 mg to 1 g may be required for some of the 2D ^{13}C experiments. Even when all the NMR experiments are successful, there may be several heteroatoms located in the carbon skeleton such that alternate structures are possible. One should exercise caution when these cases are encountered.

III. ONE-DIMENSIONAL METHODS

A. Normal Spectra

The starting point in the structural determination of an organic compound such as a fungal metabolite is the ^{1}H and ^{13}C NMR spectra. One should acquire these spectra at the highest frequency available for two reasons: (i) the greater chemical shift dispersion at high magnetic fields will permit more isolated resonances without overlap, allowing more coupling constants and chemical shifts to be obtained on a first-order basis and consequently, more assignments to be made; and (ii) the greater sensitivity at higher magnetic fields will allow the spectra to be obtained in a shorter time.

The larger chemical shift dispersion applies even for relatively small compounds. As shown in Fig. 2, although most of the protons in HT-2 toxin (**1**) can be distinguished from the 80-MHz ^{1}H spectrum, the overlap of the remaining resonances is completely resolved in the 300-MHz spectrum (Fig. 2B). An example of a more complex molecule, roseotoxin B (**2**) is given in Fig. 3. Few of

(**1**)

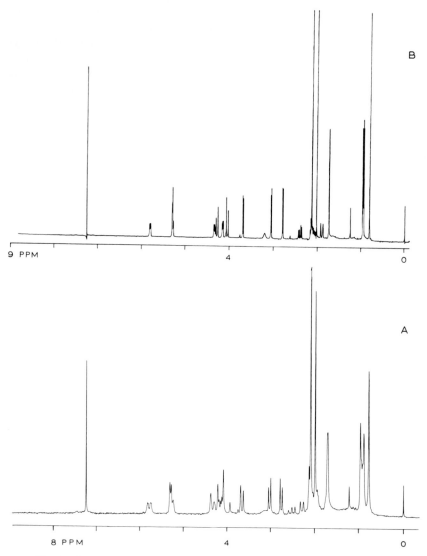

Fig. 2. ^{1}H NMR spectra of HT-2 toxin in CDCl$_3$ (A, 80 MHz and B, 300 MHz).

the protons give well-resolved resonances in the 80-MHz spectrum, whereas most of the protons give well-resolved resonances in the 300-MHz spectrum. Chemical shift dispersion is also important in ^{13}C NMR spectra. Although the chemical shift range is larger in ^{13}C NMR spectra compared to ^{1}H NMR spectra (200 ppm versus 10 ppm), obtaining the ^{13}C spectrum at higher magnetic fields

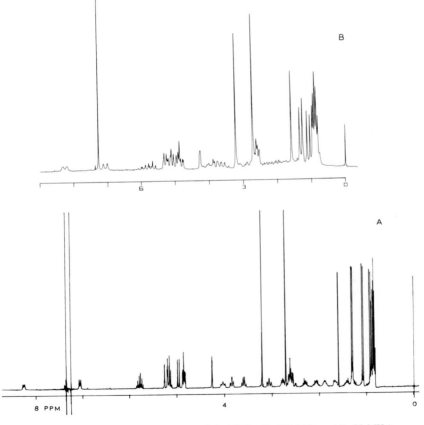

(2)

Fig. 3. ^{1}H NMR spectra of roseotoxin B in CDCl$_3$ (A, 300 MHz and B, 80 MHz).

reduces the chance of peak overlap and makes one more secure in using the normal ^{13}C NMR spectrum to determine the number of carbons present in the molecule.

Because most fungal metabolites are isolated in small amounts, the greater sensitivity of higher field instruments will allow one to obtain the spectra in a shorter time. Although initially thought to be a consideration for only ^{13}C spectra, improved sensitivity allows one to acquire the data necessary for the two-dimensional (2D) experiments using both ^1H and ^{13}C data in a shorter time. With the carbon–carbon connectivity experiments described later, one needs all the sensitivity available.

After the initial spectra have been obtained and possible assignments made, several options are available regarding what additional spectra are necessary to answer the questions posed. Several of these experiments are discussed in more detail below.

B. Decoupling

Homonuclear spin decoupling is a well-established technique for determining the coupling pathways in complex molecules. With most modern NMR spectrometers one can set both the frequency and power level of the decoupler to a high degree of accuracy. In addition, one can set up the experiment such that several decoupling experiments may be carried out sequentially under computer control.

One of the difficulties of using homonuclear spin decoupling with complex molecules is how does one ''see'' the decoupling effects in complex, overlapping multiplets. With computer-controlled spectrometers, one can subtract a controlled spectrum from a decoupled spectrum and generate a decoupling difference spectrum (DDS). The difference spectrum should reveal resonances in overlapped regions of the spectrum that were affected by the decoupling. A difficulty with this technique is the Bloch–Siegert effect, which shifts the resonances of protons close to the decoupling frequency from their respective position in the absence of the decoupling frequency (25). Thus, when the control and decoupled spectrum are subtracted, cancellation of peaks not affected by the decoupling frequency may not be effected. Fortunately, the homonuclear correlated 2D experiment (26) provides all the coupling connectivities in one experiment and usually does not require a significant increase in experiment time compared to several decoupling experiments. This technique is discussed in more detail in a later section.

Single-frequency, selective heteronuclear decoupling can be used as an aid in the assignment of the ^{13}C spectrum. If the ^1H spectrum can be assigned, one can carry out a series of ^{13}C NMR experiments where individual protons are selectively decoupled, thereby allowing a correlation between assignments in the ^1H

and ^{13}C spectra. This type of experiment is easily carried out under computer control, with several frequencies being irradiated in an overnight or weekend experiment. Furthermore, single-frequency, selective decoupling experiments can be used to aid in the assignment of long-range, carbon–hydrogen coupling constants observed in the proton-coupled ^{13}C spectrum. These coupling constants are becoming more useful as an aid in assignments and structural determinations as we learn more about the magnitude and stereochemical dependence of these couplings (27).

C. ^{1}H Relaxation Times

Proton relaxation times are becoming more important as a parameter for use in structural determinations. In dilute solution the dominant proton relaxation mechanism is the dipole–dipole relaxation mechanism. The rate equation for dipolar relaxation of a proton i with a neighboring proton j is given by $R_i \alpha \tau_c \Sigma r_{ij}^{-6}$ where τ_c is a rotational correlation time and r_{ij}'s are the internuclear distances between proton i and other protons (j) in the molecule. For more or less rigid molecules, τ_c should be effectively the same for all protons located on the backbone of the molecule. Therefore, methylene protons having a geminal partner should relax faster than methine protons with only vicinal nearest protons (r^{-6} dependence). By varying t in the standard inversion–recovery relaxation experiment ($180°$-t-$90°$-acquisition), one may be able to resolve signals within a complex group of overlapping resonances. For example, it should be possible to null the methylene resonances and observe methine resonances within a complex group of resonances. An example of this technique is shown in Fig. 4 for cytochalasin H (3). Notice that most of the methylene proton signals have been nulled into the baseline, leaving a spectrum consisting of the methine resonances. This technique should prove more useful in the future for resolving complex multiplets.

(3)

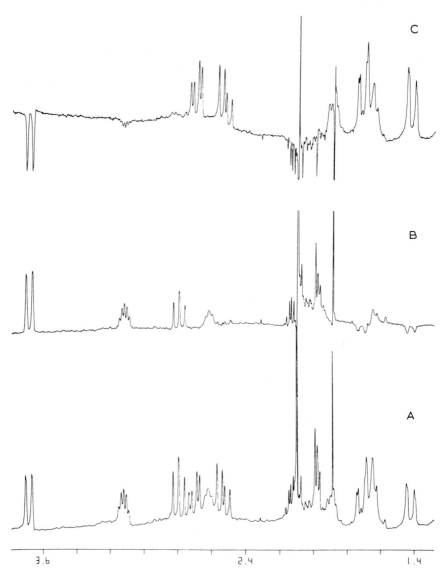

Fig. 4. (A) Normal 300-MHz ^1H NMR spectrum of cytochalasin H in CDCl$_3$. (B) The ^1H spectrum obtained with a 180°–t–90° pulse sequence, where t was adjusted to null the methylene protons. The spectrum is presented 180° out of phase to show the methine signals upright. (C) The ^1H spectrum obtained to highlight the methylene proton signals.

Further resolution enhancement may be obtained in certain cases using a $90°$-t-$180°$-t-acquisition pulse sequence (the spin-echo Fourier transform or SEFT experiment) (28). At $t = \frac{1}{2}J$, doublets and quartets are inverted in intensity whereas singlets and triplets remain positive. If the coupling constants in the multiplets are similar, one may be able to generate subspectra as an aid in the assignment of the spectrum using this technique.

D. Nuclear Overhauser Effect

The nuclear Overhauser effect (NOE) is defined as a change in intensity of the multiplet due to one proton (i) when the resonance of another proton or group of protons (j) is saturated. This effect results from dipole–dipole relaxation (see above) and is dependent on the distance separating the i and j protons (r_{ij}^{-6}) (17). The maximum enhancement for the proton–proton case is 50%. In the past the NOE was observed by comparing the integral of a multiplet with and without saturation of another resonance. Because of problems associated with accurately measuring the integrals, NOEs less than 5% could not be determined on a routine basis.

However, with the development of modern NMR spectrometers it is now possible to detect NOEs of less than 1% on a routine basis using NOE difference spectra (NOEDS) (24). A control spectrum in which no NOE enhancements are present is subtracted from a spectrum where the NOE enhancement is present, resulting in a spectrum showing only the enhancements resulting from saturating the resonance, the NOE. The detection limit depends on the signal:noise ratio and the efficiency of subtraction to yield a flat baseline in the difference spectrum. Furthermore, the decoupler used to saturate the resonance is switched off before data acquisition begins, such that Bloch–Siegert effects are eliminated.

An example of the NOEDS technique is shown in Fig. 5 for cytochalasin H (**3**). Saturation of the methyl group on C-5 results in a NOE of the methylene proton on C-12 that is syn to the methyl group, thereby allowing assignment of the C-12 methylene protons.

E. Miscellaneous Techniques

1. Deuterium Exchange

The most widely used application of deuterium exchange is that of identification of resonances due to labile protons in the ^1H spectra. The most common example is the identification of -OH resonances in the spectra of alcohols or phenols. After obtaining the ^1H spectrum in a solvent such as $CDCl_3$, a couple of drops of D_2O are added to the NMR tube, and the tube is shaken to allow D to exchange for the labile hydrogens. The spectrum is then rerun, and those resonances which are not present are due to the hydroxyl or phenolic resonances. An

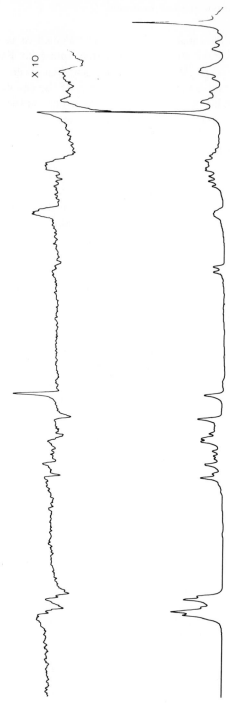

Cyto H

X 10

Fig. 5. Partial ¹H NMR spectrum (bottom) of cytochalasin H and a NOEDS (top) obtained by saturation of the 11-methyl group.

additional peak due to HOD will be present in the spectrum. In many cases with the initial spectrum of alcohols, particularly with dry solvents, the hydroxyl proton does not exchange and in the initial spectrum one may observe a multiplet for the hydroxyl proton instead of a broad singlet. A deuterium exchange experiment should identify those multiplets due to hydroxyl protons.

Deuterium substitution has also found application in ^{13}C NMR in biosynthetic studies. Substitution of deuterium for hydrogen bonded to a carbon usually results in the carbon signal being split into a triplet due to $^{13}C–D$ coupling. Even if the coupling is not observed, deuterium substitution results in an isotope shift (29) at the α- and β-carbons. The magnitude of the isotope shift is approximately 0.25 ppm per deuterium at the α-carbon and 0.1 ppm for the β-carbon. In most cases an upfield shift is observed.

The carbonyl carbons in peptides have been assigned using deuterium substitution (30). Exchange of the amide nitrogen proton with deuterium results in an upfield shift for the α-carbon and the carbonyl carbon. The usual procedure has been to acquire the spectrum in a 1 : 1 mixture of H_2O and D_2O. The α-carbon and carbonyl carbon are split into two resonances, where the upfield one is the one substituted with deuterium and the downfield one is the one substituted with hydrogen.

2. Lanthanide Shift Reagents

Lanthanide shift reagents (LSR) are paramagnetic complexes of lanthanide ions that, when added to a solution used to obtain an NMR spectrum, result in a shift of the resonances compared to the spectrum without the LSR (31,32). In many cases a complex spectrum can be made to appear first order by obtaining the spectrum with a LSR. The most widely used complexes are the tris-dipivaloylmethanates (DPM) and the tris-1,1,1,2,2,3,3-heptafluoro-7,7-octanedionates (FOD). The most commonly used lanthanide ions are Eu, Pr, and Yb. Complexes of these ions usually have better solubility in organic solvents and give the best compromise between the magnitude of the induced shift and the accompanying line-broadening. In general, downfield shifts are observed with EuLSR and YbLSR, and upfield shifts with PrLSR.

A series of spectra are usually obtained in which increasing amounts of the LSR have been added to the sample. For small relative concentrations of LSR, a plot of the induced shifts versus the concentration of the LSR is usually linear. The resonances in the spectra are then assigned and chemical shifts in the "unshifted" spectrum are found by extropolating back to zero concentration of LSR. The induced shifts arise from complexation of the LSR with a basic site in the compound under study. The effectiveness of various functional groups to complex with LSR generally follows the order: amine > hydroxyl > ketone > aldehyde > ether > ester > nitrile (33). For ^1H spectra, the induced shifts are interpreted in terms of the pseudo-contact shift mechanism $(3\cos^2 \Theta - 1)/r^3$.

The angle dependence is defined where Θ is the angle between the distance vector r joining the lanthanide ion and the nucleus in the complexed substrate and the crystal field axis of the complexed substrate. As the distance dependence indicates, nuclei closer to the site of complexation will undergo the larger induced shifts. Computer programs are available for fitting the induced shift data to obtain the best location of the lanthanide complex with respect to the substrate (34,35).

Chiral shift reagents based on camphor have been prepared and proven useful for resolving the spectra of enantiomeric mixtures (36,37). Mixtures of AgFOD and YbFOD are effective shift reagents for alkenes and aromatics (38,39).

3. Spectra of Protons Coupled to ^{13}C Nuclei Only

A method has been developed that allows the observation of proton signals coupled to ^{13}C with the cancellation of signals arising from protons bonded to ^{12}C nuclei (40). This procedure should prove useful in examining ^{13}C-enriched compounds obtained from biosynthetic or metabolic pathway studies. The method has recently been used to observe the protons bonded to ^{15}N nuclei in a labeled peptide (41).

IV. SPECTRAL EDITING IN ^{13}C NMR SPECTRA

It was recognized very early in the use of ^{13}C NMR spectra that further information was necessary to make assignments. Over the past several years various methods have been used to distinguish between carbon types in ^{13}C NMR spectra. Four of the more popular methods are discussed below in historical order and in order of increasing difficulty to implement on FT-NMR spectrometers.

A. Single-Frequency Off-Resonance Decoupling

A normal ^{13}C NMR spectrum is usually obtained with broad-band decoupling of the protons such that singlets are observed for the carbon resonances. It was recognized that valuable information toward making assignments could be obtained from experiments where single-frequency irradiation was used with the frequency being offset to be either to higher or lower frequency than the region of proton absorptions (SFORD). This leads to partially decoupled ^{13}C NMR spectra where methyl carbons appear as quartets, methylenes as triplets, methines as doublets, and quaternary carbons as singlets. An example is given in Fig. 6. The residual splitting, J_r, in the multiplets is given by

$$J_r = \frac{J\Delta v}{\gamma B_2}$$

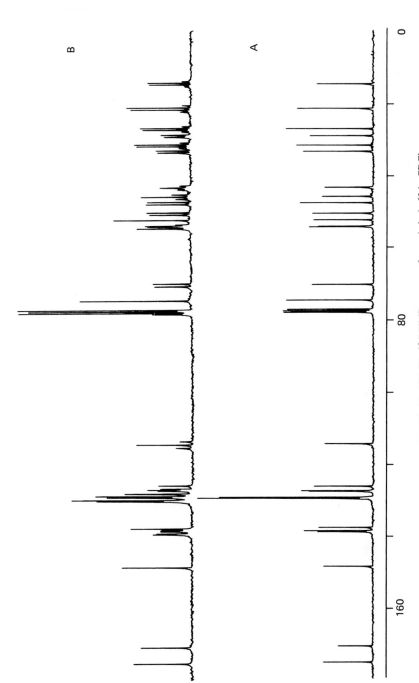

Fig. 6. Normal (A) and SFORD (B) 75-MHz ^{13}C NMR spectra of cytochalasin H in CDCl$_3$.

where J is the one-bond C–H coupling constant, γ is the frequency offset, and B_2 is the decoupler power. A graphic approach has been presented where one obtains a series of SFORD spectra where the decoupler power is held constant and the offset frequency is varied (42). By plotting the peak frequencies in the ^{13}C SFORD spectra versus the irradiation frequencies, one obtains straight lines which intersect at the proton and carbon chemical shifts. If the proton spectrum has been assigned one can use this method to correlate and assign the ^{13}C NMR spectrum.

The SFORD method of spectral editing suffers from several drawbacks, which makes it a less than ideal method. Because lines are split into multiplets, it will usually require about four times the number of pulses to achieve the same signal: noise ratio in the SFORD spectrum as in the normal decoupled spectrum. Furthermore, in complex molecules where there may be several ^{13}C resonances within a narrow frequency region, overlap of the multiplets in the SFORD spectrum may make an interpretation of the splitting patterns impossible. Second-order effects arising from strongly coupled proton spectra may also obscure the multiplet structure (43). Because of these difficulties, a number of alternative approaches have been developed and are discussed below.

B. SEFT and APT

The spin-echo Fourier transform (SEFT) technique (28,44) relies on phase modulation resulting from heteronuclear spin–spin coupling. The pulse sequence is delay-90°-t-180°-t-acquisition for the carbon frequency, with the broad-band proton decoupler being gated off during the first t time period and on the remainder of the time. The first 90° pulse tips the magnetization from along the z axis into the xy plane along the y direction. Because the decoupler is turned off at this point, there will be a vector precessing for each carbon species corresponding to each spin state of the protons attached to the carbons (i.e., two for a methine, three for a methylene, etc.). The signals become modulated because each line in a carbon multiplet precesses at a different frequency that is dependent on the one-bond C–H coupling constant. After a time t, the decoupler is turned on again and a 180° pulse is applied. This reverses the precessional direction, and after a time $2t$, an echo is acquired by the receiver. This echo is identical to a normal FID with the exception that it now contains the phase and amplitude information that existed at the time t. This phase and amplitude are different for the different types of carbons and for $t = 1/J_{CH}$, methyl and methine carbons will appear as negative signals whereas methylene carbons will appear as positive signals. Quarternary carbons are not modulated and will appear as positive signals. Thus, from this experiment methyl and methine carbons may be distinguished from methylene and quaternary carbons.

A double spin-echo technique (APT for attached proton test) has been devel-

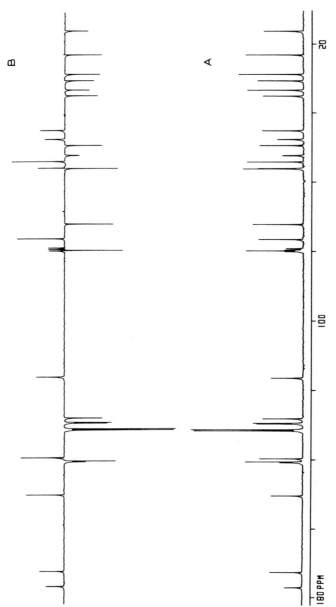

Fig. 7. Normal (A) and APT (B) 75-MHz ^{13}C NMR spectra of cytochalasin H in $CDCl_3$.

oped by Patt and Shoolery (45). Their pulse sequence is $\Theta°$-t-180°-t-v-180°-v-acquisition, with the proton decoupler being gated off during the first t period. The APT technique allows for any pulse up to 90° to be used for the first pulse, and the delays v are added to allow for complete refocusing of the magnetization as a result of less than a 90° pulse being used as the first pulse. Because the maximum amplitude of the signal appears at $t = 1/J$ and J_{CH} can vary from 125 to 180 Hz in a complex molecule, a good compromise for t is around 7 to 8 msec. Satisfactory results (Fig. 7) are obtained with a value of 1 to 3 msec for v. With either the SEFT or APT pulse sequence, a single spectrum can be acquired with essentially normal sensitivty and from which chemical shifts and multiplicity information can be obtained. Usually there is little overlap between methyl and methine carbon chemical shifts, such that these types may be distinguished. However, there may be overlap between methylene and quaternary carbons, and care should be exercised in distinguishing between these carbon types from an APT spectrum alone.

C. INEPT

The INEPT (insensitive nuclei enhanced by polarization transfer) pulse sequence was introduced to provide enhanced sensitivity in coupled ^{13}C NMR spectra (46). Modifications to this pulse sequence (47,48) allow proton-decoupled spectra to be obtained with spectral editing. The pulse sequence is given in Fig. 8. The delay D_1 is to allow for refocusing of the magnetization components and to remove frequency-dependent phase shifts and $D_2 = 1/4\,J$. Because different types of carbons are modulated differently, the delay D_2 can be chosen to discriminate different carbon types.

The usual sequence of events is first to acquire a normal spectrum. Then three INEPT spectra are acquired with different values for the delay D_1. With $D_1 = \frac{1}{6}J$, only the protonated carbon resonances appear in the spectrum. When $D_1 = \frac{1}{4}J$, the spectrum contains only the methine carbon resonances, and when $D_1 = \frac{3}{8}J$ the spectrum contains the methyl and methine carbon signals upright

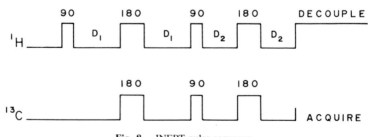

Fig. 8.   INEPT pulse sequence.

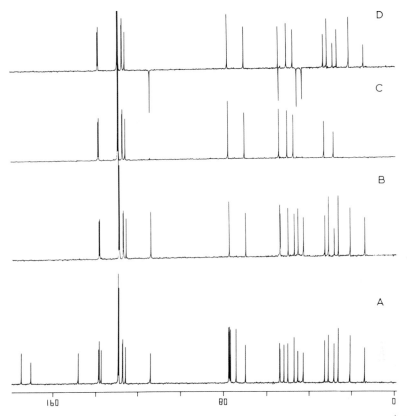

Fig. 9. Normal (A) and INEPT 75-MHz ^{13}C NMR spectra of cytochalasin H in $CDCl_3$ (B, protonated carbons: C, methine carbons; and D, methyl and methine carbons upright and methylene carbons down).

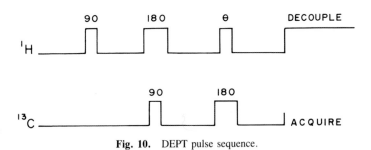

Fig. 10. DEPT pulse sequence.

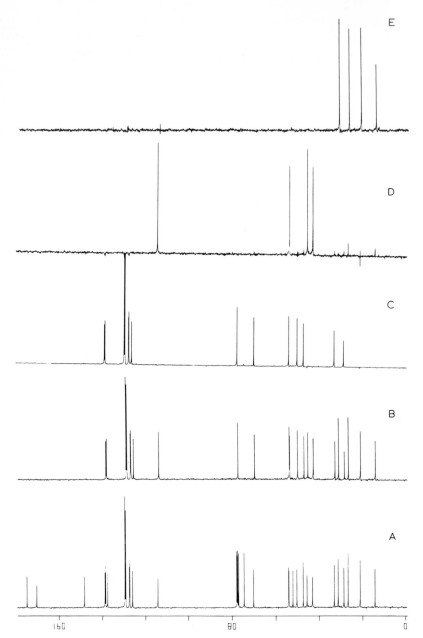

Fig. 11. Normal (A) and DEPT 75-MHz ^{13}C NMR spectra of cytochalasin H in CDCl$_3$ (B, all protonated carbons; C, methine carbons; D, methylene carbons; and E, methyl carbons).

and the methylene carbon signals inverted. An example of this editing technique is given in Fig. 9.

In the INEPT sequence the delay D_1 used to differentiate between different carbon types depends on the one-bond coupling constant J_{CH}. In a typical complex organic molecule this value may cover the range 125–180 Hz. Thus, if several types of carbons are present in the molecule the peaks due to methyl and methylene carbons may not be completely suppressed in the methine-only spectrum. This may lead to some ambiguity in the spectra of complex molecules.

D. DEPT

The recently proposed DEPT (distortionless enhancement by polarization transfer) experiment (49) appears to be the method of choice for routine work in determining the multiplicities of all resonances in a ^{13}C NMR spectrum. The pulse sequence for the DEPT technique is given in Fig. 10. The differentiation between the various carbon types is based on different values for the Θ-pulse rather than delay times as in the previous experiments. If composite pulses (50) are used for ^1H and ^{13}C 180° pulses, errors in setting the pulse angles are not as critical as in the previous editing sequences. Furthermore, the residual intensities resulting from having a range of J_{CH} values in the molecule are smaller in the DEPT sequence than in the INEPT sequence.

The procedure after acquiring a normal broad-band decoupled ^{13}C spectrum is to acquire four DEPT spectra with values for Θ of (i) 45°, (ii) 90°, (iii) 90°, and (iv) 135°. The multiplet differentiated subspectra are generated from these four spectra as follows: methine (spectrum 2 + spectrum 3); methylene (spectrum 1 − spectrum 4); and methyl (spectrum 1 + spectrum 4)-constant × (spectrum 2 + spectrum 3). The two 90° spectra are necessary to provide the same signal:noise ratios in all subspectra. The constant for multiplication to generate the methyl subspectrum can be found by trial and error add–subtract procedures. Spectrum 1 contains all the protonated carbons, and comparison of this spectrum with the normal spectrum serves to identify the quaternary carbons. An example of the DEPT spectra of cytochalasin H (3) is given in Fig. 11. A DEPT sequence has been proposed recently that will provide a quaternary-only subspectrum (51), thereby allowing complete spectral editing with the DEPT sequence. With most modern spectrometers, the entire process of acquiring the spectra and addition and subtraction to provide the individual subspectra can be automated.

V. TWO-DIMENSIONAL METHODS

A. Homonuclear Two-Dimensional *J* Spectroscopy

The homonuclear 2D *J* spectroscopy experiment uses the following pulse sequence: delay-90°-$t_{1/2}$-180°-$t_{1/2}$-acquisition (26). After an initial 90° pulse, a

180° pulse is applied in the middle of the evolution period. The FIDs are collected as a function of t_1, yielding the time domain spectra, $S(t_1,t_2)$. On double Fourier transformation, one obtains a spectrum containing chemical shift information along the F_2 axis and homonuclear coupling constant information along the F_1 axis for spin systems that are weakly coupled. The spectrum is then "tilted" or "rotated" by 45° to align all the components of a given proton multiplet at one frequency, the chemical shift value. Because all of the multiplet intensity from each proton is on lines parallel to the F_1 axis, one can generate individual cross sections from each proton yielding multiplets for the splitting of that particular proton. This experiment separates the multiplets, even though they may overlap extensively in the normal spectrum, and permits a first-order analysis of the coupling constant information. A projection along the F_2 axis yields a spectrum in which each proton appears at its chemical shift value as a singlet. Thus, one can obtain a "proton-decoupled" proton NMR spectrum. If the spectrum is not first order, peaks may appear at a chemical shift value halfway between the shifts of the coupled protons in the projected spectrum.

The resolution in the F_1 dimension (coupling constant information) is determined by the reciprocal of twice the increment in t_1, that is, $\frac{1}{2} \times N \times t_1$, where N is the number of FIDs collected and t_1 is the increment in t_1 value between each FID. Collection of 128 FIDs where t_1 is 10 msec will therefore yield a resolution of 0.39 Hz, and 256 FIDs would yield a resolution of 0.19 Hz in the F_1 axis. This should be more than adequate to resolve most long-range couplings.

An example of a projection to yield a "decoupled" proton spectrum of cytochalasin H (3) is given in Fig. 12. The advantage of the homonuclear 2D J experiment is that it provides both chemical shifts and multiplets from which one can extract coupling constants in one experiment. The multiplets for each proton are obtained even though there may be extensive overlap of several multiplets, provided the overlapping multiplets are weakly coupled. Several examples of the application of this method to obtain chemical shifts and coupling constants have appeared recently (52–54).

B. Homonuclear Correlated Two-Dimensional Spectroscopy

The homonuclear correlated 2D experiment allows one to establish which protons are spin-coupled to each other (26). The following pulse sequence is used: delay-90°-t_1-90°-acquisition, where the delay is the time between each set ·of pulses and t_1 is varied. Phase cycling, digital filtering, and quadrature detec-

---

Fig. 12.   (A) Normal 300-MHz ^1H NMR spectrum of cytochalasin H in $CDCl_3$ and (B) projection of the homonuclear 2D J spectrum to yield the "decoupled" spectrum.

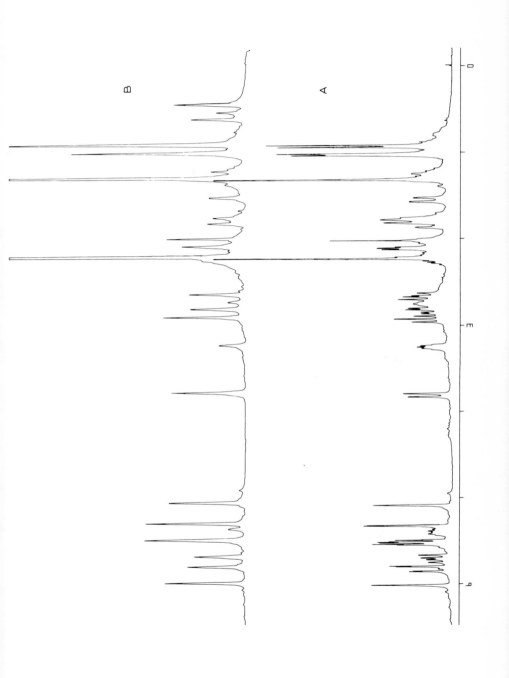

tion are used (55). The magnetization components are labeled with their characteristic precession frequencies during the evolution period t_1, and the second 90° pulse causes coherence transfer of magnetization components among all those transitions which belong to the same coupled spin system. The experiment is repeated for a set of equidistant t_1 values yielding the time domain spectrum $S(t_1,t_2)$. Fourier transformation is carried out using an equal number of data points in each dimension, yielding a square array $S(F_1,F_2)$. The data are normally presented in the form of a contour plot where the normal 1D spectrum lies along the diagonal. Cross peaks or off-diagonal peaks resulting from proton–proton coupling appear symmetrically with respect to the diagonal. Thus, one can easily establish which protons are coupled to each other.

Although the same information could in principle be obtained from a series of single-frequency decoupling experiments, the advantage of the homonuclear correlated 2D experiment is that it provides the information in one experiment. Furthermore, it avoids the problems associated with strong coupling, or those encountered because of missetting the decoupling frequency or from limited selectivity of setting the decoupling frequency in crowded spectral regions in single-frequency decoupling experiments.

An example of this experiment performed on cytochalasin H (**3**) is given in Fig. 13. The normal 1D spectrum is given along the top. Several of the coupling connectivities are illustrated. For example, although it is not clear from the normal spectrum, the homonuclear correlated 2D experiment clearly shows that both of the H-12 protons are coupled to the H-5 and H-7 protons. Note also that the region between 1.5 and 2.2 ppm is "resolved" such that one can establish the assignment of the resonances within this region. In a similar experiment carried out on the peracetylated derivative of cytochalasin H, it was readily established that the C-11 methyl group is shifted upfield by ~0.5 ppm with respect to the parent compound (56).

Several modifications of the homonuclear correlated 2D experiment exist. If the second pulse is changed to a 45° pulse (i.e., delay-90°-t_1-45°-acquisition), then the peaks located symmetrically about the diagonal have a "tilt" associated with them (57,58). By examining the slope of the off-diagonal peaks one can determine the relative signs of the coupling constants. This can be a useful diagnostic tool for distinguishing between geminal and vicinal proton–proton coupling constants, which normally have opposite signs. With another pulse sequence, broad-band decoupling in the F_1 dimension can be achieved (57,59). The pulse sequence used is delay-90°-$t_{1/2}$-180°-(t_d-$t_{1/2}$)-45°-(t_d-$t_{1/2}$)-acquisition, where phase cycling of the 90° and 45° pulses and the receiver are used, and the delay t_d remains fixed. With this experiment, a projection onto the F_1 axis yields a decoupled spectrum. The normal spectrum is present on the diagonal of the 2D spectrum, and off-diagonal peaks indicate coupling between the respective protons.

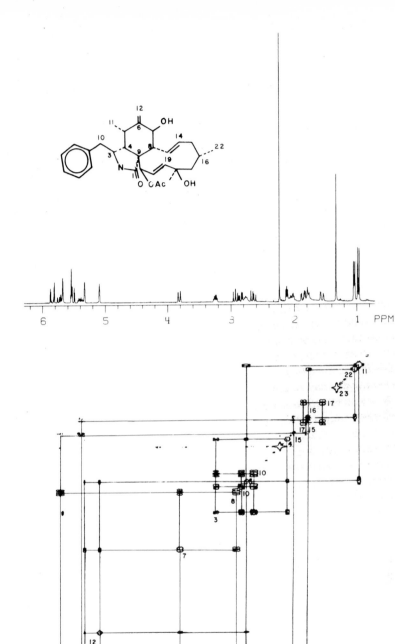

Fig. 13. Contour plot of 300-MHz ^{1}H homonuclear correlated 2D spectrum of cytochalasin H in CDCl$_3$ (1K × 1K) with a normal spectrum on top.

C. Two-Dimensional Nuclear Overhauser Enhancement Spectroscopy

A complete set of NOEs between nearby protons in a molecule may be obtained in one experiment using 2D nuclear Overhauser enhancement spectroscopy (NOESY) (60). Three nonselective $90°$ proton pulses are used in the following pulse sequence: $90°$-t_1-$90°$-t_m-$90°$-acquire. As in the homonuclear 2D J spectroscopy experiment above, the various magnetization components are frequency labeled during the evolution period t_1. During the mixing time t_m, cross-relaxation leads to exchange of magnetization between nearby protons through mutual dipolar interactions. After acquisition of the FIDs for a set of equidistant t_1 values with t_m being fixed, Fourier transformation produces a 2D frequency spectrum. One observes peaks on the diagonal disecting the two-frequency axis corresponding to magnetization components which do not exchange with other components during the mixing time t_m. Symmetrical off-diagonal peaks appear where there is magnetization transfer due to dipole–dipole cross-relaxation during the mixing time.

The 2D spectrum is usually presented as a contour plot. The mixing time can be varied to selectively detect relaxation (NOE) as a function of distance. With short mixing times of approximately 100 msec, only NOEs between protons separated by short distances will be observed. With longer mixing times of between 300 and 400 msec, NOEs between protons separated by a larger distance will be observed. By systematically varying the mixing time t_m, the buildup rates of the NOEs may be determined (61). Chemical exchange can also be examined using the 2D NOE experiment (62). In addition to yielding all the NOEs in a single experiment, this method avoids the errors associated with limited selectivity of preirradiation in crowded spectral regions in the one-dimensional NOE experiment.

D. Heteronuclear Correlated Two-Dimensional Spectroscopy

The heteronuclear correlated 2-D experiment utilizes the coupling between two nuclei (i.e., C–H) to establish a correlation between the two nuclei. Thus, the experiment can be very useful for assigning the ^{13}C NMR spectrum if one can assign the ^1H spectrum or vice versa (63,64). The pulse sequence used to obtain the heteronuclear correlated spectrum is given in Fig. 14. After the $90°$ proton pulse, the spins precess for a time $t_{1/2}$ when a $180°$ carbon pulse is applied. After another time $t_{1/2}$, the proton spins are then refocused at the end of the evolution time. After a time Δ_1 $(1/2J)$, a $90°$ pulse is applied to both the protons and carbons, resulting in a polarization transfer from the protons to the carbons. After another time Δ_2 $(1/3J)$, proton broad-band decoupling is applied and the carbon signal is acquired.

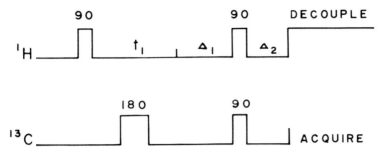

Fig. 14.   Pulse sequence for the heteronuclear correlated 2D NMR experiment.

After Fourier transformation, one obtains a spectrum where the carbon chemical shifts lie along one dimension and the proton shifts and homonuclear couplings lie along the other dimension. The chemical shifts are correlated such that a connectivity between the carbons and directly attached protons is established. Because the pulse sequence involves a polarization transfer, there is an enhancement of sensitivity and the time between pulses is governed by the proton relaxation rates and not the carbon relaxation rates, which may be several seconds longer.

An example of the heteronuclear correlated 2D spectrum of cytochalasin H (**3**) is shown in Fig. 15. The contour peaks appear where the frequency in the carbon shifts and the frequency in the proton shifts intersect and thus, establish the C–H connectivities. The correlation between some of the proton and carbon peaks is highlighted to illustrate the connectivities in Fig. 15. As can be seen from this example, if you can assign the proton spectrum, this experiment allows you to assign the protonated carbons, or vice versa.

One might question why not just run a series of selectively decoupled carbon spectra where each proton multiplet is selectively irradiated rather than utilize the heteronuclear correlated 2D experiment. For small molecules where both the proton and carbon spectra are well resolved, this approach may be faster. However, for complex molecules there are two primary reasons for using the 2D experiment. In complex molecules the proton spectrum is often overlapped to such an extent that it would not be possible to irradiate each proton frequency selectively. Second, when nonequivalent methylene groups are present in the molecule, the selective decoupled spectrum may not provide a unique answer, as second-order effects may be observed.

Although the above discussion deals with connectivities between protons and carbons, the heteronuclear correlated 2D experiment can be applied to protons and any spin ½ nuclei. A recent example has appeared where the amide protons were used to assign the ^{15}N spectrum of a peptide (65). Furthermore, one may adjust the mixing time so that long-range couplings are used to establish the

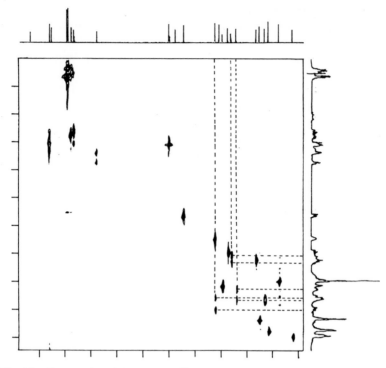

Fig. 15. Contour plot of the 75-MHz ^{13}C heteronuclear correlated 2D NMR spectrum of cytochalasin H in $CDCl_3$. The normal ^{13}C spectrum is given along the top and the ^1H spectrum along the side. The aliphatic methylene carbons are highlighted for illustrative purposes.

connectivities. A recent example has appeared where this technique has been used to aid in the structural determination of a complex molecule (66).

VI. CARBON–CARBON CONNECTIVITY

Several NMR parameters may be used to deduce the structure of an organic molecule, including proton chemical shifts and coupling constants and ^{13}C chemical shifts and proton–carbon coupling constants. Because the basic skeleton of a molecule is made up largely of carbon–carbon bonds, it might be argued that carbon–carbon coupling constants would provide much needed information about the carbon skeleton. Two major limitations have precluded the use of this type of data in the past. First, only 1 molecule in 10,000 has two ^{13}C nuclei adjacent at natural abundance levels. Enriched samples might overcome this problem, but the synthetic effort required precludes its use. Second, the strong peak resulting from molecules with only one ^{13}C nucleus would fill up the computer memory before the ^{13}C satellites could be observed. In addition, the satellite peaks may be obscured by overlap with the strong central peak in crowded spectra. Fortunately, recent experiments have shown that the strong peak at natural abundance may be suppressed by use of the special phase properties of double-quantum coherence such that $^{13}C-^{13}C$ coupling constants can be observed (67).

A. INADEQUATE

The one-dimensional experiment is called the INADEQUATE experiment (68,69). The basic idea is to create double-quantum coherence and then convert this into detectable transverse nuclear magnetization by means of a radiofrequency pulse. The strong central peak is suppressed by taking advantage of the fact that a 90° phase difference exists between the magnetization due to the central peak and the magnetization due to the satellites from $^{13}C-^{13}C$ coupling. The pulse sequence used is 90°-t-180°-t-90°-t_1-90°-acquisition, where $t = 1/4J_{cc}$ and t_1 is a constant, usually about 10 μsec. Broad-band proton decoupling is applied throughout. By using the INADEQUATE experiment, the strong central peaks are suppressed and the weaker ^{13}C satellite spectrum is revealed as AX or AB doublets.

An example of the INADEQUATE experiment with menthol (4) is shown in Fig. 16. One-half of the AX or AB doublet is upright and the other half is down due to the phase properties. The usual procedure would be to determine the $^{13}C-^{13}C$ couplings (splittings) at each carbon and then match the couplings to determine which carbons are bonded to which carbon and hence, determine the carbon skeleton. A computer program has been written that analyzes the satellite data to sort the matching pairs of coupling constants (70).

Several problems may arise in the use of the INADEQUATE experiment. If two or more of the carbon–carbon coupling constants are equal, then it may not be possible to define a unique carbon skeleton. When the spectrum is non-first order (AB type), some of the splittings may not be observable (71). If both single

(4)

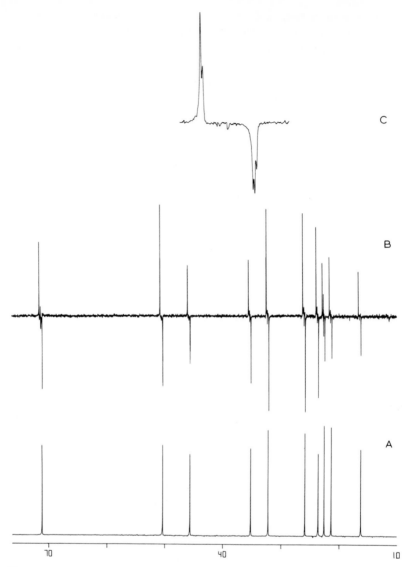

Fig. 16. Normal (A) and INADEQUATE (B) spectrum of menthol in CD_3CN at 75 MHz. An expansion of the peak at ~51 ppm is shown in (C) to illustrate the C–C couplings.

and double bonds are present in the molecule (J_{cc} = 35 and 60 Hz, respectively), then the delay t will be different for the two types of carbons and it may not be possible to suppress the central peak of both types of carbons with a single delay, $1/4J$. In these cases it may be necessary to run two experiments, one with $J = 35$ and the other with $J = 60$, in order to observe all of the satellite splittings. Finally, because of the weak satellite spectra, concentrated solutions (0.3–1.0 M) are required along with long acquisition times (overnight or over the weekend). Despite the above problems, this may be the only way to determine the structure of a compound when either it is a liquid and suitable solid derivatives cannot be made or it is a solid and suitable crystals cannot be obtained for single-crystal X-ray crystallography.

B. Two-Dimensional Double-Quantum Nuclear Magnetic Resonance

Many of the problems mentioned above for the INADEQUATE experiment may be overcome by the use of a 2D version of this experiment (72). The most useful pulse sequence for the 2D experiment appears to be (73) $90°\text{-}t\text{-}180°\text{-}t\text{-}90°\text{-}t\text{-}135°$-acquisition, where $t = 1/4J$ and the delay t_1 is a variable. Fourier transformation yields a two-dimensional spectrum in which the "normal" ^{13}C frequencies appear in the F_2 dimension and the double-quantum frequencies in the F_1 dimension. The spectrum is usually displayed as an intensity contour plot as shown in Fig. 17. The four-line AX or AB patterns between coupled carbons appear along the vertical axis as a function of the double-quantum frequencies. By determining which carbons are coupled to which carbons, the carbon backbone of the molecule may be determined.

Because this experiment depends on the double-quantum frequencies, carbon connectivities are established even if some of the carbon couplings are equal. Furthermore, the problems with varying carbon coupling constants, non-first-order spectra, and suppression of the strong central peak in the INADEQUATE experiment above are not as critical as in the two-dimensional version. Determination of the carbon–carbon connectivities and hence, determination of the carbon skeleton can be undertaken without prior knowledge. A major disadvantage of this 2D version compared to the INADEQUATE experiment is that longer acquisition times and/or more material is required for the 2D version of the carbon–carbon connectivity experiment. A problem occurs with both the 1D and 2D double-quantum connectivity experiments when heteroatoms are part of the skeleton. Heteroatoms cause a break in the carbon–carbon connectivity, and if several are present in the skeleton, one may determine several carbon fragments and not be able to determine how to connect them. While it may be possible to use the carbon chemical shifts to determine which carbons are attached to the heteroatoms, one still has several possibilities for connecting the carbon frag-

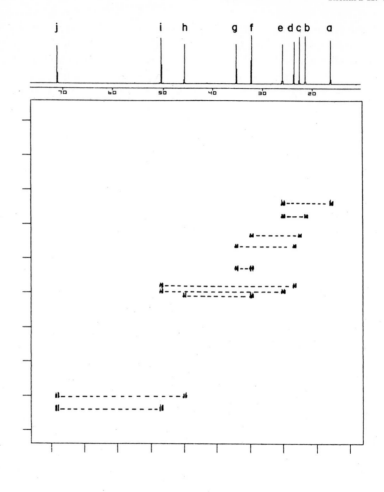

Fig. 17. Contour plot of a 75-MHz ^{13}C 2D double-quantum carbon–carbon connectivity experiment on menthol in CD$_3$CN. A normal spectrum is given along the top, and the determined connectivity of the carbons is given at the bottom.

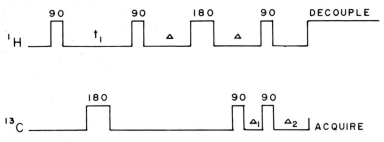

Fig. 18. Pulse sequence for the relayed-coherence 2D NMR experiment.

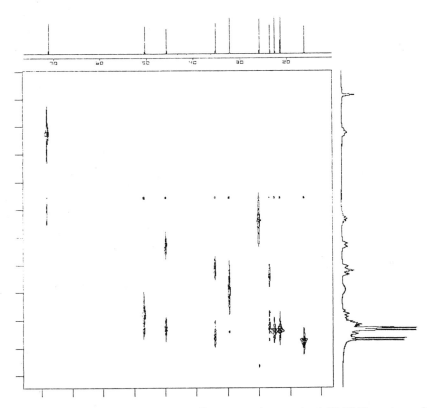

Fig. 19. Contour plot of the 75-MHz ^{13}C heteronuclear correlated 2D NMR spectrum of menthol in CD$_3$CN. The normal ^{13}C spectrum is given along the top and the ^{1}H spectrum along the side.

ments if two or more of the heteroatoms are indicated. One should be aware of
this problem if heteroatoms are present.

C. Relayed-Coherence Transfer Nuclear Magnetic Resonance

The INADEQUATE methods above provide a direct determination of carbon–
carbon connectivities. Methods have been proposed (relayed-coherence transfer
NMR) (74–76) in which information from the proton resonances are used in
assigning the carbon spectrum and in establishing their connectivities. Information is "relayed" between nuclei that are not coupled together, but possess a
common coupling partner, to yield a spectrum whereby connectivities may be
established.

An improved version of the pulse sequence used for relayed-coherence trans-

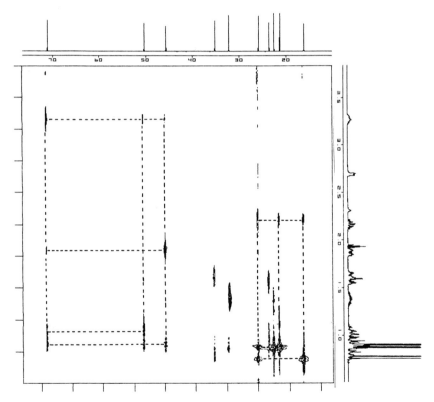

Fig. 20. Contour plot of the 75-MHz ^{13}C relayed-coherence 2D NMR spectrum of menthol in
CD$_3$CN. Several of the "relayed peaks" are highlighted to illustrate the carbons which are adjacent
to the carbons in question.

fer 2D NMR (77,78) is given in Fig. 18. During the pulse sequence magnetization transfer occurs between protons similar to the homonuclear correlation 2D experiment and between proton and carbons similar to the heteronuclear correlated 2D experiment (63,64). As a result, one obtains a two-dimensional spectrum containing both types of information such that connectivities may be established.

The usual procedure is to acquire a heteronuclear correlated 2D spectrum as shown in Fig. 19 to establish H_A-C_A connectivities. A relayed-coherence transfer experiment is then acquired and presented as a contour plot with the proton spectrum along one axis and the carbon spectrum along the other axis (Fig. 20). The additional contours (signals) appearing in the relayed experiment compared to the heteronuclear correlated experiment arise from proton–proton magnetization transfer and identify the proton–proton couplings. In the fragment $H_A-C_A-C_B-H_B$ with a proton–proton coupling of J_{AB}, magnetization transfer occurs along two pathways, $H_A-H_B-C_B$ and $H_B-H_A-C_A$. This leads to two peaks at δH_A C_B and δH_B C_A in the relayed experiment in addition to the peaks at δH_A C_A and δH_B C_B that one observes in the heteronuclear correlated experiment. These four signals appear at the corners of a rectangle in the relayed experiment and establish that the two carbons must be in the immediate vicinity within the molecule. By going through a relayed spectrum and determining the pattern of peaks, one can establish the connectivities of both the protons and carbons and hence, determine the structure of the compound.

The relayed-coherence transfer 2D NMR experiment is useful in those cases where sufficient sample is not available to establish the carbon connectivities directly using the 1D or 2D INADEQUATE experiments. In addition, the relayed spectrum can be used to assign CH carbons attached to NH or OH fragments provided there is coupling between the protons. This is not possible with the INADEQUATE experiments.

VII. BIOSYNTHETIC STUDIES

No chapter on the application of NMR to studies of fungal metabolites would be complete without a brief discussion of biosynthetic studies. Biosynthetic investigations have benefited greatly from advances made in NMR in recent years, particularly those with ^{13}C NMR. Compared to previous studies using ^{14}C-labeled precursors where extensive chemical degradation and isolation and identification of the location of the label in fragments was required, using ^{13}C precursors one simply has to identify those carbons in the ^{13}C NMR spectrum that show increased intensities over that observed for the natural abundance (1.1%) signals. A number of ^{13}C precursors including monolabeled and dilabeled acetate, sugars, and amino acids are now commercially available at a

reasonable cost. Other materials including labeled carbon dioxide or a precursor
are available to prepare specific precursors synthetically.

Implicit in the use of ^{13}C NMR for biosynthetic investigations is one's ability
to assign uniquely the spectrum of the natural abundance material. A combination of the techniques discussed in this chapter combined with comparisons of
data on structurally related compounds should serve this purpose. Incorporation
of the labeled precursor at specific sites is manifested by increased intensity at
those locations. In certain situations, one can also make use of carbon–carbon
coupling constants in biosynthetic studies. For example, when the precursor is
dilabeled acetate, the incorporation of intact acetate units is confirmed by the
observation of carbon–carbon coupling between the two carbons derived from
the acetate unit. If incorporation levels are high, one can also observe carbon–
carbon coupling when using monolabeled precursors if rearrangements occur to
place two labeled carbons adjacent in the molecule.

A number of reviews have appeared in recent years on the use of ^{13}C NMR in
biosynthetic studies (79,80). One should also consult the excellent book on
fungal metabolites by Turner and Aldridge (82) and the book on the biosynthesis
of mycotoxins edited by Steyn (81) for examples of the use of ^{13}C NMR in
biosynthetic studies. Two examples will be sited here to illustrate the technique.

Both C-2 labeled acetate and dilabeled acetate were used to investigate the
biosynthesis of morticin (**5**), a phytotoxic metabolite of *Fusarium* spp. of the
group *Mortiella* (83). The ^{13}C spectrum of **5** derived from dilabeled acetate
established that all carbons with the exception of the methoxyls were enriched.
The signals for C-1 through C-14 also exhibited satellites due to carbon–carbon
coupling demonstrating the incorporation of intact acetate units. The signals for
C-15, C-16, and C-17 were also increased in intensity, and the carbon coupling
constants indicated that C-16 and C-17 were incorporated as an intact acetate
unit. However, the enrichment observed for C-15, C-16, and C-17 was less than
that observed for the other carbons, suggesting that this three-carbon fragment
was added to the molecule in a separate step. Experiments with C-2 labeled
acetate showed that C-15 and C-16 were derived from C-2 of acetate. These

(5)

(6)

results suggested that **5** is derived from the addition of a C_3 Krebs cycle precursor to a polyheptaketide.

The biosynthesis of cytochalasin A (84), cytochalasin B, and cytochalasin D (85), and chaetoglobosin A (86) have been reported. The labeling pattern in cytochalasin B (**6**) is shown above where the ● and ■ indicate carbons derived from C-1 and C-2 of acetate and ▲ indicates carbons derived from the *S*-methyl group of methionine. Phenylalanine is incorporated intact with C-1 of phenylalanine becoming C-4 of cytochalasin B. Similar results were observed for chaetoglobosin A with the exception that the amino acid incorporated is tryptophan. Other points of interest are the acetate head-to-head condensation at C-8 and C-9 and the tail-to-tail condensation at C-4 to C-5.

REFERENCES

1. J. Jenner, Amp're International Summer School 11, Basko Polje, Yugoslavia, 1971.
2. A. Kumar, D. Welti, and R. R. Ernst, *J. Magn. Reson.* **18,** 69 (1975).
3. J. A. Pople, W. G. Schneider, and H. J. Bernstein, "High Resolution NMR Spectroscopy." Pergmon, Oxford, 1965.
4. J. W. Emsley, J. Feeney, and L. H. Sutcliffe, "High Resolution NMR Spectroscopy." Pergamon, Oxford, 1965.
5. L. M. Jackman and S. Sternhell, "Applications of Nuclear Magnetic Resonance Spectroscopy in Organic Chemistry." Pergamon, Oxford, 1969.
6. T. C. Farrar and E. D. Becker, "Pulse and Fourier Transform NMR." Academic Press, New York, 1971.
7. J. B. Stothers, "Carbon-13 NMR Spectroscopy." Academic Press, New York, 1972.
8. G. C. Levy and G. L. Nelson, "Carbon-13 Nuclear Magnetic Resonance for Organic Chemists." Wiley, New York, 1972.
9. F. W. Wehrli and T. Wirthlin, "Interpretation of Carbon-13 NMR Spectra." Heyden, London, 1976.
10. T. Axenrod and G. A. Webb, eds., "NMR Spectroscopy of Nuclei Other Than Protons." Wiley (Interscience), New York, 1974.

11. R. K. Harris and B. E. Mann, "NMR and the Periodic Table." Academic Press, New York, 1978.
12. R. Benn and H. Gunther, *Angew. Chem., Int. Ed. Engl.* **22,** 350 (1983).
13. R. Freeman and G. A. Morris, *Bull. Magn. Reson.* **1,** 5 (1979).
14. R. Freeman, *Proc. R. Soc. London, Ser. A* **373,** 149 (1980).
15. A. Bax, "Two-Dimensional Nuclear Magnetic Resonance in Liquids." Reidel Publ., Dordrecht. Netherlands, 1982.
16. L. M. Jackman and F. A. Cotten, eds., "Dynamic Nuclear Magnetic Resonance Spectroscopy." Academic Press, New York, 1975.
17. J. H. Noggle and R. E. Schirmer, "The Nuclear Overhauser Effect." Academic Press, New York, 1971.
18. R. G. Shulman, ed., "Biological Applications of Magnetic Resonance." Academic Press, New York, 1979.
19. J. S. Waugh, ed., "Advances in Magnetic Resonance." Academic Press, New York.
20. J. W. Emsley, J. Feeney, and L. H. Sutcliffe, eds., "Progress in NMR Spectroscopy." Academic Press, New York.
21. E. F. Mooney, ed., "Annual Reports on NMR Spectroscopy." Academic Press, New York.
22. G. C. Levy, ed., "Topics in Carbon-13 NMR Spectroscopy." Wiley, New York.
23. M. L. Martin, J.-J. Delpueck, and G. J. Martin, "Practical NMR Spectroscopy." Heyden, London, 1980.
24. L. D. Hall and J. K. M. Sanders, *J. Am. Chem. Soc.* **102,** 5703 (1980).
25. W. A. Anderson and R. Freeman, *J. Chem. Phys.* **37,** 85 (1962).
26. W. P. Aue, E. Bartholdi, and R. R. Ernst, *J. Chem. Phys.* **64,** 2229 (1976).
27. J. L. Marshall, "Carbon–Carbon and Carbon–Proton NMR Couplings: Applications to Organic Stereochemistry and Conformational Analysis." Verlag Chemie, Weinheim, 1983.
28. D. L. Rabenstein and T. T. Nakashima, *Anal. Chem.* **51,** 1465A (1979).
29. R. A. Bell, C. L. Chan, and B. G. Sayer, *J. Chem. Soc. D* p. 67 (1972).
30. J. Feeney, P. Partington, and G. C. K. Roberts, *J. Magn. Reson.* **13,** 268 (1974).
31. F. Inagaki and T. Miyazawa, *Prog. Nucl. Magn. Reson. Spectrosc.* **14,** 67 (1981).
32. R. E. Sievers, ed., "Nuclear Magnetic Resonance Shift Reagents." Academic Press, New York, 1973.
33. J. K. M. Sanders and D. H. Williams, *J. Am. Chem. Soc.* **93,** 641 (1971).
34. M. R. Willcott, III, R. E. Lenkinski, and R. E. Davis, *J. Am. Chem. Soc.* **94,** 1742 (1972).
35. R. E. Davis and M. R. Willcott, III, *J. Am. Chem. Soc.* **94,** 1744 (1972).
36. G. M. Whitesides and D. W. Lewis, *J. Am. Chem. Soc.* **92,** 6979 (1970).
37. H. L. Goering, J. N. Eikenberry, and G. S. Koermer, *J. Am. Chem. Soc.* **93,** 5913 (1971).
38. T. J. Wenzel, T. C. Bettes, J. E. Sadlowski, and R. E. Sievers, *J. Am. Chem. Soc.* **102,** 5903 (1980).
39. T. J. Wenzel and R. E. Sievers, *Anal. Chem.* **53,** 393 (1981).
40. M. R. Bendall, D. T. Pegg, D. M. Doddrell, and J. Field, *J. Am. Chem. Soc.* **103,** 934 (1981).
41. D. G. Reid, D. M. Doddrell, K. R. Fox, S. A. Salisbury, and D. H. Williams, *J. Am. Chem. Soc.* **105,** 5945 (1983).
42. B. Birdsall, N. J. M. Birdsall, and J. Feeney, *Chem. Commun.* p. 316 (1972).
43. J. B. Grutzner, *Chem. Commun.* p. 64 (1974).
44. D. W. Brown, T. T. Nakashima, and D. L. Rabenstein, *J. Magn. Reson.* **45,** 302 (1982).
45. S. L. Patt and J. Shoolery, *J. Magn. Reson.* **46,** 535 (1982).
46. G. A. Morris and R. Freeman, *J. Am. Chem. Soc.* **101,** 760 (1979).
47. G. A. Morris, *J. Am. Chem. Soc.* **102,** 428 (1980).
48. D. P. Burnam and R. R. Ernst, *J. Magn. Reson.* **39,** 163 (1980).
49. D. M. Doddrell, D. T. Pegg, and M. R. Bendall, *J. Magn. Reson.* **48,** 323 (1982).

50. M. H. Levitt and R. Freeman, *J. Magn. Reson.* **43,** 65 (1981).
51. V. Rutar, *J. Magn. Reson.* **48,** 155 (1982).
52. L. D. Hall and J. K. M. Sanders, *Org. Chem.* **46,** 1132 (1981).
53. J. Wernly and J. Lauterwein, *Helv. Chim. Acta* **66,** 1576 (1983).
54. G. A. Morris and L. D. Hall, *J. Am. Chem. Soc.* **103,** 4703 (1981).
55. A. Bax, R. Freeman, and G. A. Morris, *J. Magn. Reson.* **43,** 333 (1981).
56. R. H. Cox H. G. Cutler, R. E. Hurd, and R. J. Cole, *J. Agric. Food Chem.* **31,** 405 (1983).
57. A. Bax and R. Freeman, *J. Magn. Reson.* **44,** 542 (1981).
58. A. Bax and R. Freeman, *J. Magn. Reson.* **45,** 177 (1981).
59. W. P. Aue, J. Karhan, and R. R. Ernst, *J. Chem. Phys.* **64,** 4226 (1976).
60. A. Kumar, R. R. Ernst, and K. Wuthrich, *Biochem. Biophys. Res. Commun.* **95,** 1 (1980).
61. A. Kumar, G. Wagner, R. R. Ernst, and K. Wuthrich, *J. Am. Chem. Soc.* **103,** 3654 (1981).
62. C. Bosch, A. Kumar, R. Baumann, R. R. Ernst, and K. Wuthrich, *J. Magn. Reson.* **42,** 159 (1981).
63. A. A. Maudsley, A. Kumar, and R. R. Ernst, *J. Magn. Reson.* **28,** 463 (1977).
64. G. Bodenhausen and R. Freeman, *J. Magn. Reson.* **28,** 471 (1977).
65. G. A. Gray, *Org. Magn. Reson.* **21,** 111 (1983).
66. F. J. Schmitz, S. K. Agarwal, S. P. Gunasekera, P. G. Schmidt, and J. N. Shoolery, *J. Am. Chem. Soc.* **105,** 4835 (1983).
67. G. Bodenhausen, *Prog. Nucl. Magn. Reson. Spectrosc.* **14,** 113 (1980).
68. A. Bax, R. Freeman, and S. P. Kempsell, *J. Am. Chem. Soc.* **102,** 4849 (1980).
69. A. Bax, R. Freeman, and S. P. Kempsell, *J. Magn. Reson.* **41,** 349 (1980).
70. R. Richarz, W. Ammann, and T. Wirthlin, *J. Magn. Reson.* **45,** 270 (1981).
71. A. Bax and R. Freeman, *J. Magn. Reson.* **41,** 507 (1980).
72. A. Bax, R. Freeman, T. A. Frenkiel, and M. H. Levitt, *J. Magn. Reson.* **43,** 478 (1981).
73. T. H. Mareci and R. Freeman, *J. Magn. Reson.* **48,** 158 (1982).
74. P. H. Bolton, *J. Magn. Reson.* **48,** 336 (1982).
75. P. H. Bolton and G. Bodenhausen, *Chem. Phys. Lett.* **89,** 1319 (1982).
76. G. Eich, G. Bodenhausen, and R. R. Ernst, *J. Am. Chem. Soc.* **104,** 3731 (1982).
77. A. Bax, *J. Magn. Reson.* **53,** 149 (1983).
78. H. Kessler, M. Bernd, H. Kogler, J. Zarbock, O. W. Sorensen, G. Bodenhausen, and R. R. Ernst, *J. Am. Chem. Soc.* **105,** 6944 (1983).
79. A. G. McInnes and J. L. C. Wright, *Acc. Chem. Res.* **8,** 313 (1975).
80. A. G. McInnes, J. A. Walter, J. L. C. Wright, and L. C. Vining, *Top. Carbon-13 NMR Spectrosc.* **2,** 123 (1976).
81. P. S. Steyn, ed., "The Biosynthesis of Mycotoxins." Academic Press, New York, 1980.
82. W. B. Turner and D. C. Aldridge, "Fungal Metabolites II." Academic Press, New York, 1983.
83. J. E. Holenstein, H. Kern, A. Stoessl, and J. B. Stothers, *Tetrahedron Lett.* **24,** 4059 (1983).
84. C. Tamm, *in,* "Biosynthesis of Mycotixins" (P. S. Steyn, ed.), p. 269. Academic Press, New York, 1980.
85. A. Probst and C. Tamm, *Helv. Chim. Acta* **64,** 2065 (1981).
86. S. Sekita, K. Yoshihira, and S. Natori, *Chem. Pharm. Bull.* **31,** 490 (1983).

6

X-Ray Crystallography as a Tool for Identifying New Mycotoxins

JAMES P. SPRINGER

Merck Institute of Therapeutic Research
Department of Biophysics
Rahway, New Jersey 07065

I. INTRODUCTION

Single-crystal X-ray diffraction is one of the most powerful methods of structural determination. It is capable of defining a three-dimensional model for an unknown compound independently of other analytical and spectroscopic techniques. However, the combination of other experimental techniques with the results of a crystal structure determination are even more useful to the chemist or biochemist. This chapter discusses some of the principal advantages and limitations of using X-ray crystallography to define the structure of new mycotoxins.

163

Most of the points discussed may be applied in general to natural products; however, all the present examples will be from the mycotoxin literature. This chapter is mainly concerned with the practical aspects of a crystal structure determination, but it starts with a short section on the history and theoretical aspects to give the reader a perspective on the reasons that the usefulness of this technique has increased dramatically in recent years.

II. HISTORY AND DESCRIPTION OF THE SINGLE-CRYSTAL X-RAY DIFFRACTION EXPERIMENT

X Rays were first described by Roentgen in 1895, and in the next year Wiechert and Stokes concluded that X rays must be electromagnetic pulses of wavelength on the order of 1 Å. In 1912 von Laue, extrapolating from Lord Rayleigh's ideas about diffraction and scattering, showed that crystals acted as three-dimensional diffraction gratings when put in an X-ray beam. This observation piqued the interest of W. H. Bragg and his son W. L. Bragg. The younger Bragg concluded later that year that this diffraction could be thought of as a reflection of X-rays from the lattice planes of a crystal. The two Braggs were soon successful in determining the structures of a number of simple compounds from their diffraction photographs, and X-ray diffraction as a method of structural determination was born.

While instrumentation and data analysis have become increasingly sophisticated in the intervening years, the fundamental single-crystal X-ray diffraction experiment has changed very little. This experiment basically involves allowing a collimated beam of monochromatic X radiation to impinge upon a single crystal of a compound. Most of the X radiation goes right through the crystal without any interaction. However, a portion of the X rays interact with the electrons of each atom, and this interaction causes X rays of the same wavelength to be reemitted in all directions. Because the crystal consists of a three-dimensional array of molecules held in well-ordered symmetrical positions, diffraction takes place. That is, X rays come off each atom in all directions. However, because of the ordering of the molecules, constructive and destructive interference takes place, and as a consequence diffracted X rays are only seen outside the crystal at discrete points in space and not in a continuum. The net effect is such that if the diffracted X rays were stopped by a fluorescent sphere and if one were small enough to sit on top of the crystal, one would have the impression of being inside a planetarium. The main differences would be that the diffracted beams are in a more regular arrangement than the star images and the beams are not visible all at once, but would become observable as the crystal was rotated. The planeterium analogy would hold in that the bright points are separated by dark regions and that the bright points have different intensities. In the diffraction experiment one measures the intensity and position of these diffracted beams and then attempts to reconstruct the arrangement of atoms that could give rise to such a diffraction pattern. It has not been proven theoretically that a given diffraction pattern can only arise from one arrangement of atoms, but in practice this is what has been observed.

Mathematically what one is calculating is shown in Eq. (1),

$$\rho(x,y,z) = \frac{1}{V} \underset{h}{\Sigma} \underset{k}{\Sigma} \underset{l}{\Sigma} F_{hkl} e^{-2\pi i(hx + ky + lz)} \tag{1}$$

where the electron density ρ at position x,y,z is equal to a constant $(1/V)$ times a triple sum over the set of three integers h, k, and l which covers all the diffracted beams in three dimensions. V is the volume of the unit cell; the unit cell is the minimal volume of the crystal which, when translated in three dimensions, generates the whole crystal lattice. The term e is the base of the natural logarithm (i.e., $\ln e = 1$), and $i = \sqrt{-1}$. This leaves only the coefficients of the series, the structure factors, F_{hkl}. If one could experimentally measure F_{hkl} the calculation of the electron density in three dimensions and hence the solution of the crystal structure would be a trivial computational problem. However, the determination is more complicated. A compact way to expand F_{hkl} is shown in Eq. (2),

$$F_{hkl} = |F_{hkl}| e^{-2\pi i \alpha_{hkl}} \tag{2}$$

where $|F_{hkl}|$ is the structure amplitude which is proportional to I_{hkl}, the observed diffracted intensity. The final term involves α_{hkl}, the phase of the diffracted beam. Each diffracted beam comes off the crystal with a different phase. The basic problem within a single-crystal X-ray diffraction experiment is to determine what α_{hkl} is; this is what is known as the phase problem. No general experimental techniques have been developed which allow one to measure the phases directly. Various mathematical methods have been developed which allow one to determine approximate phases from the intensities of the diffracted beams, which can then be refined by least-squares techniques. These approximate phases must be fairly close to the correct phases, since the least-squares techniques have a small radius of convergence. Several excellent, readable books are available which describe these techniques as well as other aspects of X-ray diffraction (Glusker and Trueblood, 1972; Ladd and Palmer, 1977; Stout and Jensen, 1968).

In the intervening years since X-ray diffraction was first described, substantial progress has been made in the methods for the collection and analysis of these data and in particular the solving of the phase problem. The recent use of computers in general and specifically minicomputers in the laboratory has resulted in an explosion of the use of this technique to almost a routine analytical level. Minicomputers are used to drive the instruments collecting the data as well as analyzing and displaying the results. As a consequence, whereas in the past a person might have spent years determining the structure of one or two compounds, a similar analysis now takes only days. This has allowed the technique to be applied beyond simple minerals to complicated organic molecules such as mycotoxins, as well as even to proteins with molecular sizes in the tens of thousands of daltons. While not as routine as a nuclear magnetic resonance (NMR) experiment, many graduate chemistry departments are now requiring that students perform X-ray analyses themselves instead of enlisting the assistance of a full-time crystallographer.

It is important to remember that a solved crystal structure is a model of a compound consisting of regions of electron density corresponding to the positions of each of the atoms. In a normal room temperature experiment one does not see the bonding electrons; therefore, bonds are implied if the distances between two atoms are short enough. Also the multiplicity of a bond—that is, whether it is single, double, or triple—is defined by its distance and geometry as well as the valency requirement of the atoms involved.

III. ADVANTAGES AND LIMITATIONS OF USING X-RAY DIFFRACTION

This section covers some the major advantages and limitations of using single-crystal X-ray diffraction to determine the structure of complex organic molecules such as mycotoxins. Single-crystal X-ray diffraction's greatest strength is its usefulness in determining the structure of totally new compounds. Most other analytical structural techniques [e.g., NMR spectroscopy, infrared (IR) spectroscopy, ultraviolet (UV) spectroscopy, and mass spectrometry (MS)] rely on the use of model compounds. That is, one knows from past experience that a given arrangement of atoms will give a certain spectral result. However, often there are subtle, poorly understood and/or long-range effects which hinder the application of this past experience to a new problem. One of the powers and limitations of X-ray diffraction is that the solution of each new problem is independent of previous work. The power lies in the fact that one does not need previous information to determine a new structure; the limitation is that one has only a restricted number of ways of using this information anyway. As experience is gained with the other experimental techniques, the above advantage will presumably have less significance. However, at the present time, because mycotoxins have such diverse structures, this advantage is important. Also, it is important to remember that because of the fundamental nature of diffraction, the determination of an X-ray structure is very much an all-or-none proposition. That is, even if one is interested in only one portion of the molecule, the structure of the total molecule must be determined.

Many ways exist to describe a chemical entity. First of all, one can characterize a compound by what a human can sense directly: color, smell, taste, and texture. Another level is its chemistry and biology: what happens when various chemical reactions are carried out on a compound or when this compound interacts with various biological systems. On another level, a compound can be characterized by its uninterpreted spectroscopic and chromatographic results (e.g., IR, UV, NMR, HPLC, solubility, melting point). The analysis of the above data by a human brain can result in a chemical structure possibly even in three dimensions. Finally, once a molecule's structure is known it can be described mathematically by its quantum mechanical wave functions. But the most generally useful characterization of a compound in determining what its chemistry and biochemistry will be is its three-dimensional structure, which is exactly what a completed single-crystal X-ray diffraction analysis gives.

While only the relative configuration of a molecule routinely comes from a single-crystal analysis, under certain circumstances the complete absolute stereochemistry can be obtained. This is important, since often only one enantiomer has biological activity. X-Ray diffraction can be used to determine the absolute stereochemistry in one of two ways. First, since single-crystal X-ray diffraction normally determines only the relative configuration of the molecule, if one already knows the stereochemistry in an absolute sense at some stereochemical center, then from the addition of the results of the X-ray analysis one will know the absolute configuration of the whole molecule. The second method is by using the anomalous scattering of X rays. Basically, this phenomenon results from the fact that crystals of two enantiomers would give rise to very similar but not identical diffraction intensities. Consequently, one can exploit this small difference to determine which enantiomer fits the experimental data better. This technique normally works best when the structure includes atoms with atomic numbers higher than oxygen; nevertheless, it has been applied to some problems containing only "light" atoms (Engel, 1972).

It is generally agreed on by natural products chemists that two methods exist for confirming the structure of a compound: one is by total synthesis and the other is by single-crystal X-ray diffraction. If a single crystal can be formed the determination of a structure of a new mycotoxin takes on the order of several days to a week, whereas the synthesis of a new natural product may take weeks or months. The reason that X-ray diffraction is so powerful by itself is that the experiment produces from 3 to 10 times the number of data points compared to the parameters needed to fit the data. This overdetermination means that only a correct model will fit the data closely. Also one can calculate a single number, named the residual or R factor, as a measure of this fit. The smaller this residual value is, the better the model. Well-determined crystal structures generally have residuals in the range of 3 to 10%, which may seem high until one remembers the overdetermination factor.

From a practical point of view the most severe limitation to applying crystallography to mycotoxin problems is getting a suitable single crystal in the first place. To obtain a crystal of any size and quality, one needs to have the material reasonably pure (as a rough guide at least 90%). Other spectroscopic techniques can get interpretable data from impure samples, but X-ray diffraction cannot. One needs to have a crystal of sufficient size and internal order to see diffraction in the first place. It is important that the single crystal has a mass of at least 1 μg, which means that its minimum dimension will be on the order of 0.05 to 0.10 mm. But even if the native compound proves to be impossible to crystallize, one may be able to get suitable crystals of a compound by forming a chemical derivative. Because of theoretical advances in data analysis, however, it is no longer necessary to have a "heavy" atom (e.g., bromine, iodine) in the crystal to solve its X-ray structure. In the past "heavy" atom derivatives were often made to aid in the structural solution. Examples of this are shown in the next section.

One criticism of single-crystal X-ray diffraction is that the solved structure is of a compound in a rigid lattice. One would usually wish to know the structure of a mycotoxin in its active state, which is normally in solution or bound to its biological target. To a certain degree a single crystal of a compound is an artificial arrangement, since in nature a compound is rarely found in this state. But the situation is not as bleak as it may appear. Generally, no chemical bonds break or arise upon formation of a crystal. Sometimes tautomerization takes place, such as an aldehyde ($-CH_2CHO$) existing in solution while the enol ($-CH=CHOH$) exists in the crystal. The conformation in the crystal lattice is generally at a low energy minimum of the molecule, but is influenced to some degree by interactions with adjacent molecules in the lattice.

The equipment necessary to undertake single-crystal X-ray diffraction analysis should include the X-ray source, a computer-controlled diffractometer, and a computer to process the data. Computer software is available from several manufacturers.

IV. X-RAY DIFFRACTION OF MYCOTOXINS

This section gives specific examples from the mycotoxin literature to illustrate some of the points of the previous sections. No attempt has been made to include even a significant proportion of mycotoxins examined by X-ray analysis. However, the following problems serve to demonstrate the power and generality of the X-ray diffraction technique.

A. Trichodermol

The first example is from the trichothecane family of compounds. The first member of this sesquiterpene family, trichothecin, was isolated from *Trichothecium roseum* by Freeman and Morrison and shown to be an isocrotonic ester of an alcohol named trichothecolone (Freeman and Morrison, 1948). On the basis of spectroscopic work, structure (**1**) was proposed for trichothecolone (Freeman *et al.*, 1959; Fishman *et al.*, 1960). Godtfredsen and Vangedal (1964) isolated a new member of the family named trichodermin, which was shown to be a acetyl derivative of an another alcohol called trichodermol. Vigorous oxidation converted trichodermin into trichothecolone acetate (Godtfredsen and Vangedal, 1964, 1965). However, the chemistry of trichodermin, based on **1** for trichothecolone, was not consistent with its proposed structure. Therefore, a single-crystal X-ray analysis was performed on the *p*-bromobenzoate of trichodermol (Abrahamsson and Nilsson, 1964, 1966). At the time the presence of a "heavy" atom, bromine, was necessary to solve the structure. Based on its structure, the structure of trichothecolone was redefined to **2**.

In this case X-ray diffraction was used to correct a structure, and then other members of the family, both those already isolated as well as those isolated subsequently, were related to this structure.

B. Verrucarin A

A second example from the trichothecanes is verrucarin A (**3**), whose structure was proposed from spectroscopic and chemical methods (Tamm and Gutzwiller, 1962; Gutzwiller and Tamm, 1965) and then confirmed by X-ray analysis of its *p*-iodobenzene sulfonate (McPhail and Sim, 1966).

C. Verrucarin B

Finally, a third example from the trichothecane family is that of verrucarin B (**4**), in which the epoxide moiety was shown to have the $17S,18R$ configuration.

3

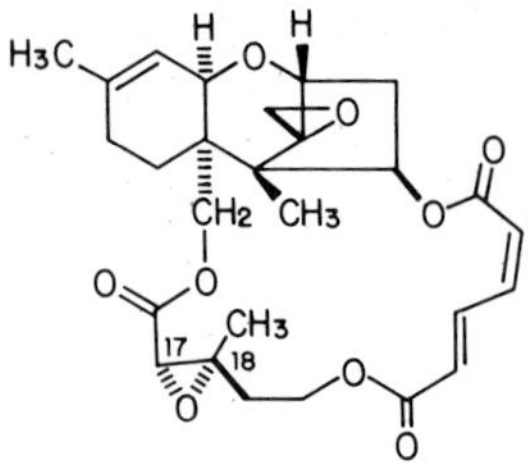

4

This particular configuration was used to help explain the biosynthesis of verrucarin B (Breitenstein *et al.*, 1979).

D. Fumitremorgin A

Switching to another class of mycotoxins, a two-dimensional structure of fumitremorgin A (**5**), a compound causing severe tremors and convulsions in mice with an ED_{50} value of 0.177 mg/kg IP, was proposed on the basis of spectroscopic work (Yamazaki *et al.*, 1975, 1980). A full three-dimensional structure showing the complete relative stereochemistry was defined by Eickman *et al.* (1975) on the basis of an X-ray diffraction analysis.

E. Tryptoquivaline

Another class of tremor-producing compounds whose structure was first established by single-crystal work is that of the tryptoquivalines and tryptoquivalones. When tryptoquivaline (**6**) was treated with methanolic HCl, a deacetylated rearrangement product was formed. Suitable crystals for an X-ray analysis of the rearrangement product resulted from the formation of a *p*-bromophenylurethane derivative (**7**) (Clardy *et al.*, 1975). The presence of the bromine atom not only

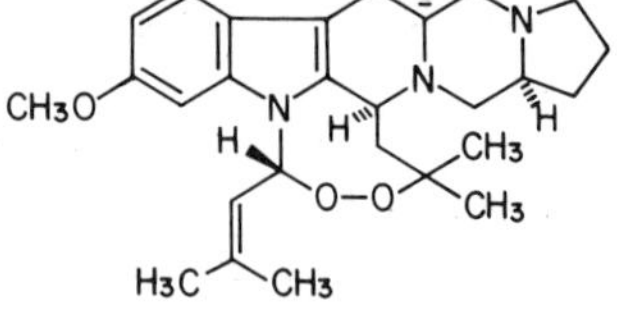

5

6   7

aided the solving of the phase problem but also allowed the determination of the absolute configuration of the compound.

F. Cytochalasin A and B

The structures of the large family of cytochalasins and chaetoglobosins have also benefited from single-crystal X-ray analysis. Many of the structural features of the first members of the family were determined by chemical and spectroscopic means. Subsequently, X-ray analysis was used to define fully all the stereochemistries including the absolute stereochemistry. One of the earliest structures looked at by X-ray analysis was cytochalasin B (8) (Sim *et al.*, 1970), and one of the most recent is cytochalasin A (9) (Griffin *et al.*, 1982).

For many mycotoxins it is not known what the biochemical site of action is. However, the cytochalasins are known to interfere with glucose transport, and an interesting proposal concerning the three-dimensional structure of these compounds and their activity has recently appeared (Griffin *et al.*, 1982).

8   R=
9   R=O

10

G. Moniliformin

Many of the examples discussed above in which X-ray analysis has played a substantial role in the structural elucidation are relatively complicated molecules with a number of stereochemical centers. An example of a structurally simple molecule in which X-ray analysis was crucial was that of moniliformin (**10**) (Springer *et al.*, 1974). Because it was a salt and had only one proton, characterization by other spectroscopic methods was unable to provide a structure.

H. Austin

Many of the previous examples have been compounds heavily characterized by chemical and spectroscopic techniques before an X-ray analysis was done. One reason for this is that many of the compounds were first isolated before X-ray diffraction could be as easily applied as it is now. Also, many of the newly discovered compounds are members of old families, and often it is straightforward to relate the new compounds to previously characterized compounds. Austin (**11**), a mycotoxin isolated from *Aspergillus ustus*, has as yet no close analogs, but its structural determination was carried out by X-ray analysis in a straightforward manner (Chexal *et al.*, 1976).

I. Gliotoxin Analog

We saw previously in the cytochalasins an example of a family of compounds in which the structures were analyzed in terms of their ability to inhibit a specific

11

12

biochemical reaction. Gliotoxin and related compounds are known to have anti-viral activity. One of the enzymes inhibited by these compounds is RNA-directed DNA polymerase (reverse transcriptase). A compound (**12**) modeled after glio-toxin was synthesized and shown to have potent inhibitory activity (Ottenheijm *et al.*, 1978). As more work goes into determining the specific site of action of mycotoxins—that is, which protein or which sequence of nucleic acids the mycotoxin reacts with—the amassed structural data on these compounds will be able to be analyzed in terms of structure–activity correlations. This will increase the depth of our understanding of biochemical processes. Considerable effort has already been put into this approach from the standpoint of understanding the important structural features of pharmaceuticals.

The above examples present only a small number of structural problems in which single-crystal X-ray diffraction techniques have been applied. However, these examples illustrate both the generality and the power of the technique in its application to a structurally diverse group of compounds such as the mycotoxins.

V. CONCLUSIONS

When single-crystal X-ray analysis was first starting to be applied to the identification of new natural products, it was considered ''unsporting'' by practitioners of other analytical techniques as well as those that employed chemical reactions to degrade a molecule and thereby make its structural determination easier. It was felt that this new method that rapidly determined previously unknown structures could be successfully used by people knowing only a minimal amount of organic chemistry. It was further claimed that organic chemistry had lost a source of new reactions, since some chemical reactions were first discovered by applying reagents to natural products to degrade them. But the study of natural products in general and mycotoxins specifically is difficult enough even employing all the structural techniques at one's disposal. Therefore, one should apply whatever techniques solve the problem in an efficient manner. Even though further progress can be made in the development of structural techniques, structural determination can no longer be thought of as an end in itself. The more routine application of single-crystal X-ray analysis as well as the other techniques in previous chapters of this volume allows us to concentrate on even more difficult questions concerning the chemistry, biochemistry, and biology of mycotoxins.

REFERENCES

Abrahamsson, S., and Nilsson, B. (1964). Direct determination of the molecular structure trichodermin. *Proc. Chem. Soc., London* p. 188.

Abrahamsson, S., and Nilsson, B. (1966). The molecular structure of trichodermin. *Acta Chem. Scand.* **20**, 1044–1052.

Breitenstein, W., Tamm, C., Arnold, E. V., and Clardy, J. (1979). The absolute configuration of the fungal metabolite verrucarin B. Biosynthetic consequences. *Helv. Chim. Acta* **62**, 2699–2705.

Chexal, K. K., Springer, J. P., Clardy, J., Cole, R. J., Kirksey, J. W., Dorner, J. W., Cutler, H. G., and Strawter, B. J. (1976). Austin, a novel polyisoprenoid mycotoxin from *Aspergillus ustus. J. Am. Chem. Soc.* **98**, 6748–6750.

Clardy, J., Springer, J. P., Buchi, G., Matsuo, K., and Wightman, R. (1975). Tryptoquivaline and trypoquivalone, two tremorgenic metabolites of *Aspergillus clavatus, J. Am. Chem. Soc.* **97**, 663–665.

Eickman, N., Clardy, J., Cole, R. J., and Kirksey, J. W. (1975). Structure of fumitremorgin A. *Tetrahedron Lett.* pp. 1051–1054.

Engel, D. W. (1972). Determination of absolute configuration and f″ values for light-atom structures. *Acta Crystallogr., Sect. B* **B28**, 1496–1509.

Fishman, J., Jones, E. R. H., Lowe, G., and Whiting, M. C. (1959). Structure and biogenesis of trichotecin. *Proc. Chem.Soc., London* pp. 127–128.

Freeman, G. G., and Morrison, R. I. (1948). Trichothecin-antifungal metabolic product of *Trichothecium roseum. Nature (London)* **162**, 30.

Freeman, G. G., Gill, J. E., and Waring, W. S. (1959). The structure of trichothecin and its hydrolysis products. *J. Chem. Soc.* pp. 1105–1132.

Glusker, J. P., and Trueblood, K. N. (1972). "Crystal Structure Analysis." Oxford Univ. Press, London and New York.

Godtfredsen, W. O., and Vangedal, S. (1964). Trichodermin, a new fungal antibiotic related to trichothecin. *Proc. Chem. Soc., London* pp. 188–189.

Godtfredsen, W. O., and Vangedal, S. (1965). Trichodermin, a new sesquiterpene antibiotic. *Acta Chem. Scand.* **19**, 1088–1102.

Griffin, J. F., Rampal, A. L., and Jung, C. Y. (1982). Inhibition of glucose transport in human erythrocytes by cytochalasins: A model based on diffraction studies. *Proc. Natl. Acad. Sci. U.S.A.* **79**, 3759–3763.

Gutzwiller, J., and Tamm, C. (1965). Verrucarine and roridine. V. The structure of verrucarine A. *Helv. Chim. Acta* **48**, 157–176.

Ladd, M. F. C., and Palmer, R. A. (1977). "Structure Determination by X-ray Crystallography." Plenum, New York.

McPhail, A. T., and Sim, G. A. (1966). Fungal metabolites. Part VI. The structure of verrucarin A: X-ray analysis of verrucarin A. *J. Chem. Soc. C* pp. 1394–1406.

Ottenheijm, H. C. J., Herscheid, J. D. M., Tijhuis, M. W., Nivard, R. F. J., DeClerq, E., and Prick, P. A. J. (1978). Gliotoxin analogues as inhibitors of reverse transcriptase. 2. Resolution and X-ray crystal structure determination. *J. Med. Chem.* **21**, 799–804.

Sim, G. A., McLaughlin, G. M., Kiechel, J. R., and Tamm, C. (1970). The absolute stereochemistry of phomin: X-ray analysis of the phomin-silver fluoroborate complex. *Chem. Commun.* pp. 1398–1399.

Springer, J. P., Clardy, J., Cole, R. J., Kirksey, J. W., Hill, R. K., Carlson, R. M., and Isidor, J. L. (1974). Structure and synthesis of moniliformin, a novel cyclobutane microbial toxin. *J. Am. Chem. Soc.* **96**, 2267–2268.

Stout, G. H., and Jensen, L. H. (1968). "X-ray Structure Determination." Macmillan, New York.

Tamm, C., and Gutzwiller, J. (1962). Verrucarins and roridins. II. Partial structure of verrucarin A. *Helv. Chim. Acta* **45**, 1726–1731.

Yamazaki, M., Fujimoto, H., and Kawasaki, T. (1975). Structure of a tremorgenic metabolite from *Aspergillus fumigatus,* fumitremorgin A. *Tetrahedron Lett.* pp. 1241–1244.

Yamazaki, M., Fujimoto, H., and Kawasaki, T. (1980). Chemistry of tremorgenic metabolites. I. Fumitremorgin A from *Aspergillus fumigatus. Chem. Pharm. Bull.* **28**, 245–254.

7

Application of Biosynthetic Techniques in the Structural Studies of Mycotoxins

PIETER S. STEYN AND ROBERT VLEGGAAR

National Chemical Research Laboratory
Council for Scientific and Industrial Research
Pretoria 0001, Republic of South Africa

I. INTRODUCTION

The constituents of microorganisms can be classified either as compounds which are of primary metabolic concern or otherwise as secondary metabolites which are apparently nonessential to the producing organism. Eminent scientists have described different hypotheses for the role of both secondary metabolism and its metabolites. Fungi have a marvelous capacity to produce secondary metabolites with an array of diverse and often novel molecular structures. Some of the compounds show beneficial activity (antibiotics, fungicides, and enzyme inhibitors) and others potent toxicity (mycotoxins). This chapter is devoted to biosynthetic approaches in relation to the structural studies of mycotoxins. Several monographs have been published recently on the biosynthesis of secondary metabolites (Steyn, 1980; Packter, 1973; Haslam, 1974; Herbert, 1981; Mann, 1978).

MODERN METHODS IN THE ANALYSIS
AND STRUCTURAL ELUCIDATION
OF MYCOTOXINS

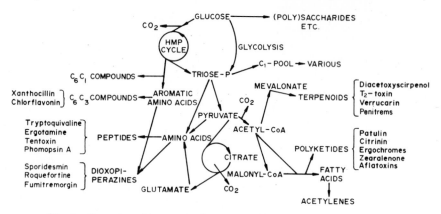

Fig. 1. Formation of mycotoxins from a primary pool of primitive precursors.

Mycotoxins do not constitute a single well-defined structural category and have as such no molecular features in common. These toxins are formed from a primary pool of primitive precursors (e.g., acetate, pyruvate, malonate, mevalonate, shikimate, and amino acids) as shown in Fig. 1. The reactions involved in the formation of these compounds include condensation, oxidation, reduction, alkylation, and halogenation, and lead to the formation of unique molecular structures. A survey of the secondary metabolites in biosynthetic terms indicates that variety within a particular group is generated by the multiple branching of the biosynthetic pathways.

The cardinal problems associated with the study of secondary metabolism involve the identification of the source(s) in primary metabolism from which the secondary metabolites originate and the identification of the mechanism and types of intermediates by which the end products are formed. The structure of a primary metabolite does not necessarily yield any immediate clues to its mode of biosynthesis, whereas the structure of a secondary metabolite often provides accurate information on its origins and mechanism of formation (Herbert, 1981). In this chapter the opposite approach is emphasized by highlighting the structural information that can be obtained from well-planned biosynthetic labeling experiments using mainly stable isotopes. The concept of biosynthetic architecture in fungal metabolites is based on the polyketide biosynthesis, the isoprene rule, and the recognition of amino acids. These approaches have contributed immensely to structural elucidation.

II. TECHNIQUES

The main steps involved in biosynthetic labeling experiments involve the administration of labeled likely precursors to cultures of a specific fungus. The product is isolated from the cultures, and the incorporation of the precursor and the location of the label in the product are determined. Precursors labeled with the radioactive isotopes ^{14}C, ^{3}H, and ^{35}S are frequently used in biosynthetic studies, as very low levels of incorporation of the precursor in the product can be assayed by liquid scintillation counting. Unfortunately, this approach suffers from two drawbacks in that safety precautions are necessary in handling radioactive compounds and specific laborious degradative reactions may be required to determine the position of the labeled nucleus in the metabolite. A severe complication in this regard is the possibility that unexpected rearrangements may occur in the course of these reactions. The application of nuclear magnetic resonance (NMR) spectroscopy, particularly at very high magnetic field strengths, obviates these problems as most nuclei detected by NMR spectroscopy are not radioactive (one notable exception is ^{3}H), and the technique detects the labeling site in a compound by simpler nondestructive means. The inherent higher sensitivity of radiolabeling, however, should not be overlooked.

A chemically defined medium or in exceptional cases a solid medium has to be formulated for the production of a specific metabolite as a prelude to the biosynthetic feeding experiments. Details of many liquid media are recorded in the Catalog of Strains I of the American Type Culture Collection (1982), in the Merck Handbook of Culture Media (1981), and in the Difco Manual (1966). Cultivation of the fungal cultures in shake culture (using a gyratory shaker) enhances fungal growth and is the method of choice for biosynthetic studies. However, we observed that most mycotoxins were produced in adequate yield in stationary culture. The following step is the determination of the yield of the end product against time, the so-called production curve. Next, specifically labeled precursors are added at specified times to growing cultures. The dilution value for a radiolabeled precursor is defined in Eq. (1).

$$\text{Dilution value} = \frac{\text{Specific activity (precursor)}}{\text{Specific activity (product)}} \times \frac{m \text{ (product)}}{n \text{ (precursor)}} \tag{1}$$

where m and n are the appropriate number of labeled sites. The absolute incorporation is given by Eq. (2).

$$\text{Incorporation (\%)} = \frac{S_m}{S_p} \times \frac{M_m}{M_p} \times 100 \tag{2}$$

where S_m and S_p designate the molar specific activity of the metabolite and precursor, respectively, and M_m and M_p are number of moles of isolated metabolite and administered precursor, respectively (Pita Boente, 1981). A dilution value of <80 is necessary for meaningful single-label ^{13}C-labeling experiments.

Addition of precursors to fungal cultures is usually done during the logarithmic phase, steps a to b of metabolite production (see Fig. 2). The precursor can be added under sterile conditions to the fermentation either as a single dose, as daily additions over a period of time, or continuously by using a peristaltic

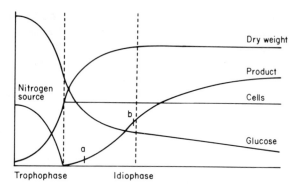

Fig. 2. Schematic representation of a nitrogen-limited fermentation.

pump. The metabolite in question is isolated when its concentration has reached a certain level as indicated by the production curve.

A prerequisite for biosynthetic studies using ^{13}C and ^2H isotopes is the unambiguous assignment of the resonances in the natural abundance ^1H and ^{13}C NMR spectra of the substance. The techniques commonly used in the assignment of ^{13}C resonances have been summarized by Steyn et al. (1980) and by Wirthlin (1982). The chemical shift values of the various ^{13}C resonances are obtained from the broad-band proton-decoupled ^{13}C NMR spectrum. The multiplicities of each resonance can be derived from a comparison of this spectrum with either an off-resonance proton-decoupled or single-frequency coupled ^{13}C spectrum. In the assignment procedure, use can be made of selective proton-decoupling and selective population inversion (SPI) experiments, shift reagents, solvent effects, and nuclear Overhauser effect (NOE) experiments.

Newer experimental methods for multiplet selection involve fundamentally different techniques such as cross-polarization via scalar coupling, J-modulated refocusing of spin echoes (APT, Attached Proton Test), or polarization transfer experiments such as INEPT (Insensitive Nuclei Enhanced by Polarization Transfer) and DEPT (Distortionless Enhancement by Polarization Transfer). Two-dimensional (2D) NMR spectroscopy, using heteronuclear 2D-J resolved and heteronuclear 2D-shift correlated experiments, is a powerful tool for assignment purposes.

Spin–spin coupling between ^{13}C nuclei is of great importance for structural elucidation, since this phenomenon yields information directly related to the carbon skeleton of the metabolite. The necessary condition for the observation of $(^{13}C,^{13}C)$ coupling is the presence of two anisochronous ^{13}C atoms in a molecule. This condition is, however, met for only a few molecules in a sample ($\approx 0.011\%$) as a result of the low natural abundance of ^{13}C (1.1%). Line splittings as a result of $(^{13}C,^{13}C)$ couplings are present in the ^{13}C NMR spectrum only as very weak satellites of the intense signals of molecules at the natural abundance level. The detection of these satellites at the natural abundance level can be achieved through detection of double-quantum coherence in an INADE-QUATE experiment (Incredible Natural Abundance Double-QUAntum Transfer Experiment). In the case of biosynthetic experiments, structural information can be obtained from the observed coupling between contiguous ^{13}C-enriched carbon atoms. This situation can be realized by addition of ^{13}C-doubly labeled precursors (e.g., [1,2-^{13}C$_2$]acetate or [1,2,3-^{13}C$_3$]malonate). The lack of a one-bond (C,C) coupling in these instances is an indication of a rearrangement involving a bond cleavage in the biosynthetic sequence.

The supplementation of fungal cultures with, for example, sodium [1-^{13}C]acetate and [2-^{13}C]acetate, and [2-^{13}C]mevalonic acid leads to so-called singly labeled end products. A comparison of the intensity of the resonances observed in the broad-band proton-decoupled ^{13}C NMR spectrum of the enriched samples with that of the natural abundance sample identifies the enrichment sites, provided the dilution values are not too high. It is of importance to note, however, that for comparison purposes, the data should be recorded under virtually identical conditions (e.g., solvent, temperature, concentration, pulse width, and relaxation delay between pulses). The number of enriched sites is indicative of the number of acetate or mevalonate units involved in the biogenesis. Thus, in the case of averufin, $C_{20}H_{16}O_7$ incorporation of either [1-^{13}C]acetate or [2-^{13}C]acetate led to the enrichment of 10 carbon atoms, and 10 acetate units are therefore involved in the biosynthesis of this metabolite. The observation of satellite resonances due to one-bond (^{13}C,^{13}C) coupling in these experiments is relatively rare and is usually associated with skeletal rearrangements.

Biosynthetic studies usually follow the successful elucidation of a particular structure. However, in this chapter attention is directed mainly at the information to be gained from biosynthetic experiments carried out in conjunction with structural elucidation studies. This approach was followed in our studies on the penitrems, phomopsin A, and the asticolorins.

III. BIOSYNTHESIS OF POLYKETIDES

The polyketide biosynthetic pathway leading to secondary metabolites is the metabolic route most characteristic of the fungi, and it is therefore not surprising that a large number of mycotoxins are produced by this route. A close relationship exists between the biosynthesis of fatty acids and that of polyketides. In the case of the fatty acids the original oxygen atoms in the polyketide chain, with the exception of those of the terminal carboxyl group, are lost as a result of reduction and water elimination, whereas in the case of the polyketide pathway these oxygen atoms are mostly retained.

Polyketides are formed by the condensation of acetyl-coenzyme A ($\equiv$ acetyl-SCoA) and malonyl-SCoA. The latter is formed in turn from hydrogen carbonate and acetyl-SCoA via an enzyme-mediated carboxylation process which requires biotin as cofactor. In the condensation reaction of acetyl-SCoA with malonyl-SCoA a concerted loss of carbon dioxide accompanies attack at the carbonyl group of acetyl-SCoA, as shown in Fig. 3. Experiments using ^{14}C-labeled carbon dioxide have shown that the carbon atom lost as carbon dioxide in the condensation reaction is the same as the one derived initially from hydrogen carbonate.

In the formation of polyketides acetyl-SCoA serves as the starter unit and the carbon chain is lengthened by condensation with a number of malonyl-SCoA units [three in the case of orsellinic acid (1)] (see Fig. 3). It has been suggested

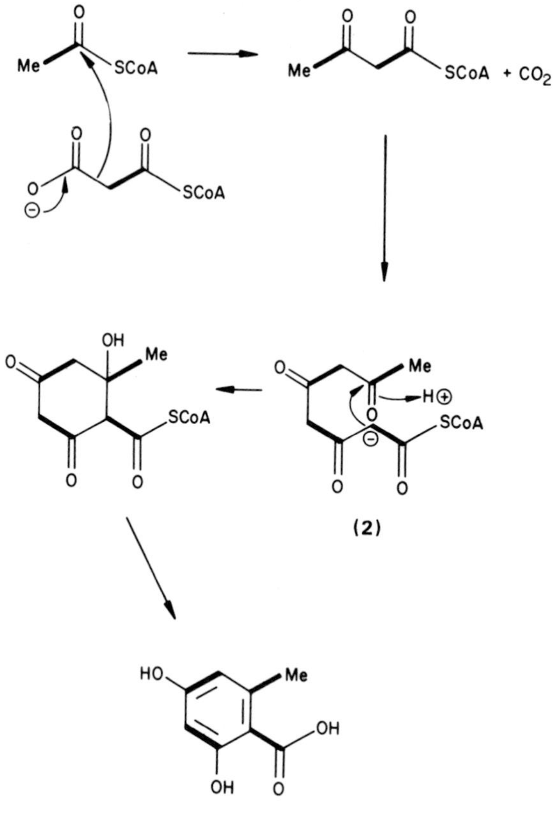

Fig. 3.   Biosynthesis of orsellinic acid (1) from $[1,2\text{-}^{13}C_2]$acetate.

that the growing β-polyketide thioester chain is stabilized by hydrogen bonding to a multienzyme complex or by chelation of an enolate tautomer with a metal atom held by the enzyme (Mann, 1978). At this stage some of the carbonyl groups may be reduced and alkylation of some of the activated methylene groups by S-adenosylmethionine could also occur. Coiling of the polyketide chain can lead to intramolecular aldol or Claisen condensations. This process is shown for the formation of orsellinic acid (**1**) via a tetraketide (**2**). It is evident that feeding experiments with labeled acetate would find wide application in the study of polyketide biosynthesis.

A. Aflatoxin B₁

The aflatoxins are as a result of their worldwide occurrence and potent hepatocarcinogenicity the most important group of mycotoxins. It is therefore not surprising that the biogenesis of these hazardous compounds has attracted so much attention (Steyn et al., 1980; Townsend and Christensen, 1983).

The polyketide nature of the aflatoxins is based on the results obtained by Biollaz et al. (1970) from extensive degradation studies on aflatoxin B₁ (**3**) derived from [1-^{14}C]acetate and [2-^{14}C]-acetate, and [methyl-^{14}C]methionine (see Fig. 4). Studies on aflatoxin biosynthesis using stable isotopes will be used as a model for the structural and biosynthetic information to be gained from such investigations despite the fact that the structures were known prior to the biosynthetic studies. These studies and the time course experiments of Zamir and Hufford (1981) indicated the following series of intermediates in the pathway leading to aflatoxin B₁: [acetyl-SCoA + malonyl-SCoA] → polyketide → norsolorinic acid (**4**) → averantin → averufin (**5**) → versiconal acetate (**6**) → versicolorin A (**7**) → sterigmatocystin → aflatoxin B₁ (**3**). Townsend and Christensen (1983) have investigated the sequence of events leading to averufin and presented a biosynthetic pathway in which nidurufin plays a key role.

The main source of structural information in feeding experiments is the one-bond (^{13}C,^{13}C) coupling constants. The results from our experiments on versiconal acetate (Steyn et al., 1979), averufin (Gorst-Allman et al., 1977), versicolorin A (Gorst-Allman et al., 1978), austocystin D (**8**) (Horak et al., 1983), and aflatoxin B₁ (Pachler et al., 1976) using [1,2-^{13}C₂]acetate are shown in Fig. 4. Norsolorinic acid (Steyn et al., 1981) and versiconal acetate (Steyn et al., 1979) were converted to their methyl ethers [(**9**) and (**10**), respectively] prior to the recording of the ^{13}C NMR spectra. From a quick perusal of the data it is evident that this information has both biosynthetic and structural implications in that the J(CC) value is characteristic for a specific chemical environment. The magnitude of the coupling constants increases with increasing s-character of the bonds. The data also facilitate the assignment procedure, that is, the assignment of a specific resonance to a particular carbon atom. The main structural value is

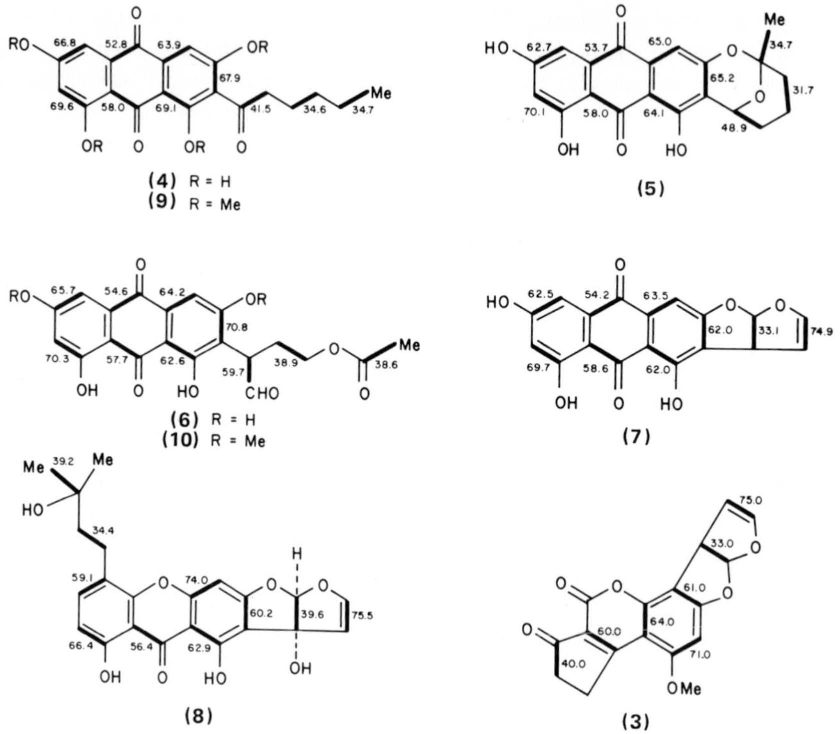

Fig. 4. One-bond (^{13}C,^{13}C) coupling constants (in hertz) for a number of mycotoxins (**4–10**) enriched with [1,2-^{13}C$_2$]acetate.

based on the fact that two-carbon units can be recognized in an organic substance.

In the case of aflatoxin B$_1$ feeding experiments with [1-^{13}C]acetate and [2-^{13}C]acetate established the enrichment of nine and seven carbon atoms, respectively. The remaining carbon atom, the methoxy group, is derived from methionine. The observed one-bond (^{13}C,^{13}C) coupling constant of 34 Hz for C-5 and C-6 (Hsieh *et al.*, 1975), and of 44 Hz for C-11 and C-14 in the spectra of [1-^{13}C]acetate- and [2-^{13}C]acetate-derived aflatoxin B$_1$ (Pachler *et al.*, 1976), respectively, show that rearrangements involving bond fission occur during the biosynthetic process. This information is structurally also important, as it unambiguously defines the C-5–C-6 and C-11–C-14 linkages. The two-bond (^{13}C,^{13}C) coupling [^2J(CC) 21 Hz] between C-2 and C-4 in the proton-decoupled ^{13}C NMR spectrum of aflatoxin B$_1$ derived from [2-^{13}C]acetate has similar structural implications.

The most useful biosynthetic and structural information was obtained from the proton-decoupled ^{13}C NMR spectrum of $[1,2\text{-}^{13}C_2]$acetate-derived aflatoxin B_1. The observed one-bond coupling constants indicated that C-2–C-6, C-3–C-4, C-7–C-12, C-8–C-9, C-10–C-11, C-13–C-14, and C-15–C-16 orginate from seven intact acetate units. The other two carbon atoms, C-1 and C-5, are derived from the carboxyl carbon atom of two separate acetate units, which each lost the methyl carbon in the biosynthetic pathway leading to aflatoxin B_1 (3), and appear as enhanced resonances in the spectrum.

The intermediacy of averufin in aflatoxin B_1 biosynthesis (de Jesus *et al.*, 1980; Simpson *et al.*, 1982a) was confirmed by feeding a mixture of doubly labeled averufins to mycelial pellets of *Aspergillus parasiticus* (Townsend and Davis, 1983). The ^{13}C NMR spectrum of the formed aflatoxin B_1 showed satellite signals, due to one-bond $(^{13}C,^{13}C)$ coupling, for the signals at δ 117.3 (C-2) and 201.2 (C-3) (J 57 Hz), as well as an enhanced resonance at δ 29.0 (C-5) (see Fig. 5). A unique finding by Townsend and Christensen (1983) is the incorporation of $[1\text{-}^{13}C]$hexanoate into averufin, a result which indicates that a linear C_6 starter unit produced by a fatty acid synthetase is involved in averufin biosynthesis.

B. The Asticolorins

The fate of the hydrogen atoms in the biosynthesis of a metabolite can be studied using both ^2H- and $^{13}C,^2H$-labeled precursors (Garson and Staunton, 1979). Both direct (^2H NMR) and indirect methods (^{13}C NMR) can be used to determine the subsequent incorporation of deuterium in the metabolite. The incorporation of deuterium located β to a ^{13}C atom in a precursor is detected by a small characteristic upfield β-isotope shift in the resonance position of the ^{13}C nucleus in the ^{13}C NMR spectrum of the enriched metabolite. The number of ^2H atoms located β to a particular ^{13}C atom can be deduced from the value of the β-isotope shift (Abell and Staunton, 1981). This technique, which has proved most useful in biosynthetic studies (Simpson *et al.*, 1982b; Simpson and Stenzel, 1982; de Jesus *et al.*, 1983b; Gorst-Allman *et al.*, 1983), can of course also be applied to structural studies (e.g., the asticolorins).

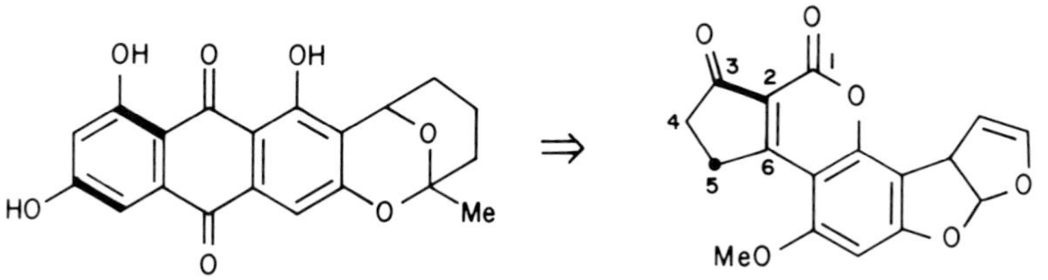

Fig. 5. Conversion of a mixture of $[5,6\text{-}^{13}C_2]$averufin and $[8,11\text{-}^{13}C_2]$averufin into aflatoxin B_1.

The asticolorins, toxic metabolites isolated from cultures of *Aspergillus multicolor,* are characterized by the novel way in which a mevalonate-derived 3,3-dimethylallyl group is utilized to link two dibenzofuran moieties. The problem of the structural elucidation of the asticolorins A (**11**), B (**12**), and C (**13**), was solved in two ways. The one method involves X-ray crystallography of asticolorin A (**11**), and the other is based on NMR spectroscopy and biosynthetic studies of asticolorin C (**13**) as outlined below (Rabie *et al.,* 1984; Steyn *et al.,* 1984).

The assignment of the resonances in the ^{13}C NMR spectrum of asticolorin C (**13**) provided a wealth of structural information and proceeded as follows. First-order analysis of the signals in the 500.13-MHz ^{1}H NMR spectrum yielded the values of the proton chemical shifts and coupling constants. From the values of the coupling constants as corroborated by extensive ^{1}H-[^{1}H] homonuclear decoupling experiments, the (^{1}H,^{1}H) connectivity pattern for asticolorin C could be constituted. The residual splittings observed in a series of off-resonance proton-decoupled ^{13}C NMR experiments enabled us to correlate all the signals of proton-bearing carbon atoms with specific proton resonances. Extensive heteronuclear ^{13}C-[^{1}H] selective population inversion experiments (Pachler and Wessels, 1973, 1977) established the two- and three-bond (C,H) connectivity pattern for asticolorin C. The method, however, does not allow us to differentiate between the resonances of ring A and H carbon atoms. This ambiguity was resolved by the observation of one-bond (C,C) couplings for C-11–C-12 (58.9 Hz) and C-20–C-21 (54.7 Hz) in the broad-band proton-decoupled ^{13}C NMR spectrum of asticolorin C derived from [2-^{13}C]acetate. In addition the spectrum showed enhancement of the signals of 19 carbon atoms, whereas that of asticolorin C derived from [1-^{13}C]acetate showed only 14 enhanced signals.

(**11**)	$R^1 = R^2 = H, R^3 = OH$
(**12**)	$R^1 = R^2 = H, R^3 = \!=\!O$
(**13**)	$R^1 = OH, R^2 = H, R^3 = \!=\!O$

The arrangement of intact acetate units in asticolorin C was studied by addition of [1,2-^{13}C$_2$]acetate to the culture medium. The signals of a number of carbon atoms exhibited, as a result of one-bond (C,C) couplings, two pairs of satellite peaks of equal intensity in the broad-band proton-decoupled ^{13}C NMR spectrum of this enriched asticolorin C. This phenomenon, observed for the corresponding carbon atoms in rings A, C, F, and H, indicates the existence of two different arrangements of intact acetate units in each of these rings (see Fig. 6) and proves the intermediacy of four molecules of a symmetrical intermediate orcinol, in the biosynthesis of the asticolorins. Both the arrangement of intact acetate units, each of which defines two contiguous carbon atoms, and the magnitude of the one-bond coupling constants, ^1J(CC), provide important information about the structure of the metabolite.

A part of the two-bond (C,H) connectivity pattern for asticolorin C was established, in a study on the fate of the hydrogen atoms in the biosynthesis of this metabolite, by addition of [1-^{13}C,2-^{2}H$_3$]acetate to cultures of *Aspergillus multicolor*. A number of resonances in the proton-decoupled ^{13}C NMR spectrum of asticolorin C enriched with [1-^{13}C,2-^{2}H$_3$]acetate exhibited signals shifted upfield as a result of one or more deuterium atoms two bonds removed (i.e., the so-called β-isotope shift). The labeling pattern is shown in Fig. 7, and the values of the β-isotope shifts are collated in Table I. The presence of three isotopically shifted signals for C-19, for example (see Fig. 8), due to the incorporation of one

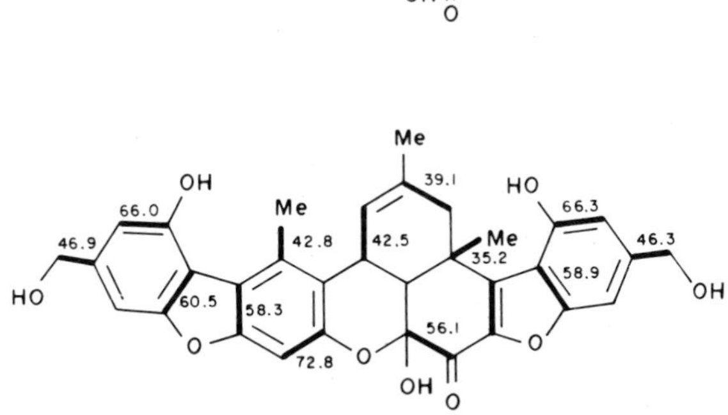

Fig. 6. Arrangement of intact acetate units in asticolorin C derived from [1,2-^{13}C$_2$]acetate. Values of ^1J(CC) in hertz.

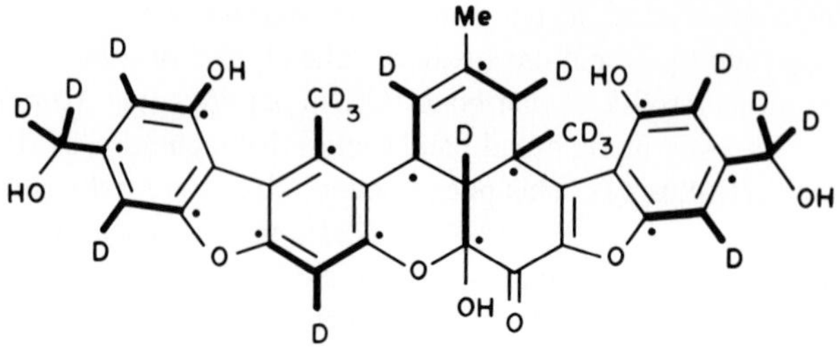

Fig. 7. Labeling pattern of asticolorin C derived from [1-^{13}C,2-^{2}H$_3$]acetate.

TABLE I

β-Isotope Shifts Observed for Asticolorin C Enriched by [1-^{13}C,2-^{2}H$_3$]Acetate

Carbon atom	δ_C/ppm[a]	$\Delta\delta$/ppm[b]
2	149.82	0.044
4	154.16	0.046
6	157.40	0.049
8	142.35	0.053
		0.052
10	151.37	0.036
13	131.88	0.034
		0.038
		0.039
15	29.56	0.082
17	130.73	0.072
19	34.22	0.069
		0.067
		0.069
22	152.76	0.036
24	146.62	0.052
		0.056
26	158.21	0.044
30	93.13	0.089

[a] Recorded on a Bruker WM-500 spectrometer at 125.76 MHz.
[b] ^{13}C-^{1}H Upfield β-shift.

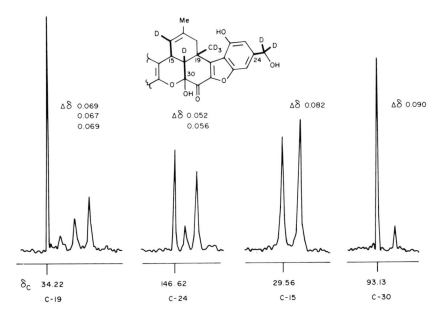

Fig. 8. Part of the broad-band proton-decoupled ^{13}C NMR spectrum of asticolorin C derived from [1-^{13}C, 2-$^2H_3$]acetate showing the isotopically shifted ^{13}C signals. $\Delta\delta$ values are in parts per million.

to three deuterium atoms into the C-35 methyl group, confirms that this carbon atom is part of the "starter" acetate unit of the original polyketide chain of one orcinol unit in asticolorin C. Experimental evidence has shown that the methyl hydrogens of acetyl-SCoA are incorporated into fatty acids in varying degrees; the predominant species (~80%) at the terminal methyl group (i.e., the methyl group of the starter unit) is $^{13}C^2H_3$ (White, 1980; McInnes *et al.*, 1979). Little loss of deuterium from the precursor occurs before the incorporation (McInnes *et al.*, 1979). The deuterium losses observed for the C-35 methyl group in asticolorin C arise as the result of a limited conversion by an acetyl-SCoA $\rightleftarrows$ malonyl-SCoA equilibrium.

C. The Penitrems

The penitrems constitute a group of highly complex substances which induce a neurotoxic syndrome in vertebrates. The neurotoxicosis is characterized by sustained tremors, limb weakness, ataxia, and convulsions. All the fungal isolates involved in the production of penitrem A (**14**) were identified by Pitt (1979) as *Penicillium crustosum*.

Penitrem A was discovered in 1968, but structural studies on the penitrems

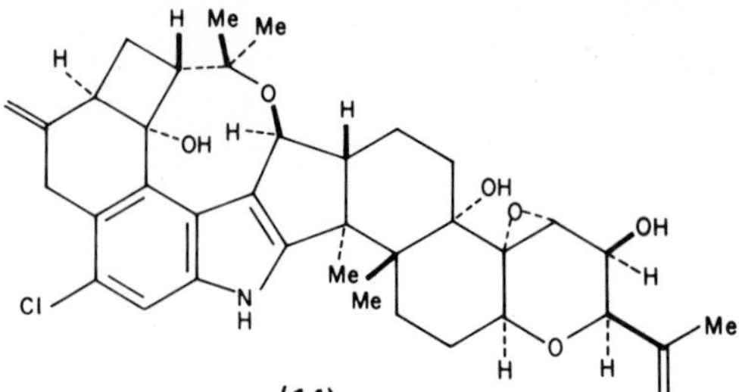

(14)

were initially hampered by the acid sensitivity of the substances as well as the inadequate quantities of material available to investigators. We approached the structural elucidation of penitrem A (14) in a concerted manner by the simultaneous application of physical techniques [very high-field NMR spectroscopy, ultraviolet (UV) spectroscopy, circular dichroism, and mass spectrometry (MS)] and biosynthetic studies (de Jesus et al., 1983a,b,c). The great value of the complementary usage of these techniques cannot be overemphasized. The assigned ^{13}C NMR spectrum of penitrem A (14) and the one-bond (C,H) couplings are shown in Fig. 9.

Absorptions at λ_{max} 295 and 233 nm (ϵ 11,600 and 37,000, respectively) in the UV spectrum of penitrem A (14) indicated the presence of a substituted indole moiety in the molecule. Feeding experiments employing different labeled tryptophans substantiated this supposition. In separate experiments (2S)-[3-^{14}C]-tryptophan and (2RS)-[benzene-ring-U-^{14}C]tryptophan were added to growing cultures of Penicillium crustosum. The ring-labeled precursor gave an eightfold higher incorporation (0.16%) of radioactivity than the side chain-labeled tryptophan. The indole part of tryptophan thus contributes the aromatic part of the

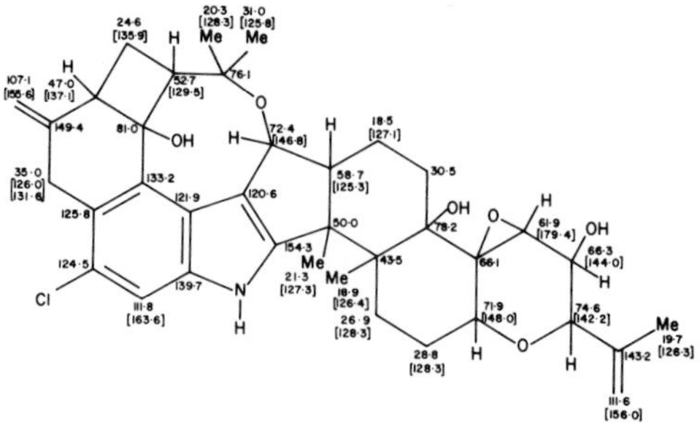

Fig. 9. The ^{13}C NMR assignments for penitrem A. Values in brackets are the one-bond (C,H) coupling constants (in hertz).

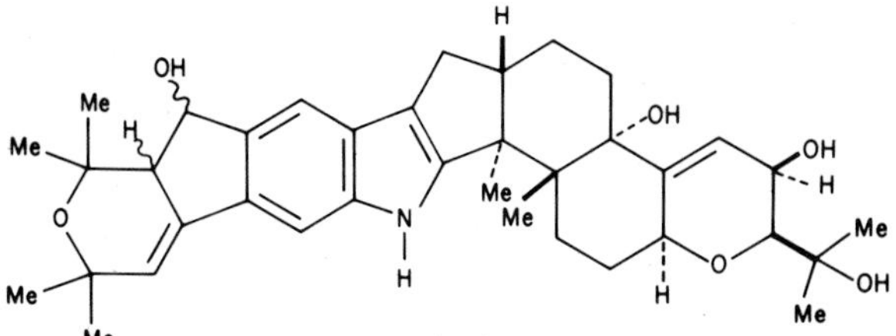

(15)

penitrems. This observation was extended in a separate study by supplementing cultures of the penitrem-producing fungus with (2RS)-[indole-2-^{13}C,2-^{15}N]-tryptophan. The proton-decoupled ^{13}C NMR spectrum of this penitrem A showed enhancement only of the signal at δ 154.36, which was assigned to C-2. The above experiments accounted for 8 of the 37 carbon atoms of penitrem A.

The broad-band proton-decoupled ^{13}C NMR spectrum of [1-^{13}C]acetate-derived penitrem A showed enhancement of the signals of 12 carbon atoms, subsequently assigned to C-11, C-13, C-15, C-16, C-18, C-21, C-23, C-25, C-29, C-31, C-32, and C-37 (see Fig. 10). This result unambiguously pointed to the involvement of six isoprene (C$_5$) units in the bioconstruction of penitrem A. This information is of great structural importance, since it enables the chemist to explore the location of the isoprene units in the molecular structure of the substance. In addition, one-bond (C,C) couplings were observed for some signals. A one-bond (C,C) coupling of 37.1 Hz was observed for the two quaternary carbon atoms which resonate at δ 43.55 and 50.08 (later identified as C-31 and C-32). The two carbon atoms must be contiguous and a 1,2-shift must occur in the course of the biogenesis. Very low-intensity one-bond (C,C) couplings were observed for C-16 and C-34 (39.4 Hz) and C-18 and C-19 (39.7 Hz). These couplings were ascribed to coupling between the ^{13}C-enriched carbon atoms

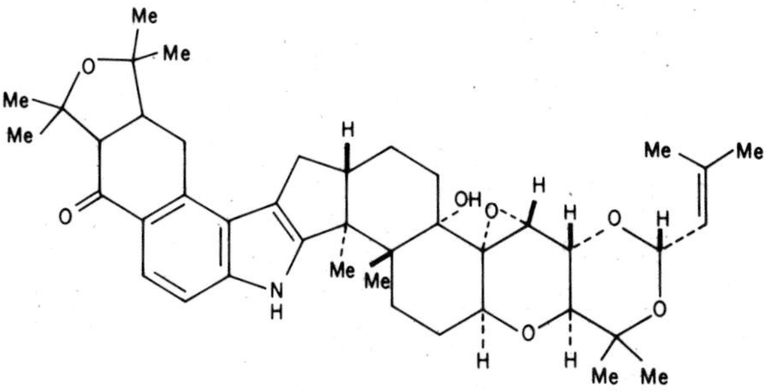

(16)

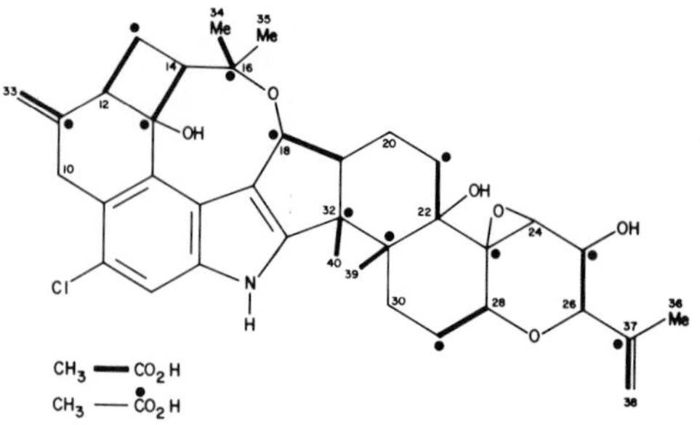

Fig. 10. Labeling pattern of penitrem A derived from [1-^{13}C]acetate and the arrangement of intact acetate units in penitrem A derived from [1,2-^{13}C$_2$]acetate.

(C-16 and C-18) and the adjacent carbon atoms (C-34 and C-19) present at the natural abundance level (1.11%).

The 125.76 MHz broad-band proton-decoupled ^{13}C NMR spectrum of [2-^{13}C]acetate-derived penitrem A showed the enhancement of 17 carbon atoms, namely C-10, C-12, C-14, C-19, C-20, C-22, C-24, C-26, C-28, C-30, C-33, C-34, C-35, C-36, C-38, C-39, and C-40. The bioorigins of all the 37 carbon atoms have therefore been accounted for. Six mevalonate units would contribute 30 carbon atoms, but the observation of only 17 [2-^{13}C]acetate-derived enriched carbon atoms indicates that a single carbon atom derived from C-2 of acetate, and originally present as a methyl group of a mevalonate unit, is lost.

A closer scrutiny of the data derived from the broad-band proton-decoupled ^{13}C NMR spectrum of [2-^{13}C]acetate-derived penitrem A revealed a further wealth of biosynthetic and structural information. Many of the carbon signals in the spectrum exhibited one-bond (C,C) couplings (see Fig. 11). The intensities of the satellite peaks vary from one signal to the next and reflect the different probabilities of the biosynthetic processes. The magnitude of the ^1J(C,C) values showed the presence of 11 intact acetate units with an arrangement similar to that found for [1,2-^{13}C$_2$]acetate-derived penitrem A (see Fig. 10). The formation of [1,2-^{13}C$_2$]-acetate from [2-^{13}C]acetate during the fermentation is the result of the frequent cycling of [2-^{13}C]acetate in the Krebs citric acid cycle. These one-bond (C,C) couplings occur whenever two ^{13}C nuclei are located on adjacent sites, as in [1,2-^{13}C$_2$]acetate-derived metabolites, leading to intraacetate (C,C) coupling, or between the ^{13}C atoms of two different acetate units furnishing an interacetate coupling.

In the broad-band proton-decoupled ^{13}C NMR spectrum of [2-^{13}C]acetate-derived penitrem A, interacetate (C,C) coupling was evident between C-19 and C-20 [^1J(CC) 35.2 Hz] due to a rearrangement involving bond migration. Furthermore, multiple labeling between adjacent labeled acetate units and between adjacent labeled mevalonate units leads to interacetate coupling. These couplings

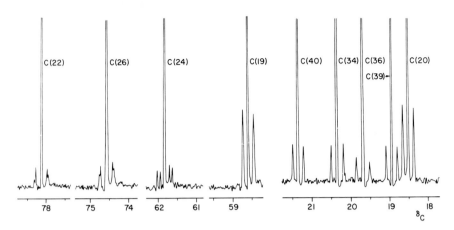

Fig. 11. A part of the broad-band proton-decoupled ^{13}C NMR spectrum of penitrem A derived from [2-^{13}C]acetate.

give valuable structural information and were readily detected in the functionalized rings of penitrem A. In these cases the significant differences in the $^1J(C,C)$ values allowed the resolution of one or more such couplings at a particular site. The ability to observe intermevalonate coupling exemplifies the high resolution obtained by sophisticated high-field NMR spectrometers.

The (C,C) coupling constants measured for [1,2-$^{13}C_2$]acetate-derived penitrem A showed the presence of 11 intact acetate units (Fig. 10). The carbon

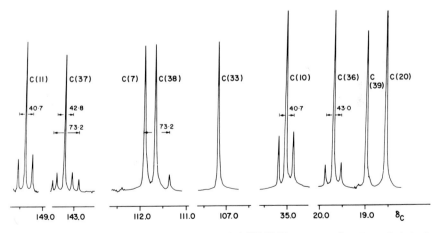

Fig. 12. Part of the broad-band proton-decoupled ^{13}C NMR spectrum of penitrem A derived from [2,3-$^{13}C_2$]mevalonate.

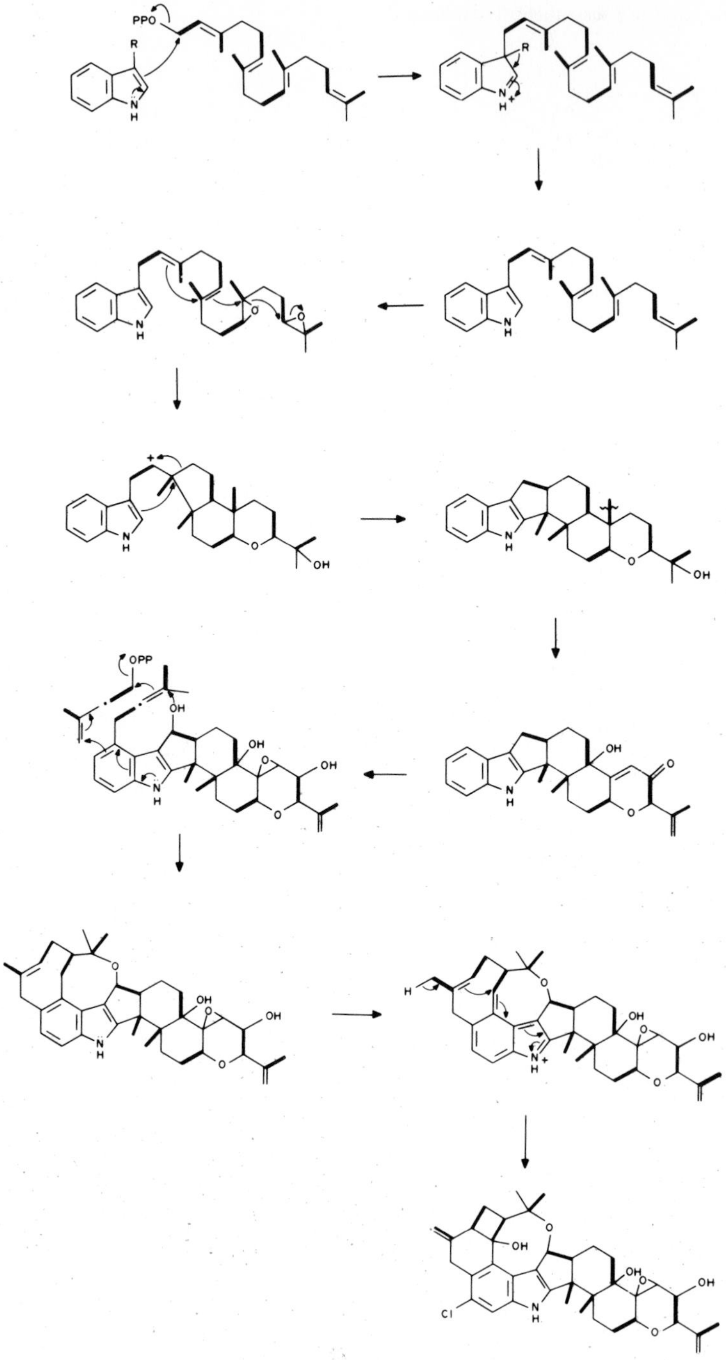

Fig. 13. Proposed biosynthetic pathway for penitrem A.

atoms constituting the isopropylidene moiety exhibited an interesting phenomenon, in that the C-37 resonance exhibited two pairs of satellite peaks of equal intensity due to coupling with C-38 [^1J(CC) 73.2 Hz] and C-36 [^1J(CC) 42.7 Hz]. In the ^{13}C NMR spectrum of penitrem A derived from (3RS)-[2-^{13}C]mevalonolactone, seven carbon signals were enhanced, namely, C-10, C-20, C-24, C-30, C-35, C-36, and C-38. Supplementation of the growing culture with (3RS)-[2,3-^{13}C$_2$]mevalonolactone gave penitrem A containing four intact two-carbon units (see Fig. 12), namely, C-10–C-11, C-16–C-35, C-23–C-24 and C-30–C-31. Again, C-36 (42.7 Hz) and C-38 (73.2 Hz) displayed equal-intensity satellite peaks. This observation evidences for the cleavage of the 2,3-bond of mevalonolactone by a 1,2-bond migration as C-20, which is derived from C-2 of mevalonolactone, shows no one-bond (C,C) coupling. As a result of this rearrangement a coupling is observed between C-19 and C-20 in the spectrum of [2-^{13}C]acetate-derived penitrem A as well as between C-31 and C-32 in that of [1-^{13}C]acetate-derived penitrem A.

A biosynthetic pathway leading to penitrem A could be postulated (see Fig. 13). Subsequent experiments employing deuterium labeling were performed to study the mechanism of the formation of penitrem A.

Our prior knowledge of the biosynthesis and structure of penitrem A (de Jesus *et al.*, 1983a) and penitrems B–F (de Jesus *et al.*, 1983b) enabled us to recognize the involvement of six and seven mevalonate units in the bioformation and structures of the vitally important tremorgenic mycotoxins, the janthitrems including janthitrem E (**15**) (de Jesus *et al.*, 1984) and lolitrem B (**16**) (Gallagher *et al.*, 1984), respectively.

IV. AMINO ACIDS IN THE BIOSYNTHESIS OF MYCOTOXINS

The secondary metabolism of amino acids is notable for the diversity of the pathways that are utilized (Mann, 1978). Many different types of metabolites are produced, particularly in higher plants, and the alkaloids comprise the largest and most exotic class (see Fig. 14). Most amino acid molecules, which are synthesized by all organisms, are utilized in the biosynthesis of essential proteins. It should furthermore be emphasized that the aromatic amino acids, such as phenylalanine and tryptophan, are formally shikimate-derived metabolites.

The great variety of mycotoxins which is generated within a group by the branching of biosynthetic pathways was discussed by Bu'Lock (1980). Figure 15 illustrates the structurally diverse mycotoxins which can be formed from the dimethylallylation of tryptophan and its derivatives.

The amino acids play an important role in the bioformation of many fungal metabolites. They are readily recognized in the dioxopiperazines, formed from the condensation of two amino acids, for example echinulin (Casnati *et al.*,

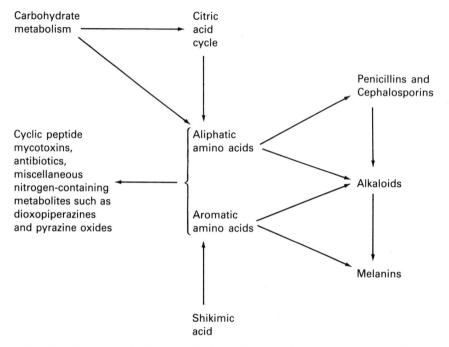

Fig. 14. The central role of amino acids in the bioformation of many secondary metabolites. [Adapted from Mann (1978).]

1962; Birch and Farrer, 1963), mycelianamide (Birch *et al.,* 1956), brevianamides (Birch and Wright, 1970), and austamides (Steyn, 1973).

Birch, who pioneered many of the concepts of secondary fungal metabolism and particularly the polyketide route and the principles of biogenetic architecture, used labeling experiments with (2RS)-[3-^{14}C]tryptophan and (2S)-[4-^{14}C]proline to establish the intermediacy of these amino acids in the formation of the ψ-indoxyl brevianamide A (Birch and Wright, 1970). At this stage the reader is referred to the remark by Birch and Wright (1970): ''We wish to reemphasize the part which biogenetic considerations and experiments played in determining the structures. The physical evidence is too complex to lead directly to complete structures, and the biogenetic basis permits possible structures to be drawn which can be tested out against the available physical evidence.'' This statement is still true, although in the last few years, with the advent of very high-field NMR spectroscopy (500 MHz for ^{1}H) and routine X-ray crystallography, structural elucidation has become much more facile, and more effort has been directed to solving biosynthetic problems. It is of importance to note that many total syntheses of alkaloids were done by so-called biomimetic syntheses before the actual biosynthetic pathways were estab-

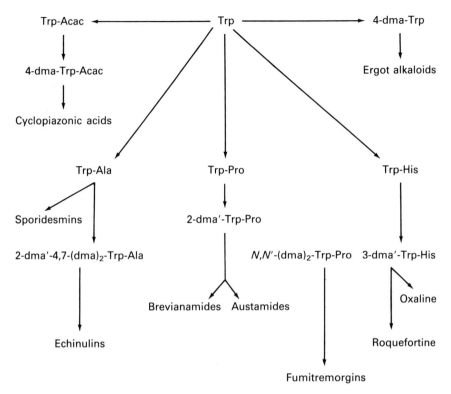

Fig. 15. Branching pathways to mycotoxins involving dimethylallylation of tryptophan and its derivatives. Trp, Tryptophan; Acac, acetoacetyl; Ala, alanine; Pro, proline; His, histidine; dma, 3,3-dimethylallyl; dma′, 1,1-dimethylallyl. [From Bu'Lock (1980).]

lished. The success of these approaches evidenced that biosynthesis might proceed via standard chemical reactions, albeit enzyme catalyzed.

Many mycotoxins, by incorporating within their structures the biogenetic subunits of two (or more) metabolic pathways, have a mixed biosynthetic origin. α-Cyclopiazonic acid (**17**) is such a compound containing tryptophan (amino acid pathway), one C_5 subunit (mevalonate pathway), and two C_2 units (polyketide pathway) (Holzapfel and Wilkins, 1971).

Amino acids play an obvious role in the biosynthesis of the following structurally diverse mycotoxins: ochratoxins (**18**) (phenylalanine) (Steyn et al., 1970); the tetramic acids, tenuazonic acid (**19**) (isoleucine) (Stickings and Townsend, 1961); erythroskyrine (valine) (Shibata et al., 1966) and cyclopiazonic acid (**17**) (tryptophan) (see above); the cytochalasins (**20**) (phenylalanine) (Tamm, 1980); the chaetoglobosins (**21**) (tryptophan) (Sekita et al., 1983); and the xanthocillins (**22**) (tyrosine and the non-amino acid p-hydroxyphenylpyruvic acid) (Achenbach and König, 1972). The structural features of some of these toxins (17–22) which contain one amino acid are shown in Fig. 16.

Very few mycotoxins, with the exception of the cyclic peptides (see later), the pyrazine oxides (see later), and the epipolythiodioxopiperazines contain amino acids as the only biosynthetic building units. The epipolythiodioxopiperazines (Kirby and Robins, 1980) are an important group of mycotoxins, including gliotoxin (23) (phenylalanine and alanine) and sporidesmin (24) (tryptophan and alanine); the latter compound is the causative agent of facial eczema, a photosensitization syndrome among sheep. Kirby and Varley (1974) established the intermediacy of cyclo-(S)-[3-³H]alanyl-(S)-[3-¹⁴C]tryptophanyl in the biosynthesis of sporidesmin. The biogenetic role of the amino acids shown in parentheses has been established by the relevant feeding experiments.

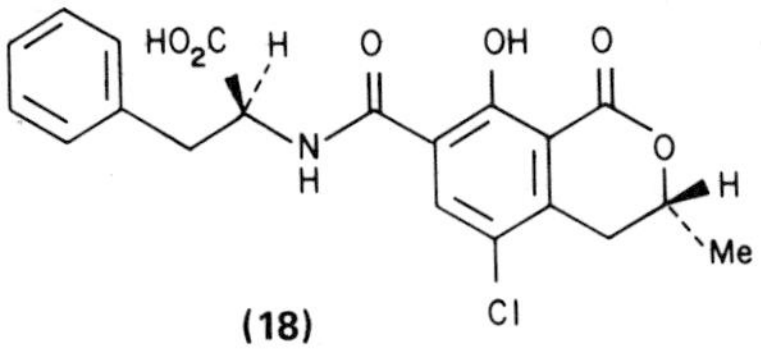

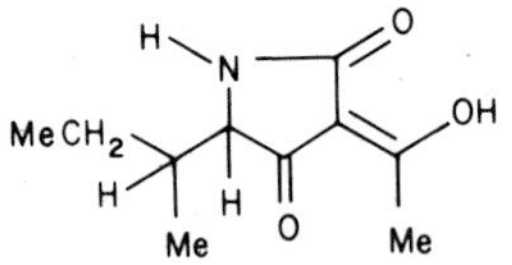

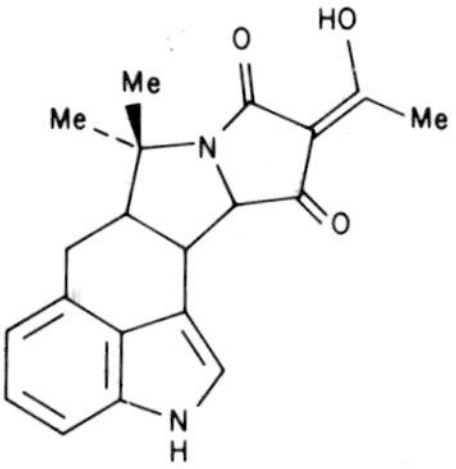

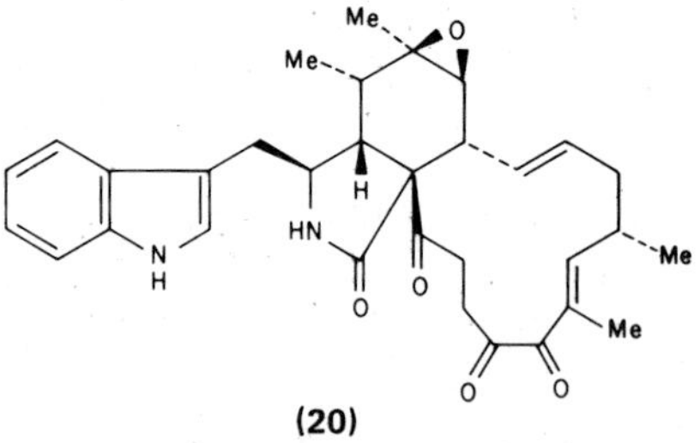

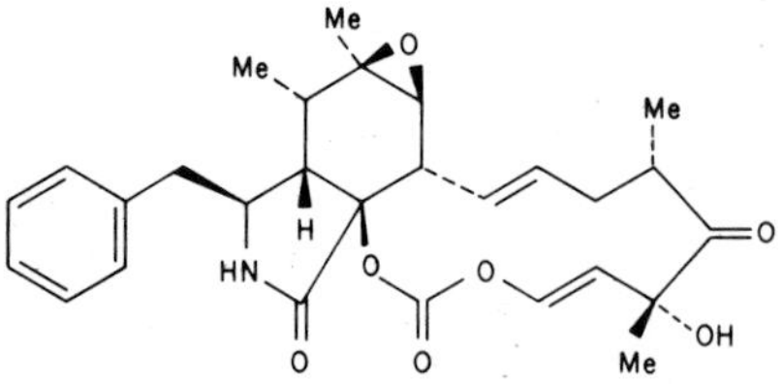

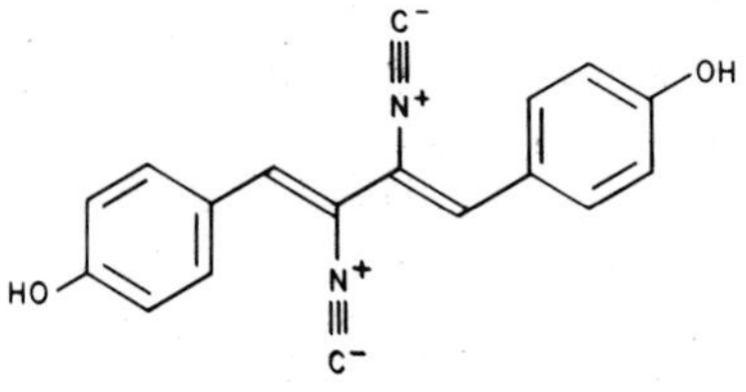

Fig. 16.   Structures of some mycotoxins (17–22) containing one amino acid.

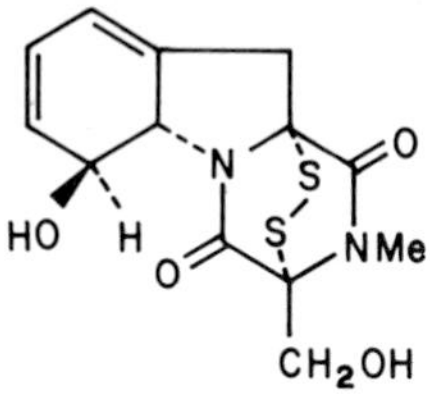

(23) 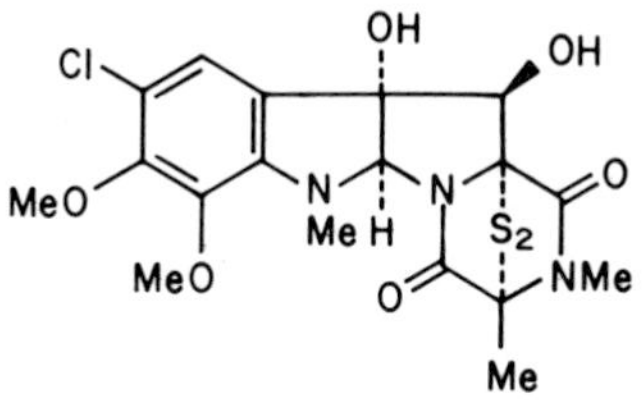   (24)

Noteworthy experiments were recently done on roquefortine (**25**) and its inter-mediacy established in the formation of oxaline (**26**). Roquefortine, a metabolite of *Penicillium roqueforti,* possesses neurotoxic properties and is a natural con-taminant of blue cheese (Scott and Kennedy, 1976). Steyn and Vleggaar (1983) confirmed the bioorigin of roquefortine by incorporation of (2*S*)-[*ring*-2-^{14}C]histidine and (2*S*)-[3-^{14}C]tryptophan into roquefortine by fungal cultures. Both biosynthetic and structural information was obtained by feeding (2*RS*)-[*indole*-2-^{13}C,2-^{15}N]tryptophan to cultures of *Penicillium ox-alicum.* The proton-decoupled ^{13}C NMR spectrum of the enriched oxaline point-ed to the presence of contiguous ^{13}C and ^{15}N atoms, as the resonance at δ_C 101.60 ppm (C-2 of roquefortine) exhibited a one-bond (^{13}C,^{15}N) coupling of 10.1 Hz. This experiment showed that the amino group of tryptophan was retained during the formation of roquefortine. In subsequent experiments, the role of roquefortine as an intermediate in the biosynthesis of the complex al-kaloid oxaline was established by Steyn and Vleggaar (1983), shown in Fig. 17 (**25,26**).

Research on tremorgenic mycotoxins led to the isolation and characterization of a number of nitrogen-containing (amino acid) metabolites (Yamazaki, 1980). An inspection of the structures of the neurotropic mycotoxins, fumitremorgin A (Yamazaki *et al.,* 1980), and of verruculogen (Foyes *et al.,* 1974), indicated the

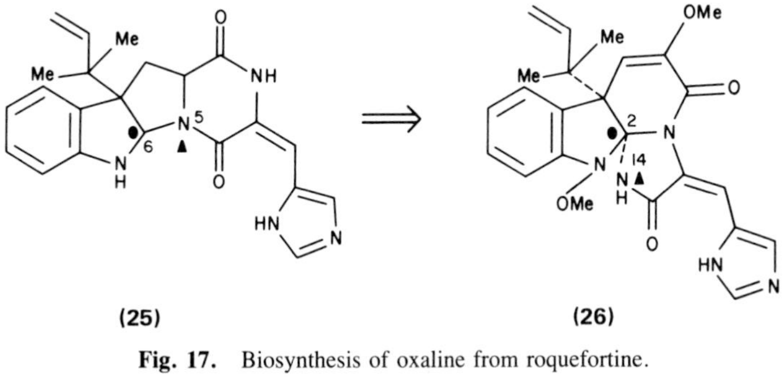

(25)   (26)

Fig. 17.   Biosynthesis of oxaline from roquefortine.

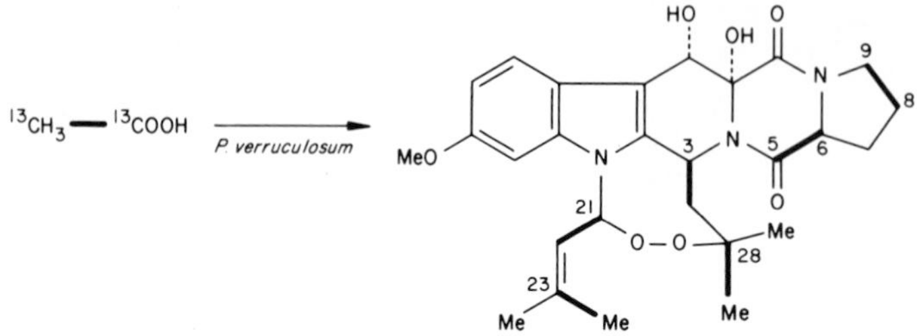

Fig. 18. Incorporation of [1,2-^{13}C$_2$]acetate into verruculogen.

role of mevalonic acid and of the amino acids tryptophan and proline as precursors of these complex dioxopiperazines. This supposition was established by Yamazaki *et al.* (1975). The intermediacy of acetate in the formation of these substances was shown by the addition of sodium [1,2-^{13}C$_2$]acetate to the culture medium of *Penicillium verrucolosum*. The isolated verrucologen contained six intact acetate units, namely, C-5–C-6, C-8–C-9, C-21–C-22, C-23–C-24, C-28–C-29, and C-3–C-31. Two acetate units were incorporated into the proline moiety through de novo synthesis of proline: acetate → TCA cycle → glutamic acid → proline. The other four acetate units were incorporated into the two isoprenyl groups in the verruculogen molecule (Uramoto *et al.*, 1977), as shown in Fig. 18.

The tryptoquivalines are complex tremorgenic metabolites produced by *Aspergillus clavatus* (Clardy *et al.*, 1975) and *Aspergillus fumigatus* (Yamazaki *et al.*, 1976). It is of importance to note that cultures of *A. fumigatus* produced both the fumitremorgins and the tryptoquivalines. The tryptoquivalines are most likely derived from the four amino acids tryptophan, anthranilic acid, valine, and alanine [or methylalanine in the case of tryptoquivaline (**27**)].

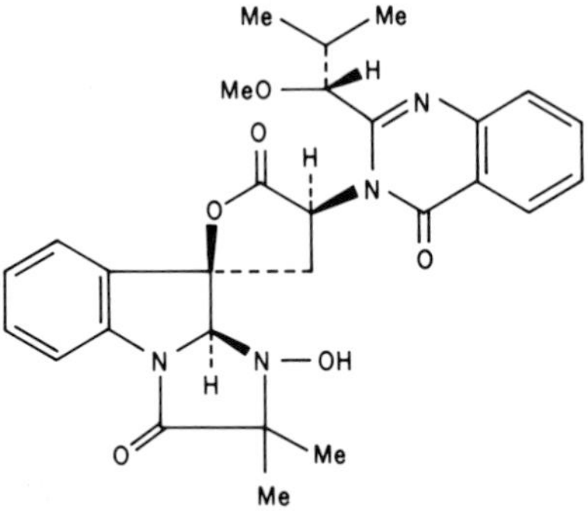

(**27**)

(28) R = Me
(29) R = CH(Me)$_2$

Aspergillic acid and the related pyrazine oxides form an important group of mycotoxins. Inspection of the structures of these substances evidences for the precursor role of amino acids in their bioformation. Labeling experiments showed that aspergillic acid was formed from (2S)-leucine and (2S)-isoleucine (Yamazaki, 1980).

The ergot toxins are certainly the best known fungal metabolites which affect the central nervous system of various species, and have also been associated with one of the most severe human mycotoxicoses, namely ergotism. The biosynthesis of these substances has been studied in great detail, particularly that of (R)-lysergic acid (Floss and Anderson, 1980). The structural features of the cyclol-type peptide ergot alkaloids indicates the involvement of the amino acids tryptophan, alanine, proline, and phenylalanine in the formation of ergotamine (**28**), whereas ergochristine (**29**) is formed from tryptophan, valine, proline, and phenylalanine.

The biologically active substance tentoxin, a metabolite of *Alternaria tenuis*, is one of the best-studied cyclic peptides (Meyer *et al.*, 1975). Tentoxin is unique in containing dehydrophenylalanine as a constituent amino acid.

Steyn *et al.* (1983) recently isolated a novel cyclic heptapeptide rhizonin (**30**) from cultures of *Rhizopus microsporus*. Problems are still encountered in produc-

(30)

ing rhizonin on a chemically defined medium, and no investigation on its bioorigin has therefore been undertaken. Its structure is based on extensive NMR and mass spectral data and single-crystal X-ray crystallography. Rhizonin contains *allo*-isoleucine, leucine, *N*-methylalanine, valine, *N*-methyl-3-(3-furyl)alanine (1 : 1 : 1 : 2 : 2).

Phomopsin A (**31**), a mycotoxin produced by *Phomopsis leptostromiformis*, is the causative agent of lupinosis, a severe liver disease of sheep and cattle grazing lupins (Frahn *et al.*, 1983). The structure of phomopsin A was recently established by application of chemical degradation and derivatization studies, using very high-field ^{1}H and ^{13}C NMR studies and biosynthetic experiments (Culvenor *et al.*, 1983).

Phomopsin A (**31**), $C_{36}H_{43}ClN_6O_{11}$, is a substance with a high nitrogen content, and among fungal secondary metabolites this could indicate the presence of amino acid residues, albeit in modified form. This supposition was supported by the presence of seven peaks in the carbonyl region (δ 163.96–168.72 ppm) and four methine carbon signals in the δ 67–56 ppm region. Subsequent studies showed that phomopsin A contained a dehydroaspartyl moiety and that the signal at δ 167.28 was attributable to its free carboxyl carbon atom. Phomopsin A therefore contains four "normal" amino acids and two dehydro amino acids. Supplementation of cultures of the phomopsin-producing fungus with (2*S*)-[U-^{14}C]valine, (2*S*)-[U-^{14}C]isoleucine, (2*S*)-[U-^{14}C]phenylalanine and (2*S*)-[U-^{14}C]proline led to the isolation of phomopsin A containing reasonable levels of radioactivity (Payne, 1983).

In a highly significant experiment, stable isotopes in the form of (2*S*)-[3-^{13}C]phenylalanine were administered to cultures of *Phomopsis leptostromiformis*. The site of enrichment in the isolated phomopsin A was determined by ^{13}C NMR spectroscopy. Only the oxygen-bearing benzylic carbon

(31)

atom (δ 69.56) was enriched. This single experiment unambiguously established the presence of a phenylalanyl moiety in the structure of phomopsin A.

An acid hydrolysate of phomopsin A gave glycine (0.63), sarcosine (0.11), 3,4-dehydrovaline (0.03), valine (0.18), two β,γ-dehydroisoleucines (0.20 and 0.44), and 3,4-dehydroproline (1.00). However, prior catalytic reduction (PtO_2-H_2) of phomopsin A, followed by an acid hydrolysis, led to the formation of valine, isoleucine, proline, aspartic acid (1 : 1 : 1 : 1) and the β,γ-dehydroisoleucines, as observed earlier. The data on the constituent amino acids were of vital importance in the eventual construction of the phomopsin molecule. The amino acid sequence was unambiguously established by extensive ^{13}C-[^1H] selective population experiments (Culvenor et al., 1983).

From the foregoing it must be evident that the principles of biosynthetic architecture and labeling experiments using radioactive and stable isotopes can find extensive application in the structural elucidation of fungal metabolites.

REFERENCES

Abell, C., and Staunton, J. (1981). The use of ^2H n.m.r. spectroscopy and β-isotopic shifts in the ^{13}C n.m.r. spectrum to measure deuterium retention in the biosynthesis of the polyketide 6-methylsalicylic acid. *J.C.S. Chem. Commun.* p. 856.

Achenbach, H., and König, F. (1972). Die Frage der biogenetischen Gleichwertigkeit der beiden Xanthocillin-Hälften. *Chem. Ber.* **105**, 784.

Biollaz, M., Büchi, G., and Milne, G. (1970). The biosynthesis of the aflatoxins. *J. Am. Chem. Soc.* **92**, 1035.

Birch, A. J., and Farrer, K. R. (1963). Studies in relation to biosynthesis. Part XXXIII. Incorporation of tryptophan into echinulin. *J. Chem. Soc.* p. 4277.

Birch, A.J., and Wright, J. J. (1970). Studies in relation to biosynthesis. XLII. The structural elucidation and some aspects of the biosynthesis of the brevianamides-A and -E. *Tetrahedron* **26**, 2329.

Birch, A. J., Massey-Westropp, R. A., and Rickards, R. W. (1956). Studies in relation to biosynthesis. Part VIII. The structure of mycelianamide. *J. Chem. Soc.* p. 3717.

Bu'Lock, J. D. (1980). Mycotoxins as secondary metabolites. *In* "The Biosynthesis of Mycotoxins" (P. S. Steyn, ed.), p. 1. Academic Press, New York.

Casnati, G., Cavalleri, R., Piozzi, F., and Quilico, A. (1962). Echinulina.—Nota XI. (XVII di rierche chimiche nel gruppo dell'*Aspergillus glaucus*.) *Gazz. Chim. Ital.* **92**, 105–128.

Clardy, J., Springer, J. P., Büchi, G., Matsuo, K., and Wightman, P. (1975). Tryptoquivaline and tryptoquivalone, two tremorgenic metabolites of *Aspergillus clavatus*. *J. Am. Chem. Soc.* **97**, 663.

Culvenor, C. C. J., Cockrum, P. A., Edgar, J. A., Frahn, J. L., Gorst-Allman, C. P., Jones, J. A., Marasas, W. F. O., Murray, K. E., Smith, L. W., Steyn, P. S., Vleggaar, R., and Wessels, P. L. (1983). Structure elucidation of phomopsin A, a novel cyclic hexapeptide mycotoxin produced by *Phomopsis leptostromiformis*. *J.C.S. Chem. Commun.* p. 1259.

de Jesus, A. E., Gorst-Allman, C. P., Steyn, P. S., Vleggaar, R., Wessels, P. L., Wan, C. C., and Hsieh, D. P. H. (1980). The conversion of averufin derived from [1,2-^{13}C]acetate into aflatoxin B_1. *J.C.S. Chem. Commun.* p. 389.

de Jesus, A. E., Steyn, P. S., van Heerden, F. R., Vleggaar, R., Wessels, P. L., and Hull, W. E.

(1983a). Tremorgenic mycotoxins from *Penicillium crustosum*. Isolation of penitrems A-F and the structure elucidation and absolute configuration of penitrem A. *J.C.S. Perkin I* p. 1847.

de Jesus, A. E., Steyn, P. S., van Heerden, F. R., Vleggaar, R., Wessels, P. L., and Hull, W. E. (1983b). Tremorgenic mycotoxins from *Penicillium crustosum*. Structure elucidation of penitrems B-F. *J.C.S. Perkin I* p. 1857.

de Jesus, A. E., Gorst-Allman, C. P., Steyn, P. S., van Heerden, F. R., Vleggaar, R., Wessels, P. L., and Hull, W. E. (1983c). Tremorgenic mycotoxins from *Penicillium crustosum*. Biosynthesis of penitrem A. *J.C.S. Perkin I* p. 1863.

de Jesus, A. E., Steyn, P. S., van Heerden, F. R., and Vleggaar, R. (1984). Structure elucidation of the janthitrems, novel tremorgenic mycotoxins from *Penicillium janthinellum*. *J.C.S. Perkin I* p. 697.

Floss, H. G., and Anderson, J. A. (1980). Biosynthesis of ergot toxins. *In* "The Biosynthesis of Mycotoxins" (P. S. Steyn, ed.), p. 17. Academic Press, New York.

Foyes, J., Lokensgard, D., Clardy, J., Cole, R. J., and Kirksey, J. W. (1974). Structure of verruculogen, a tremor producing peroxide from *Penicillium verruculosum*. *J. Am. Chem. Soc.* **96,** 6785.

Frahn, J. L., Jago, M. V., Culvenor, C. C. J., Edgar, J. A., and Jones, A. J. (1983). Chemical and biological properties of phomopsin, a toxic metabolite of *Phomopsis leptostromiformis*. *Toxicon* **21,** Suppl. 3, 149.

Gallagher, R. T., Hawkes, A. D., Steyn, P. S., and Vleggaar, R. (1984). Tremorgenic neurotoxins from perennial ryegrass causing ryegrass staggers disorder of livestock: structure elucidation of lolitrem B. *J.C.S. Chem. Commun.* p. 614.

Garson, M. J., and Staunton, J. (1979). Some new n.m.r. methods for tracing the fate of hydrogen in biosynthesis. *Chem. Soc. Rev.* **8,** 539.

Gorst-Allman, C. P., Pachler, K. G. R., Steyn, P. S., Wessels, P. L., and Scott, De B. (1977). Carbon-13 nuclear magnetic resonance assignments of some fungal C_{20}-anthraquinones; their biosynthesis in relation to that of aflatoxin B_1. *J.C.S. Perkin I* p. 2181.

Gorst-Allman, C. P., Steyn, P. S., Wessels, P. L., and Scott, De B. (1978). Carbon-13 nuclear magnetic resonance assignments and biosynthesis of versicolorin A in *Aspergillus parasiticus*. *J.C.S. Perkin I* p. 961.

Gorst-Allman, C. P., Steyn, P. S., and Vleggaar, R. (1983). Biosynthesis of diplosporin by *Diplodia macrospora*. Part 2. Investigation of ring formation using stable isotopes. *J.C.S. Perkin I* p. 1357.

Haslam, E. (1974). "The Shikimate Pathway." Wiley, New York.

Herbert, R. B. (1981). "The Biosynthesis of Secondary Metabolites." Chapman & Hall, London.

Holzapfel, C. W., and Wilkins, D. C. (1971). On the biosynthesis of cyclopiazonic acid. *Phytochemistry* **10,** 351.

Horak, R. M., Steyn, P. S., and Vleggaar, R. (1983). Biosynthesis of austocystin D in *Aspergillus ustus*. A carbon-13 n.m.r. study. *J.C.S. Perkin I* p. 1745.

Hsieh, D. P. H., Seiber, J. N., Reece, C. A., Fitzell, D. L., Yang, S. L., Dalezios, J. I., La Mar, G. N., Budd, D. L., and Motell, E. (1975). ^{13}C Nuclear magnetic resonance spectra of aflatoxin B_1 derived from acetate. *Tetrahedron* **31,** 661.

Kirby, G. W., and Robins, D. J. (1980). The biosynthesis of gliotoxin and related epipolythiodioxopiperazines. *In* "The Biosynthesis of Mycotoxins" (P. S. Steyn, ed.), p. 301. Academic Press, New York.

Kirby, G. W., and Varley, M. J. (1974). Synthesis of tryptophan stereoselectively labelled with tritium and deuterium in the β-methylene groups; the steric course of hydroxylation in sporidesmin biosynthesis. *J.C.S. Chem. Commun.* p. 883.

McInnes, A. G., Walter, J. A., and Wright, J. L. C. (1979). Differential hydrogen exchange during the biosynthesis of fatty acids in *Anacystis nidulans:* the incorporation of $[2,2,2-^2H_3, 2-^{13}C_{0,1}]$acetate. *Tetrahedron Lett.* p. 3245.

Mann, J. (1978). "Secondary Metabolism," pp. 171–235. Oxford Univ. Press, London and New York.

Meyer, W. L., Templeton, G. E., Grable, C. I., Jones, R., Kuyper, L. F., Lewis, R. B., Sigel, C. W., and Woodhead, S. H. (1975). Use of ¹H nuclear magnetic resonance spectroscopy for sequence and configuration analysis of cyclic tetrapeptides. The structure of tentoxin. *J. Am. Chem. Soc.* **97**, 3802.

Pachler, K. G. R., and Wessels, P. L. (1973). Selective population inversion (SPI). A pulsed double resonance method in FT NMR spectroscopy equivalent to INDOR. *J. Magn. Reson.* **12**, 337.

Pachler, K. G. R., and Wessels, P. L. (1977). Sensitivity gain in a progressive-saturation selective population inversion NMR experiment. *J. Magn. Reson.* **28**, 53.

Pachler, K. G. R., Steyn, P. S., Vleggaar, R., Wessels, P. L., and Scott, De B, (1976). Carbon-13 nuclear magnetic resonance assignments and biosynthesis of aflatoxin B₁ and sterigmatocystin. *J.C.S. Perkin I* p. 1182.

Packter, N. R. (1973). "Biosynthesis of Acetate-Derived Compounds." Wiley, New York.

Payne, A. L. (1983). Biosynthesis of radiolabelled phomopsin by *Phomopsis leptostromiformis*. *Appl. Environ. Microbiol.* **45**, 389.

Pita Boente, M. J. (1981). Ph.D. Thesis, Univ. of Glasgow.

Pitt, J. I. (1979). *Penicillium crustosum* and *P. simplicissimum* the correct names for two common species producing tremorgenic mycotoxins. *Mycologia* **71**, 1166.

Rabie, C. J., Simpson, T. J., Steyn, P. S., van Rooyen, P. H., and Vleggaar, R. (1984). Structure and absolute configuration of the asticolorins, toxic metabolites from *Aspergillus multicolor*. *J.C.S. Chem. Commun.* p. 764.

Scott, P. M., and Kennedy, B. P. C. (1976). Analysis of blue cheese for roquefortine and other alkaloids from *Penicillium roqueforti*. *J. Agric. Food Chem.* **24**, 865.

Sekita, S., Yoshihira, K., and Natori, S. (1983). Chaetoglobosins, cytotoxic 10-(indol-3-yl)-[13]cytochalasans from *Chaetomium* spp. IV. C-nuclear magnetic resonance spectra and their application to a biosynthetic study. *Chem. Pharm. Bull.* **31**, 490–498.

Shibata, S., Sankawa, U., Taguchi, H., and Yamazaki, K. (1966). Biosynthesis of natural products. III. Biosynthesis of erythroskyrine, a coloring matter of *Penicillium islandicum* Sopp. *Chem. Pharm. Bull.* **14**, 474–478.

Simpson, T. J., and Stenzel, D. J. (1982). Application of ²H β-isotopic shifts in ¹³n.m.r. spectra to biosynthetic studies. Incorporation of [1-¹³C, 2-²H₃]acetate into *O*-methylasparvenone in *Aspergillus parvulus*. *J.C.S. Chem. Commun.* p. 1074.

Simpson, T. J., de Jesus, A. E., Steyn, P. S., and Vleggaar, R. (1982a). Biosynthesis of aflatoxins. Incorporation of [4′-²H₂]averufin into aflatoxin B₁ by *Aspergillus flavus*. *J.C.S. Chem. Commun.* p. 631.

Simpson, T. J., de Jesus, A. E., Steyn, P. S., and Vleggaar, R. (1982b). Biosynthesis of aflatoxins. Incorporation of [2-²H₃]acetate and [1-¹³C, 2-²H₃]acetate into averufin. *J.C.S. Chem. Commun.* p. 632.

Steyn, P. S. (1973). The structures of five diketopiperazines from *Aspergillus ustus*. *Tetrahedron* **29**, 107.

Steyn, P. S. (1980). "The Biosynthesis of Mycotoxins." Academic Press, New York.

Steyn, P. S., and Vleggaar, R. (1983). Roquefortine, an intermediate in the biosynthesis of oxaline in cultures of *Penicillium oxalicum*. *J.C.S. Chem. Commun.* p. 560.

Steyn, P. S., Holzapfel, C. W., and Ferreira, N. P. (1970). The biosynthesis of the ochratoxins, metabolites of *Aspergillus ochraceus*. *Phytochemistry* **9**, 1977.

Steyn, P. S., Vleggaar, R., Wessels, P. L., and Scott, De B. (1979). Biosynthesis of versiconal acetate, versiconol acetate and versiconol, metabolites from a blocked mutant of *Aspergillus parasiticus*. The role of versiconal acetate in aflatoxin biosynthesis. *J.C.S. Perkin I* p. 460.

Steyn, P. S., Vleggaar, R., and Wessels, P. L. (1980). The biosynthesis of aflatoxin and its congeners. *In* "The Biosynthesis of Mycotoxins" (P. S. Steyn, ed.), p. 105. Academic Press, New York.

Steyn, P. S., Vleggaar, R., and Wessels, P. L. (1981). A ^{13}C n.m.r. study of the biosynthesis of norsolorinic acid. *S. Afr. J. Chem.* **34,** 12.

Steyn, P. S., Tuinman, A. A., van Heerden, F. R., van Rooyen, P. H., Wessels, P. L., and Rabie, C. J. (1983). The isolation, structure, and absolute configuration of the mycotoxin, rhizonin A, a novel cyclic heptapeptide containing *N*-methyl-3-(3-furyl)alanine, produced by *Rhizopus microsporus. J.C.S. Chem. Commun.* p. 47.

Steyn, P. S., Vleggaar, R., and Simpson, T. J. (1984). Stable isotope labelling studies on the biosynthesis of asticolorin C by *Aspergillus multicolor.* Evidence for a symmetrical intermediate. *J.C.S. Chem. Commun.* p. 765.

Stickings, C. E., and Townsend, R. J. (1961). Studies in the biochemistry of micro-organisms. 108. Metabolites of *Alternaria tenuis* auct.: the biosynthesis of tenuazonic acid. *Biochem. J.* **78,** 412.

Tamm, C. (1980). The biosynthesis of the cytochalasans. *In* "The Biosynthesis of Mycotoxins" (P. S. Steyn, ed.), p. 269. Academic Press, New York.

Townsend, C. A., and Christensen, S. B. (1983). Stable isotope studies of anthraquinone intermediates in the aflatoxin pathway. *Tetrahedron* **39,** 3575.

Townsend, C. A., and Davis, S. G. (1983). The regiochemistry of A-ring-labelled averufin incorporation into aflatoxin B_1. *J.C.S. Chem. Commun.* p. 1420.

Uramoto, M., Cary, L., Tanabe, M., Hirotau, K., and Clardy, J. (1977). *Proc. Annu. Meet., Kanto Branch, Agric. Chem. Soc. Tokyo.*

White, R. H. (1980). Stoichiometry and stereochemistry of deuterium incorporated into fatty acids by cells of *Escherichia coli* grown on [*methyl*-^{2}H$_3$]acetate. *Biochemistry* **19,** 9.

Wirthlin, T. (1982). Assignment techniques in ^{13}C n.m.r. Classical and new methods. *Varian FT-NMR Semin., 2nd,* Pretoria, South Africa.

Yamazaki, M. (1980). The biosynthesis of neurotropic mycotoxins. *In* "The Biosynthesis of Mycotoxins" (P. S. Steyn, ed.), p. 193. Academic Press, New York.

Yamazaki, M., Fujimoto, H., Kawasaki, T., Okuyama, E., and Kuga, T. (1975). *Proc. Symp. Chem. Nat. Prod., 19th, Hiroshima.*

Yamazaki, M., Fujimoto, H., and Okuyama, E. (1976). Structure determination of six tryptoquivaline-related metabolites from *Aspergillus fumigatus. Tetrahedron Lett.* p. 2861.

Yamazaki, M., Fujimoto, H., and Kawasaki, T. (1980). Chemistry of tremorgenic metabolites. I. Fumitremorgen A from *Aspergillus fumigatus. Chem. Pharm. Bull.* **28,** 245.

Zamir, L. O., and Hufford, K. P. (1981). Precursor recognition by kinetic pulse-labeling in a toxigenic aflatoxin B_1-producing strain of *Aspergillus. Appl. Environ. Microbiol.* **42,** 168.

8

Immunoassays for Mycotoxins

F. S. CHU

Department of Food Microbiology and Toxicology
University of Wisconsin-Madison
Madison, Wisconsin 53706

I. INTRODUCTION

The presence of mycotoxins in foods and feeds has been considered to be a potential hazard to human and animal health (Cole and Cox, 1981; Mirocha *et al.*, 1979; Rodricks *et al.*, 1977; Stoloff, 1980). Since the mycotoxin problem is difficult to avoid, the most effective measure for their control depends on a

207

MODERN METHODS IN THE ANALYSIS
AND STRUCTURAL ELUCIDATION
OF MYCOTOXINS

rigorous program to monitor their presence in foods and feeds. Consequently, sensitive and accurate methods for analysis of mycotoxins in foods are essential for decreasing the risk of human exposure to mycotoxins (Mirocha *et al.*, 1979; Rodricks *et al.*, 1977; Stoloff, 1980). The importance of monitoring mycotoxins in foods and feeds has been treated by Cole elsewhere in this book and will not be discussed here.

Since the discovery of the aflatoxins (Afla) as potent carcinogens, extensive investigations have been made to develop new analytical methods for mycotoxins. However, three main problems are generally associated with mycotoxin analysis:

1. Due to the diversity in chemical structures and thus their chemical and physical properties, individual methods must be developed for each group of mycotoxins. Some mycotoxins, including aflatoxins, ochratoxins, and zearalenone, have defined spectral and fluorescence properties, which permit direct chemical analysis after separation of different structurally related compounds and other contaminants. Other mycotoxins, including the trichothecenes and rubratoxins, do not have well-defined absorption maxima; hence, alternate methods must be employed. Methods used to analyze these compounds generally have low sensitivity and need expensive instrumentation.

2. Apart from the detection problem, the separation of structurally related analogs within a group of mycotoxins is also a difficult task. Because only trace amounts of toxin are present in a very complex matrix such as foods or feeds, it is also necessary to remove large amounts of impurities before any analysis can be done. Hence, extensive cleanup is necessary for most analyses, making each analysis very time-consuming. In addition, since the types of contaminants may vary considerably for different samples, a different cleanup method may be needed for each commodity (Horwitz, 1980).

3. Mycotoxins are unevenly distributed in agricultural commodities. To decrease statistical variance, it is necessary to analyze a large number of samples for a specific lot of commodity (Davis *et al.*, 1980).

Therefore, in attempting to develop new methods for mycotoxin analyses, efforts should be made to overcome all three major problems. A good method should be very *specific, sensitive,* and relatively *simple* to operate. Preferably, it should be adaptable to automation. Recent investigations on analytical methods for the mycotoxins have led to several new techniques containing many of these features.

One method which has gained great popularity for analysis of mycotoxins as well as other small molecular weight compounds is high-performance liquid chromatography (HPLC). As discussed elsewhere in this volume (Chapter 11), HPLC methods combined with fluorescence and ultraviolet (UV) detection systems have shown good sensitivity and excellent reproducibility. However, there are still some shortcomings: (a) because of the high sensitivity, samples must first be subjected to an extensive cleanup treatment before separation by HPLC; (b) since only one sample can be analyzed at a time instrumentally, even if an automatic injection system is available, a large number of samples cannot be analyzed in a short time; and (c) instrument maintenance and solvents are expensive, and specially trained personnel are needed for such analyses.

Recent investigations in our laboratory as well as others have led to the development of highly sensitive, specific, and simple methods for detecting mycotoxins (Chu, 1983, 1984a,b). These techniques are immunoassays which involve the interaction of mycotoxins with specific antibodies against the homologous mycotoxins. Two assay systems have been developed: a radioimmunoassay (RIA) and an enzyme-linked immunosorbent assay (ELISA). Details on the methods for production of specific antibodies against mycotoxins and the application of these two techniques are discussed in this chapter.

II. PRODUCTION OF ANTIBODY AGAINST MYCOTOXINS

Unlike many bacterial toxins, which are highly immunogenic, mycotoxins are secondary fungal metabolites of low molecular weight. Hence, they are not immunogenic and must be conjugated to a protein or polypeptide carrier before immunization. Conjugation of mycotoxin to a protein carrier is complicated by the functional groups present in the molecule. Whereas some mycotoxins, including ochratoxins, patulin, and penicillic acid, have a reactive group that can be used for direct-coupling reaction, most mycotoxins, including aflatoxins and the trichothecenes, lack a reactive group. In the latter case, a reactive carboxyl or other reactive group must be first introduced into the toxin molecule. During the last few years, methods for preparation of mycotoxin conjugates have been developed, and specific antibodies against aflatoxins (Afla) B_1 (Biermann and Terplan, 1980; Chu and Ueno, 1977; Langone and Van Vunakis, 1976; Sizaret *et al.*, 1980), B_{2a} (Gaur *et al.*, 1981; Lau *et al.*, 1981b; Lawellin *et al.*, 1977b), M_1 (Harder and Chu, 1979), G_{2a} (Chu *et al.*, 1985), Q_1 (Fan *et al.*, 1984a), B_1 diol (Pestka and Chu, 1984a) and B_1-DNA adduct (Groopman *et al.*, 1982; Haugen *et al.*, 1981; Hertzog *et al.*, 1982), diacetoxyscirpenol (DAS) (Chu *et al.*, 1984a), deoxyverrucarol (DOVE) (Chu *et al.*, 1984b), kojic acid (Abdalla and Grant, 1981), ochratoxin A (OA) (Aalund *et al.*, 1975; Chu *et al.*, 1976; Morgan *et al.*, 1982, 1983; Worsaae, 1978), rubratoxin B (Davis and Stone, 1979), T-2 toxin (Chu *et al.*, 1979), sterigmatocystin (Li and Chu, 1984), and zearalenone (Thouvenot and Morfin, 1983) were produced. The approaches that have been used for coupling mycotoxins to a protein carrier for antibody production are summarized in Table I. Details of these approaches are discussed in the following sections.

TABLE I

**Methods Used for Conjugation of
Mycotoxin to Proteins**[a]

Mycotoxins	Derivatives prepared	Methods of conjugation
Afla M_1, B_1	CMO	WSC, MA
Afla B_1	Dichloride	Direct
Afla B_{2a}, B_1 diol	—	RA, TAB
Afla B_{2a}	HG	WSC, MA
Afla G_{2a}	—	RA
Afla Q_1	HS	WSC, MA
Afla Q_{2a}	—	RA
DAS	HG	WSC, MA
DOVE	HS	WSC, MA
Kojic acid	—	TAB
Ochratoxin A	—	WSC, MA
Rubratoxin B	—	WSC
Sterigmatocystin	HA	WSC
T-2 Toxin	HS, HG	WSC, MA
Zearalenone	CMO	WSC, MA

[a] CMO, Carboxymethyloxime; DAS diacetoxy-scirpenol; DOVE, deoxyverrucarol; HA, hemi-acetal; HG, hemiglutarate; HS, hemisuccinate; MA, mixed-anhydride method; RA, reductive alkylation method; TAB, tetrazobenzidine; WSC, water-soluble carbodiimide method.

A. Preparation of Mycotoxin Derivatives

1. Carboxymethyloxime of Mycotoxins

Mycotoxins which contain a carbonyl group are generally conjugated to the protein after the formation of a carboxymethyloxime (CMO). O-Carboxymethyloxime of Afla B_1 (Chu *et al.*, 1977; Langone and Van Vunakis, 1976), M_1 (Harder and Chu, 1979), and zearalenone (Thouvenot and Morfin, 1983), and an Afla analog, 5,7-dimethoxycyclopentenone[2,3-*c*]coumarin (Langone and Van Vunakis, 1976), have been prepared. This approach is very simple and the reaction is almost quantitative. In a typical experiment, CMO of Afla B_1 was obtained after refluxing Afla B_1 with ethylhydroxylaminohemi-HCl for 4 hr in the presence of pyridine (Chu *et al.*, 1977). Other bases such as NaOH have also been used (Langone and Van Vunakis, 1976).

2. Hemisuccinate or Hemiglutarates of Mycotoxins

To introduce a carboxylic acid group into mycotoxins containing one or more reactive hydroxyl groups, the bifunctional acid anhydrides are generally used. Acylation reagents such as succinic and glutaric anhydrides are commonly used in conjunction with a catalyst such as pyridine or 4,N,N-dimethylaminopyridine. Hemisuccinate (HS) and hemiglutarate (HG) of Afla B_{2a} (Lau et al., 1981b) and Q_1 (Lau et al., 1981a), T-2 toxin (Chu et al., 1979), DAS (Chu et al., 1984a), and DOVE (Chu et al., 1984b) have been prepared. Useful antibodies were obtained after immunizing rabbits with proteins conjugated with these hemiacylated derivatives. However, because the stability of these acylated derivatives varies considerably, the rate of spontaneous hydrolysis of these two derivatives in aqueous solution should be tested. For example, the hemisuccinate of Afla B_{2a} hydrolyzed rapidly in neutral solution, but the hemiglutarate of B_{2a} was very stable, so the HG was selected for subsequent conjugation (Lau et al., 1981b). In addition, because the chain length (i.e., the spacer chain between protein and mycotoxin) may affect the orientation of the hapten in the protein molecule and thus affect the immunogenicity of the conjugate, the efficacy of HS-mycotoxin-protein or HG-mycotoxin-protein conjugate for antibody production should be experimentally studied.

B. Conjugation of Mycotoxins to Proteins

1. Water-Soluble Carbodiimide and Mixed-Anhydride Methods

Generally, a water-soluble carbodiimide (WSC) or mixed-anhydride method (MA) is used to conjugate a mycotoxin to a protein carrier. In the WSC method, the mycotoxin or its derivatives are coupled to the protein in the presence of a WSC such as 1-ethyl-3,3-dimethylaminopropylcarbodiimide (Chu and Ueno, 1977; Chu et al., 1976; Langone and Van Vunakis, 1976), whereas in the MA method the mycotoxins were generally activated to their corresponding anhydride by isobutyl chloroformate in dry tetrahydrofuran and triethylamine and simultaneously reacted with the protein (Chu et al., 1982; Lau et al., 1981b; Worsaae, 1978). Although bovine serum albumin (BSA) was most frequently employed as the carrier, other proteins such as γ-globulin (Aslund et al., 1975), hemocyanin (Hertzog et al., 1982; Morgan et al., 1982), human serum albumin (Biermann and Terplan, 1980), ovalbumin, lysozyme, and polylysine (Chu et al., 1976; Langone and Van Vunakis, 1976) have also been used as the carriers. Ochratoxin A (Aalund et al., 1975; Chu et al., 1976; Worsaae, 1978) and rubratoxin B (Davis and Stone, 1979) were conjugated to a protein carrier directly without any derivatization. Other mycotoxins have been conjugated to protein after introducing a carboxylic acid group into the molecules using one of the methods described before. Although both WSC and MA methods have been used for the conjugation of mycotoxins to proteins, the former method is milder and is generally used for conjugation of mycotoxins or their derivatives to an enzyme for ELISA. The MA method usually is capable of coupling more mycotoxin to a protein molecule.

The efficiency of conjugation of mycotoxin derivatives to BSA could be improved by using a modified BSA in which the carboxylic group was blocked by ethylenediamine (Chu *et al.*, 1983). It must be pointed out, however, that the presence of a large number of hapten groups in a protein molecule may not necessarily give a good immune response (Fan *et al.*, 1984a). The amount of mycotoxin coupled to the protein should be determined. The efficacy of immunogens which contain different amounts of mycotoxin in each protein molecule for antibody production also should be studied.

2. Conjugation of Mycotoxins to Protein via Other Methods

An alternative for conjugation of a mycotoxin to a protein molecule is the reductive alkylation method which involves the dihydrofuran moiety in the aflatoxins B_{2a} (Gaur *et al.*, 1981), G_{2a} (Chu *et al.*, 1985), Q_{2a} (Fan *et al.*, 1984a), diol (Pestka and Chu, 1984a), and the hemiacetal of sterigmatocystin (Li and Chu, 1984). Above pH 7, the furan ring opens and exposes two free aldehyde groups which can readily form a Schiff base with protein. Subsequent reduction with sodium borohydride results in a stable covalent bond (Ashoor and Chu, 1975). Conjugation of Afla to protein was also achieved by direct reaction of Afla B_1 2,3-dichloride (or 2,3-dibromide) with protein (Sizaret *et al.*, 1980).

A cross-linking bifunctional reagent such as tetraazobenzidine has been used for conjugation of Afla B_{2a} and kojic acid to protein, and useful antibodies have been obtained from rabbits immunized with these conjugates (Lawellin *et al.*, 1977a,b).

For the production of antibodies against mycotoxin-DNA adducts, the principle of strong electrostatic interaction between methylated protein and DNA was employed (Plescia, 1982). Thus, when Afla-DNA adduct mixed with either methylated BSA (Groopman *et al.*, 1982; Haugen *et al.*, 1981) or methylated hemocyanin (Hertzog *et al.*, 1982) was used as an immunogen in BALB/c mice in which the spleen cells were isolated and used in the fusion studies, monoclonal antibodies against the Afla–DNA adducts were obtained.

C. Production of Antibodies against Mycotoxins in Rabbits

Once the conjugate was prepared, antibody generally was produced in rabbits by injecting a mixture of conjugate and complete Freund's adjuvant (Chu and Ueno, 1977) at multiple sites on the backs of rabbits. Antibodies having sufficient titer were generally obtained 5–7 weeks after the initial immunization.

Subsequent booster injections were made into the thigh once every month using incomplete adjuvant as the dispersion agent (Gaur *et al.*, 1980). Although goats have been used for the production of antibody against Afla, the titers were lower than those obtained in rabbits (Gaur *et al.*, 1980). Production of antibody against zearalenone has been done in pigs (Thouvenot and Morfin, 1983).

D. Production of Monoclonal Antibodies

The successful production of antibodies against mycotoxins in rabbits and other animal systems as well as the use of such antibodies in the immunoassay of mycotoxins have led to a great demand for specific antibodies against several important mycotoxins. Also, because of recent rapid progress in the hybridoma technology area, specific monoclonal antibodies from selected hybridoma cell lines could be generated in virtually unlimited amounts. This technique has been applied to produce antibodies against aflatoxins. Haugen *et al.* (1981) produced specific monoclonal antibody against Afla-DNA adduct by fusion of mouse myeloma cells with spleen cells isolated from BALB/c mice that had been immunized with Afla B_1-DNA adducts incorporated into methylated BSA. A similar technique was used by Hertzog *et al.* (1982) to generate antibodies against guanine imidazole ring-opened Afla B_1-DNA, except that the mice were immunized with guanine imidazole ring-opened Afla B_1-DNA coupled electrostatically to methylated keyhole limpet hemocyanin. Monoclonal antibody against Afla B_1 which has cross-reactivity with Afla B_{2a} and M_1 was prepared using spleen cells obtained from BALB/c mice which had been immunized with Afla B_{2a}-BSA conjugate (Lubet *et al.*, 1983; Sun *et al.*, 1983). Groopman *et al.* (1984) have used Afla B_1–bovine γ-globulin as an antigen to generate monoclonal antibody against Afla B_1. The antigen was prepared through activation of Afla B_1 with *m*-chloroperoxybenzoic acid. Monoclonal antibody against Afla M_1 was prepared using spleen cells of BALB/c mice which had been immunized with CMO of Afla M_1-BSA conjugate (Woychik *et al.*, 1984). Monoclonal antibody against T-2 toxin has also recently been obtained (Hunter *et al.*, 1985).

III. CHARACTERIZATION OF ANTIBODY

The specificity of an antibody is determined primarily by the antigens used in antibody production. Before conducting any RIA or ELISA tests, the specificity of an antibody preparation must be tested thoroughly by a competitive binding assay in which the antibody is incubated with a constant amount of radioactive-labeled mycotoxin or mycotoxin–enzyme conjugate together with different kinds of unlabeled mycotoxin analogs of varied concentrations. The relative concentration needed to displace the radioactivities from the antibodies by different ana-

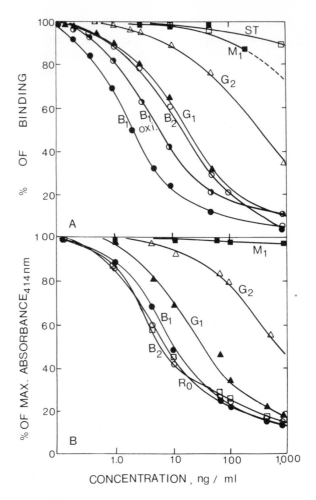

Fig. 1. Cross-reactivity of aflatoxin analogs with Afla B_1 antibody as determined by RIA (A) and competitive direct ELISA (B). In the RIA, an equilibrium dialysis method was used (Ueno and Chu, 1977). ^3H-Afla B_1, 0.3–0.5 ml (15,000–20,000 cpm/ml), together with a constant amount of IgG (30 µg/ml), was dialyzed against an equal volume of different concentrations of unlabeled Afla analogs. The extent of binding of ^3H-Afla B_1 with IgG in the absence of unlabeled toxin (i.e., dialyzed against buffer alone) was considered as 100%. In the ELISA, an Afla B_1–horseradish peroxidase conjugate was used as an enzyme indicator (Pestka *et al.*, 1980). An antiserum at 1:400 dilution was coated to a microplate for the assay. The figure is in semilog scale, where the *x* axis is in log scale. In the ELISA, 25 µl of sample was used in each analysis. Therefore, the amount in each assay was 40 times less on the ng/ml basis. The letters B_1 (●), B_2 (○), G_1 (▲), G_2 (△), M_1 (■), and R_0 (aflatoxicol, □) represent different aflatoxins; ST (□), sterigmatocystin.

logs is then determined. A typical example for the determination of the specificity of antibodies against Afla B$_1$ by the RIA and ELISA is shown in Fig. 1A and B, respectively.

A. Specificity of Different Aflatoxin Antibodies

Since antibody specificity is determined primarily by the specific group exposed on the protein molecule, different analogs of a specific mycotoxin can be used in conjugation. Alternatively, the reactive group could be introduced at a different side chain of the same toxin molecule for subsequent conjugation. Extensive studies on aflatoxins have been made both by our laboratories and by others, and the approaches are summarized in Fig. 2.

The relative specificities of different antibodies obtained from rabbits after immunizing with different Afla analogs is summarized in Table II. When rabbits are immunized with Afla conjugated through the cyclopentenone portion of the molecule, such as the CMO of Afla B$_1$ (Biermann and Terplan, 1980; Chu and Ueno, 1977; Langone and Van Vunakis, 1976) and M$_1$ (Harder and Chu, 1979), the antibodies generally recognize the dihydrofuran portion of the molecule. Specific antibodies against Afla B$_1$ and M$_1$ were produced in this manner. When conjugates are prepared through the dihydrofuran portion of the Afla molecule, such as Afla B$_{2a}$ (Gaur et al., 1981; Lau et al., 1981b), Afla B$_1$ diol (Pestka and Chu, 1984a), and Afla B$_1$ chloride (Sizaret et al., 1980), the antibody has a specificity directed toward the cyclopentenone ring. When both the cyclopentenone ring and the dihydrofuran moiety of Afla are exposed, as in the case of BSA-Afla Q$_1$-HS as an immunogen, the antibody cross-reacts with most B-type aflatoxins (Fan et al., 1984a). In view of the diversity of antibody specificities, different antibody preparations should be used for analysis of different Afla metabolites.

It is interesting to note that the monoclonal antibodies for Afla B$_1$ and Afla M$_1$ were not as specific as polyclonal antibodies. The relative cross-reactivity of the monoclonal antibody for Afla B$_1$ with different aflatoxins, as determined by the amount of Afla necessary to cause 50% inhibition of maximal absorbance for the monoclonal antibody, was 1.0, 1.2, 1.6, and >100 for Afla B$_1$, B$_2$, M$_1$, and aflatoxicol, respectively. No cross-reactivity toward Afla G$_1$ and G$_2$ was found (Lubet et al., 1983). The cross-reactivity of monoclonal antibody for Afla M$_1$ with different aflatoxins was 1, 12, >40, 12, and >40 for Afla M$_1$, B$_1$, B$_2$, G$_1$, and G$_2$, respectively (Woychik et al., 1984). Thus, the B$_1$ monoclonal antibody had a specificity toward the cyclopentenone ring portion of Afla molecules, whereas M$_1$ monoclonal antibody had a specificity toward the dihydrofuran moiety of the Afla molecule. Since the type(s) of clone(s) selected depended on the marker ligand used in the assay system, it is very important to select an adequate ligand in the initial screen system. It is not known whether the mono-

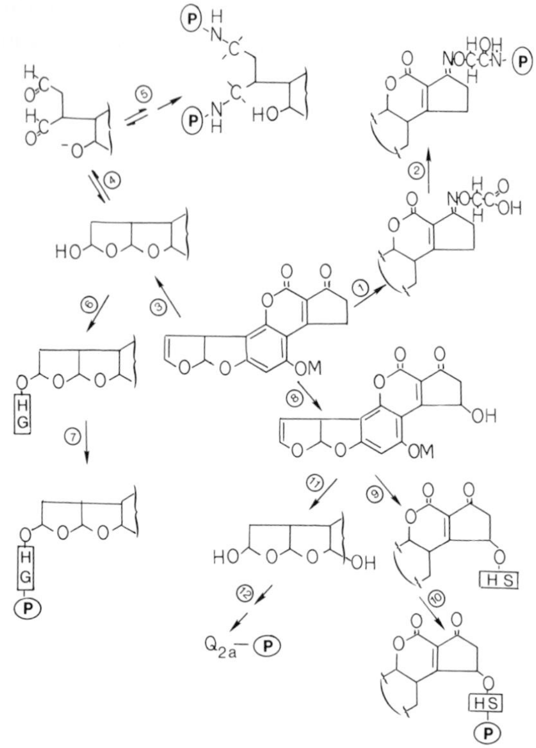

Fig. 2. Approaches used in the conjugation of Afla B_1 to protein. Afla B_1 (1) or M_1 may convert to its O-carboxymethyloxime, which is then conjugated to protein (2); or may convert to B_{2a} (3), in which the furan ring opens (4) at $\geq$pH 7.0 and subsequently forms a Schiff's base with protein. The bond is stabilized by reduction with sodium borohydride (5). Acylation of Afla B_{2a} with glutaric anhydride (6) results in a hemiglutarate which is then conjugated to protein (7). Afla B_1 may also chemically convert Afla Q_1 (8), which is then either acylated to hemisuccinate (9) for conjugation (10) or converts to Q_{2a} (11) for subsequent coupling to protein (12) via the reductive alkylation route such as those for B_{2a}.

clonal antibody for Afla B_1 indeed has a diversified specificity or the isolated hybridoma contained a mixed clone because of the problems encountered in the selection of the clone.

Because it has been postulated that the carcinogenic and acute toxic effects of Afla B_1 are due to the metabolic activation of this toxin and subsequent binding to macromolecules (Miller, 1978), efforts have been made in several laboratories to produce specific antibody against Afla-DNA adducts and its related metabolites. In our laboratory, we found that the Afla B_{2a} (Pestka et al., 1982) and Afla

TABLE II

Specificity of Antibodies Obtained from Rabbits after Immunization with Different Aflatoxin Conjugates[a]

Conjugates used	Aflatoxin analogs								
	B_1	B_2	G_1	G_2	M_1	B_{2a}	P_1	Q_1	R_0
BSA-CMO-B_1	100	11.1	9.1	0.6	<0.2	—[b]	—	5.9	2.5
	_100_	_100_	_33.3_	_1.3_	_<0.9_	—	—	—	_100_
PLL-CMO-B_1	100	43.4	15.1	4.5	—	1.1	0.2	8.3	—
PLL-CMO-B_1	100	18.9	1.6	0.3	—	0.8	0.2	1.4	—
BSA-B_{2a}	100	12.5	3.1	<0.3	6.3	8.3	—	—	0.5
	_100_	_25_	_<1.0_	_<0.1_	_1.1_	_125_	—	—	_1.4_
BSA-B_1Cl[c]	100	238	3.0	2.0	2.7	—	—	—	—
BSA-B_{2a}-HG	100	10	6.3	0.2	2.0	11.1	—	—	0.4
BSA-G_{2a}	_2.1_	_1.6_	_8.0_	_15.4_	_2.0_	—	—	—	—
BSA-CMO-M_1	55.6	<1.0	<1.0	<1.0	100	<1.0	—	1.9	1.7
	_0.6_	_<0.6_	_<0.6_	_<0.6_	_100_	_<0.6_	—	—	_<0.6_
BSA-Q_1-HS	100	1.4	0.1	1.4	—	—	—	2.9	33.3
BSA-Q_{2a}	_<0.3_	—	—	—	—	_1.0_	—	_1.0_	—

[a] Expressed as percentage of reactivity relative to Afla B_1 except BSA-G_{2a}, BSA-CMO-M_1, and BSA-Q_{2a} antibodies, which are relative to Afla G_{2a}, M_1, and Q_{2a}, respectively. The data were obtained from both the RIA and ELISAs (underlined values).

[b] Not determined.

[c] BSA-B_1Cl, Afla B_1 dichloride conjugated to BSA directly.

diol (Pestka and Chu, 1984a) antibodies cross-react with the Afla-DNA adduct and related metabolites. Monoclonal antibodies against two types of Afla-DNA adducts have also been prepared. The cross-reactivity of these antibodies with different Afla-DNA-related metabolites is summarized in Table III. It is apparent that these monoclonal antibodies prepared against Afla B_1-guanine modified DNA (Groopman et al., 1982; Haugen et al., 1981) and Afla B_1–ring-opened modified DNA (i.e., 2,3-dihydro-2-(N-5-formyl)-2',5',6'-triamino-4'-oxy-N-pyrimidyl-3-hydroxyaflatoxin B_1) (Hertzog et al., 1982) showed virtually no cross-reaction with Afla analogs such as Afla B_1, B_{2a}, diol, and Afla-guanine. Instead, these antibodies have high degree of recognition for Afla-DNA adducts regardless of whether it is Afla B_1-DNA or Afla G_1-DNA adduct. These results suggested that the monoclonal antibodies had high specificity to the altered DNA molecule rather than to either Afla B_1 or DNA itself. In contrast, the cyclopentenone ring of Afla B_1 was the primary epitope for the Afla B_{2a} and diol antibodies (Pestka et al., 1982; Pestka and Chu, 1984a).

TABLE III

Specificity of Different Antibodies against Aflatoxin-DNA Adduct and Its Analogs[a]

	Conjugates used			
Afla analogs	$BSA-B_{2a}$	$BSA-B_1$ diol	$MHC-B_1-FAPyr-DNA$	$MBSA-B_1-DNA$[b]
B_1	80	200	—	11,000
B_{2a}	100	100	—	10,000
Diol	80	100	<0.1	2,000
G_1	<1.0	6	—	260
M_1	<1.0	130	—	5,100
Q_1	—	2	—	—
Aflatoxicol	1.4	20	—	—
B_1-DNA	47.6	32	52.6	5–12
B_1-FAPyr-DNA	40	—	100	—
B_1-Gua	17.5	0.6	<0.1	2,100
B_1-FAPyr	22.3	—	<0.1	2,100
G_1-DNA	—[c]	—	3.3	—
G_1-FAPyr-DNA	—	—	23.3	—
G_1-Gua	—	—	—	—
G_1-FAPyr	—	—	—	—
DNA	<0.1	<0.1	0.01	—
RNA	—	—	<0.01	—

[a] Expressed as percentage of reactivity relative to Afla B_{2a} for the $BSA-B_{2a}$ and $BSA-B_1$ diol antibodies, which were obtained from rabbits, and to B_1-FAPyr-DNA for $MHC-B_1$-FAPyr-DNA (methylated hemocyanin-ring opened Afla-DNA adduct) antibody, which was obtained from one hybridoma line (Hertzog *et al.*, 1982).

[b] Monoclonal antibody obtained from another hybridoma line. The BALB/c mice were immunized with Afla B_1-DNA adduct complexed to methylated BSA. Data expressed in this column are concentrations (picamoles) that resulted in a 50% inhibition of binding of antibody to Afla B_1-DNA coated to the plate by free B_1-DNA and to <2% inhibition by all other analogs (Groopman *et al.*, 1982; Haugen *et al.*, 1981).

[c] Not determined.

B. Specificity of Antibodies against Trichothecene Mycotoxins

When rabbits were immunized with BSA-T-2 toxin hemisuccinate (Chu *et al.*, 1979), the antibody elicited was highly specific for T-2 toxin, less specific for HT-2 toxin, and least specific for T-2 triol. Weak cross-reactions were observed for neosolaniol, T-2 tetraol, and 8-acetylneosolaniol. Practically no cross-reactions were observed for diacetoxyscirpenol, deoxynivalenol, trichodermin, verrucarin, verrucarol A, and roridin (Chu *et al.*, 1977; Fontelo *et al.*, 1983). The relative concentrations necessary for the displacement of 50% of the radioactive T-2 toxin bound to this antiserum by different trichothecenes were found to be 1,

5.7, 46, 453, 1428, and 5370 for T-2, HT-2, T-2 triol, neosolaniol, T-2 tetraol, and 8-acetylneosolaniol, respectively (Chu *et al.*, 1979). Similar results were obtained in an indirect ELISA (Peters *et al.*, 1982). In addition, this antiserum had a higher affinity to acetyl-T-2 and T-2 HS than to T-2 toxin. The relative concentrations for the inhibition of binding of the antibody to polylysine-T-2 conjugate in an indirect ELISA system by acetyl-T-2, T-2 HS, and T-2 toxin were found to be 0.1, 0.5, and 1, respectively (Peters *et al.*, 1982).

Antibodies against DAS were recently obtained from rabbits in our laboratory after immunizing the animals with DAS-HG-BSA (Chu *et al.*, 1984a). These antibodies also showed high specificity toward DAS itself. The cross-reactivity with monoacetoxyscirpenol was ~80 times weaker than DAS, and practically no cross-reaction with scirpenol triol, T-2 toxin, deoxynivalenol, and verrucarin A was detected.

Attempts to elicit antibodies that have a broad specificity for different trichothecenes have been made in our laboratory by immunizing a group of rabbits with a hemisuccinate derivative of deoxyverrucarol (Chu *et al.*, 1984b). This compound has only one hydroxylic group in the trichothecene skeleton, and the -OH group was modified by succinylation for conjugation purposes. Nonetheless, the antibody showed high specificity against the deoxyverrucarol and only had weak cross-reaction with DAS and verrucarol A. Again, the antibody did not cross-react with T-2 toxin, deoxynivalenol, and verrucarin A.

From the above studies it appears that both the trichothecene skeleton and the side chain groups in the trichothecene molecule are important in determining the specificity of the antibodies elicited. Modification of the side chain in the trichothecene molecule produces antibodies which are completely different from each other.

C. Specificity of Antibodies against Other Mycotoxins

The antibody for ochratoxin A (OA) was most specific for OA and ochratoxin C (OC), and is not specific for ochratoxins B (OB), α and other coumarins (Chu *et al.*, 1976). The relative cross-reactivity of the antibody, as analyzed by both RIA (Chu *et al.*, 1976) and ELISA (Pestka *et al.*, 1981c) for OA, OC, and OB, was 1, 2, and 1000, respectively. Chu *et al.* (1976) suggested that the amide bond, chlorinated dihydroisocoumarin ring, and a portion of the phenylalanine ring all contribute to the immunogenicity of the toxin.

The specificity of antibody against sterigmatocystin (ST), which was prepared by immunizing rabbits with the hemiacetal-ST-BSA, was determined by an ELISA test (Li and Chu, 1984). The antibody was most specific for ST, and had little or no cross-reaction with Afla B_1, B_{2a}, G_1, G_2, and M_1, and O-methyl-ST, The dehydro-ST was ~16 times less reactive with the antibody than ST. The antibody thus has an affinity to the anthraquinone moiety of the ST molecule (Li and Chu, 1984). Those results are similar to those of Afla B_{2a} antibody where Afla B_2-BSA conjugated by the reductive method was used as an immunogen for antibody production (Gaur *et al.*, 1981).

When pigs were immunized with 6'-CMO of zearalenone conjugated to BSA, the antibodies were most specific for zearalenone, but they could not distinguish zearalenone, zearalanone, and low melting point (LMP) zearalenol as analyzed by RIA (Thouvenot and Morfin, 1983). Other analogs such as *cis*-zearalenone, high melting point (HMP) zearalanol, HMP zearalenol, LMP zearalanol, and zearalane have only 70–31% of the cross-reactivity as compared with zearalenone. Compounds such as dibenzylzearalenone, Afla B_1, β-resorcylic acid, resorcinol, and estradiol-17β had less than 0.02% cross-reactivity. Thus, Thouvenot and Morfin (1983) concluded that the whole resorcylic acid lactone ring including the two phenolic hydroxyls, the lactone, and the 10'-chiral center, are all necessary for binding with the antibody.

IV. RADIOIMMUNOASSAY FOR MYCOTOXINS

A. General Considerations

Although a number of immunochemical methods have been used for the analysis of small molecular weight biological substances (Langone and Van Vunakis, 1981; Van Vunakis and Langone, 1980), only the RIA and ELISA have been developed for the analysis of mycotoxins. Both techniques are based on the competition of binding between the unlabeled toxin in the sample and the labeled toxin in the assay system for the specific binding sites of antibody molecules. In the RIA (Yalow, 1978) a radiolabeled toxin is used, whereas a toxin–enzyme conjugate is used in the ELISA (Maggio, 1980; Schuurs and Van Weemen, 1977).

The RIA procedure involves simultaneous incubation of a solution of unknown sample or known standard dissolved in phosphate buffer with a constant amount of labeled toxin and specific antibody. Free toxin and bound toxin are then separated by an appropriate technique, and the radioactivity in each fraction is then determined. The toxin concentration of the unknown sample is determined by comparing the results to a standard curve, which is established by plotting the ratio of radioactivity in the bound fraction and free fraction versus log of concentration of unlabeled standard toxin. A typical standard curve for Afla B_1 and OA analysis is shown in Fig. 3. The specific activity of radioactive ligands plays an essential role in the sensitivity of RIA. Currently, methods for the preparation of tritiated OA (Chang and Chu, 1976), T-2 toxin (Wallace *et al.*, 1977), and zearalenone (Thouvenot and Morfin, 1983) of high specific activity are available, and tritiated Afla B_1 of high specific activity can be obtained

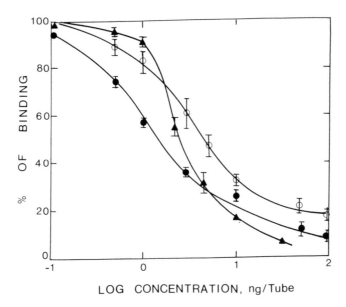

LOG CONCENTRATION, ng/Tube

Fig. 3.   Standard curves for RIA of Afla B$_1$ (●), OA (○), and T-2 toxin (▲). For Afla B$_1$ and OA, a solid-phase RIA was used for the separation of free and bound toxin (Sun and Chu, 1977). In the assay, 0.2 ml of IgG–sepharose gel was incubated with 0.2 ml of ^{3}H-Afla B$_1$ (36,500 dpm) or ^{3}H-OA, and 0.5 ml of different unlabeled Afla B$_1$ or OA at appropriate concentrations in a Quick-Sep column at room temperature for 30 min, then 6°C overnight (Sun and Chu, 1977). In the T-2 toxin analysis, 50 μl of ^{3}H-T-2 toxin (20,000 dpm) and 50 μl of diluted antiserum preparation were incubated with 100 μl of different concentrations of unlabeled T-2 toxin at 4°C for 1 hr. The free and bound toxin was then separated by albumin-coated charcoal (Fontelo *et al.*, 1983).

commercially (Chu and Ueno, 1977; Langone and Van Vunakis, 1976). Tritiated T-2 was prepared by reducing the 3-dehydro–T-2 toxin with high specific radioactivity tritiated sodium borohydride (Wallace *et al.*, 1977). Three other radioactive mycotoxins were prepared by the tritium exchange method (Chang and Chu, 1976; Langone and Van Vunakis, 1976; Thouvenot and Morfin, 1983). To eliminate the nonspecific binding, the purity of all the radioactive ligands should be analyzed by one or more methods prior to use. Because of the tritium exchange problem which may occur during storage, the specific activity and purity of the compounds should also be frequently analyzed.

B. Techniques for Separation of Free and Bound Mycotoxins

Several methods have been used for the separation of free and bound toxins. The ammonium sulfate precipitation method is simplest and has been used successfully for Afla (Sizaret *et al.*, 1980) and for T-2 toxin analysis (Lee and Chu, 1981a,b). Analysis of Afla has also been achieved either by the double-antibody technique (Langone and Van Vunakis, 1976) or by a solid-phase RIA method in which the immunoglobulin G (IgG) was conjugated to CNBr-activated Sepharose gel (Sun and Chu, 1977). Thus, after incubation of the standard or sample solution and radioactive toxin with the IgG gel, the free and bound forms are separated by a simple filtration step. A dextran-coated charcoal column, which has been used commonly for RIA of other compounds (Yalow, 1978), has also been adapted to the analysis of Afla (Thorpe and Yang, 1979; Tsuboi *et al.*, 1984), T-2 toxin (F. S. Chu *et al.*, unpublished observations), and zearalenone (Thouvenot and Morfin, 1983). Most recently, Fontelo *et al.* (1983) have successfully used albumin-coated charcoal to separate the free and bound toxin in the RIA of T-2 toxin.

C. Applications

The sensitivity of RIA for different mycotoxins in foods and biological fluids is summarized in Table IV. In general, RIA can detect as low as 0.25 to 0.5 ng of purified mycotoxin in a standard preparation. The lower limits for mycotoxin detection in food or feed samples is about 2 to 5 ppb (μg/kg) for samples that have not been subjected to a cleanup treatment. In these assays, Afla B_1 was extracted from corn, wheat, and peanut butter according to the AOAC procedures (El-Nakib *et al.*, 1981; Langone and Van Vunakis, 1976) or other methods (Sun and Chu, 1977), and the extracts were then concentrated, redissolved in dimethylsulfoxide or methanol, diluted with buffer, and then subjected to RIA by either the double-antibody method (Langone and Van Vunakis,

TABLE IV

Sensitivity of Radioimmunoassay for Mycotoxins[a]

Mycotoxins	Standard range (ng)	Detection limits (ppb)	Recovery (%)					
			C	P	W	M	S	U
Afla B_1	0.5–5.0	5.8	64	104	97	—	—	—
Afla M_1	5–50	5.0	—	—	—	108[b]	—	—
T-2	0.2–2.5	1.0	—	—	86	—	103	—
T-2	0.2–2.5	2.5	103	—	100	117	—	98
T-2	0.2–2.5	5.0	102	—	—	100	—	102
Zearalenone	0.25–10	5.0	—	—	—	—	101	—

[a] C, corn; M, milk; P, peanut butter; S, serum; U, urine; W, wheat.

[b] Values with an underline indicate that the samples have been subjected to a cleanup treatment prior to RIA.

1976) or the solid-phase RIA (Sun and Chu, 1977). The recovery for the toxin added to the samples was generally around 60 to 100% when no cleanup chromatographic procedure was used.

Using the Afla B_1 dichloride antibodies, Sizaret et al. (1980) have demonstrated the "Aflatoxin-like" metabolites in rat urine after the animal ingested Afla by RIA. The RIA has also been used for the detection of Afla in human serum (Tsuobi et al., 1984). In this study, antibodies obtained from rabbits after immunizing with a CMO derivative of Afla B_1 conjugated to protein was used. Separation of free and bound toxin in this assay was achieved by the dextran-coated charcoal method. Human serum samples obtained from subjects 2–3 hr after lunch and from the fasting subjects in two regions of Japan were analyzed. Since the antibody is specific for Afla B_1, 5 of 20 samples in the fasting group had 33.4 ± 14.6 (SD) ppt of Afla B_1, whereas 29 of the 80 samples had a level of 218.1 ± 268.3 ppt of B_1. The presence of Afla B_1 in those human serum samples was also confirmed by HPLC as well as by mass spectral analysis.

The sensitivity of RIA can be improved by a simple cleanup procedure after extraction. For example, after a Sep-Pak treatment, the minimum detection level for T-2 toxin in corn and wheat by the RIA decreased from 2.5 and 5 ppb to 1 and 2.5 ppb, respectively (Lee and Chu, 1981a,b). Because radioactive Afla M_1 of high specific activity is not available, tritiated Afla B_1 was used in the RIA of Afla M_1 in milk (Harder and Chu, 1979; Pestka et al., 1981b). This resulted in the low sensitivity for RIA of Afla M_1 in milk. Nonetheless, Yasaei et al. (1983) demonstrated that the detection limits of RIA for Afla B_1 and M_1 in liver tissue improved considerably when sample extracts were subjected to Sep-Pak C_{18} reversed-phase cartridge two times under different solvent systems. No detectable interference was observed in extracts of ≤2 g of liver. Since the minimum detection limits for the standard was 20 pg/assay, this corresponded to a detection limit of 10 ppt. In view of the extensive cleanup necessary with RIA, it is not presently the best method for detecting Afla M_1 in milk.

V. ENZYME-LINKED IMMUNOSORBENT ASSAY FOR MYCOTOXINS

A. General Considerations

Two types of ELISA are generally used for analysis of biologically active substances (Maggio, 1980; Schuurs and Van Weemen, 1977). One type is called a homogeneous ELISA in which an enzyme–toxin conjugate must first be prepared. In this assay, the enzyme activity is altered after it is bound to specific antibodies. Thus, it is not necessary to separate the free and bound form of the enzyme–ligand conjugate in this assay. This type of ELISA has not been used for mycotoxin analysis. The other type is called heterogeneous competitive ELISA in which the toxin must first be conjugated to an enzyme or to a protein. The enzyme activity remains unaltered after binding with the antibody; and separation of the free and bound enzyme–ligand is necessary. The ELISA used for mycotoxin analysis is almost exclusively the latter type, which constitutes two different systems. One system, direct competitive ELISA, involves the use of mycotoxin–enzyme conjugate (Biermann and Terplan, 1980; Lawellin et al., 1977b; Lee and Chu, 1984; Pestka et al., 1980, 1981a,b,c, 1982), and the other system, indirect competitive ELISA (Fan and Chu, 1984; Fan et al., 1984b; Morgan et al., 1982; Peters et al., 1982), involves the use of a protein–mycotoxin conjugate and a secondary antibody to which an enzyme has been conjugated. Horseradish peroxidase (HRP) has been used in all the ELISA tests except in the indirect ELISA for Afla B_1-DNA adduct (Groopman et al., 1982; Haugen et al., 1981; Hertzog et al., 1982) and OA (Morgan et al., 1982) systems, which used alkaline phosphatase and β-galactosidase (Hertzog et al., 1982). Although several methods have been used in the conjugation of mycotoxin to HRP, water-soluble carbodiimide is milder than others and thus has been most frequently used.

B. Direct Competitive ELISA Using Microtiter Plate

In this assay, specific antibodies are first coated onto a microtiter plate (Pestka et al., 1980). Among several methods tested for coating of antibody to the microtiter plate, the glutaraldehyde and bicarbonate methods appear to be most suitable for the assay (Biermann and Terplan, 1980; Lee and Chu, 1984; Pestka et al., 1980). Once the antibody is dried on the plate, it is stable for 3 to 6 months. The plate is washed with buffer before use, and the sample solution or standard toxin is generally incubated simultaneously with enzyme conjugate (Pestka et al., 1980, 1981b) or incubated separately in two steps (Biermann and Terplan, 1980). The plate is washed again, and the residual enzyme bound to the plate is determined by incubation with a substrate solution containing hydrogen peroxide and appropriate oxidizable chromogens. The resulting color is measured spectrophotometrically or by visual comparison with the standards. The oxidizable chromogens for HRP that have been used for ELISA of mycotoxins include o-toluidine (Lawellin et al., 1977b), 5-aminosalicylic acid (Biermann and Terplan, 1980), and 2,2'-azinodi-3-ethylbenzthizoline6-sulfonate (ABTS) (Pestka et al., 1980, 1981b). ABTS has been shown to be the most convenient chromogen because of its high sensitivity and stability. A typical ELISA plate for the analysis of Afla M_1 is shown in Fig. 4.

The sensitivity of ELISA for mycotoxin analysis is summarized in Table V. In general, ELISA is approximately 10–100 times more sensitive than RIA when purified mycotoxins are used. As low as 2.5 pg of pure mycotoxin can be

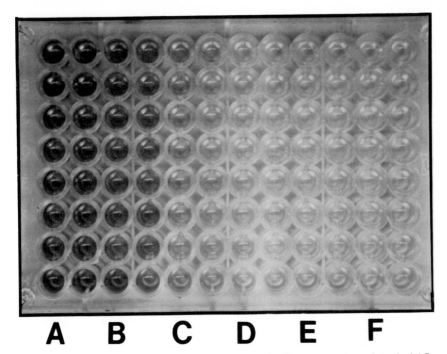

A B C D E F

Fig. 4. Representative ELISA plate for Afla M_1 analysis. The concentrations of standard Afla M_1 in rows A, B, C, D, E, and F were 0, 2.5, 12.5, 25, 125, and 250 pg/assay, respectively.

measured in certain instances (Pestka *et al.*, 1981a). However, when different commodities containing added toxins were tested, the lower limit for detection was comparable to or only slightly better than RIA. One of the major advantages of ELISA is that it does not use radioactive substances, and thus avoids expensive instrumentation and problems associated with disposal of radionuclides. The ELISA method is also more advantageous when the radioactive toxin is not available. For example, as low as 0.25 ppb of Afla M in milk can be readily detected by the ELISA technique, whereas with RIA the minimum detection level was 5 ppb (Pestka *et al.*, 1981b). Since milk samples can be used directly in the ELISA, a cleanup step is not necessary; hence, many samples can be analyzed within a relatively short period (2–3 hr). When milk or urine samples were subjected to a simple cleanup treatment with a C_{18} reversed-phase Sep-Pak cartridge (Hu *et al.*, 1984), the sensitivity of ELISA for Afla M_1 reached the 10–25 ppt level. Similar approaches have been tested for other dairy products (Fremy and Chu, 1984) and have a sensitivity of 10 to 50 ppt. The microplate ELISA appears to be the most specific, effective, and simplest method for analysis of Afla M_1 in milk relative to other methods that have been developed.

TABLE V

Sensitivity of ELISA for Mycotoxins[a]

Mycotoxins	Standard range (pg)	Detection limits (ppb)	Recovery (%)					
			C	P	W	M	S	U
Direct ELISA								
Afla B$_1$	25–1000	3	70	72	62	—	—	—
		5.8	73	97	81	—	—	—
Afla M$_1$	25–1000	0.015	—	—	—	87[b]	—	80
		0.25	—	—	—	120–130	—	—
		0.5	—	—	—	96–120	—	—
OA	25–500	1–2	—	—	85	—	—	—
T-2	2.5–200	1.0	124	—	86	—	—	—
		2.5	146	—	96	—	—	—
		5.0	109	—	83	—	—	—
Indirect ELISA								
Afla B$_1$	20–1000	5	96	68	—	—	—	—
		10	99	68	—	—	—	—
OA	10–1000	0.5[c]	—	—	—	—	—	—
T-2	2.0–200	0.2	—	—	—	80	51	79
		1.0	—	—	—	83	55	77
		5.0	—	—	—	83	95	82

[a] C, corn; M, milk; P, peanut butter; S, serum; U, urine; W, wheat.

[b] Values with an underline indicate that the samples have been subjected to a cleanup treatment prior to ELISA.

[c] The recovery of OA added to the barley sample was reported between 95 and 98% in the range 0.5–48 ppb.

We have evaluated the potential of the ELISA technique for the detection of low levels of Afla M$_1$ in naturally contaminated samples in collaboration with other laboratories. Approximately 100 milk samples naturally contaminated at levels < 0.5 ppb were received from several laboratories. The amount of Afla M$_1$ present in these samples was determined by other laboratories using either HPLC or TLC methods. We have found excellent agreement between the ELISA method and results obtained by other methods (J. J. Pestka and F. S. Chu, unpublished observations). Since 1980, we have also participated in the mycotoxin check sample program, which was sponsored by the International Agency for Research on Cancer. The program involved analyzing aflatoxins in corn, peanut, defatted peanut meal, and milk. Again the ELISA data agreed well with those obtained by other methods. For example, in the 1980 mycotoxin check sample program, the amounts of Afla B$_1$ found in corn, peanut, and defatted peanut meal by all other methods were 15, 231, and 49 ppb, respectively (Friesen and Garren, 1982a), whereas the ELISA data were 14, 270, and 35.5, respectively. The mean value for Afla M_1 in the milk sample obtained by other laboratories was 7.6 ppb (Friesen and Garren, 1982b), and the mean value with ELISA was 8.6 ppb. The coefficient of variation (CV) for other methods within laboratories were in the range of 11 to 23%, while the ELISA gave a CV of 9 to 21%. The disadvantage of ELISA in these collaborative studies was its inability to monitor aflatoxins B_2, G_1, and G_2 in the samples.

C. Direct Competitive ELISA Using Other Solid Phases

In addition to the use of the microtiter plate as a solid phase, the polystyrene tubes have been used as the solid phase for antibody coating. Lawellin *et al.* (1977b) used this technique for direct ELISA of Afla B_1 and B_{2a}, but the sensitivity was not as good as with the microtiter plate. Comparable results were obtained by Biermann and Terplan (1980) for the ELISA of Afla B_1 when they tested both the tube assay and the microtiter plate assay. Nylon beads and Terisaki plate were tested for the ELISA of Afla B_1, M_1, and T-2 toxin by Pestka and Chu (1984b). Both methods had detection limits comparable to that for mycotoxin microtiter plate ELISAs. The ELISA competition curve for Afla B_1, M_1, and T-2 toxin exhibited linear response between 1.0 and 100, 0.1 and 100, and 0.1 and 10.0 ng/ml, respectively, in the nylon beads assay. Response ranges for the Terisaki plate ELISAs for Afla B_1, M_1, and T-2 toxin were 1.0–50, 0.05–0.5, and 0.5–1.0 ng/ml, respectively. These procedures did not require specialized instrumentation and may be adaptable as an economical screening method for mycotoxins in the fields.

D. Indirect ELISA

In the indirect ELISA (or sandwich ELISA), instead of using the mycotoxin–enzyme conjugate, which may involve the enzyme stability problem, a mycotoxin–protein (or polypeptide) conjugate is prepared and coated to the microplate before assay. The plate is then incubated with specific rabbit antibody in the presence or absence of the homologous mycotoxin. The amount of antibody bound to the plate coated with mycotoxin–protein conjugate is determined by reaction with goat anti-rabbit IgG–enzyme complex and by subsequent reaction with the substrates. Both HRP and alkaline phosphatase conjugated to the goat anti-rabbit IgG have been used in the indirect ELISAs.

The effectiveness of this technique has been tested for Afla B_1 (Cerny *et al.*, 1983; Fan and Chu, 1984; Fan *et al.*, 1984a; Martin *et al.*, 1984), OA (Morgan *et al.*, 1982, 1983), and T-2 toxin (Fan *et al.*, 1984b; Peters *et al.*, 1982). In the ELISA of aflatoxin, Afla B_1-DNA, afla B_1-ovalbumin (Cerny *et al.*, 1983), Afla B_1 dichloride-ovalbumin (Martin *et al.*, 1984), and Afla B_1-polylysine (Fan and Chu, 1984) were used as the primary coating ligands. Aflatoxin B_1 itself also was shown to be an effective ligand when the microplate was first coated with polylysine (Cerny *et al.*, 1983; Lubet *et al.*, 1983). The detection limit was approximately 250 pg/assay for Afla B_1 and M_1 when Afla B_1-DNA was coated to the plate. A biotin–avidin system consisting of biotinylated second antibody (anti-mouse IgG, in the case of using monoclonal antibody) and HRP-avidin D has been used for the analysis of milk samples. The detection limit of this system for the milk sample which had been treated with Sep-Pak C_{18} reversed-phase cartridge was 0.05 ppb (Cerny *et al.*, 1983).

When plates coated with Afla B_1-polylysine were used in the assay, the detection limit for Afla B_1 was found to be approximately 25 pg/assay (Fan and Chu, 1984). Without a sample cleanup treatment, the recovery of the toxin added to the samples in the range of 5 to 40 ppb was 80–99% and 70–97% for the cornmeal and peanut butter samples, respectively. The sensitivity of this method was comparable to or slightly better than the direct ELISA, with the added advantage that much less antibody was required. For example, in the direct ELISA, an antibody dilution of 1:400 was necessary for the analysis of Afla B_1 (Pestka *et al.*, 1980), whereas similar sensitivity and specificity could be achieved at antibody dilution of 1:40,000 using the indirect ELISA (Fan and Chu, 1984). Therefore, this assay is ideal for monitoring the antibody titers of hybridoma culture fluids for the screening of monoclonal antibody-producing cells. The disadvantage of the method is that an additional incubation step is necessary for the assay. Thus, it is more time-consuming. In addition, the background is sometimes higher than that obtained from the direct ELISA.

Afla B_1 dichloride antibodies obtained from rabbits also have been tested as a reagent for monitoring Afla exposure by an indirect ELISA (Martin *et al.*, 1984). The rationale was that those antibodies, like the Afla B_{2a} antibodies, have a wide spectrum of specificity; thus one can monitor different types of Afla metabolites in a single run. The data obtained in such an analysis might be used as an indicator for Afla exposure. An Afla B_1-ovalbumin conjugate, prepared by reacting Afla B_1 dichloride or dibromide with ovalbumin, was coated to the microtiter plate. Although the lower detection limit for the indirect ELISA using the B_1 dichloride antibodies in estimating the "overall" Afla metabolites level in urine was ~10 ppt, these investigators observed that undiluted human urine from normal subjects sometimes inhibited the binding of antibodies to the plate resulting in false positive data. The nonspecific inhibition could be prevented by either dilution or by a cleanup procedure using C_8 reversed-phase treatment.

In the OA assay, OA-hemocyanin was coated onto the plate and alkaline phosphatase-labeled goat anti-rabbit IgG was used in the assay (Morgan *et al.*, 1982, 1983). The detection limit was found to be 10 pg/assay. Two approaches have been used in the indirect ELISA for T-2 toxin. Peters *et al.* (1982) coated the polylysine onto the microplate followed by adding HS of T-2 toxin and WSC to form the polylysine-T-2 HS directly in the microplate. In contrast, Fan *et al.*

(1984b) coated the polylysine-T-2 HS conjugate directly onto the plate. The latter investigators found that the detection limits for T-2 toxin in the serum, urine, and milk samples were 1, 0.2, and 0.2 ppb, respectively, using indirect ELISA in combination with a Sep-Pak C_{18} reversed-phase cartridge cleanup step.

VI. IMMUNOHISTOCHEMICAL METHODS AND OTHER STUDIES

With the availability of different antibodies showing diverse specificity, immunological methods have great potential for uses other than monitoring toxins in foods. For example, Lawellin *et al.* (1977a) have used an enzyme-linked immunocytochemical method to visualize the deposition of Afla B_1 in the hyphae of *Aspergillus parasiticus*. Elling (1977) has used the immunofluorescence microscopy method to locate OA in rat kidneys. Antibodies against Afla B_{2a} have been useful in monitoring Afla-DNA adducts (Pestka *et al.*, 1982), and antibodies against Afla B_1 and M_1 have been used to study the kinetics of transformation of Afla B_1 to M_1 in mice (Li and Chu, 1982). We have also demonstrated the possible use of immunization as a prophylactic method against aflatoxicosis (Ueno and Chu, 1978; Adekunle and Chu, 1979). Sizaret *et al.* (1982) found that the mutagenicity of Afla B_1 in the Ames test was inhibited by specific antibodies obtained from rabbits that had been immunized either with CMO-Afla B_1-BSA conjugate or with Afla B_1 diol-BSA conjugate. A novel approach to using antibodies as an affinity reagent in the cleanup step for analysis of Afla M_1 in human urine has been reported by Wu *et al.* (1983). These investigators conjugated the monoclonal antibody against Afla M_1 to a solid phase and subsequently used the immobilized antibody column for purification and concentration of Afla M_1 in human urine. After this treatment, trace amounts of Afla M_1 in human urine can be detected by either the HPLC method or ELISA method without any interference substances.

In addition, immunohistochemical methods have been established for monitoring Afla B_1, OA, and T-2 toxin in different organs of rats and mice which have been fed the toxins. Afla B_{2a} antibody was used by Pestka *et al.* (1983) in an indirect immunoperoxidase method to localize Afla B_1 bound to rat liver. The tissue sections after fixing and inactivating the endogenous peroxidase were incubated with the rabbit antiserum and then with goat anti-rabbit IgG–peroxidase (GARP) complex, followed by incubation with diaminobenzidine–hydrogen peroxide substrate. A positive reaction yielded a dark red-brown precipitate visible under the light microscope. In the Afla study, bound toxin could be identified in excellent detail in tissues fixed with periodate–lysine–paraformaldehyde and embedded in glycol methacrylate, but poor resolution was observed when unfixed cryostat sections were used. Strong peroxidase-positive reactions were observed in the nuclei of hepatocytes of Afla B_1-treated rats (Fig. 5).

Another immunohistochemical method is the peroxidase–antiperoxidase (PAP) technique, in which the slides are first treated with antiserum against

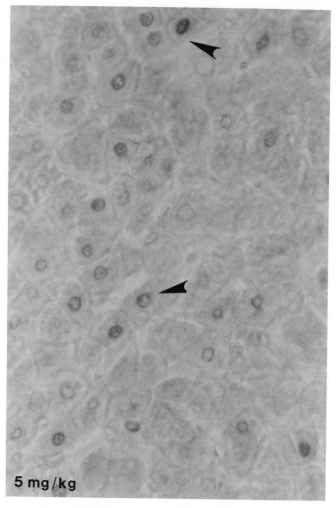

5 mg/kg

Fig. 5. Immunohistochemical localization of Afla B_1 in rat hepatocytes. The peroxidase-positive areas are confined to the nuclear portion of hepatocytes 2 hr after an IP injection of a rat with 5 mg/kg of Afla B_1. Glycol methacrylate 4-μm sections, unstained. Magnification: ×640.

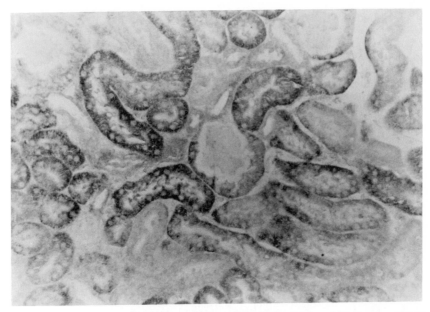

Fig. 6. Peroxidase and antiperoxidase (PAP) method for localization of OA in mice. Significantly greater PAP-positive areas can be seen within cells of the proximal convoluted tubules rather than the distal tubular cells 40 min after mice received 25 mg/kg OA orally. Paraffin section 5–7 μm, unstained. Magnification: ×500.

mycotoxin, then with GARP, and finally with rabbit antiperoxidase–peroxidase complex. Lee *et al.* (1984a) found that the PAP method was more sensitive than the indirect immunoperoxidase method. This method has been used successfully to study the fate of OA in mice (Lee *et al.*, 1984a) and to localize T-2 toxin (Lee *et al.*, 1984b) in different tissues and organs of mice after the animal received the toxin. Immunohistochemical staining for OA was most intense in the GI tract, intermediate in the kidney (Fig. 6), and least in the liver. In the T-2 toxin study, the toxin was demonstrable in the esophagus from 5 to 27 hr postdosing (Fig. 7). Positive stains were also observed in the cytoplasm of intact and injured epithelial cells of stomach, epithelium, and phagocytic elements of the duodenal lamina propria, and weak stain in the villous tip epithelial cells of jejunum and medulla of kidney. T-2 Toxin was never demonstrable in any of the hepatic tissue examined. These studies indicate that the antibodies could be very useful as a specific immunohistochemical reagent for diagnosis of certain mycotoxicoses. The use of antibody against mycotoxins for diagnosis of mycotoxicoses has recently been discussed by the author (Chu, 1986).

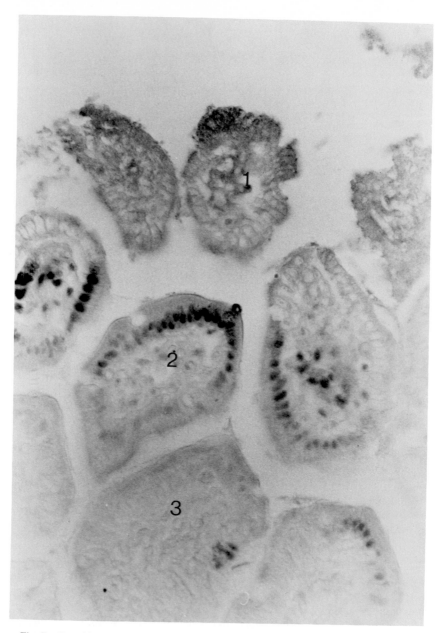

Fig. 7. Peroxidase and antiperoxidase (PAP) method for localization of T-2 toxin mice. High-power view of injured and recovered duodenal villous tips of mice 3 hr after orally receiving 11 mg/kg of T-2 toxin PO. Three types of villous tips are demonstrated: (1) necrotic, villous tips, and (2) partially necortic villous tips. Note cytoplasmic and nuclear binding of T-2 by epithelial cells as well as by phagocytic cells and fibroblasts of duodenal lamina propria; (3) normal, intact villous tips containing some T-2 toxin. Paraffin section 4–μm, unstained. Magnification: ×500.

VII. CONCLUDING REMARKS

The development of new immunochemical methods for mycotoxin analyses have progressed rapidly in recent years. Preliminary results have shown that the RIA and ELISA techniques have some advantages over the other chemical and biological methods and can be used for routine mycotoxin analyses, possibly for automation. However, as with any new method, it must be tested rigorously by others and extensive collaborative studies should be made to validate the assay's accuracy and reproducibility. Since the recoveries were low (60%) and the CV between each analysis was high (9–30%) in some immunoassays, additional research is necessary for improvement of both methods. Immunochemical methods for mycotoxin studies are not limited to the analysis of toxins in foods, feeds, and biological samples. We have demonstrated that they could be used as a histochemical tool for diagnosis of mycotoxicoses as well as to monitor the metabolism of mycotoxins.

The improvement and the extensive use of RIA and ELISA procedures for detecting mycotoxins in foods depends on the availability of antibodies. Thus, there is a need for developing a more efficient method for antibody production. In view of the recent success of hybridoma techniques, including the production of monoclonal antibody against Afla B_1, M_1 (Lubet *et al.*, 1983; Goopman *et al.*, 1984; Woychik *et al.*, 1984), and an Afla- DNA adduct (Groopman *et al.*, 1982; Haugen *et al.*, 1981; Hertzog *et al.*, 1982), efforts should be made for production of monoclonal antibodies against other mycotoxins. Additional efforts should also be directed to develop methods for production of antibodies against mycotoxins that are not currently available. Optimal conditions for production of antibodies in large animals such as goats should also be studied.

ACKNOWLEDGMENTS

This work was supported by the College of Agricultural and Life Sciences, the University of Wisconsin, Madison (NC-129). Part of the work described in this chapter was supported by a Public Health Service grant (CA 15064) from National Cancer Institute, FDA contracts 223-79-2270 and 223-79-2288, and an U.S. Army contract DAMD-C-2021.

REFERENCES

Aalund, O., Brundeldt, K., Hald, B., Krogh, P., and Poulsen, K. (1975). A radioimmunoassay for ochratoxin A: a preliminary investigation. *Acta Pathol. Microbiol. Scand., Sect. C* **83,** 390–392.

Abdalla, A. E., and Grant, D. W. (1981). An immunoanalysis of kojic acid. *Sabouradia* **18,** 191–196.

Adekunle, A. A., and Chu, F. S. (1979). Immunization against aflatoxicosis. *IRCS Med. Sci.* **7,** 452.

Ashoor, S. H., and Chu, F. S. (1975). Reduction of aflatoxin B_{2a} with sodium borohydride. *J. Agric. Food Chem.* **23,** 445–447.

Biermann, V. A., and Terplan, G. (1980). Nachweis von Aflatoxin B_1 mittels ELISA. *Arch. Lebensmittelhyg.* **31,** 51–57.

Cerny, M. E., Qian, G. S., and Yang, G. C. (1983). Rapid extraction and determination of aflatoxin M_1 in milk with an enzyme-linked imunosorbent assay. *Abstr. AOAC Annu. Meet., 97th, Washington, D.C.* p. 70.

Chang, F. C. C., and Chu, F. S. (1976). Preparation of ^{3}H-ochratoxins. *J. Labelled Compd. Radiopharm.* **12,** 231–238.

Chu, F. S. (1983). Immunochemical methods for mycotoxin analysis. *Proc. Int. Symp. Mycotoxin, Cairo, 1981* pp. 177–194.

Chu, F. S. (1984a). Immunochemical studies of mycotoxins. *In* "Toxigenic Fungi—Their Toxins and Health Hazard" (H. Kurata and Y. Ueno, eds.), p. 234. Kodansha, Tokyo and Elsevier, Amsterdam.

Chu, F. S. (1984b). Immunoassays for analysis of mycotoxins. *J. Food Prot.* **47,** 562–569.

Chu, F. S. (1986). Immunochemical methods for diagnosis of mycotoxicoses. *In* "Diagnosis of Mycotoxicoses" (J. L. Richard and J. R. Thurston, eds.), pp. 163–176. Martinus Nighoff Publishers, Dordrecht/Boston/Lancaster.

Chu, F. S., and Ueno, I. (1977). Production of antibody against aflatoxin B_1. *Appl. Environ. Microbiol.* **33,** 1125–1128.

Chu, F. S., Chang, F. C. C., and Hinsdill, R. D. (1976). Production of antibody against ochratoxin A. *Appl. Environ. Microbiol.* **31,** 831–835.

Chu, F. S., Hsia, M. T. S., and Sun, P. (1977). Preparation and characterization of aflatoxin B_1-O-carboxymethyloxime. *J. Assoc. Off. Anal. Chem.* **60,** 791–794.

Chu, F. S., Grossman, S., Wei, R. D., and Mirocha, C. J. (1979). Production of antibody against T-2 toxin. *Appl. Environ. Microbiol.* **37,** 104–108.

Chu, F. S., Lau, H. P., Fan, T. S., and Zhang, G. S. (1982). Ethylene diamine modified bovine serum albumin as protein carrier in the production of antibody against mycotoxins. *J. Immunol. Methods* **55,** 73–78.

Chu, F. S., Liang, M. Y., and Zhang, G. S. (1984a). Production and characterization of antibody against diacetoxyscirpenol. *Appl. Environ. Microbiol.* **48,** 777–780.

Chu, F. S., Zhang, G. S., William, M. D., and Jarvis, B. B. (1984b). Production and characterization of antibody against deoxyverrucarol. *Appl. Environ. Microbiol.* **48,** 781–784.

Chu, F. S., Steinert, B. W., and Graur, P. K. (1985). Production and characterization of antibody against aflatoxin G_1. *J. Food Safety* **7,** 161–170.

Cole, R. J., and Cox, R. H. (1981). "Handbook of Toxic Fungal Metabolites." Academic Press, New York.

Davis, N. D., Dickens, J. W., Freie, R. L., Hamilton, P. B., Shotwell, O. L., and Wyllie, T. D. (1980). Protocols for surveys, sampling, post-collection handling, and analysis of grain samples involved in mycotoxin problems. *J. Assoc. Off. Anal. Chem.* **63,** 95–102.

Davis, R. M., and Stone, S. S. (1979). Production of anti-rubratoxin antibody and its use in a radioimmunoassay for rubratoxin B. *Mycopathologia* **67,** 29–33.

Elling, F. (1977). Demonstration of ochratoxin A in kidney of pigs and rats by immunofluorescence microscopy. *Acta Pathol. Microbiol. Scand., Sect. A* **85,** 15–156.

El-Nakib, O., Pestka, J. J., and Chu, F. S. (1981). Determination of aflatoxin B_1 in corn, wheat, and peanut butter by enzyme-linked immunosorbent assay and solid phase radioimmunoassay. *J. Assoc. Off. Anal. Chem.* **64,** 1077–1082.

Fan, T. S. L., and Chu, F. S. (1984). An indirect enzyme-linked immunosorbent assay for detection of aflatoxin B_1 in corn and peanut butter. *J. Food Prot.* **47,** 263–266.

Fan, T. S. L., Zhang, G. S., and Chu, F. S. (1984a). Production and characterization of antibody against aflatoxin Q_1. *Appl. Environ. Microbiol.* **47,** 526–532.

Fan, T. S. L., Zhang, G. S., and Chu, F. S. (1984b). An indirect enzyme-linked immunosorbent assay for T-2 toxin in biological fluids. *J. Food Prot.* **47,** 964–967.

Fontelo, P. A., Beheler, J., Bunner, D. L., and Chu, F. S. (1983). Detection of T-2 toxin by an improved radioimmunoassay. *Appl. Environ. Microbiol.* **45,** 640–643.

Fremy, J. M., and Chu, F. S. (1984). Direct ELISA for determining aflatoxin M_1 at picogram levels in dairy products. *J. Assoc. Off. Anal. Chem.* **67,** 1098–1101.

Friesen, M. D., and Garren, L. (1982a). International mycotoxin check sample program Part I. Report on the performance of participating laboratories for the analysis of aflatoxin B_1, B_2, G_1, and G_2. *J. Assoc. Off. Anal. Chem.* **4,** 855–863.

Friesen, M. D., and Garren, L. (1982b). International mycotoxin check sample program Part II. Report on the performance of participating laboratories for the analysis of aflatoxin M_1 in milk. *J. Assoc. Off. Anal. Chem.* **4,** 864–868.

Gaur, P. K., El-Nakib, O., and Chu, F. S. (1980). Comparison of antibody production against aflatoxin B_1 in goats and rabbits. *Appl. Environ. Microbiol.* **40,** 678–680.

Gaur, P. K., Lau, H. P., Pestka, J. J., and Chu, F. S. (1981). Production and characterization of aflatoxin B_{2a} antiserum. *Appl. Environ. Microbiol.* **41,** 478–482.

Groopman, J. D., Haugen, A., Goodrich, G. R., Wogan, G. N., and Harris, C. C. (1982). Quantitation of aflatoxin B_1-modified DNA using monoclonal antibodies. *Cancer Res.* **42,** 3120–3124.

Groopman, J. D., Trudel, L. J., Marshak-Rothstein, A., and Wogan, G. N. (1984). High affinity monoclonal antibodies recognizing aflatoxins and aflatoxin-DNA adducts. *Abstr. AACR Annu. Meet. Toronto.* p. 98.

Harder, W. O., and Chu, F. S. (1979). Production and characterization of antibody against aflatoxin M_1. *Experientia* **36,** 1104–1105.

Haugen, A., Groopman, J. D., Hsu, I. C., Goodrich, G. R., Wogan, G. N., and Harris, C. C. (1981). Monoclonal antibody to aflatoxin B_1-modified DNA detected by enzyme immunoassay. *Proc. Natl. Acad. Sci. U.S.A.* **78,** 4124–4127.

Hertzog, P. J., Linday Smith, J. R., and Garner, R. C. (1982). Production of monoclonal antibodies to guanine imidazole ring opened aflatoxin B_1 DNA, the persistent DNA adduct *in vivo*. *Carcinogenesis* **3,** 825–828.

Horwitz, W. (1980). "Official Methods of Analysis," 13th Ed. Assoc. Off. Anal. Chem., Arlington, Virginia.

Hu, W. J., Woychik, N., and Chu, F. S. (1984). ELISA of picogram quantities of aflatoxin M_1 in urine and milk. *J. Food Prot.* **47,** 126–127.

Hunter, K. W., Jr., Brimfield, A. A., Miller, M., Finkelman, F. D., and Chu, F. S. (1985). Preparation and characterization of monoclonal antibodies to the trichothecene mycotoxin T-2. *Appl. Environ.* Microbiol. **49,** 168–172.

Langone, J. J., and Van Vunakis, H. (1976). Aflatoxin B_1: Specific antibodies and their use in radioimmunoassay. *J. Natl. Cancer Inst.* **56,** 591–595.

Langone, J. J., and Van Vunakis, H., eds. (1981). "Immunochemical Techniques, Parts B and C," Methods in Enzymology, Vols. 73 and 74. Academic Press, New York.

Lau, H. P., Fan, T., and Chu, F. S. (1981a). Preparation and characterization of aflatoxin Q_1-hemisuccinate. *J. Assoc. Off. Anal. Chem.* **64,** 681–683.

Lau, H. P., Gaur, P. K., and Chu, F. S. (1981b). Preparation and characterization of aflatoxin B_{2a}-hemiglutarate and its use for the production of antibody against aflatoxin B_1. *J. Food Saf.* **3,** 1–13.

Lawellin, D. W., Grant, D. W., and Joyce, B. K. (1977a). Aflatoxin localization by enzyme-linked immunocytochemical technique. *Appl. Environ. Microbiol.* **34,** 88–93.

Lawellin, D. W., Grant, D. W., and Joyce, B. K. (1977b). Enzyme-linked immunosorbent analysis of aflatoxin B_1. *Appl. Environ. Microbiol.* **34,** 94–96.

Lee, S., and Chu, F. S. (1981a). Radioimmunoassay of T-2 toxin in corn and wheat. *J. Assoc. Off. Anal. Chem.* **64,** 156–161.

Lee, S., and Chu, F. S. (1981b). Radioimmunoassay of T-2 toxin in biological fluids. *J. Assoc. Off. Anal. Chem.* **64,** 684–688.

Lee, S. C., Beery, J. T., and Chu, F. S. (1984a). Immunohistochemical fate of ochratoxin A in mice. *Toxicol. Appl. Pharmacol.* **72,** 218–227.

Lee, S. C., Beery, J. T., and Chu, F. S. (1984b). Immunoperoxidase localization of T-2 toxin. *Toxicol. Appl. Pharmacol.* **72,** 228–235.

Lee, S. S., and Chu, F. S. (1984). Enzyme-linked immunosorbent assay of ochratoxin A in wheat. *J. Assoc. Off. Anal. Chem.* **67,** 45–49.

Li, Y. K., and Chu, F. S. (1982). Kinetics of transformation of aflatoxin B_1 to aflatoxin M_1 in lactating mouse, an ELISA analysis. *Experientia* **38,** 842–843.

Li, Y. K., and Chu, F. S. (1984). Production and characterization of antibody against sterigmatocystin. *J. Food Saf.* **6,** 119–126.

Lubet, M. T., Olson, D. F., Yang, G., Ting, R., and Steuer, A. (1983). Use of a monoclonal antibody to detect aflatoxin B_1 and M_1 in enzyme immunoassay. *Abstr. AOAC Annu. Meet., 97th, Washington, D.C.* p. 71.

Maggio, E. T. (1980). "Enzyme-Immunoassay." CRC Press, Boca Raton, Florida.

Martin, C. N., Garner, R. C., Tursi, F., Garner, J. V., Whittle, H. C., Ryder, R. W., Sizaret, P., and Montesano, R. (1984). An ELISA procedure for assaying aflatoxin B_1. *In* "Monitoring Human Exposure to Carcinogenic and Mutagenic Agents" (A. Berlin, M. Draper, K. Hemminki, and H. Vainio, eds.), Sci. Publ. No. 59. IARC, Lyon.

Miller, E. C. (1978). Some current perspectives on chemical carcinogenesis in human and experimental animals: presidential address. *Cancer Res.* **38,** 1479–1496.

Mirocha, C. J., Pathre, S. V., and Christensen, C. M. (1979). Mycotoxins. *Adv. Cereal Sci. Technol.* **3,** 159–224.

Morgan, M. R. A., Matthew, J. A., McNerney, R., and Chan, H. W. S. (1982). The immunoassay of ochratoxin A. *Proc. Int. IUPAC Symp. Mycotoxins Phycotoxins, 5th, Vienna* pp. 32–35.

Morgan, M. R. A., McNerney, R., and Chan, H. W. S. (1983). Enzyme-linked immunosorbent assay of ochratoxin A in barley. *J. Assoc. Off. Anal. Chem.* **66,** 1481–1484.

Pestka, J. J., and Chu, F. S. (1984a). Aflatoxin B_1 dihydrodiol antibody: Production and specificity. *Appl. Environ. Microbiol.* **47,** 472–477.

Pestka, J. J., and Chu, F. S. (1984b). Enzyme-linked immunosorbent assay of mycotoxins using nylon bead and Terasaki plate solid phases. *J. Food Prot.* **47,** 305–308.

Pestka, J. J., Gaur, P. K., and Chu, F. S. (1980). Quantitation of aflatoxin B_1 antibody by an enzyme-linked immunosorbent microassay. *Appl. Environ. Microbiol.* **40,** 1027–1031.

Pestka, J. J., Lee, S. S., Lau, H. P., and Chu, F. S. (1981a). Enzyme-linked immunosorbent assay for T-2 toxins. *J. Am. Oil Chem. Soc.* **58,** 940A–944A.

Pestka, J. J., Li, Y. K., Harder, W. O., and Chu, F. S. (1981b). Comparison of a radioimmunoassay and an enzyme-linked immunosorbent assay for the analysis of aflatoxin M_1 in milk. *J. Assoc. Off. Anal. Chem.* **64,** 294–301.

Pestka, J. J., Steinert, B. W., and Chu, F. S. (1981c). An enzyme-linked immunosorbent assay for detection of ochratoxin A. *Appl. Environ. Microbiol.* **41,** 1472–1474.

Pestka, J. J., Li, Y. K., and Chu, F. S. (1982). Reactivity of aflatoxin B_{2a} antibody with aflatoxin B_1-modified DNA and related metabolites. *Appl. Environ. Microbiol.* **44,** 1159–1165.

Pestka, J. J., Beery, J. T., and Chu, F. S. (1983). Indirect immunoperoxidase localization of aflatoxin B_1 in rat liver. *Food Chem. Toxicol.* **21,** 41–48.

Peters, H., Dierich, M. P., and Dose, K. (1982). ELISA for detection of T-2 toxin. *Hoppe-Seyler's Z. Physiol. Chem.* **363,** 1437–1447.

Plescia, O. J. (1982). Preparation and assay of nucleic acids as antigens. *In* "Nucleic Acid" (L. Grossman and K. Moldave, eds.), Methods in Enzymology, Vol. 22, Part B, pp. 893–899. Academic Press, New York.

Rodricks, J. V., Hesseltine, C. W., and Mehlman, M. A. (1977). Mycotoxins in Human and Animal Health." Pathotox, Park Forest South, Illinois.

Schuurs, A. H. W. M., and Van Weemen, B. K. (1977). Enzyme-immunoassay. *Clin. Chim. Acta* **81,** 1–40.

Sizaret, P., Malaveille, C., Montesano, R., and Frayssinet, C. (1980). Detection of aflatoxins and related metabolites by radioimmunoassay. *J. Natl. Cancer Inst.* **69,** 1375–1380.

Sizaret, P., Malaveille, C., Bruhn, G., Aquelon, A. M., and Toussaint, G. (1982). Inhibition by specific antibodies of the mutagenicity of aflatoxin B_1 in bacteria. *Oncodevl. Biol. Med.* **3,** 125–134.

Stoloff, L. (1980). Aflatoxin control: Past and present. *J. Assoc. Off. Anal. Chem.* **63,** 1067–1073.

Sun, P., and Chu, F. S. (1977). A simple solid-phase radioimmunoassay for aflatoxin B_1. *J. Food Saf.* **1,** 67–75.

Sun, T., Wu, Y., and Wu, S. (1983). Monoclonal antibody against aflatoxin B_1 and its potential applications. *Chin. J. Oncol.* **5,** 401–405.

Thorpe, C. W., and Yang, G. C. (1979). Comparison of two HPLC methods and an RIA procedure for the determination of aflatoxin in cornmeal. *Abstr. Annu. Meet. AOAC, 93rd, Washington, D.C.* pp. 46.

Thouvenot, D., and Morfin, R. F. (1983). A radioimmunoassay for zearalenone and zearalanol in human serum: production, properties and use of porcine antibodies. *Appl. Environ. Microbiol.* **45,** 16–23.

Tsuboi, S., Nakagawa, T., Tomita, M., Seo, T,, Ono, H., Kawamura, K., and Iwamura, N. (1984). Detection of aflatoxin B1 in serum samples of male Japanese subjects by radioimmunoassay and high performance liquid chromatography. *Cancer Res.* **44,** 1231–1234.

Ueno, I., and Chu, F. S. (1978). Modification of hepatotoxic effect of aflatoxin B_1 in rabbits by immunization. *Experientia* **34,** 85–86.

Van Vunakis, H., and Langone, J. J., eds. (1980). "Immunochemical Techniques, Part A," Methods in Enzymology, Vol. 70, p. 525. Academic Press, New York.

Wallace, E. M., Pathre, S. V., Mirocha, C. J., Robison, T. S., and Fenton, S. W. (1977). Synthesis of radiolabeled T-2 toxin. *J. Agric. Food Chem.* **25,** 836–838.

Worsaae, H. (1978). Production of an ochratoxin A antigen with high hapten/carrier molar ratio. *Acta Pathol. Microbiol. Scand., Sect. C* **86,** 203–204.

Woychik, N. A., Hinsdill, R. D., and Chu, F. S. (1984). Production and characterization of monoclonal antibody against aflatoxin M_1. *Appl. Environ. Microbiol.* **48,** 1096–1099.

Wu, S., Yang, G., and Sun, T. (1983). Studies on immuno-concentration and immunoassay of aflatoxins. *Chin. J. Oncol.* **5,** 81–84.

Yalow, R. S. (1978). A probe for the fine structure of biologic systems. *Science* **200,** 1236–1245.

Yasaei, P., Qian, G. S., and Yang, G. C. (1983). Radioimmunoassay detection of aflatoxin B_1 and M_1 in liver tissue extracts using disposable column. *Abstr. 97th, AOAC Annu. Meet., Washington, D.C.* p. 71.

9

Thin-Layer Chromatography/High-Performance Thin-Layer Chromatography as a Tool for Mycotoxin Determination

STANLEY NESHEIM AND MARY W. TRUCKSESS
Food and Drug Administration
Division of Chemistry and Physics
Washington, D. C. 20204

MODERN METHODS IN THE ANALYSIS
AND STRUCTURAL ELUCIDATION
OF MYCOTOXINS

I. INTRODUCTION

The use of thin-layer chromatography (TLC) for separation and analysis of complex mixtures grew rapidly after E. Stahl standardized the procedure in 1958 and showed its wide applicability (Stahl, 1969). It is therefore no coincidence that the isolation, structural elucidation, and development of TLC methods of analysis for aflatoxins occurred at about that time. The aflatoxins were discovered in 1960, as fluorescent blue and green spots on TLC. Pure aflatoxins for structural determination were prepared by preparatory TLC, and quantitative methods were developed. In some cases the aflatoxins served as the model compounds for development of reliable methods to measure trace levels (few micrograms per kilogram) in foods and feeds. More recently many other mycotoxin TLC methods have been developed (Romer, 1984; Scott, 1982). With advances in technique, TLC is becoming for some mycotoxins the method of choice. The separation efficiency is comparable to other modes of chromatography, that is, 5000 theoretical plates for 5-cm migration in high-performance TLC (HPTLC) or 10 cm in TLC (Rogers, 1984), compared with 5000 for a 20-cm-long HP liquid chromatographic (HPLC) column.

The mycotoxins include a broad range of types of compounds (Pohland and Thorpe, 1982; Betina, 1984). TLC is ideal for handling a great variety of compounds, through judicious choice of mobile phases, multiple development with the same or different solvents or solvent mixtures, or use of multidirectional (two-dimensional) elution.

Determination of mycotoxins using TLC has many advantages. It is simple and economical. Generally it is nondestructive, so that compounds may be recovered for further analysis. The adsorbent can be impregnated with a variety of chemicals so as to achieve specific separations. Quantitation can be carried out either visually or instrumentally, and a number of reagents are available (usually applied as sprays) to achieve visualization of specific compounds. The capacity of one-dimensional TLC can be as many as 25 discrete sample extract spots, and for two-dimensional TLC it can be 100–400 spots. Until recently the precision of measurement of the spots, expressed as coefficients of variation (CVs) of relatively crude estimations made visually or using TLC scanners, was in the range of 5 to 35%. Instrumentation that greatly improves this precision has become available. The absorbance of aflatoxin spots can now be determined automatically with CVs less than 1%. A high-speed digital microdensitometer drum scanner, which can scan 3 cm^2/sec (20 × 20 cm plate in 2 min) with a repeatability of 0.07 to 0.5%, is commercially available. This apparatus has not yet been routinely applied to TLC or to mycotoxin determination.

The term "high performance," adopted from high-performance liquid chromatography, originally was used to describe the use of small, uniform 5- to 20-μm particle adsorbents, in contrast to the 37- to 70-μm adsorbents previously available. More recently, the definition has been broadened to include all the techniques used to improve performance, such as multidimensional, multisolvent, multidevelopment, and totally instrumentalized TLC: from sample extract application and densitometric scanning of the plate to collection and evaluation of the data by computer. Considerable progress continues to be made in mycotoxin determination by TLC. Cleanup techniques are being simplified and made more efficient. With new instrumentation, better precision and accuracy and lower determination limits are becoming possible.

II. THIN-LAYER CHROMATOGRAPHIC PROCEDURES FOR MYCOTOXINS

This chapter, without going into extraction and cleanup details, presents the best presently available TLC quantitation and identification procedures for selected mycotoxins. Comprehensive reviews of TLC methods have been published for several mycotoxins. Some of the more useful methods are listed in Table I. Many of the reviews detail extraction and cleanup techniques as well.

As for most other separation and determinative techniques, the analyte must be extracted and concentrated. The most appropriate solvent for complete extraction and minimum matrix carryover is chosen for simplicity of the subsequent cleanup. Acetonitrile, chloroform, acetone, ethyl acetate, and methanol, or mixtures of these with water, are used. Acids or bases are added if the toxins are acidic or basic. The most polar organic solvents are used for water-soluble toxins such as patulin and some of the trichothecenes.

Efficient extraction and cleanup is critical to successful TLC. To determine trace levels of mycotoxins, the analyte has traditionally been concentrated and purified by solvent partition, metal complexing, and precipitation, or by adsorption column chromatography on silica gel, alumina, magnesium silicate (Florisil), or charcoal. Improvements made in this area include changes from large (>10 g) to smaller (0.25–2 g) columns, the use of materials with smaller and more uniform particles, and the use of commercially available prepacked disposable plastic columns. The columns are eluted with vacuum or pressure.

A. Thin-Layer Chromatographic Plates

The general principles and recent progress in TLC have been reviewed (Bertsch *et al.*, 1980; Fenimore and Davis, 1981; Hauck, 1981; Hauck and Jost, 1983; Kaiser, 1982; Rogers, 1984; Sherma and Fried, 1982; Touchstone, 1982; Zlatkis and Kaiser, 1977). There is no doubt that present knowledge and techniques applied to mycotoxin trace determination would vastly improve the quality of analytical data, so that they would equal and surpass data obtained by most

TABLE I

TLC of Mycotoxins on Silica Gel

Mycotoxins	Commodity	Spotting solvent	Mobile phase[a]	Detection method and limit (ng/g)[b]	Reference
Aflatoxins	Peanuts, peanut products, corn, nuts	C_6H_6–CH_3CN (98 + 2)	Acetone–$CHCl_3$ (1 + 9) C_6H_6–EtOH–water (40 + 6 + 3 trough), (4 + 27 + 20 tank)	Fl. blue B_1, B_2, Fl. green G_1, G_2 ex. 366 nm, em. 420 nm B_1, G_1 (5); B_2, G_2 (1)	Stoloff and Scott (1984)
	Mixed feed		Acetone–$CHCl_3$ (1 + 9)	B_1 (2)	Shannon et al. (1983)
	Cottonseed oil		Et_2O–MeOH–water (96 + 3 + 1)	B_1 (2)	M. W. Trucksess (1984 unpublished)
	Ginger	$CHCl_3$	Acetone–$CHCl_3$ (1 + 9)	B_1, G_1 (2) B_2, G_2 (0.6)	Trucksess and Stoloff (1980)
	Eggs	$CHCl_3$	Hexane–THF–EtOH (70 + 20 + 10)	B_1 (0.05)	Trucksess and Stoloff (1984)
Aflatoxin M_1	Milk and milk products	C_6H_6–CH_3CN (9 + 1)	i-PrOH–acetone–$CHCl_3$ (5 + 10 + 85)	Fl. blue (0.05); ex. 366 nm, em. 420 nm	Stubblefield (1979)
	Milk	C_6H_6–CH_3CN (9 + 1)	Et_2O–hexane–MeOH–water (85 + 10 + 4 + 1)		Price et al. (1981)
	Blood, urine, tissue	$CHCl_3$	Et_2O–MeOH–Water (96 + 4 + 1)		Trucksess et al. (1982)
Aflatoxicol	Urine	$CHCl_3$	$CHCl_3$–acetone–i-PrOH (85 + 10 + 5) equil. tank	Fl. blue, ex. 365 nm (1 ng/spot)	Lovelace et al. (1982)
Ochratoxin A	Corn, barley, tissue	C_6H_6–HOAc (99 + 1)	C_6H_6–HOAc–MeOH (90 + 5 + 5)	Fl. blue, NH_3–MeOH vapor, ex. 366 nm, em. 420 nm; grains (10), tissue (5)	Nesheim et al. (1984)
Sterigmatocystin	Barley, wheat, cheese, corn	$CHCl_3$	C_6H_6–HOAc–MeOH (90 + 5 + 5)	$AlCl_3$ Spray (20 g/100 ml EtOH), heat 5 min 110°C; silicone–Et_2O spray (15 + 82); fl. yellow, ex. 365 nm (5–20), cheese (2)	Francis et al. (1985)

242

Compound	Extraction	Solvent	TLC solvent system	Detection	Reference
Patulin	Apples, pears, juices, jams	CHCl₃	Touene–EtOAc–CHCl₃–HCOOH (105 + 45 + 48 + 3)	MBTH spray (0.5 g/100 ml H₂O), heat 15 min 130°C; fl. yellow-brown, ex. 366 nm (120–130)	Gimeno and Martins (1983)
Citrinin	Apples, pears, juices, jams	CHCl₃	Toluene–EtOAc–CHCl₃–HCOOH (90 + 45 + 50 + 5)	Silica gel–oxalic acid (95 + 5 w/w) AlCl₃ spray heat 5 min 105°C; fl. blue, ex. 366 nm (30–40)	Gimeno and Martins (1983)
Citrinin	Corn, barley	CHCl₃	Toluene–EtOAc–CHCl₃–HCOOH (70 + 50 + 50 + 20) or (80 + 50 + 69 + 6)	Silica gel–glycolic acid, AlCl₃ spray (20 g/100 ml EtOH), heat 5 min 105°C; fl. blue, ex. 366 nm (15–20)	Gimeno (1984)
Penicillic acid	Corn, oats, barley, dried beans	CH₂Cl₂	Toluene–EtOAc–HCOOH (6 + 3 + 1)	p-Anisaldehyde spray (0.5 ml/85 ml MeOH–HOAc–H₂SO₄ (70 + 10 + 5), heat 8 min 130°C; fl. blue, ex. 366 nm (50)	Thorpe (1982)
Cyclopiazonic acid	Corn, peanuts	CHCl₃	EtOAc–i-PrOH–NH₄OH (50 + 15 + 10)	p-Dimethylaminobenzaldehyde spray (1 g/75 ml EtOH + 25 ml HCl), fl. blue (125)	Lansden (1984)
Alternariol, alternariol methyl ether	Tomatoes, catsup, cherry juice	CHCl₃	C₆H₆–HOAc–MeOH (90 + 5 + 5)	Fl. blue, ex. 365 nm (5)	Stack (1984)
Altertoxin I and II				I₂ Vapor, fl. purple-brown (2 µg/spot)	
Tenuazonic acid	Tomatoes	CHCl₃	C₆H₆–HOAc–MeOH (90 + 5 + 5)	Fl. quenching, ex. 254 nm (2 µg/spot)	Stack (1984)
Mycophenolic acid	Cheese	CHCl₃	CHCl₃–Et₂O–HCOOH (60 + 20 + 0.1) or CHCl₃–EtOAc–HCOOH (60 + 40 + 0.2)	Ethylamine spray or NH₃ vapor; fl. blue, ex. 366 nm (25)	Lafont and Debeaupuis (1982)
Luteoskyrin	Rice	MeOH	Benzene–acetone, (4 + 1) or upper-phase hexane–acetone–water (5 + 5 + 3.5)	Silica gel–oxalic acid, fl. yellow (20)	Ueno (1982)
Xanthomegnin	Green coffee beans	CHCl₃	C₆H₆–HOAc–MeOH (90 + 5 + 5)	Fl. yellow	Stack et al. (1983)
Viomellein				Fl. brown	

(continued)

243

TABLE I (*Continued*)

Mycotoxins	Commodity	Spotting solvent[a]	Mobile phase[a]	Detection method and limit (ng/g)[b]	Reference
Vioxanthin Rubratoxin B	Corn	EtOAc	CHCl₃–MeOH–HOAc (80 + 20 + 2)	Fl. pink (1 μg/spot) Fl. quenching, ex.; 254 nm, em. 540 nm	Sandor (1982)
T-2, HT-2, T-2 tetraol, T-2 triol, Diacetoxyscirpenol	Standards	C₆H₆–CH₃CN (98 + 2)	Toluene–EtOAc–HCOOH (5 + 4 + 1)	10% Chromotropic acid–63% H₂SO₄ (1 + 5) spray, heat 5–15 min 110°C; fl. blue, ex. 365 nm (50 ng/spot) (100 ng/spot)	Baxter *et al.* (1983)
Zearalenone	Mixed feed	CHCl₃	CHCl₃–MeOH (97 + 3)	Fl. blue, ex. 250 nm (10)	Howell and Taylor (1981)
Deoxynivalenol (vomitoxin)	Corn, wheat	CHCl₃–CH₃CN (8 + 2)	CHCl₃–i-PrOH–acetone (8 + 1 + 1)	Silica gel–AlCl₃, heat 7 min 120°C; fl. blue, ex. 313 nm, em. 420 nm (40)	Trucksess *et al.* (1984b)

[a] Mobile phase (v/v): HOAc, acetic acid; Et₂O, ethyl ether; EtOAc, ethyl acetate; HCOOH, formic acid (90%); i-PrOH, 2-propanol; MeOH, methanol; THF, tetrahydrofuran.

[b] Fl., Fluorescence; ex., optimum excitation wavelength; em., optimum emission wavelength; MBTH, 3-methyl-2-benzothiazolinone hydrazone hydrochloride.

other methods. To do this it is necessary to have homogeneous, representative samples, well-characterized standards, chromatographic systems under control, and an error- or bias-free measurement.

The TLC of the aflatoxins has received the most attention over the years; consequently it is the most refined and generally serves as a model for the other mycotoxins. Silica gel is used as the adsorbent almost exclusively. Most analysts prefer commercially prepared plates because of the durability and homogeneity of the adsorbent layer. The preferred layer thickness is 0.25–0.5 mm. Commercial plates as received must be cleaned for quantitative work by prewashing the plate with methanol or with a solvent mixture somewhat more polar than the one to be used for the analysis, and then drying at 102°C for 15 min. A more efficient method is to layer the plates with adsorbent pads or filter paper and wash with solvent. Besides being clean, the silica gel plate must meet the performance test of separating aflatoxins B_1, B_2, G_1, and G_2 (Stoloff and Scott, 1984, Sect. 26.031). Commercially precoated plates that have been found satisfactory include silica gel 60 or Alufoil (E. Merck) and Unisil G (Analtech). The Alufoil (aluminum-backed) plates are popular because they can easily be cut to different sizes, and spots can be cut out for extraction during preparatory TLC.

Plates are easily prepared in the laboratory, and for many analysts this is the most economical method. The silica gel can be any of a number of commercially available materials incorporating 10–15% $CaSO_4$ as a binder. The following four were judged to be satisfactory for determination of aflatoxins: Silic AR TLC-7G or Silic AR TLC-4G (Mallinckrodt), silica gel G HR (Macherey & Nagel), Adsorbosil-Plus-2 (Applied Science). The properties of silica gel vary from manufacturer to manufacturer and from batch to batch. Convenient slurry-spreading apparatuses include the Instrutec self-glass-leveling apparatus and the Camag automatic spreader. Single-strength window glass is strong enough to be reused many times. Laboratory-prepared plates are very soft and require gentle handling. Their performance must also be monitored by a B_1, B_2, G_1, and G_2 separation test.

To prepare smooth TLC plates of even thickness, the adsorbent is shaken by hand for 30 sec with enough water to give a slurry with a creamy or paintlike consistency. If the slurry is too thin, less uniform layers are produced as the gel settles on the TLC plate and forms water puddles. If the slurry is too thick, the gel flows unevenly, and layers with ridges are formed. The slurry is spread at a rapid and constant speed and allowed to harden undisturbed for 15 to 30 min; then the plate is dried in an 80°C oven for 2 hr. If the relative humidity is less than 60%, the plates can be dried overnight at room temperature. The Camag automatic spreader, which requires much less skill to operate than the hand spreader and saves time, is particularly useful when many plates are needed. The dried and cleaned plate must be stored in a sealed storage cabinet.

The solvent for spotting should be volatile. It must readily dissolve the mycotoxin but must not spread the sample extract excessively as it is applied. Suggested solvents are listed in Table I. Compounds of interest must be stable in the solvent. Aflatoxin M_1, for example, is less stable in chloroform than in benzene–acetonitrile (9 + 1 v/v) and decomposes in a few days in methanol. Some compounds of interest are sensitive to ultraviolet (UV) light, making it necessary to apply the sample extract under subdued incandescent light. For qualitative work, hand application with a 10-μl syringe or Drummond micropipettes can be used. A transparent template and a desktop hood are also essential.

For ease in applying the sample extract by hand a comfortable sitting position should be assumed. The sample extract should be applied rapidly and carefully without damaging the layer but slowly enough to avoid flow over the surface. The application should be as small as possible. For precise work, a mechanical spotter such as the Linomat III (Camag) should be used. The sequence of sample extracts and standard solutions applied across the TLC plate should be planned before spotting, and should be recorded in the notebook. For good quantitation, it is recommended that standard solutions be applied at three concentrations, bracketing the ranges of the expected concentrations of the sample extracts. If the sample extract shows interfering spots near the mycotoxin spots, an additional intermediate concentration is applied, and an approximately equal quantity of mycotoxin standard is superimposed as an internal standard. The range of concentration of mycotoxins should be within the linear range of a standard curve. The linear range depends on the toxin, amount and kind of residual solvent on the TLC plate, and the densitometric instrument used.

B. Chromatogram Development

Many factors in development can contribute to poor resolution and therefore to unsatisfactory quantitation. Humidity plays a major role in adsorption chromatography (normal-phase chromatography) using organic or nonaqueous solvents. Relative humidity outside the range of 40 to 60% usually presents problems in the TLC of aflatoxins. Increased moisture in the silica gel layer deactivates it, giving rise to high R_f values, and in the extreme case, to total loss of resolution. The plates must be reactivated immediately before use and spotted under nitrogen. If the moisture content is too low, tailing may occur. In this case, water is added to the developer or placed in the bottom of the chamber. Silica gel 60-precoated plates (E. Merck) have a polyamide binder and are not affected as much by humid conditions.

Solvent selectivity, of course, is very important to resolution. Table I lists references to methods and preferred solvents for separation of individual mycotoxins in different types of extracts. In addition, the sections on individual mycotoxins give reviews of many useful solvents. The solvent composition as described may have to be adjusted to compensate for variation in silica gel activity to bring the R_f values of the mycotoxin of interest into the desired range of 0.3 to 0.8. A change in the R_f value decreases or increases the separation between components in a chromatographic mixture but not their relative position. To separate incompletely resolved toxins and/or interferences, different mobile phases of different selectivities should be evaluated. The TLC separation process on silica gel is complex. Water and mobile-phase vapors absorbed above the solvent front modify the adsorbent properties of the silica gel, and the mobile phase is separated to form a gradient as the plate is developed. Therefore, adsorption, partition, and ionic chromatography, or mixtures of these, can take place. Mycotoxin TLC includes examples of all modes. The most efficient approach to solving a resolution problem is to test a few mobile phases with different selectivities operating in different chromatographic modes (neutral–normal phase, acidic–normal phase, and reversed phase). To develop the TLC plates with similar mobile phases, however, serves little purpose except perhaps to gain a false assurance of positive chemical identification.

The standard metal development chamber for 20-cm plates is commonly used. To minimize temperature gradients the chamber must be insulated (not necessary for glass) and be well sealed to avoid uneven development across the plate. In laboratories where temperature is poorly controlled, it may be necessary to build a special TLC room or to place the development chamber in an incubator. Masking tape can be used to seal a poorly fitting cover. A nonequilibrated tank is used for mycotoxins that are difficult to separate, such as the aflatoxins. Equilibration speeds up development and gives more uniform R_f values at the expense of resolution. Standard plate development should take no more than 90 min. For HPTLC (small-particle, short-time, and short-distance development), small chambers with a suitable mobile phase and/or with controlled-chamber atmosphere must be used.

Other factors which affect R_f value and resolution include uniformity of particle size distribution and thickness of the adsorbent layer; volatility of mobile phases; the size of the chamber; temperature; and proximity of the spots to the side of the plate or irregular score lines which affect lane evaporation rates.

C. Multidimensional Development

When simple development fails to resolve the mycotoxins from interferences (e.g., in the analysis of eggs, cheese, milk, spices, and certain mixed feeds for aflatoxins), the following special techniques can be useful:

1. Two-solvent development: The plate is developed first with a solvent that removes the interferences and then with the usual development solvent in the same direction. An example is the determination of aflatoxin in corn. Ethyl ether

[cannot be used for silica gel 60 or Adsorbosil 1 plates (Trucksess, 1974)] is used to move interferences to the top of the plate. When silica gel with an aluminum support is used, the top of the plate is cut off and the plate is developed again in the same direction. With a glass plate, this may be accomplished by scoring the silica gel below the interference and developing again in the same direction as the first time.

2. Two-dimensional development: The sample extract is spotted in one corner with reference standard solutions on two sides. The plate is developed in one direction, dried in an oven for 2 min at 50°C, and then developed in a second direction (90° from first direction) with a different mobile phase (Stoloff and Scott, 1984, Sect. 26.074).

D. Screening Methods

TLC is well suited for use in screening methods. It is very flexible and simple, and many samples can be chromatographed at the same time under the same or different conditions. Screening methods serve many purposes, including the following:

1. These tests eliminate large numbers of negative samples with minimum effort. Usually each spot represents one sample extract. The amount of extract used must not exceed the TLC adsorbent binding capacity. It need only be large enough to contain an amount of mycotoxin at a level above the limit of detection. For determination of the strongly fluorescent mycotoxins, a good starting point is to spot the amount of extract equivalent to 0.25 g of original sample and an amount of standard one to five times the smallest determinable quantity. For sample extracts with complex chromatographic patterns, a useful technique is to apply two spots and superimpose the standard (internal standard) on one spot to facilitate identification of the mycotoxin.

2. Such tests allow a preliminary estimate of concentration for use in dilution of the sample extract to match its mycotoxin concentration to that of the standard for subsequent quantitation. A good spotting volume range for the standard and/or sample is 1–20 μl.

3. Such tests allow determination of more than one mycotoxin on the same plate. Many multitoxin methods have been published (Gimeno, 1980; Howell and Taylor, 1981; Johann and Dose, 1983; Nowotny et al., 1983; Patterson, 1983; Stoloff et al., 1971; Takeda et al., 1979; Wilson et al., 1976). In these methods extracts prepared by one or several techniques are applied to several TLC plates, with different mycotoxins grouped according to chromatographic properties. The plates may be developed with different solvents appropriate to the group, or all may be developed with the same solvent but treated with different visualization reagents.

E. Quantitative Evaluation

TLC is a very popular chromatographic method, especially for qualitative analysis and preparatory separations. However, in some cases it has not been accepted for quantitative work. The reason often given is the poor precision of many of the reported data. This opinion is not justified. Several basic problems have contributed to variability and lack of precision: unawareness of sources of errors (Novacek, 1973); variability in extract cleanup, technique, materials, equipment, and sample extract application; poor control of the many undefined parameters of chromatography (development); and, finally, differences in performance of the several types of instruments and lack of quality control and evaluation of the data.

Quantitation can be based on measurement of fluorescence, color, or UV adsorption of the toxin or derivative of the toxin. Fluorescence determination is preferred for several reasons: It is more specific than other techniques; amounts in the low or fraction of a nanogram range can be measured; and the linear range is large ($\geq$10-fold). For instrumental quantitation the following procedures apply: Each plate must have its own standards. Volatile mobile phases must be evaporated totally or to a constant level. If solvent is not to be removed, it can be sealed in the layer by covering the layer with another clean plate or by spraying with mineral oil. The concentration and kind of solvent present on the silica gel affect fluorescence intensity and must be controlled. This effect varies for different mycotoxins. For example, aflatoxin spots on a plate developed in acetone–chloroform fluoresce more strongly than ones developed with ether–methanol–water. Ochratoxin fluorescence is enhanced approximately 10-fold after exposure to methanol–ammonia, more so than with ethanol–ammonia or ammonia alone; the excitation maximum is shifted from 333 to 370 nm. Sealing the layer by spraying the plate with paraffin oil in hexane after development cuts down fluorescent light loss due to scattering. For scanning, the optimum excitation and emission wavelengths giving maximum response and least interference are chosen. In Table I are listed appropriate wavelengths useful for determining different mycotoxins.

No general settings for instrumental parameters can be recommended when using a densitometer. These parameters are generally different for each type of instrument. The parameters to be adjusted include type of filters and lamp, length and width of slits, sensitivity, amplification and damping, recorder damping, and/or methods of data handling. The chromatographic pattern must be carefully aligned with the scanning stage of the instrument. Correct alignment is critical to an analysis. Instrumental parameters and initial trial scans are best optimized by using one of the higher concentration standards. Settings should not be changed during the scan of the rest of the TLC plate. The sample extract spots should be scanned parallel to the direction of development and from low to high background (baseline). After the plate is scanned, one or two of the first sample extract spots should be rescanned to check for response stability. The response should be within 1% of the value originally obtained. A change in response >1% may indicate spot fading or instrumental error. The optimum instrumental parameters developed in our laboratory for the Camag TLC scanner using the high-pressure mercury lamp are as follows for ochratoxin A (OT) after development in NH_4OH–MeOH, aflatoxins (AT), and deoxynivalenol (DON) after spraying with $AlCl_3$ solution: excitation 366 nm (AT, OT), 313 nm (DON); emission 420 nm. For 5-cm development, 6-mm band application using Linomat III (Camag) sample applicator: slit 3×0.3 mm (Micro position) for sample extract applied with 10-μl syringe; for 10-cm development, slit 6×0.6 mm; sensitivity 8 on scale 1–12; scanning speed 0.5 mm/sec (slowest of settings, highest is 4 mm/sec); settings for the Spectra Physics 4100 integrator: attenuation 64, peak threshold 1000, chart speed 4 cm/min and peak width 3 sec.

F. Calculations

Common practice is to use peak area, rather than peak height, as a measure of concentration. In our experience with the Schoeffel SD 3000 and Camag TLC scanner, the measured aflatoxin peak heights are not proportional to concentration. This should be expected because peak size and shape are influenced by many factors, such as volume and amount of material applied, origin area, and distance of spot travel. Automatic integrators such as the Spectra Physics 4100 can be programmed to evaluate peaks by different modes, that is, where to start and stop integrating, and how to determine the baseline. The method giving the best signal:noise ratio for particular commodities and response linear to quantity of standard should be selected. For most sample extract spots, integration from peak trough to peak trough gives the best results. The readings from each sample extract spot on the plate should be evaluated for precision and accuracy before they are used in calculating results. A simple method for doing this is to calculate the average response per unit volume R_{av}/μl of the three or more standard spots and also for the two or more spots of each sample as follows:

$$R_{av}/\mu l = [R_1/V_1 + R_2/V_2 + R_3/V_2]/3$$
$$\% \, R_s = R_1/R_{av} \times 100$$

$R_{av}/\mu l =$ Average response per microliter
$R_1, R_2,$ and $R_3 =$ Response for each of the three standard spots
$V_1, V_2,$ and $V_3 =$ Microliter volumes of the respective spots
$\% \, R_s =$ Response for each spot 1, 2, 3, as % of the average response

Any spot which is more than $\pm$ 5% different from the average (or $\pm 1\%$ in HPTLC) should not be used for quantitative evaluation. Data for sample extract spots should not be extrapolated outside the range of standards covered on the TLC plate. A similar calculation should be made for the two or more different amounts of the sample extract applied.

G. Quantitation by Two-Dimensional Thin-Layer Chromatography

Quantitative analysis by two-dimensional TLC is the only available method in some cases. For several reasons, however, it is more time-consuming. Only one sample extract is applied to each TLC plate. Three or four different concentrations of standards, a sample extract, and a sample extract plus standard (internal standard) solution are applied as reference materials in two perpendicular margins. Replicate analyses, if needed, are done on separate TLC plates. The best quantitation is obtained when the standards are developed in the second dimension. Both sets of standards are each developed in only one direction (Stoloff and Scott, 1984, Sect. 26.106). Aligning the spots with the optics of the scanner is of critical importance. On hard plates, the spot can be marked either above or below with a pencil and the marks used to align the spots manually. Coordinates can also be assigned from a benchmark on the side of the plate and used to align the sample extract spot. On soft plates, a clean cover plate placed over the layer can be used to mark the spot. Before final scanning, the plate is manually adjusted in two directions until maximum response is obtained in the direction that gives a minimum rising baseline.

H. Confirmation of Identity by Thin-Layer Chromatography

Any discussion of chromatography leads naturally to the question of confirmation of identity of the compound measured. The methods available depend on the chemical characteristics of the mycotoxin. The formation of derivatives with changed chromatographic properties is probably the simplest way. The identity of some mycotoxins can be confirmed with various spray reagents (e.g., H_2SO_4, $AlCl_3$, and p-anisaldehyde) or use of multiple different mobile phases. The chemical methods for confirmation of identity are not absolute when compared with mass spectrometry (MS) (Nesheim and Brumley, 1981). Confirmation of identity of mycotoxins by means of MS analyses requires additional steps for sample extract cleanup.

After TLC separation, the TLC spot is eluted from the adsorbent by various devices. In our laboratory, we use the Camag Eluchrom system to elute six spots simultaneously. However, when the mycotoxin spot is separated by two-dimensional TLC or when confirmation of only one sample extract spot is needed, we locate and scrape the spot from the plate and transfer the adsorbent into a syringe filter (Millipore SR 0.2 μm) or onto a glass filter placed in a small filtering funnel. A solvent slightly more polar than the developer is used to elute the mycotoxin, which is collected in a conical centrifuge tube, and the solvent is evaporated. The sample extract residue is dissolved in acetone and is analyzed by either direct-probe MS (Brumley *et al.*, 1981) or gas chromatography–mass spectrometry (GC-MS) (Trucksess *et al.*, 1984a). This preparation is time-consuming, and the analyte is sometimes only partly recovered (Park *et al.*, 1985). Contamination of the isolated mycotoxin with impurities can also be a problem.

Recent advances in MS techniques avoid some of these problems. A TLC scanner interfaced with a mass spectrometer has been developed to transport the compound into the ion source directly from the TLC plate without prior elution (Ramaley *et al.*, 1983). The desorption energy is provided by either a tungsten filament incandescent lamp or a small CO_2 laser. Methane, a chemical ionization reagent gas, was used to sweep desorbed analytes through an isolation valve and down a glass-lined, stainless-steel transfer line to the ion source. The $(M + H)^+$ peak was determined. The efficiency of this technique is dependent on volatility and polarity of the analytes, chamber temperature, and TLC surface. Compounds of over 20 carbon atoms are difficult to detect by these techniques.

Another method for obtaining direct mass spectra of TLC spots is the use of fast-atom-bombardment (FAB) MS (Chang *et al.*, 1984). Nonvolatile or thermally labile compounds can be ionized directly while on the adsorbent. The probe tip, covered with a strip of double-faced masking tape (adhesive on both sides), is pressed against the TLC spot to pick up silica gel and adsorbed compound. Solvent and FAB matrix liquid are added, and the probe tip is inserted into the mass spectrometer. The technique has not been applied routinely to mycotoxins.

Recently the analyses of crude extracts for deoxynivalenol, zearalenone, and aflatoxin B_1 by MS-MS have been reported (Plattner, 1984). For enhanced sensitivity and specificity, it would be advantageous to combine this technique with a preliminary concentration or purification by TLC.

III. ANALYSIS OF INDIVIDUAL MYCOTOXINS

A. Aflatoxins

The aflatoxins occur in a great many commodities (Table I). The simplest method should be tried first; Table I can be used as a guide. The more complex methods are needed for commodities with many interferences, notably ginger root, spices, liver tissue, eggs, and complex feeds. Hundreds of solvent mixtures have been proposed (Nesheim, 1980; Scott, 1982), but for most problems one or more of the following are adequate: (a) acetone–chloroform (9 + 1 v/v), (b) ether–methanol–water (96 + 3 + 1 v/v/v), (c) benzene–ethanol–water (40 + 6 + 3 v/v/v), (d) dichloromethane–trichloroethylene–n-amyl alcohol–formic acid (80 + 15 + 4 + 1 v/v/v/v), and (e) benzene–acetic acid–methanol (90 + 5 + 5 v/v/v). If one system shows that the R_f values of standards and unknown are nonidentical, one could conclude that the spots are not aflatoxins. If all five systems separate aflatoxins B_1, B_2, G_1, and G_2, and the R_f values of standard and unknown spots are identical, one could still not conclude absolutely that the spots are aflatoxins, but one could be fairly sure that further efforts with TLC would probably be useless. In (a) through (c) there is a gradation of adsorption toward partition chromatography. In (d) the resolved order of the aflatoxins changes from B_1, B_2, G_1, G_2, to B_1, G_1, B_2, G_2, and in (e) the acidic inter- ferences are separated from the mycotoxins.

The major difficulties in aflatoxin TLC analyses have been spot fading and false positive results. A chemically clean laboratory environment, proper light- ing, and low humidity are essential to avoid spot fading (Nesheim, 1980). To eliminate false positive results, two-dimensional development with two solvent mixtures of widely different selectivities is most useful. Chemical confirmation and MS analysis are also being used.

Confirmation of identity of aflatoxins is performed by means of either MS analyses or chemical derivatization. Procedures for preparation and isolation of analyte prior to MS analyses are indicated in an earlier part of this chapter. Trifluoroacetic acid (TFA) is the most common reagent used for chemical deri- vatization of the aflatoxins. Aflatoxin B_1 or G_1 reacts with water across the double bond of the vinyl ether when TFA is used as the catalyst. This reaction can be carried out directly on a spot of the extract at the origin of a thin-layer plate; 2 μl TFA is superimposed at the origin on one of the mixed standard spots and on the sample extract spots containing the presumed aflatoxins. After 5 min at room temperature, the excess TFA is evaporated by heating in a 40°C oven for 10 min. The plate is then developed with $CHCl_3$–acetone–2-propanol (85 + 10 + 5 v/v/v), a solvent more polar than the one used for the parent compounds. The fluorescent water addition products of B_1 and G_1 are observed at R_f values of ~0.2 and ~0.15, while the unmodified B_1 and G_1 have R_f values of 0.9 and 0.8.

The TFA derivatives can also be prepared from the B_1 and G_1 spots used for the original TLC analysis as follows: After quantitation, locate under UV light the B_1 and G_1 spots originating from both sample extract and standard solutions. With a pencil make two small marks on the silica gel beside each of the toxin spots. Apply 1 ng B_1 and 1 ng G_1 on a horizontal line with the presumed B_1 and G_1, respectively. Use the marks as a guide to apply 2 μl TFA to each of the toxin spots; then heat as before at 40°C for 10 min. Score the silica gel across the plate about 2 cm below the G_2 spot to eliminate the slow-migrating interferences, and place the plate in a developing tank, using a 10-cm-deep trough for the mobile phase as described elsewhere (Trucksess and Stoloff, 1979). Cut the plate 3 cm below G_2 when silica gel on aluminum is used. Develop the TLC plate, and compare the R_f values of the analyte derivatives with those of the standards. The coincidence of R_f values is confirmation of identity of the aflatoxins.

The identity of aflatoxin M_1 can be confirmed in a similar manner. Aflatoxin M_1 in the sample extract spot and in a standard spot is overspotted with TFA. The spotted plate is covered with a clean glass plate, then heated in a 70°C oven for 8 min (Stubblefield, 1979). The plate is developed with $CHCl_3$–acetone–2-propanol (85 + 10 + 7 v/v/v). The R_f values of the M_1 derivative and the unreacted M_1 are ~0.3 and ~0.7, respectively. The structure of the derivative has yet to be determined.

It is important that analysts who do routine aflatoxin determinations periodically compare their analytical results with those from other laboratories to ensure constant quality control of each step of the methodology. International check samples are available through the International Agency for Research on Cancer and the International Smalley Aflatoxin Check Sample Program of the American Oil Chemists' Society. All the results with statistical analyses are published periodically (Friesen and Garren, 1982; McKinney, 1984). The evaluations so far have indicated poor accuracy and great variability between analyst results. This is partly attributable to analyst and method errors. Poor TLC conditions are also contributing factors.

The application of HPTLC as described earlier in this chapter to aflatoxin determination has provided excellent results (Bicking et al., 1985; Huff and Sepaniak, 1983; Lee et al., 1980; Moi and Schmutz, 1983; Tosch et al., 1984). Instrumental quantitation is used instead of visual estimation, which at best is semiquantitative. A range of instruments with varying prices are commercially available. Among the least expensive is one described by Dickens et al. (1980).

B. Ochratoxins

During TLC analysis on silica gel with neutral mobile phases, ochratoxins A and B (pK_a 7.2 and 5.3, respectively) streak badly, and R_f values increase with the amount applied. The mobile phase must therefore include an acid modifier. The acid in the mobile phase separates during chromatography as a secondary solvent front. If ochratoxin A migrates with the secondary solvent front, the mobile phase must be adjusted. If the R_f is high, the acid component should be decreased; if it is low, the polar-neutral component should be increased. If the ochratoxin A spot is at the secondary solvent front, it will be badly distorted and not at its true R_f (it is traveling with the front of the acid) and will be heavily contaminated with other acidic compounds also traveling with this front. An acid-modified mobile phase affects quantitation in several ways. Although this has been known for more than 10 years, little has been done about it, and this no doubt has contributed to the lack of precision of TLC methodology for ochratoxin. In the past, ochratoxins usually have been estimated directly on the TLC plate after development with acid-modified solvents, and the concentration of residual acid and solvent has not been controlled. The fluorescence intensity can change by as much as 10-fold when ochratoxin A is exposed to NH_3–methanol vapor. The magnitude of the change is influenced by residual mobile phase. In the method given in Table I, the TLC plate is exposed to NH_3–methanol vapor and then is covered with another glass plate to prevent evaporation of the NH_3–methanol. If the NH_3–methanol does escape and fluorescence intensity drops, it can be restored by reexposure to fresh NH_3–methanol. The fluorescent spots under these conditions are stable for several days, whereas they occasionally fade in minutes on acidic plates. The method cited in Table I is recommended for most commonly contaminated commodities. It has been further evaluated (Nesheim *et al.*, 1984) in preparation for a planned collaborative study sponsored by the Association of Official Analytical Chemists (AOAC) and the International Union of Pure and Applied Chemistry (IUPAC). The method includes a confirmatory step. Methyl esters are prepared, with BF_3 as a catalyst. The esters are identified by comparing the R_f values of the standard and analyte derivatives. The esters have also been prepared directly on the TLC plate (Kleinau, 1981).

For special purposes and problems, methods other than these cited above might be more suitable and may be chosen from the numerous ones described in several excellent reviews (Harwig *et al.*, 1983; Nesheim, 1976).

C. Citrinin

Citrinin has been found occurring naturally in grains and fruits. Much effort has been expended over the years to perfect procedures for the TLC of this toxin. The major problems have been that (a) it is weakly yellow fluorescent and difficult to see at low levels against an often yellow-fluorescing sample background; (b) it tails badly in normal-phase TLC on silica gel, making it difficult to detect nanogram amounts; and (c) it is unstable.

Improvements have partially overcome the first two problems. More intense fluorescence and easier detection against the yellow background were accomplished with $AlCl_3$ spray followed by heating, which changes the yellow fluorescence to blue (Gimeno and Martins, 1983). On the supposition that citrinin streaking might be a result of its chelating with heavy metals in the silica, EDTA was incorporated in the silica but was not effective.

A second improvement has been the incorporation of an acid in the silica gel to reduce tailing. Oxalic acid was used first, but more recently glycolic acid was found to be better because of reduced diffusion of the citrinin spots and hence enhanced detectability (Gimeno, 1984). No suitable TLC method of confirmation is presently available.

D. Sterigmatocystin

Sterigmatocystin in microgram quantities fluoresces brick red. This fluorescence is too weak to be useful for determining sterigmatocystin in the sample at nanograms per gram levels. However, when the TLC plate is sprayed with $AlCl_3$, a much more intense yellow fluorescence is produced, allowing measurement of low nanogram amounts (Stoloff and Scott, 1984, Sect. 26.132). A problem with the yellow fluorescence is its instability. Under some conditions it appears to fade in a matter of minutes but can be restored by heating the TLC plate.

A solution to this problem has recently been reported (Francis et al., 1985). The yellow fluorescent sterigmatocystin–$AlCl_3$ complex is prepared as usual by spraying with 20% ethanolic $AlCl_3$ solution and heating to 110° for 5 min. While still warm, the plate is sprayed with an ethyl ether–silicone solution (DC-200, 12,500 centistokes; No. 08107, Applied Science Laboratories, Inc.), dried 3 min at 110°C, sprayed a second time, and finally dried 3 min again. This treatment enhances the fluorescence about 10-fold and stabilizes it. The spots remain unchanged while unprotected from light and moisture for at least 2 weeks. Sterigmatocystin is estimated either visually or fluorodensitometrically (excitation 360 nm, emission 500 nm; Johann and Dose, 1983).

The identity of sterigmatocystin can be confirmed chemically as follows: The toxin is isolated by preparatory TLC and is converted to the acetate derivatives, which are identified by TLC (Stoloff and Scott, 1984, Sect. 26.138). However, a more efficient approach is the same as for aflatoxin B_1 and G_1, that is, to add 4 μl TFA–benzene (1 + 1 v/v) to the TLC spots from sample extract and standard solutions to form the water adducts, which are then identified by comparison of R_f values (Shannon and Shotwell, 1976). The derivatization can also be accomplished on the TLC plate by spraying with benzene–TFA (4 + 1 v/v) (van Egmond et al., 1980) or by overspotting the sterigmatocysin spot with TFA anhydride (Thurm et al., 1979).

E. Zearalenone

The method for zearalenone cited in Table I is a modification of the interlaboratory-tested AOAC method (Stoloff and Scott, 1984, Sect. 26.139). The zearalenone is detected by fluorescence (250-nm excitation). The minimum detectable amount on TLC is about 10 ng; much more toxin is required against a sample extract background. Use of a Fast Violet spray (Scott et al., 1978;

Gimeno, 1983) lowers the detectability to 5 to 10 ng standard zearalenone as pink spots that turn purple and fluoresce when sprayed with sulfuric acid. Use of HPTLC plates greatly reduces the time needed for development and improves resolution, so that sample extract interferences are separated from zearalenone [not separable on regular TLC plates (Swanson *et al.*, 1984)]. HPTLC plates with a preadsorbent zone are preferred because no special spotting apparatus, such as an automatic spotter, is needed. Zearalenone can be measured on the TLC plate by fluorodensitometry [excitation 313 nm, emission 443 nm (Shotwell *et al.*, 1976)]. However, this technique is still not as sensitive as HPLC techniques (Bennett *et al.*, 1985; Trenholm *et al.*, 1984).

F. Trichothecenes

The trichothecenes now number more than 60. Not all of these have been found to occur naturally; many were produced in laboratory culture or were synthesized. They are difficult to determine as a group because of the large variations in chemical and physical properties, especially solubility and chromatographic behavior. Most of them also lack functionalities detectable by UV or fluorescence detectors. Basically, the compounds have a cyclic C_{15} triterpenoid nucleus. Most of the naturally occurring trichothecenes include a hindered, and therefore relatively unreactive, epoxide substituent. Destruction of this epoxide group also eliminates toxicity. In addition, the trichothecenes have one or more macrocyclic, hydroxy, ester, or keto functions. A recent book summarizes the discovery, chemistry, mycology, and toxicology of trichothecenes (Ueno, 1983). Over the years Ueno and others have divided the trichothecenes into four categories (A–D) based on their structure.

The category A trichothecenes fluoresce when sprayed and heated with sulfuric acid. The limit of determination of these is about 50 to 100 ng. T-2 Toxin is an example of this group, on which much work has been published. Category A and B compounds can be determined on the same TLC plate. To do this, the plate is first sprayed and heated with $AlCl_3$ to visualize the 8-keto compounds (category B); then the plate is sprayed and heated with sulfuric acid. The 8-keto compounds turn into dark red, UV-absorbing spots, and the category A compounds become fluorescent.

Those in category B have a keto substituent and fluoresce blue on TLC plates when sprayed and heated with $AlCl_3$. About 20 to 50 ng can be determined. Deoxynivalenol is a good example of category B. Its methodology has received the most attention recently because of its widespread occurrence in corn and wheat in 1982 and 1983 (Eppley *et al.*, 1984). The AOAC has conducted a collaborative study of the method (Table I), and it was adopted as official.

Category C consists of those compounds which have a large-ring substituent. Some of these fluoresce naturally, but others are detected as fluorescent spots with sulfuric acid and heating. Group D consists of all the rest of the tri-chothecenes which do not fit into the other three categories. Microgram quan-tities are found as dark spots when heated with sulfuric acid. All the epoxy-substituted trichothecenes can be converted to fluorescent derivatives by treat-ment on the TLC plate with nicotinamide and 2-acetylpyridine (Sano *et al.*, 1982). Many solvent mixtures have been used for TLC, but the most useful ones consist of 3 to 10% 2-propanol in chloroform and sometimes include up to 10% acetone.

The fluorescent trichothecene complexes or derivatives can be measured by densitometry (Takitani *et al.*, 1979): for example, the AlCl$_3$ complex of DON (Trucksess *et al.*, 1984b), sulfuric acid-induced fluorescence of T-2 (Schmidt *et al.*, 1981), or nicotinamide–2-acetylpyridine derivatives of epoxy-substituted trichothecenes (Sano *et al.*, 1982).

Good methods for determining identity, purity, and concentration of standards are needed. Presently, standards are purified as much as possible and weighed accurately to prepare standard solutions.

TLC methods for confirmation of identity of the trichothecenes are limited (Stack and Eppley, 1982). The recommended method is to isolate the toxin by preparative TLC and identity it by MS as described for DON (Trucksess *et al.*, 1984b). Some degree of confirmation can be achieved by two-dimensional TLC or by using other spray reagents, for example, nicotinamide and 2-acetylpyridine (Sano *et al.*, 1982) or chromotropic acid (Baxter *et al.*, 1983). Preparation of derivatives by reaction at the hydroxyl functions is not very useful because of the possibility of reaction at one or more of the several hydroxyl sites and formation of many products.

G. Patulin

Patulin is found in fruits and grains. The pure compound is not stable as a thin film; therefore, standards are prepared and distributed either as the bulk material or as dilute solutions. It is moderately stable in apple juice (Scott, 1984) and ground grains. Patulin is not fluorescent and in the earliest methods was deter-mined by quenching on 254-nm fluorescent TLC plates. About 40 μg/kg grain was required (Pohland and Allen, 1970).

The AOAC patulin method studied and accepted by AOAC in 1977 uses a 0.5% 3-methyl-2-benzothiazolinone hydrazone hydrochloride (MBTH)–water TLC spray and heating to 130°C for 15 min. The patulin appears as a yellow spot which also fluoresces yellow-brown (Scott and Kennedy, 1973; Scott, 1974). The determination limit in juice is 20–25 μl/liter.

Scopoletin and 5-(hydroxymethyl)furfural and other materials present in some apple juices interfere with TLC analysis. To overcome this, it has been proposed that the patulin be converted to its dinitrophenylhydrazone and then analyzed by

TLC (Stinson *et al.*, 1977). According to this procedure a patulin sample is passed over 2,4-dinitrophenylhydrazine 66%–H_3PO_4 on Celite in a minicolumn from which it is recovered with dichloromethane. The derivative is purified by preparatory TLC (scraping and eluting from the plate and determining by UV spectrophotometry at 375 nm).

A second method for the elimination of interference occurring in grape and wine extracts is the use of two-dimensional TLC (Ough and Corison, 1980) using toluene–ethyl acetate–formic acid (5 + 4 + 1 v/v/v) and $CHCl_3$–acetone (90 + 10). The separated patulin spot is suitable for fluorescence TLC densitometry after visualization with MBTH: excitation 300–400 nm, emission 485 nm.

H. Penicillic Acid

Penicillic acid exists in an uncyclized form in polar solvents such as aqueous and basic solutions. It reverts to the cyclic form in neutral or acidic solvents. The cyclic form is more stable than the linear one. Exposure to basic conditions must therefore be minimized.

Penicillic acid is not as stable as many of the other mycotoxins. It tends to disappear from materials on standing after grinding. This has been shown to result from reactions with sulfhydryl groups. It also decomposes as dry films in the pure state. Standards are therefore best stored in weakly acidic solution.

The limited work in TLC has been thoroughly reviewed (Ehnert *et al.*, 1981). Excellent quantitation by TLC densitometry has also been demonstrated (Reimerdes *et al.*, 1975).

TLC depends on the conversion of penicillic acid, on the TLC plate after development, to fluorescent derivatives by spraying with one of the following reagents: 1% *p*-tolualdehyde or 1% dimethylaminobenzaldehyde in H_2SO_4–HOAc–MeOH (5 + 10 + 70 v/v/v) (Neelakantan *et al.*, 1978); 1% diphenylboric acid–2-ethanolamine in methanol spray (Johann and Dose, 1983); and 0.5 ml *p*-anisaldehyde–85 ml MeOH–HOAc–H_2SO_4 (70 + 10 + 5 v/v/v) (Thorpe, 1982). The *p*-anisaldehyde is most widely used. The exposure of the developed plate to ammonia vapor also produces a blue fluorescent derivative. However, the fluorescence of the compound is not as stable as that obtained with the *p*-anisaldehyde spray. The plate can be resprayed to intensify the fluorescence again if spot fading occurs.

IV. CONCLUSIONS

We have attempted to present the state of the art of TLC mycotoxin determinations through selection of the best and most recent methods. Within the limitations of time and space we have tried to emphasize information we regard as most useful to the analyst. The techniques used in TLC have been changing rapidly, and this method will continue to evolve into what is being called high-performance TLC.

We have discussed some of the mycotoxins that are currently considered significant and to which TLC is applicable. The significance of the various mycotoxins changes as we gain greater knowledge concerning occurrence and toxicology of the known and as yet undiscovered mycotoxins. A group of toxins which has received much interest recently is the *Alternaria* toxins (Schade and King, 1984). Relatively few TLC methods have been reported for these compounds.

With the advance in HPLC techniques, the automated, unattended HPLC analysis for routine materials might be preferred. However, TLC is often the method of choice by virtue of its simplicity, versatility, and economy. With the improved cleanup techniques, better laboratory quality control, and advances in instrumentation, TLC can be expected to provide data with coefficients of variation from the TLC step of less than 5% and be as accurate and precise as quantitation by other techniques.

REFERENCES

Baxter, J. A., Terhune, S. J., and Qureshi, S. A. (1983). Use of chromotropic acid for improved thin-layer chromatographic visualization of trichothecene mycotoxins. *J. Chromatogr.* **15,** 130–133.

Bennett, G. A., Shotwell, O. L., and Kwolek, W. F. (1985). Determination of α-zearalenol and zearalenone in corn by high performance liquid chromatography: collaborative study. *J. Assoc. Off. Anal. Chem.* **68,** 958–961.

Bertsch, W., Hara, S., Kaiser, R. E., and Zlatkis, A., eds. (1980). "Instrumental HPTLC." Huthig Verlag, New York.

Betina, V., ed. (1984). "Mycotoxins Production, Isolation, Separation and Purification." Elsevier, New York.

Betina, V. (1985). Thin-layer chromatography of mycotoxins. *J. Chromatogr.* **334,** 211–276.

Bicking, M. K. L., Kniseley, R. N., and Svec, H. J. (1985). Determination of aflatoxins in air samples of refuse-derived fuel by thin-layer chromatography with laser-induced fluorescence spectrometric detection. *Anal. Chem.* **55,** 200–204.

Brumley, W. C., Nesheim, S., Trucksess, M. W., Trucksess, E. W., Dreifuss, P. A., Roach, J. A. G., Andrzejewski, D., Eppley, R. M., Pohland, A. E., Thorpe, C. W., and Sphon, J. A. (1981). Negative ion chemical ionization mass spectrometry of aflatoxins and related mycotoxins. *Anal. Chem.* **53,** 2003–2006.

Chang, T. T., Lay, J. O., Jr., and Francel, R. J. (1984). Direct analysis of thin layer chromatography spots by fast atom bombardment mass spectrometry. *Anal. Chem.* **56,** 109–111.

Dickens, J. W., McClure, W. F., and Whitaker, T. B. (1980). Densitometric equipment for rapid quantitation of aflatoxins on thin layer chromatograms. *J. Am. Oil Chem. Soc.* **57,** 205–208.

Ehnert, M., Popken, A. M., and Dose, K. (1981). Quantitative Bestimmung der Penicillinsaure in pflanzlichen Lebensmitteln. *Z. Lebensm.-Unters. Forsch.* **172,** 110–114.

Eppley, R. M., Trucksess, M. W., Nesheim, S., Thorpe, C. W., Wood, G. E., and Pohland, A. E.

(1984). Deoxynivalenol in winter wheat: thin layer chromatographic method and survey. *J. Assoc. Off. Anal. Chem.* **67,** 43–45.

Fenimore, D. C., and Davis, C. M. (1981). High performance thin-layer chromatography. *Anal. Chem.* **43,** 252A–266A.

Francis, O. J., Ware, G. M., Carman, A. S., and Kuan, S. S. (1985). Thin layer chromatographic determination of sterigmatocystin in cheese. *J. Assoc. Off. Anal. Chem.* **68,** 643–645.

Friesen, M. D., and Garren, L. (1982). International mycotoxin check sample program: part 1. Report on laboratory performance for determination of aflatoxins B_1, B_2, G_1, and G_2 in raw peanut meal, deoiled peanut meal, and yellow corn meal. *J. Assoc. Off. Anal. Chem.* **65,** 855–863.

Gimeno, A. (1980). Improved method for thin layer chromatographic analysis of mycotoxins, *J. Assoc. Off. Anal. Chem.* **63,** 182–186.

Gimeno, A. (1983). Rapid thin layer chromatographic determination of zearalenone in corn, sorghum, and wheat. *J. Assoc. Off. Anal. Chem.* **66,** 565–569.

Gimeno, A. (1984). Determination of citrinin in corn and barley on thin layer chromatographic plates impregnated with glycolic acid. *J. Assoc. Off. Anal. Chem.* **67,** 194–196.

Gimeno, A., and Martins, M. L. (1983). Rapid thin layer chromatographic determination of patulin, citrinin, and aflatoxin in apples and pears, and their juices and jams. *J. Assoc. Off. Anal. Chem.* **6,** 85–91.

Harwig, J., Kuiper-Goodman, T., and Scott, P. M. (1983). Microbial food toxicants: ochratoxins. *In* "CRC Handbook of Foodborne Diseases of Biological Origin" (M. Rechcigl, Jr., ed.), pp. 193–238. CRC Press, Boca Raton, Florida.

Hauck, H. E. (1981). HPTLC: New and further developments in the area of HPTLC precoated plates. *Swiss Chem.* **3**(16), 41–46.

Hauck, H. E., and Jost, W. (1983). Characterization and application of reversed phase TLC. *Am. Lab.* **15**(8), 72–77.

Howell, M. V., and Taylor, P. W. (1981). Determination of aflatoxins, ochratoxin A and zearalenone in mixed feeds, with detection by thin layer chromatography or high performance liquid chromatography. *J. Assoc. Off. Anal. Chem.* **64,** 1356–1363.

Huff, P. B., and Sepaniak, M. J. (1983). Laser fluorometric detection for thin-layer chromatography. *Anal. Chem.* **55,** 1994–1996.

Johann, H., and Dose, K. (1983). Multianalytical method for the routine determination of the aflatoxin B_1, B_2, G_1 and G_2 and citrinin, ochratoxin A, patulin, penicillic acid and sterigmatocystin in molded foods. *Fresenius' Z. Anal. Chem.* **314,** 139–142.

Kaiser, R. E., ed. (1982). "Instrumental High Performance Thin-Layer Chromatography." Inst. Chromatogr., Bad Durkheim, West Germany.

Kleinau, G. (1981). Zur Derivatisierung von Ochratoxin A auf der Dunnschichtplatte. *Nahrung* **25,** K9–K10.

Lafont, P., and Debeaupuis, J. P. (1982). Analytical method—thin layer chromatographic assay of mycophenolic acid. *In* "Environmental Carcinogens Selected Methods of Analysis. Vol. 5: Some Mycotoxins" (H. Egan, L. Stoloff, M. Castegnaro, P. Scott, J. K. O'Neill, H. Bartsch, and W. Davis, eds.), pp. 389–398. IARC, Lyon.

Lansden, J. A. (1984). Cyclopiazonic acid analysis by thin layer chromatographic densitometry. *Abstr. AOAC Annu. Int. Meet., 98th, Washington, D.C.* No. 124.

Lee, K. Y., Poole, C. F., and Zlatkis, A. (1980). Simultaneous multimycotoxin determination by high performance thin-layer chromatography. *Anal. Chem.* **52,** 837–842.

Lovelace, C. E. A., Njapau, H., Salter, L. F., and Bayley, A. C. (1982). Screening method for the detection of aflatoxin and metabolites in human urine: aflatoxins B_1, G_1, M_1, B_{2a}, G_{2a}, aflatoxicol I and II. *J. Chromatogr.* **227,** 256–261.

McKinney, J. K. (1984). Analyst performance with aflatoxin methods as determined from AOCS

Smalley check sample program: short-term and long-term views. *J. Assoc. Off. Anal. Chem.* **67,** 25–32.

Moi, J., and Schmutz, H. R. (1983). Instrumental thin-layer chromatography. *Am. Lab.* **15**(1), 54–60.

Neelakantan, S., Balsubramanian T., Balasaraswathi, R., Jasmine, G. I., and Swaminathan, R. (1978). Detection of penicillic acid in foods. *J. Food Sci. Technol.* **15,** 1–2.

Nesheim, S. (1976). The ochratoxins and other related compounds. *In* "Mycotoxins and Other Fungal Related Food Problems" (J. V. Rodricks, ed.), pp. 276–295. Am. Chem. Soc., Washington, D.C.

Nesheim, S. (1980). Factors affecting the thin layer chromatography of aflatoxins. *In* "Thin Layer Chromatography Quanitative Environmental and Clinical Applications" (J. C. Touchstone and D. Rogers, eds.), pp. 194–240. Wiley, New York.

Nesheim, S., and Brumley, W. C. (1981). Confirmation of identity of aflatoxins. *J. Am. Oil Chem. Soc.* **58,** 945A–949A.

Nesheim, S., Trucksess, M. W., and Thorpe, C. W. (1984). Improved method for determination of ochratoxin A in barley, corn, and pig tissues using high performance thin layer chromatography. *Abstr. AOAC Annu. Int. Meet., 98th, Washington, D.C.* No. 126.

Novacek, V. M. (1973). Errors in quantitative analysis of thin-layer chromatograms. *Am. Lab.* **5**(3), 85–91.

Nowotny, P., Baltes, W., Kronert, W., and Weber, R. (1983). Screening method for the determination of 22 mycotoxins in moldy foods. *Proc. Int. IUPAC Symp. Mycotoxins Phycotoxins, 5th, Vienna, 1982* pp. 36–39.

Ough, C. S., and Corison, C. A. (1980). Measurement of patulin in grapes and wines. *J. Food Sci.* **45,** 476–478.

Park, D. L., DiProssimo, V., Abdel-Malek, E., Trucksess, M. W., Nesheim, S., Brumley, W. C., Sphon, J. A., Barry, T. L., and Petzinger, G. (1985). Negative ion chemical ionization MS procedure for confirmation of the identity of aflatoxin B_1: collaborative study. *J. Assoc. Off. Anal. Chem.* **68,** 636–640.

Patterson, D. S. P. (1983). Screening procedures for mycotoxins. *Proc. Int. Symp. Mycotoxins, 1982* pp. 167–175.

Plattner, R. D. (1984). Identification and quantitation of mycotoxins by MS/MS. *Finnigan Spectra* **9,** 25–28.

Pohland, A. E., and Allen, R. (1970). Analysis and chemical confirmation of patulin in grains. *J. Assoc. Off. Anal. Chem.* **53,** 686–687.

Pohland, A. E., and Thorpe, C. W. (1982). Mycotoxins. *In* "Handbook of Carcinogens and Hazardous Substances" (M. C. Bowman, ed.), pp. 303–390. Dekker, New York.

Price, R. L., Jorgensen, K. V., and Billotte, M. (1981). Citrus artifact interference in aflatoxin M_1 determination in milk. *J. Assoc. Off. Anal. Chem.* **64,** 1383–1385.

Ramaley, L., Nearing, M. E., Vaughan, M. A., Ackman, R. G., and Jamieson, W. D. (1983). Thin-layer chromatographic plate scanner interfaced with a mass spectrometer. *Anal. Chem.* **55,** 2285–2289.

Reimerdes, E. H., Engel, G., and Behnert, J. (1975). Untersuchungen zur bildung von Mykotoxinen und deren quantitative Bestimmung I. Die Bildung von Penicillinsaure durch *Penicillium cyclopium. J. Chromatogr.* **110,** 361–368.

Rogers, D. (1984). 2-D TLC: enhanced resolution and capability. *Am. Lab.* **16**(5), 65–73.

Romer, T. (1984). Chromatographic techniques for mycotoxins. *In* "Food Constituents and Food Residues: Their Chromatographic Determination" (J. F. Lawrence, ed.), pp. 355–393. Dekker, New York.

Sandor, G. S. (1982). Thin layer chromatographic determination of rubratoxin B in corn. *In* "Environmental Carcinogens Selected Methods of Analysis. Vol. 5: Some Mycotoxins" (H. Egan,

L. Stoloff, M. Castegnaro, P. Scott, J. K. O'Neill, H. Bartsch, and W. Davis, eds.), pp. 445–455. IARC, Lyon.

Sano, A., Asabe, Y., Takitani, S., and Ueno, Y. (1982). Fluorodensitometric determination of trichothecene mycotoxins with nicotinamide and 2-acetylpyridine on a silica gel layer. *J. Chromatogr.* **235**, 257–265.

Schade, J. E., and King, A. D., Jr. (1984). Analysis of the major *Alternaria* toxins. *J. Food Prot.* **47**, 978–995.

Schmidt, R., Bieger, A., Ziegenhagen, E., and Dose, K. (1981). Bestimmung von T-2 Toxin in pflanzlichen Nahrungsmitteln. *Fresenius' Z. Anal. Chem.* **308**, 133–136.

Scott, P. M. (1974). Collaborative study of a chromatographic method for determination of patulin in apple juice. *J. Assoc. Off. Anal. Chem.* **67**, 621–625.

Scott, P. M. (1982). Mycotoxin analysis by TLC. *In* "Advances in Thin Layer Chromatography" (J. C. Touchstone, ed.), pp. 321–342. Wiley, New York.

Scott, P. M. (1984). Effects of food processing on mycotoxins. *J. Food Prot.* **47**, 489–499.

Scott, P. M., and Kennedy, B. P. C. (1973). Improved method for the thin layer chromatographic determination of patulin in apple juice. *J. Assoc. Off. Anal. Chem.* **56**, 813–816.

Scott, P. M., Panalaks, T., Kanhere, S., and Miles, W. F. (1978). Determination of zearalenone in cornflakes and other corn-based food by thin layer chromatogrpahy high pressure liquid chromatography, and gas–liquid chromatography/high resolution mass spectrometry. *J. Assoc. Off. Anal. Chem.* **61**, 593–599.

Shannon, G. M., and Shotwell, O. L. (1976). Thin layer chromatographic determination of sterigmatocystin in cereal grains and soybeans. *J. Assoc. Off. Anal. Chem.* **59**, 963–965.

Shanon, G. M., Shotwell, O. L., and Kwolek, W. F. (1983). Extraction and thin layer chromatography of aflatoxin B_1 in mixed feed. *J. Assoc. Off. Anal. Chem.* **66**, 582–586.

Sherma, J., and Fried, B. (1982). Thin-layer and paper chromatography. *Anal. Chem.* **56**, 46R–63R.

Shotwell, O. L., Goulden, M. L., and Bennett, G. A. (1976). Determination of zearalenone in corn: collaborative study. *J. Assoc. Off. Anal. Chem.* **59**, 666–670.

Stack, M. E. (1984). Isolation, Identification, and Mutagenicity of Altertoxins I, II, III, Metabolites of Alternaria. M.S. Thesis, American Univ., Washington, D.C.

Stack, M. E., and Eppley, R. M. (1982). Analysis of trichothecenes by TLC. *In* "Advances in Thin Layer Chromatography: Clinical and Environmental Applications" (J. C. Touchstone, ed.), pp. 495–502. Wiley, New York.

Stack, M. E., Mislivec, P. B., Denizel, T., Gibson, R., and Pohland, A. E. (1983). Ochratoxin A and B, xanthomegnin, viomellein and vioxanthin production by isolates of *Aspergillus ochraceus* from green coffee beans. *J. Food Prot.* **46**, 965–968.

Stahl, E. (1969). "Thin-Layer Chromatography: A Laboratory Handbook," 2nd Ed. Springer-Verlag, New York.

Stinson, E. E., Huktanen, C. N., Zell, F. E., Schwartz, D. P., and Osman, S. F. (1977). Determination of patulin in apple juice products as the 2,4-dinitrophenylhydrazone derivative. *J. Agric. Food Chem.* **25**, 1220–1222.

Stoloff, L., and Scott, P. M. (1984). Natural poisons. *In* "Official Methods of Analysis" (S. Williams, ed.), pp. 477–502. Assoc. Off. Anal. Chem., Arlington, Virginia.

Stoloff, L., Nesheim, S., Yin, L., Rodricks, J. W., Stack, M., and Campbell, A. D. (1971). A multimycotoxin detection method for aflatoxins, ochratoxins, zearalenone, sterigmatocystin, and patulin. *J. Assoc. Off. Anal. Chem.* **54**, 91–97.

Stubblefield, R. D. (1979). The rapid determination of aflatoxin M_1 in milk. *J. Am. Oil Chem. Soc.* **56**, 800–802.

Swanson, S. P., Corley, R. A., White, D. G., and Buck, W. B. (1984). Rapid thin layer chromatographic method for determination of zearalenone and zearalenol in grains and animal feeds. *J. Assoc. Off. Anal. Chem.* **67**, 580–582.

Takeda, Y., Isohata, E., Amano, R., and Uchiyama, M. (1979). Simultaneous extraction and fractionation and thin layer chromatographic determination of 14 mycotoxins in grains. *J. Assoc. Off. Anal. Chem.* **62**, 573–578.

Takitani, S., Asabe, Y., Kato, T., Suzuki, M., and Ueno, Y. (1979). Spectrodensitometric determination of trichothecene mycotoxins with 4-(p-nitro-benzyl)pyridine on silica gel thin layer chromatograms. *J. Chromatogr.* **172**, 335–342.

Thorpe, C. W. (1982). Determination of penicillic acid in foodstuffs. *In* "Environmental Carcinogens Selected Methods of Analysis. Vol. 5: Some Mycotoxins" (H. Egan, L. Stoloff, M. Castegnaro, P. Scott, J. K. O'Neill, H. Bartsch, and W. Davis, eds.), pp. 333–348. IARC, Lyon.

Thurm, V., Paul, P., and Koch, C. E. (1979). Zur hygienischen Bedeutung von Sterigmatocystin in planzlichen Lebensmitteln. 1. Mitt. analytischer Nachweis von Sterigmatocystin. *Nahrung* **23**(2), 111–115.

Tosch, D., Waltking, A. E., and Schlesier, J. F. (1984). Comparison of liquid chromatography and high performance thin layer chromatography for determination of aflatoxin in peanut products. *J. Assoc. Off. Anal. Chem.* **67**, 337–339.

Touchstone, J. C., ed. (1982). "Advances in Thin Layer Chromatography." Wiley, New York.

Trenholm, H. L., Warner, R. M., and Fitzpatrick, D. W. (1984). Rapid, sensitive liquid chromatographic method for determination of zearalenone and α- and β-zearalenol in wheat. *J. Assoc. Off. Anal. Chem.* **67**, 968–972.

Trucksess, M. W. (1974). Observation on the thin layer chromatographic development of corn extracts with ethyl ether. *J. Assoc. Off. Anal. Chem.* **57**, 1220–1221.

Trucksess, M. W., and Stoloff, L. (1979). Deep solvent trough for thin layer chromatography. *J. Assoc. Off. Anal. Chem.* **62**, 1181–1182.

Trucksess, M. W., and Stoloff, L. (1980). Thin layer chromatographic determination of aflatoxins in dry ginger root and ginger oleoresin. *J. Assoc. Off. Anal. Chem.* **63**, 1052–1054.

Trucksess, M. W., and Stoloff, L. (1984). Determination of aflatoxicol and aflatoxins B_1 and M_1 in eggs. *J. Assoc. Off. Anal. Chem.* **67**, 317–320.

Trucksess, M. W., Stoloff, L., Brumley, W. C., Wilson, D. M., Hale, O. M., Sangster, L. T., and Miller, D. M. (1982). Aflatoxicol and aflatoxin B_1 and M_1 in the tissues of pigs receiving aflatoxin. *J. Assoc. Off. Anal. Chem.* **65**, 885–887.

Trucksess, M. W., Brumley, W. C., and Nesheim, S. (1984a). Rapid quantitation and confirmation of aflatoxins in corn and peanut butter, using a disposable silica gel column, thin layer chromatography, and gas chromatography/mass spectrometry. *J. Assoc. Off. Anal. Chem.* **67**, 973–975.

Trucksess, M. W., Nesheim, S., and Eppley, R. M. (1984b). Thin layer chromatographic determination of deoxynivalenol in wheat and corn. *J. Assoc. Off. Anal. Chem.* **67**, 40–43.

Ueno, Y. (1982). Thin layer chromatographic determination of luteoskyrin in rice grains. *In* "Environmental Carcinogens Selected Methods of Analysis. Volume 5: Some Mycotoxins" (H. Egan, L. Stoloff, M. Castegnaro, P. Scott, J. K. O'Neill, H. Bartsch, and W. Davis, eds.), pp. 405–417. IARC, Lyon.

Ueno, Y., ed. (1983). "Trichothecenes: Chemical, Biological and Toxicological Aspects." Elsevier, New York.

van Egmond, H. P., Paulsch, W. E., Deijll, E., and Schuller, P. L. (1980). Thin layer chromatographic method for analysis and chemical confirmation of sterigmatocystin in cheese. *J. Assoc. Off. Anal. Chem.* **63**, 110–114.

Wilson, D. M., Tabor, W. H., and Trucksess, M. W. (1976). Screening method for the detection of aflatoxin, ochratoxin, zearalenone, penicillic acid, and citrinin. *J. Assoc. Off. Anal. Chem.* **59**, 125–127.

Zlatkis, A., and Kaiser, R. E., eds. (1977). "HPTLC High Performance Thin-Layer Chromatography," Journal of Chromatography Library, Vol. 9. Elsevier, New York.

10

Gas Chromatography in Mycotoxin Analysis

RODNEY W. BEAVER

Coastal Plain Experiment Station
Department of Plant Pathology
University of Georgia
Tifton, Georgia 31793-0748

I. INTRODUCTION

This chapter is organized into two main sections. The first describes some practical aspects of gas chromatography (GC) and the general applicability of gas chromatography to mycotoxin analysis. The second main section describes gas chromatographic methods for specific mycotoxins taken from the literature. Although space limitations made it impossible to describe completely every method

MODERN METHODS IN THE ANALYSIS
AND STRUCTURAL ELUCIDATION
OF MYCOTOXINS

for every known mycotoxin, the methods and mycotoxins chosen illustrate the general procedures and considerations necessary to develop a useful GC method for mycotoxin analysis.

II. PRACTICAL ASPECTS OF GAS CHROMATOGRAPHY

Gas chromatography refers to the process whereby analyte molecules are vaporized and separated in a long tube or column. The GC column can be packed with an inert support which has been coated with a thin, uniform film of a high molecular weight polymeric liquid or gum (packed-column GC), or the inside walls of the column can be coated with a thin film of the polymer (capillary-column GC). The thin film of polymer serves as the stationary phase. The mobile phase is a stream of gas which is forced through the column. In order for solutes to migrate through the column they must be vaporized; thus GC is typically performed with the column heated to temperatures ranging from 30° to 350°C. Separation occurs because different analyte molecules have different affinities for the stationary phase and/or different vapor pressures at the separation temperature. Strongly interacting molecules, or those with lower vapor pressures, spend a greater amount of time bound or dissolved in the stationary phase and thus spend less time in the moving gas stream. More weakly interacting molecules, or those with higher vapor pressures, spend less time in the stationary phase and are swept to the column outlet more quickly. Once eluted from the column, the analytes pass through a detector which generates a signal proportional to the number or concentration of molecules.

The following discussion describes in greater detail some of the practical aspects of packed-column and capillary-column GC and GC detectors as well as some techniques which enhance the suitability of GC for the analysis of mycotoxins. This discussion of GC techniques is necessarily general and somewhat arbitrary in the topics covered. Excellent books by Perry (1981) and Jennings (1980) cover GC theory and practice in great detail.

A. Packed Columns

1. Column Materials

Tubing for packed columns is typically of $\frac{1}{8}$- or $\frac{1}{4}$-in. o.d. and 2–6 mm i.d. This tubing may be glass, Teflon, stainless steel, or nickel. Stainless steel and nickel have the advantage of strength and ruggedness, but can possess considerable catalytic activity, especially at the temperatures typically encountered in GC. This activity can be a severe restriction in the analysis of mycotoxins which are not thermally stable. Although it has been reported that nickel is considerably less reactive than stainless steel (Fenimore et al., 1977), it should still be used with considerable caution. The opacity of the metal columns presents another disadvantage, because voids and other column faults are not visually apparent and are not detected until the chromatography has degraded.

Although Teflon tubing is reasonably inert and essentially unbreakable, it has a relatively low maximum working temperature (<250°C). There is also the potential for atmospheric oxygen to diffuse into Teflon tubing. The presence of oxygen can quickly ruin some stationary phases, especially the polyethylene glycols, which could result in sample degradation.

The preferred material for packed-column GC tubing is glass. Borosilicate glass (which is used almost exclusively) is much more inert than the metals, is transparent so that the quality of the packing can be easily observed, and is impermeable to the diffusion of atmospheric oxygen. The only disadvantage of glass relative to the other materials is its relative fragility. However, this is not a serious shortcoming, and only moderate care in handling is required to use glass columns successfully.

2. Support Materials

The purpose of the support material in packed-column GC is to provide an inert, high surface area matrix over which a thin, uniform film of the stationary phase is distributed. This purpose can be served by a variety of materials including silica, alumina, carbon, Teflon, synthetic porous polymers, or diatomaceous earth (diatomite).

Due to its high surface area, relative inertness, and very high temperature tolerance, diatomaceous earth is the most popular and widely applicable of these materials. This material is usually acid washed to remove metal impurities and is often treated with dimethyldichlorosilane or hexamethyldisilizane in order to passivate active adsorption sites (e.g., free silanol groups). Diatomite is available in various sizes according to seive-cuts, with the most popular being 60–80, 80–100, and 100–120 mesh. Although mesh size does affect the ultimate achievable efficency of a packed column, a more important consideration is probably the size uniformity of the particles. This need for a narrow distribution of particle sizes results in the one major disadvantage of diatomite as a support material. Careful handling of diatomite and columns packed with diatomite is required, since the particles are quite frangible, and rough treatment results in fractured particles and a concomitant decrease in column performance. Perry (1981) gives an excellent and complete description of support materials and their influence on GC column performance.

3. Stationary Phases

Stationary phases for packed-column GC must be thermally stable, nonreactive, and possess the necessary selectivity for the separation at hand. Although there are a myriad of phases available (one GC supplier's catalog lists over 100), many of these are redundant. Some of the more popular phases will be briefly described here; for a more complete description of stationary-phase selection and characterization, the reader is directed to the classic work by McReynolds (1970). Leary *et al.* (1973) used statistical methods to suggest a list of 12 phases which could cover the entire selectivity and polarity range of the more than 200 phases studied by McReynolds.

The most popular of the stationary-phase materials are the polysiloxanes. These silicone oils or gums are available specifically prepared for GC application and thus are highly purified and possess a relatively narrow molecular weight distribution. The silicones are exceptionally thermally stable, are tolerant of small amounts of water or oxygen in the carrier gas, and are available in a variety of modifications ranging from nonpolar to polar. A nonpolar phase results when the polymer backbone is substituted solely with methyl groups. Polarity and differences in selectivity can be obtained by substituting phenyl groups for some of the methyl substituents. Even more polarity can be obtained by introducing cyanoethyl or cyanopropyl groups. Trifluoropropyl moieties can also be used to replace methyls on the polymer backbone to yield still different selectivity. Haken (1984) has recently reviewed the composition, physical characteristics, and availability of the polysiloxanes.

Other widely used phases are the polyethylene glycols and polyesters. The polyethylene glycols are much more polar than any of the polysiloxanes and also possess hydrogen-bonding capabilities. This hydrogen bonding can lead to considerable selectivity advantages. The polyesters (e.g., diethylene glycol succinate, DEGS) yield still different polarities and selectivities. The polyethylene glycols and polyesters have considerably lower maximum temperature limits than the silicones, and each can be degraded by traces of water or oxygen in the carrier gas.

Depending on the specific application, stationary phase, and support material, packed GC columns are typically used with stationary-phase loadings ranging from 1 to 10% (w/w). The percentage loading refers to the amount of stationary phase deposited on the support. A standard method for preparing the column-packing material is to dissolve a weighed amount of stationary phase in an appropriate solvent. This mixture is then added to the appropriate amount of support material and a vacuum applied so that the solution will penetrate the pores of the support. The solvent is then removed on a rotary evaporator and the material is heated in a shallow pan in an oven to remove the last traces of solvent. The packing is then poured slowly into the column with gentle vibration. Leibrand and Dunham (1973) have described packed-column preparation in detail and offer a much more complicated procedure which yields columns superior to those prepared by the simple method outlined above.

4. Injection Techniques

Relative to capillary columns, packed-column GC injection techniques are simple. The primary concerns are plug sample introduction and the prevention of sample degradation. Several operational procedures accomplish each of these goals.

Ideally, sample should be introduced into the column as an infinitely narrow band or plug in order to minimize the contribution of sample introduction to peak band width. The best approximation of this condition is achieved by injecting a small volume of sample as rapidly as possible. The injector area of the GC should be heated sufficiently so that the sample and solvent are rapidly vaporized and sample vapors mix minimally with carrier gas before being rapidly sorbed by the stationary phase.

The high injector temperatures required for rapid sample volatization can cause problems for thermally labile compounds. Therefore, it is very important to ensure that marginally stable samples do not contact metal or other catalytic surfaces in the injector. The best way to avoid this problem is to inject directly onto the column. If the packing material is held in place (as is usual practice) by a plug of glass wool at the column inlet, it is important that the glass wool be carefully silanized. For very sensitive samples, silanized quartz wool should be used.

B. Capillary Columns

1. Column Materials

Until the late 1970s, capillary columns were most often constructed of either soft (soda lime) or borosilicate glass. Metal capillary columns have been used for some applications but are unsuitable for all but the most inert samples. Soft glass has the advantage of being easily workable (i.e., columns are easily drawn) and being slightly less prone to breakage, while some compounds tend to undergo less degradation in borosilicate columns.

Dandeneau and Zerenner (1979) first described the use of fused silica for the construction of capillary columns. Since that time, fused silica has become immensely popular due to the advantages over both soda lime and borosilicate glasses. A significant practical advantage of fused silica is its flexibility and ruggedness. Fused-silica columns can be easily inserted into injector and detector fittings and is essentially unbreakable under normal usage as long as the protective polyimide coating is not damaged. Another advantage of fused silica is its inherent inertness. The material used for columns is prepared synthetically from silicon tetrahchloride and typically contains <1 ppm total metal impurities (Lee *et al.*, 1984). This high purity results in a surface which exhibits very few active sites and little catalytic activity, thus obviating many problems of peak asymmetry and sample decomposition. It should be noted, however, that the fused-silica surface is inherently acidic, which can lead to some problems when chromatographing basic compounds.

The advantages of fused silica for capillary columns lead directly to the two major disadvantages of this material. The polyimide coating necessary to protect the column and to confer flexibility and strength is not completely transparent, so that problems within the column, such as phase separation and droplet formation, are not readily visible. Also, the very inert surface and low surface energy make coating with the more polar stationary phases difficult. However, this particular problem has been largely overcome, and high-quality columns coated with very polar polyethylene glycols are commercially available. Except in some specialized cases, the advantages of fused silica outweigh any disadvantages, and fused silica should be the nearly universal choice for routine capillary GC applications.

2. Stationary Phases

Of the many available GC stationary phases, only a few are suitable for capillary applications. The phases used are almost exclusively the polysiloxanes and polyglycols (Lee et al., 1984). Fortunately, as discussed in Section II,A,3, these phases are available to cover a wide range of polarities and selectivities.

The most widely used capillary GC technique is known as wall-coated open tubular (WCOT) capillary GC. The WCOT method will be discussed here. Jennings (1980) describes the other types of capillary techniques. In WCOT columns, the interior of the column tubing serves as the support and a thin film of the stationary phase is deposited on the tubing surface. Columns are available with film thicknesses ranging from 0.1-μm in 0.2-mm i.d. columns to 1.0-μm in 0.54-mm i.d. columns. The various implications of film thickness on chromatographic performance have been discussed by Ettre (1983) and by Ettre and co-workers (1983), but in general, thicker films yield lower efficiency and higher sample capacity, while thinner films give enhanced chromatographic efficiency with diminished sample capacities.

Due to the relatively small total amount of stationary phase present in a WCOT column, slight thermal bleed of the stationary phase or the solubility of the phase in the solvents used for sample dissolution can seriously affect column performance. In order to overcome these problems, "cross-linked" or bonded phases have been developed. In these bonded phases a small amount of an organic peroxide such as dicumyl peroxide (Gandara et al., 1984) or an azo compound such as aziosobutyronitrile (Springston et al., 1983) is added to the coating solution. After coating, the column is sealed and heated to 150° to 250°C. Decomposition of the peroxide–azo compound leads to the production of free radicals which react with the stationary phase, leading to cross-linking of the stationary-phase polymer chains. Alternatively, the coated-column stationary phase can be cross-linked by exposure to γ radiation (Hubball *et al.*, 1984). Radiation cross-linking is claimed to be advantageous, since there are no decomposition products arising from the organic initiator to contaminate the stationary phase.

Cross-linked phases exhibit essentially the same polarity and selectivity characteristics as the same phase prior to cross-linking. However, they are more thermally stable, exhibit less bleed, last longer, and are resistant to the deleterious effects of large volumes of injection solvent. Indeed, these phases are so stable that columns whose performance has declined after many injections can often be returned to nearly new condition by rinsing with a solvent such as pentane or dichloromethane.

3. Injection Techniques

There are three basic methods for delivering a sample solution into a capillary column. These are known as split, splitless, and on-column injections. Each has advantages and disadvantages, and the particular requirements of a given analysis will determine which technique is required. Each technique is described briefly here; for a complete description of injection techniques including diagrams see Jennings (1980).

In split injections, the sample is injected into a glass sleeve several centimeters long and with an i.d. of ~2 mm. The sleeve is often packed with deactivated glass beads or with standard packed-column packing material. This packing provides additional thermal mass so that the sample and solvent are quickly and uniformly vaporized. Carrier gas flow is routed through the sleeve so that some variable ratio of the vaporized sample, usually ranging from 0.1 to 2%, is delivered to the column while the remainder is vented to the atmosphere. The split-injection technique is used for concentrated samples where the amount of analyte present could overload the column or detector. Split injection is also useful when the solvent is deleterious to the column. However, if sample is limited or if the concentration of the individual analytes within the solvent is very low, it is inadvisable to throw away 50–99% of the available sample.

In the splitless-injection technique, the entire sample is vaporized and then delivered to the column. The injection port is lined with a glass sleeve in which the vaporization takes place. Unless the sample is reconcentrated at the top of the column, the widths of eluting peaks will reflect the volume of the glass sleeve. In order to achieve the necessary reconcentration, it is necessary either that the temperature of the column be ~10°–30°C below the boiling point of the injection solvent (resulting in the so-called solvent effect), or that the temperature of the column be 100°–150°C below the boiling point of the injected analytes (resulting in cold-trapping of the analytes). Splitless injection circumvents the often painful necessity of wasting a large portion of the sample, but requires careful consideration of solvent and temperature. More complete descriptions of splitless-injection techniques can be found in Jennings (1980), Grob and Grob (1974), and Grob and Grob (1972).

In both split and splitless techniques, the sample and solvent must be vaporized by contact with a hot, relatively high surface area liner. For particularly unstable samples, it may be necessary to deliver the sample solution directly onto a cold column. On-column injection is a very mild technique, but its use requires a complicated injection system because the ultrafine syringe needles required to fit inside small-diameter capillary columns are not sturdy enough to pierce a septum. Once injection has been accomplished, the conditions for suitable chromatography require either cold-trapping or a solvent effect, just as in splitless injection. Readers are referred to the manufacturer of their particular instrument for details of the operation and availability of on-column injection systems.

C. Detectors

1. Flame Ionization Detector

The flame ionization detector (FID) is a sensitive detector for GC. Its basis for operation is the production of ions when carbon-containing compounds are burned in a hydrogen–air flame as they elute from the column. The ions are collected by an electrode and produce a current which is proportional to the number of ions present at a given instant. Well-designed FIDs can detect as little as picograms per second of some compounds (Drushel, 1983).

Although the FID is a widely applicable detector, there are several disadvantages which must be addressed. The FID response is dependent on the structure of a detected compound (e.g., propane gives a higher response than acetone), thus the relative response of various compounds must be determined by the use of standards. Also, some compounds such as the fixed gases, highly halogenated compounds, carbon disulfide, and other "nonflammable" compounds, yield greatly reduced signals with an FID. This can be advantageous in cases where the solvent can be chosen so that a large solvent peak does not interfere with early-eluting peaks. Perhaps the most severe disadvantage of the FID in the analysis of mycotoxins is its universality. Since mycotoxins are typically found in complex biological matrices, many potentially interfering peaks may appear in the chromatogram.

2. Electron Capture Detector

The electron capture detector (ECD) operates by measuring a decrease in signal when a suitable compound passes through the detector. A β-radiation source (usually ^{63}Ni) ionizes carrier gas molecules as they pass through the detector. The electrons produced by this ionization are collected by an electrode and measured. The signal produced by these electrons when only carrier gas is present in the detector is known as the standing current. When a sample component with a high electron affinity passes through the detector, some of these electrons are captured by the sample component and are unavailable for collection at the sampling electrode, which decreases the standing current. The decrease in standing current is dependent on the amount of sample present in the detector. In some cases the ECD is capable of detecting on the order of hundreds of femtograms per second (Drushel, 1983).

Although the ECD can be quite sensitive, the requirements for this sensitivity are stringent. Compounds such as benzene, alkanes, and many oxygen- and nitrogen-containing compounds give little or no response. The presence of a halogen atom in a compound results in a large ECD response. The addition of a second halogen can result in an increase in response of orders of magnitude. The selectivity and sensitivity makes the ECD a very useful detector for the detection of halogenated compounds in complex matrices. However, for the advantages of the ECD to be exploited in mycotoxin analysis, the majority of common mycotoxins must be derivatized with a halogenated derivatizing reagent. The ECD is a complex and often tricky detector with optimum performance depending on a variety of operating parameters. The text by Perry (1981) thoroughly describes ECD operation and theory.

3. Other Detectors

Several other types of GC detectors are available, but in general they are less applicable to the analysis of mycotoxins than the FID or ECD. The thermal conductivity detector (TCD) is a truly universal detector except for extremely rare cases where the heat capacity of the analyte coincidentally exactly matches the heat capacity of the carrier gas, but it suffers from low sensitivty and selectivity. The TCD also has a nonnegligible volume and therefore contributes significantly to the band width of narrow (as from capillary columns) peaks.

Several detectors are available which have high specificity for certain elements such as nitrogen, phosphorus and sulfur, and for unsaturated compounds. These include the thermionic emission detector, the flame photometric detector, the photoionization detector, and the chemiluminescent nitrogen detector. Drushel (1983) gives a succinct description of the operational aspects of each of these detectors. However, these detectors have found very limited use in mycotoxin analysis.

D. Comparison of Packed-Column and Capillary-Column Techniques

The common perception of capillary GC is that it yields vastly greater efficiencies than packed-column GC. In terms of total plates available per column, this is true. However, the actual plate height obtainable with packed columns is comparable to that which can be obtained with capillary columns. Packed columns approaching 1100 plates/ft ($\sim$3300 plates/m) can be constructed (Leibrand and

Dunham, 1973) with careful technique. A very good capillary column might exhibit 3000 plates/m.

The way in which capillary columns gain the efficiency advantage is sheer length. Due to pressure considerations, space within the column oven, and thermal mass (placing a very long ¼-in. packed column into an oven causes problems with maintaining temperature equilibra and in obtaining uniform temperatures during temperature programs), packed columns are limited to a maximum length of 3 to 6 m. Capillary columns of 25 to 50 m are routinely used, and separations on 100-m capillary columns are common. It must be noted, however, that column length in excess of that which is required should be avoided, since added length carries a time penalty.

In spite of the tremendous number of plates available with capillary columns, there are two primary reasons for choosing a packed column over a capillary column: (1) instrument considerations and (2) column capacity. In order to realize the full efficiency of capillary columns, it is necessary to eliminate as much extra-column dead volume as possible. This means that injectors and detectors must be appropriately designed. Also, column connections must be secure and properly made. If other than fused-silica capillary columns are used, the column must be correctly mounted and supported in the oven and the ends must be straightened in order to fit the injector and detector fittings. Most of these details are not significant obstacles and should not prevent the use of capillary columns when the requirements of the separation mandate their use.

The question of sample capacity can obviate the use of capillary columns. Due to their small i.d., very large sample volumes cannot be injected into these columns. In cases where the analyte is present in the solvent at the parts-per-million level, large injections are required for detectability. Packed columns can handle tens or even hundreds of microliters of injected solvent. In the latter case a packed column may be preferable to a capillary column.

For most mycotoxin analyses, where considerable effort is expended on sample cleanup prior to the GC analysis, the choice of either capillary-column or packed-column GC will probably depend largely on what is available in the analyst's lab and on which technique is more familiar to him or her. In many cases, the extensive sample cleanup will eliminate the need for the great resolving power of capillary GC and at the same time concentrate the sample sufficiently so that the full capacity of packed-column GC will not be required. In cases where either technique is available and either appears to be suitable, the higher resolution of capillary columns should probably be utilized in order to minimize the possibility of peaks eluting coincidently with the analyte.

E. Derivatization

There are two primary reasons for derivatizing a sample prior to GC analysis. Many compounds of analytical interest, including mycotoxins, contain polar groups such as hydroxyls, carboxyls, or amines. Conversion of these groups to esters, ethers, amides, and so on, will enhance their suitability for GC by reducing hydrogen bonding, increasing the volatility, and sometimes improving thermal stability. Even in cases where volatility or stability is not a limitation, these same groups can be derivatized with reagents which greatly increase the detectability of the compound and/or the selectivity of the detector for that particular compound. A fringe benefit of derivatization is often the elimination of peak asymmetry due to the reduction of interactions between the polar functional groups and active sites in the column walls or packing. The text by Knapp (1979) contains a wealth of information on derivatization reactions and reagents for numerous compound classes.

Alcohols can be readily converted to their trimethylsilyl derivative with a variety of reagents. The simplest of these is trimethylsilyl chloride. Other silylating reagents such as hexamethyldisilazane and N-trimethylsilylimidazole can also be used. The various trimethyl-silylating reagents require different reaction conditions and have different trimethylsilyl-donating capabilities, so the reagent of choice for a particular application is a matter of trial and error based on similar types of compounds from the literature. The trimethylsilyl ethers of alcohols are often considerably more volatile than the free alcohol.

In order to enhance the response of hydroxyl-containing compounds to an ECD, the trichloroacetate or trifluoroacetate can be formed. These reactions are easily carried out with trichloroacetic or trifluoroacetic anhydride. Besides having the potential for orders of magnitude better detectability, the trihalo acetates are often more volatile than the parent compound. Perfluoro esters of alcohols, formed with reagents such as pentafluoropropionic anhydride or heptafluorobutyric anhydride, also give greatly enhanced ECD responses, but with these higher molecular weight esters volatility can be a concern.

Acids are often converted to the methyl ester in order to enhance volatility. An excellent reagent for this purpose is diazomethane, which reacts quickly and quantitatively, with N_2 being the only side product. However, diazomethane is a powerful and sensitive explosive and is highly toxic, so it must be handled very carefully. HCl–methanol is an alternative methylating agent but is rather harsh and can lead to sample decomposition. A mixture of the boron trichloride with methanol is an effective methylating agent that may be milder than HCl–methanol. In order to enhance ECD detectability, carboxylic acids can be reacted with pentafluorobenzyl bromide in a basic medium to form the pentafluorobenzyl esters. This reaction must be carefully considered, however, because the high molecular weight of the derivatizing agent may result in a marginally volatile derivative.

Amino compounds are easily converted to the more volatile N-trimethylsilyl derivatives by a variety of silylating agents. Amide formation, using acid anhydrides (e.g., acetic anhydride), is also easily carried out. If the perfluoro anhydrides are utilized, the perfluoroamides formed can be readily detected with an ECD.

III. USE OF GAS CHROMATOGRAPHY
IN THE ANALYSIS OF MYCOTOXINS

The following examples of the use of GC in the analysis of mycotoxins are intended to be illustrative and not to represent a comprehensive review of all known mycotoxins. The examples discussed are current for the particular mycotoxin and represent the state of the art as found in the literature.

Since mass spectrometry (MS) is discussed in another chapter of this text, studies in which GC is only ancillary to MS are not covered in this discussion. In cases where the GC method itself was an important part of the work and the mass spectrometer served primarily as a detector, the development of the GC method will be discussed in this chapter with no critical evaluation of the MS.

A. Trichothecenes

The trichothecenes are a related group of fungal metabolites characterized by a 12,13-epoxytrichothec-9-ene ring system. They are produced by *Fusarium* species as well as some other fungi and often occur in corn, wheat, and other grains (Cole and Cox, 1981).

Deoxynivalenol (DON, vomitoxin) has been determined by several groups in wheat, corn, and feeds using GC with detection by ECD and FID. These methods depend on extraction, followed by cleanup and derivatization prior to GC analysis.

Ware and co-workers (1984) extracted 25 g of wheat with 125 ml of chloroform–ethanol (80 + 20 v/v) by shaking for 60 min. The sample was cleaned up by evaporating 10 ml of sample extract, dissolving the residue in dichloromethane, and applying the sample to a centrifugable silica gel column. Nonpolar materials were eluted from the column with toluene–acetone (80 + 20 v/v), and the DON was eluted with dichloromethane–methanol (95 + 5 v/v).

The dried sample, representing 2 g of the total wheat sample, was treated with heptafluorobutyric anhydride containing dimethylaminopyridine catalyst. The derivatization was carried out for 20 min at 60°C, after which aqueous sodium carbonate was added to the derivatizing vial and 100 μl of the organic layer was transferred to a vial containing 900 μl hexane. The final sample solution represented 0.0002 g/μl underivatized wheat extract. Gas chromatography was performed on a 1.8 m × 2 mm i.d. glass column packed with 3% OV-101 (methyl silicone) on 80–100 mesh Chromosorb W HP. The carrier gas was 5% methane in argon at 60 ml/min, and the oven temperature was programmed from 175°C (10 min initial hold) to 250°C at 10°C/min. The ECD was adjusted so that 100 pg of standard gave a response of 10% of full-scale signal. The trisheptafluorobutyrate of DON eluted at ~6.5 min under the conditions described. The method was reported to be linear from 100 to 500 pg of injected standard DON derivative.

The efficacy of heptafluorobutyric anhydride for the derivatization was compared with heptafluorobutyrylimidazole, and the heptafluorobutyric anhydride was shown to give more complete reaction in less time. Using spiked samples, this method was reported to give recoveries ranging from 99.5% at the 1184-ppb level to ~88% at the 118-ppb level, with a mean recovery of 88.1% [coefficient of variance (CV) = 8.6%]. The limit of detection of DON in wheat was reported to be 118 ppb.

Terhune *et al.* (1984) determined DON in wheat, corn, and feed by extracting 25 g of sample for 1 hr with 125 ml of water. A portion of the extract (20 ml) was added to a Clin-Elut column and eluted with ethyl acetate. After evaporation of the ethyl acetate, the residue was transferred with dichloromethane to a silica Sep-Pak cartridge, and the Sep-Pak was rinsed with toluene–acetone (95 + 5 v/v). DON was then eluted with 6 ml of methanol–dichloromethane (10 + 90). This eluate was evaporated and then brought to 4 ml with toluene–acetonitrile (95 + 5 v/v). The final extract represented 1 g/ml of the initial sample. A 1-ml portion of sample solution was added to 50 μl of heptafluorobutyrylimidazole and the derivatization carried out for 1 hr at 60°C. The solution was extracted with aqueous sodium bicarbonate, and 100 μl of the organic phase was added to 900 μl of hexane to yield a solution representing 0.1 g of the initial sample. GC conditions were as follows: 3 m × 4 mm i.d. 3% OV-3 (10% phenyl methyl silicone) on 80–100 mesh Chromosorb W HP; carrier gas, 5% methane in argon at 50 ml/min; oven temperature isothermal at 185°C; ECD operated at 225°C. The trisheptafluorobutyryl-DON derivative eluted at about 16 min under these conditions with a total of 35 min required between injections to allow interfering peaks to elute prior to subsequent injections. Injections ranging from 20 to 200 pg of standard DON derivative gave linear calibration curves. Studies on spiked wheat samples showed that recoveries were in the 88–98% range for DON levels of 50 to 100 ppb. The detection limit in wheat, feed, and corn was 20 ppb.

A method by Bennett *et al.* (1983) used methanol–water (50 + 50 v/v) to extract DON from wheat. The sample was shaken with the methanol–water for 30 min, filtered, and a portion precipitated with 30% aqueous ammonium sulfate. This solution was filtered and a portion of the filtrate was extracted with ethyl acetate. The ethyl acetate extract was evaporated to dryness and the residue transferred to a silica gel column using dichloromethane. Washes with toluene–acetone (5 + 95 v/v) and then hexane removed nonpolar impurities. DON was eluted with dichloromethane–methanol (95 + 5 v/v). Derivatization was with heptafluorobutrylimidazole in toluene–acetonitrile (95 + 5 v/v) for 1 hr at 60°C. The mixture was washed with aqueous sodium bicarbonate and an aliquot of the organic phase mixed with hexane. The final derivatized sample represented 100 μg/μl of the original wheat. GC was performed using a glass 50 × 1.3 cm o.d. column packed with 3% OV-101 (methyl silicone) on Gas-Chrom Q at 160°C with 5% methane in argon carrier gas at 60 ml/min. The average recovery for

DON was given as 81.5% with a CV of 7 to 8% at the 100–1000 ppb range using an ECD. The linear range of ECD response was not given, nor was the expected retention time of the derivatized DON. The lower limit of detection of DON in wheat was not stated, but spiked samples were analyzed down to 100 ppb. This method was presented as a modification of that of Scott *et al.* (1981), with the primary advantage being a decrease in analyst time per sample.

Cohen and Lapointe (1982) described the determination of DON in wheat, oats, barley, and corn. The mycotoxin was extracted from a 10-g sample using 200 ml chloroform–ethanol (80 + 20 v/v) plus 30 ml H_2O. [For a modification to the original 1982 procedure, which is the method discussed here, see Cohen and Lapointe (1983).] The extract was filtered and dried with sodium sulfate. After removal of the organic solvent by evaporation, the residue was transferred to a silica Sep-Pak cartridge with dichloromethane. The Sep-Pak was rinsed with dichloromethane, and then the DON was eluted with methanol–dichloromethane (5 + 95 v/v). The solvent was once again removed and the residue transferred with methanol–dichloromethane to a Sephadex LH-20 column. After washing the column with methanol–dichloromethane (10 + 90 v/v), the DON was removed with methanol–dichloromethane (20 + 80 v/v). The collected eluate was evaporated, and the extract was diluted to 10 ml with toluene–acetonitrile (95 + 5 v/v), and 1 ml of diluted extract was added to a J. T. Baker Cyano extraction column. The Cyano column was rinsed with an additional portion of toluene–acetonitrile (95 + 5 v/v), and the DON was eluted with toluene–acetonitrile (80 + 20 v/v). The solvent was removed, and the DON-trisheptafluorobutyrate was prepared by reaction with heptafluorobutyrylimidazole in much the same way as described by Bennett *et al.* (1983). The GC column was a 30 m × 0.25 mm i.d. fused-silica column coated with a 0.25-μm-thick film of SE-54 (5% phenyl, 1% vinyl, methyl silicone). The column was swept with helium carrier gas at a linear velocity of 35 cm/sec. Makeup gas consisting of 5% methane in argon at a flow of 40 ml/min was added to the column effluent prior to detection by an ECD. The column was programmed from 100°C (2 min initial hold) to 250°C at 10°C/min and held for 5 min and then to 300°C at 25°C/min. DON eluted during the hold at 250°C, after which the column was rapidly heated to remove strongly retained substances. Sample solutions were injected in the splitless mode. Quantitation was by peak height with a reported linear range of 10 to 100 pg injected. The derivatized DON eluted at 18.35 min in a region of the chromatogram relatively clear of interfering peaks. Quantitation was reported to be reliable down to 200 ppb if the Cyano column cleanup step was omitted, and down to 50 ppb if the Cyano column cleanup was included. Reported recoveries in wheat and barley ranged from 87 to 97% at levels from 360 to 56 ppb with CVs of less than 10%.

A multi-fusariotoxin method by Bata *et al.* (1983) utilized a capillary column and FID detection to determine DON and several other mycotoxins in wheat.

Extraction was carried out by shaking a 10-g wheat sample with 200 ml of ethyl acetate. The wheat was filtered, reextracted with 200 ml of methanol–water (60 + 40 v/v), and both extracts were combined and dried with sodium sulfate. After evaporation of the extract, the residue was transferred to a silica gel column and the column washed with benzene to remove lipids. The mycotoxins were then eluted with benzene–acetone (50 + 50 v/v). The eluate containing the mycotoxins was evaporated and then brought to 2 ml in acetone. Derivatization was carried out by adding 500 μl of the acetone solution to 100 μl of N,O-bis (trimethylsilyl)trifluoroacetamide and heating at 60°C for 15 min. The solution of trimethylsilylated mycotoxins obtained was injected directly into the GC column without further cleanup. GC was carried out on a 12 m × 0.25 mm i.d. borosilicate glass column coated with SE-52 (5% phenyl methyl silicone). Other stationary phases, including SE-30 (methyl silicone) and OV-17 (50% phenyl methyl silicone) were examined and found to be unsuitable. The carrier gas was hydrogen maintained at an inlet pressure of 40 kPa. Detection was with an FID. Both isothermal (200°C) and programmed (180°–260°C at 4°C/min) oven temperatures were utilized. Under isothermal conditions, DON was reported to have a Kovats retention index of 2323. Using the programmed oven temperature, DON had a retention time of approximately 12 min. Although the sensitivity and selectivity of the FID used in this study is less than that of an ECD (meaning that more interfering peaks might be expected), DON and the other mycotoxins examined eluted in fairly clean areas of the chromatogram. Quantitation was by the internal standard method with T-2 tetraol serving as the internal standard, but the linear range was not given. DON could be quantitated at the 100-ppb level in feed samples. Recoveries of DON at 100 ppb in corn averaged 70% with a CV of approximately 6%.

A later report by Bata and co-workers (1984) examined the same mycotoxins as the earlier study. Cleanup, derivatization, and GC conditions were essentially the same as in the prior study. However, the detection limit for DON and the other fusaritoxins was approximately halved (to 50 ppb) by utilizing a small-i.d. capillary column (0.13 mm i.d.). The decreased i.d. of the column resulted in higher efficiencies and gave enhanced peak heights, thus effectively decreasing the minimum detectable quantity (MDQ).

Kamimura and associates (1981) extracted *Fusarium* mycotoxins from grains with methanol–water. The crude extract was initally purified on an XAD-4 column eluted with methanol. The eluted mycotoxin fraction was further purified on a Florisil column using chloroform–methanol (90 + 10 v/v) to elute the mycotoxins. Mycotoxins were converted to the trimethylsilyl derivatives using a mixture of N-trimethylsilylimidazole and trimethylchlorosilane in ethyl acetate. GC was performed on a 2 m × 3.2 mm i.d. column packed with 2% OV-17 (50% phenyl methyl silicone) on 80–100 mesh Gas-Chrom Q. Both an FID and ECD were used to detect the DON-trimethylsilyl derivatives. Using the FID,

DON levels down to 100 ppb were reportedly detected. With the ECD, levels of DON as low as 2 ppb were said to be detectable. The recoveries of DON from barley, rice, and corn were 91, 89, and 93%, respectively, when spiked at the 1-ppm level.

Stahr *et al.* (1979) extracted mycotoxins from feeds by blending with acetonitrile–water. The blended sample was filtered and the filtrate was defatted by extraction with petroleum ether. The extracted sample was treated with 10% aqueous ferric chloride solution and then extracted with chloroform. After evaporation of the solvent, the chloroform residue was chromatographed on 6-ft 3% OV-101 (methyl silicone), 10% OV-101, and Dexsil-300 (a carboborane) columns. The mycotoxins were detected without derivatization with an FID. However, the method was only applicable for levels of approximately 1 ppm DON. The fact that metal columns resulted in analyte destruction was noted, as was the need to passivate the column prior to analysis by injections of Silyl-8. Silyl-8 is a commercially available (Pierce Chemical Co.) mixture of trimethylsilylating reagents designed to be injected onto a GC column in order to react with active sites present in the injector, column, and column packing.

Bennett and co-workers (1984) extracted DON from corn, wheat, oats, rice, and barley by blending 50-g samples with 250 ml of methanol–water (50 + 50 v/v). The sample slurries were centrifuged, the centrifugate was removed, the sediment reblended with methanol–water, recentrifuged, and the combined centrifugates were extracted with ethyl acetate. After evaporating the ethyl acetate, the residue was dissolved in acetonitrile and defatted with petroleum ether. The acetonitrile was then removed under vacuum and the residue transferred with dichloromethane to a silica gel column. The column was washed with benzene, benzene–acetone (95 + 5 v/v), and chloroform–methanol (98 + 2 v/v), and then DON was eluted with chloroform–methanol (95 + 5 v/v). The DON eluate was evaporated and then dissolved in 1 ml acetone. Derivatization was carried out by mixing 200 µl of the acetone solution with 100 µl of Tri-Sil TBT, a commercial mixture (Pierce Chemical Co.) of trimethylchlorosilane, N,O-bis(trimethylsilyl)acetamide, and N-trimethylsilylimidazole, and heating for 1 hr at 60°C. The tristrimethylsilyl DON was chromatographed on a 4 ft × 2 mm i.d. glass column packed with 3% OV-1 (methyl silicone) on 100–120 mesh Gas-Chrome Q. The column was programmed from 175°C to 250°C at 5°C/min, and DON was detected by FID. Tristrimethylsilyl-DON eluted at 6.4 min under the analysis conditions. In wheat, rice, and corn, DON could be detected at levels down to 200 ppb. Due to a greater amount of interferences in barley, the detection limit was approximately 500 ppb. Recoveries from corn, oats, and wheat averaged 82%, while in barley and rice DON was recovered at 69 and 70%, respectively.

T-2 toxin is another important trichothecene toxin which is associated with *Fusarium* species and other fungi and which is often found in the same grains as DON (Cole and Cox, 1981).

Cohen and Lapointe (1984) used methanol–water (50 + 50 v/v) to extract T-2 toxin from wheat, barley, and oats. The methanol–water extract was treated with 30% aqueous ammonium sulfate and extracted with ethyl acetate. After drying with sodium sulfate, the ethyl acetate was evaporated and the residue was transferred to a silica-gel Sep-Pak using dichloromethane. The Sep-Pak was rinsed with dichloromethane, and then T-2 toxin was eluted with dichloromethane–methanol (95 + 5 v/v). The dichoromethane–methanol was evaporated and the residue dissolved in chloroform–hexane (50 + 50 v/v), which was transferred to a cyano extraction column. Washing the cyano column with chloroform–hexane (50 + 50) eluted T-2 toxin. A portion of the eluate equivalent to 2 g of the initial sample was evaporated, mixed with 100 μl of heptafluorobutyrylimidazole and 1 ml of toluene–acetonitrile (95 + 5), and heated for 1 hr at 60°C. The mixture was vortexed with 5% aqueous sodium bicarbonate, and 100 μl of the organic phase was diluted to 1 ml with heptane for GC analysis. The GC column was a 30 m × 0.32 mm i.d. fused-silica capillary column coated with a 0.25-μm-thick film of DB-5 (bonded 5% phenyl methyl silicone). The carrier gas was helium at 35 cm/sec. Oven temperature was programmed from 75°C (2 min initial hold) to 225°C at 5°C/min, then, after a 10-min hold, to 250°C at 5°C/min, and, after a 5-min hold at 250°C, to 280°C at 30°C/min. An ECD was used, with makeup gas (10% methane in argon) being added to the column effluent at 40 ml/min. Samples were injected in the splitless mode by an autosampler. T-2 toxin eluted at 49.47 min with a CV of 0.02% for 30 replicates. Detector response for T-2 toxin was linear from 100 to 1000 pg injected with a calibration curve correlation coefficient of 0.997. Recoveries of T-2 toxin in wheat ranged from 80% at 200 ppb to 87.4% at 800 ppb. In barley, recoveries of 65 and 68% were noted at levels of 800 and 2000 ppb, respectively. The recovery in oats at 2000 ppb was 82%, and at 4000 ppb recovery was 81%.

Swanson et al. (1983) determined T-2 toxin in blood plasma. The plasma (2.5 ml) was extracted with benzene (10 ml) and the benzene was washed with sodium hydroxide. A portion of the benzene extract (2 ml) was mixed with 4 ml hexane and transferred to a Florisil cleanup column. The column was washed with dichloromethane, and the T-2 toxin was eluted with chloroform–methanol (95 + 5 v/v). The eluate was evaporated and then redissolved in 1 ml toluene. Heptafluorobutyrylimidazole (50 μl) was added and, after a 30-sec reaction time, aqueous sodium bicarbonate was added. Hexane (4 ml) was added, and after centrifugation 4-μl aliquots of the organic phase were used for GC analysis. GC conditions were as follows: 1.8 m × 2 mm i.d. glass column packed with 3% OV-1 (methyl silicone) on 100–120 mesh Supelcoport; carrier gas (5% methane in argon) flow of 22 ml/min; column temperature isothermal at 230°C; detection by ECD. Under these conditions T-2 toxin eluted at 8.6 min. Excellent recoveries of T-2 toxin, ranging from 100.5% at 50 ng/ml to 96.4% at 1000 ng/ml, were reported. As little as 25 pg of standard T-2 derivative could be detected; however, in actual plasma samples, 50 pg was required due to background interferences. The MDQ was given as 25 ng/ml plasma. Detector response was said to be linear from 25 pg to 2 ng.

Romer et al. (1978) determined T-2 toxin (and diacetoxyscirpenol) in corn, livestock feeds, and pet food. Methanol–water (50 + 50 v/v) was blended or shaken with a 50-g sample. An aliquot of the filtered extract was mixed with 30% aqueous ammonium sulfate and then filtered with the aid of diatomite. A portion of this filtrate was extracted repetitively with chloroform, and the combined chloroform extracts were washed with aqueous sodium hydroxide. A portion of the chloroform extract, mixed with hexane, was added to a silica gel column. The column was washed successively with benzene and with benzene–acetone (95 + 5 v/v), and then the mycotoxins were eluted with diethyl ether. A portion of the ethyl ether eluate equivalent to 2.0 g of the original sample was evaporated, and 2 ml benzene and 100 µl heptafluorobutyrylimidazole were added to the residue. After mixing for 15 sec, the derivatization mixture was washed with aqueous sodium bicarbonate and 0.5 ml of the organic phase was mixed with 4.5 ml of pentane containing methoxychlor internal standard. A 5-µl portion of the resulting solution was subjected to GC under the following conditions: 4 ft × 2 mm i.d. glass column packed with 3% SE-30 (methyl silicone) on 100–120 mesh Gas-Chrom Q; carrier gas (unspecified type) at 24 ml/min; column temperature isothermal at 200°C; detection by ECD. The method was applied to corn, dog chow, high-protein dog chow, dairy feed, and pig feed. The average recovery for samples spiked with T-2 toxin at the 1035-ppb level was 105% with a CV of 17%. The MDQ was found to be 150 pg of injected T-2 toxin. T-2 toxin at the 100-ppb level was quantifiable by the method. Detector response was stated to be linear over the range of 150 to 2000 pg injected T-2 toxin.

Of the multi-fusariotoxin methods discussed in detail previously with respect to DON, the following results were obtained for T-2 toxin. Bata et al. (1984) reported a detection limit of 50 ppb T-2 toxin. Stahr and co-workers (1979) detected T-2 toxin at levels to 1000 ppb. Kamimura et al. (1981) detected T-2 toxin to 80 ppb and reported recoveries of T-2 toxin from barley, rice, and corn of 81, 72, and 83%, respectively.

Methods for the GC determination of various other trichothecene toxins exist. The following references are representative of some of the trichothecenes amenable to GC analysis. Diacetoxyscirpenol has been determined by Swanson and co-workers (1982). Cohen and Lapointe (1984) determined HT-2 toxin and diacetoxyscirpenol along with T-2 toxin. The method by Romer and associates (1978) for T-2 toxin described above was also applicable to diacetoxyscirpenol. Bata et al. (1984) also used the method described for DON and T-2 toxin to analyze for scirpentriol, diacetoxyscirpenol, neosolaniol, HT-2 toxin, and others. The previously described method of Kamimura and co-workers (1981) was also applied to the determination of HT-2 toxin, nivalenol, neosolaniol, fusarenon-X, and diacetoxyscirpenol.

B. Zearalenone

Zearalenone is produced by *Fusarium* and other fungi growing on corn, wheat, and other cereal grains. It has been shown to exhibit estrogenic and anabolic activities and can result in shrunken ovaries and abortion in pregnant swine. Young male swine experience atrophy of the testes and may develop enlarged mammary glands (Cole and Cox, 1981).

Mirocha and co-workers (1974) extracted 100 g of corn or barley either via Sohxlet extraction with ethyl acetate or by shaking with ethyl acetate. One-quarter of the extract, representing 25 g initial sample, was evaporated to dryness and the residue redissolved in 25 ml chloroform. The chloroform was gently extracted with two portions of 1 *M* aqueous sodium hydroxide; the phenolic and therefore acidic zearalenone was thus partitioned into the aqueous phase. The sodium hydroxide extract was acidified to pH 9.5 with phosphoric acid and extracted with chloroform to partition the now-reprotonated zearalenone into the organic phase. The chloroform was removed by evaporation and the residue from the organic phase was used directly for derivatization. An alternative cleanup procedure, reported to be less efficient but of utility when untenable emulsions resulted from the base cleanup procedure, involved evaporating a portion of the initial sample extract and dissolving the residue in acetonitrile. The acetonitrile was defatted by extraction with petroleum ether, and the acetonitrile was evaporated. Preparative thin-layer chromatography (TLC) on a 500-μm layer of silica gel G, developed with chloroform–methanol (97 + 3), was utilized as a final cleanup step. After extracting the zearalenone zone from the TLC sorbent and removal of solvent, the sample residue was derivatized by adding 50 μl of Tri-Sil BT (a commercial mixture of *N,O*-bis(trimethylsilyl)acetamide and trimethylchlorosilane) and allowing the mixture to stand at room temperature for 20 min. Gas chromatography was carried out on a 0.93 m × 3.2 mm i.d. glass or metal column packed with 3% OV-1 (methyl silicone) on 100–200 mesh Gas-Chrom Q. Nitrogen at 20 ml/min was used as the carrier gas, and detection was by FID. The oven temperature was programmed from 150°C to 275°C at 6°C/min. Under these conditions, the bistrimethylsilyl ether of zearalenone eluted at approximately 17 min. Two methods for confirming zearalenone, one involving the formation of the zearlenone dimethyl ether and the other the formation of the methyloxime of bistrimethylsilyzearalenone, were also described. The method was reported to detect less than 50 ppb zearalenone. CVs were estimated to be 28% at levels of 20 to 100 ppb, 9% at levels of 1000 to 5000 ppb, and 6% at levels of 20,000 to 34,000 ppb. Recoveries of zearalenone (concentrations not reported) were 83.5 ± 7% with the base cleanup procedure and 73% with the alternate procedure.

Thouvenot and Morfin (1979) quantitated zearalenone in corn using capillary GC. A 25-g corn sample was spiked with internal standard zearalanone and the sample extracted using the method of Mirocha described above. After the preparative TLC cleanup procedure, in which the zearalanone and zearalenone cochromatographed, the bistrimethylsilyl ethers of zearalenone were prepared by reacting the corn extract with a mixture of N,O-bis(trimethylsilyl)trifluoroacetamide and trimethylchlorosilane for 3 hr at room temperature. GC conditions were as follows: 57 m × 0.29 mm i.d. borosilicate capillary column coated with a 0.21-μm-thick film of SE-52 (5% phenyl methyl silicone); helium carrier gas at 1 ml/min (linear velocity 13.3 cm/sec); oven temperature isothermal at 280°C; detection by FID. The column was said to exhibit 97,000 theoretical plates. The bistrimethylsilyl ether of zearalenone eluted at 27.3 min (Kovats retention index = 2788), and the bistrimethylsilyl ether of the internal standard zearalanone eluted at 24.6 min (retention index = 2727). The precision of the method at the 2000-ppb level in corn was said to be ±5%. At the 100-ppb limit of detection, precision was ±15%.

Trenholm and associates (1980) described a method for the determination of zearalenone in blood serum. The plasma was applied to a porous polymer chromatographic column, and zearalenone was eluted with isopropanol–dichloromethane. The organic eluate was washed with aqueous sodium hydroxide to extract zearalenone; the aqueous base was then neutralized with aqueous hydrochloric acid and zearalenone reextracted into dichloromethane. A variety of silylating reagents was examined. The most efficient derivatization was found to occur when the residue from the evaporation of the final dichloromethane extract was dissolved in acetone and reacted at room temperature for 2 hr with N-methyl-N-trimethylsilyltrifluoroacetamide. GC was carried out on a 3.7 m × 2 mm i.d. glass column packed with 3% OV-17 on 120–140 mesh Gas-Chrom Q at an oven temperature of 262°C. An FID was utilized and quantitation was by the internal standard method with epicoprostanol as the internal standard. Detector response was linear from 500 to 3000 pg of injected zearalenone derivative. The limit of detection in blood serum was found to be 100 ng zearalenone per milliliter of serum. Studies on spiked samples showed recoveries ranging from 68% (8000 ng/ml) to 74% (1600 ng/ml).

C. Patulin

Patulin is chemically a relatively simple molecule. It is primarily produced by *Penicillium* and *Aspergillus* species. Patulin produces digestive tract hemorrhaging and is carcinogenic but, due to its instability in the digestive tract of animals, the impact of patulin toxicosis in animals is unknown. Patulin is often found in apple juice and cider vinegar (Cole and Cox, 1981).

Pohland and associates (1970) determined patulin in apple juice by extracting 50-ml samples with ethyl acetate. The extracts, after drying with sodium sulfate, were evaporated and the residue derivatized by one of three methods. The acetate derivative was formed by dissolving the residue in 175 μl tetrahydrofuran and 10 μl pyridine and adding 500 μl acetic anhydride. Reaction was carried out over-night at room temperature. The chloroacetate was prepared analogously but with chloroacetic anhydride instead of acetic anhydride. This reaction required 2 hr at room temperature. The trimethylsilyl ether was formed by reaction with 500 μl of *N,O*-bis(trimethylsilyl)acetamide in 500 μl of ethyl acetate for 1.5 hr on a steam bath. In all cases, after reaction the derivatizing solutions were evaporated to dryness and the residue was redissolved in 5 ml ethyl acetate containing 0.15 mg/ml hexadecane or octadecane internal standard. GC conditions were as fol-lows: (a) for the acetate derivative 6 ft × 4 mm i.d. glass column packed with 3% JXR (methyl silicone) on Gas-Chrom W operated at 110°C; (b) for the chloroace-tate derivative, same as in (a) except oven temperature of 130°C; (c) for the trimethylsilyl ether derivative, 10 ft × 4 mm i.d. glass column packed with 1% SE-30 (methyl silicone) on Gas-Chrom Q. Both ECD and FID detection were used for the chloroacetate; only FID detection was examined with the other derivatives. The silyl derivative was found to give linear calibration curves from 50 to 500 ng injected but was not considered suitable for quantitation due to the difficulty in achieving 100% derivatization and the instability, even under re-frigeration, of the derivative. The acetate derivative was easily formed in quan-titative yield but was not stable indefinitely. Patulin acetate gave linear calibra-tion curves from 80 to 500 ng injected. The detection limit for the acetate derivative was approximately 700 ppb. The chloroacetate derivative was stable indefinitely and give linear calibration curves from 60 to 500 ng injected patulin with an FID. With the FID, the detection limit for the chloroacetate was approx-imately 500 ppb. Using an ECD, the chloroacetate could be detected at the 150-ppb level. Recoveries from apple juice spiked at levels from 700 to 60,000 ppb patulin ranged from 91% (lower level) to 101%.

Pero and Harvan (1973) examined a variety of column packings for several mycotoxins including patulin. Columns were 5 ft × 2 mm i.d. packed with OV-11 (35% phenyl methyl silicone), OV-17 (50% phenyl methyl silicone), OV-25 (75% phenyl methyl silicone), OV-101 (methyl silicone), or Dexsil-300 (a carboborane) coated at the 3% level on Gas-Chrom Q. The GC oven was programmed from 100°C to 250°C at 8°C/min, and nitrogen at 25 ml/min was used as the carrier gas. Patulin was derivatized with a mixture of *N,O*-bis(tri-methylsilyl)acetamide–trimethylchlorosilane–pyridine (6 + 2 + 9). Approxi-mate retention times of the trimethylsilyl ether of pautlin on the various phases were as follows: OV-11, 10.1 min; OV-17, 9.8 min; OV-25, 8.3 min; OV-101, 5.9 min; Dexsil-300, 8.8 min. The purpose of the method was to develop a screening procedure for a wide range of mycotoxins likely to be found in human feedstuffs. For this purpose, the OV-17 column was found to be the most suitable with regard to peak symmetry and complete separation of the mycotoxins examined.

Suzuki *et al.* (1975) studied the efficacy of various reagents for the silylation of patulin. Of several reagents studied, the most effective was found to be N,O-bis(trimethylsilyl)acetamide, which reacted quantitatively in 20 min at room temperature. Using a 2 m × 3 mm i.d. glass column packed with 10% DC-200 (methyl silicone) and 15% QF-1 (50% trifluoropropyl methyl silicone) on Gas-Chrom Q at 175°C and an ECD, trimethylsilyl patulin was found to give satisfactory peak symmetries when 750 ng of the compound was injected. The derivative was found to be stable in benzene solution for greater than 30 days when stored at either 5° or 20°C. These findings are in contrast with those of Pohland et al. (1970) described above.

Fujimoto et al. (1975) extracted penicillic acid and patulin from rice by shaking a 20-g sample with 100 ml of methanol–5% aqueous sodium chloride (50 + 50 v/v) and 75 ml of hexane. After filtration, the aqueous methanol layer of the extract was washed with hexane, and then the methanol was removed by evaporation. The remaining aqueous solution was extracted with ethyl acetate, which was then dried with sodium sulfate and evaporated to dryness. The residue, dissolved in 0.5 ml of benzene–methanol (97 + 3 v/v), was streaked on a silica gel G TLC plate, which was then developed with benzene/methanol/acetic acid (18 + 1 + 1 v/v/v). The silica gel containing patulin was removed and extracted with benzene–methanol. After removal of the solvent, the trimethylsilyl derivative of patulin was formed in benzene using N,O-bis(trimethylsilyl)acetamide. Several GC columns were examined, but best results were obtained on a 2 m × 3 mm i.d. glass column packed with a 1 : 1 mixture of 10% DC-200 (methyl silicone) and 15% QF-1 (50% phenyl methyl silicone) on 80–100 mesh Gas-Chrom Q operated at isothermal temperatures ranging from 205° to 160°C. Patulin could be detected at the 20-ppb level in rice. Recoveries for the method described were between 70 and 80% in rice, flour, and soybeans at the 50-ppb level. An alternative extraction procedure gave recoveries of patulin ranging from 85 to 92% (see the original citation).

D. Penicillic Acid

Penicillic acid is a mycotoxin produced by several *Penicillium* and *Aspergillus* species. It is toxic and is a suspected carcinogen (Cole and Cox, 1981).

Thorpe and Johnson (1974) reported a method for the analysis of penicillic acid in corn, dried beans, and apple juice. Corn and beans (60 g) were blended with ethyl acetate, while apple juice (50 ml) was extracted with ethyl acetate to remove the penicillic acid. The organic extractant was then partitioned three times with aqueous sodium bicarbonate. After acidifying to pH 3 with aqueous hydrochloric acid, the penicillic acid was back-extracted into ethyl acetate. The ethyl acetate was removed using a Kuderna-Danish concentrator, and the residue was transferred to a silica gel column using hexane–ethyl acetate–formic acid (75 + 25 + 1 v/v/v). Penicillic acid was then eluted with the same solvent and the eluate concentrated with a Kuderna-Danish apparatus. The residue was derivatized with 200 μl of trifluoroacetic anhydride (room temperature for 15 min). Excess reagent was then removed and the sample dissolved in 1.0 ml isooctane. GC conditions were as follows: 6 ft × $\frac{1}{4}$ in. i.d. glass column packed with 3% OV-101 (methyl silicone) on 100–120 mesh Gas-Chrom Q operated isothermally at 140° to 160°C; nitrogen carrier gas at a flow rate of 70 to 80 ml/min; detection with an ECD. The method was sensitive enough to quantitate as little as 4 ppb penicillic acid, with the detection limit said to be considerably lower. Recoveries at levels greater than 15 ppb were greater than 80%, and recoveries at levels below 15 ppb ranged from 55 to 80%. The loss at the lower levels was believed to be due to the volatility of the penicillic acid trifluoroacetate, which evaporated when excess reagent was removed during the derivatization step.

Penicillic acid was determined by Fujimoto et al. (1975) (using the method described in Section II,C) in rice, flour, and soybeans. Recoveries from these commodities were 88.8, 87.6, and 91.6%, respectively. The MDQ was found to be approximately 20 ppb.

Phillips and co-workers (1981) described an interesting derivative for the confirmation of penicillic acid. Reaction of penicillic acid with diazomethane was found to proceed in a two-step reaction via the expected acid methyl ester to a pyrazoline derivative. The pyrazoline was presumed to form when diazomethane added to the double bond of the δ-methylene group of methyl penicillate. Using a 6 ft × 2 mm i.d. glass column packed with 1.5% SP-2250 (50% phenyl methyl silicone) and 1.95% SP 2401 (trifluoropropyl silicone) on 100–120 mesh Supelcon at 140°C, methyl penicillate eluted at 1.83 min, and the pyrazoline derivative eluted at 2.92 min. With a 3% OV-101 (methyl silicone) column, the pyrazoline eluted at 70 sec and the FID gave a linear response in the range of 10 to 1000 ng injected.

E. Slaframine and Swainsonine

Slaframine and swainsonine are chemically similar compounds produced by the fungus Rhizoctonia leguminicola. When forage infested by this fungus is consumed by livestock, a condition known as "the slobbers" results. Slaframine stimulates secretions from exocrine glands such as the salivary glands and pancreas. Swainsonine disrupts glycoprotein synthesis. It is currently thought that livestock slobber syndrome could be a result of ingestion of both slaframine and swainsonine (Broquist et al., 1984).

Hagler and Behlow (1981) extracted red clover in a Sohxlet apparatus with 95% ethanol. The acidified extract was partitioned against chloroform and then, after adjusting to pH 10, slaframine was extracted into chloroform. After concentration, the extract was streaked onto a silica gel 60 TLC plate. The plate was developed with chloroform–methanol–ammonium hydroxide (90 + 10 + 1 v/v/v), and the slaframine zone was removed and eluted from the silica gel with acetone. A variety of slaframine derivatives were prepared and used for qualitative GC identification of slaframine with the following GC conditions: 1 m × 3.2 mm i.d. nickel column packed with 3% OV-17 (50% phenyl methyl silicone) on 100–120 mesh Gas-Chrom Q programmed from 100° to 200°C at 8°C/min; nitrogen carrier gas at 40 ml/min; FID. The GC retention times of these derivatives were found to be as follows: (a) slaframine, 5.3 min; (b) *O*-deacetylslaframine, 3.4 min; (c) trimethylsilyl slaframine, 6.0 min; (d) trimethylsilyl-*O*-deacetyl-*N*-acetylslaframine, 9.6 min; (e) *N*-acetylslaframine, 11.4 min; (f) *N*-acetyl-*O*-deacetylslaframine, 10.0 min; (g) trimethylsilyl-*O*-deacetylslaframine, 3.9 min; (h) heptafluorobutyrylslaframine, 13.5 min.

Broquist and co-workers (1984) extracted swainsonine from red clover hay as described above for slaframine. The crude extract was purified by percolation through a Bio-Rad AG 50W-X8 ion exchange column eluted with aqueous sodium hydroxide and sodium bicarbonate. The column eluate containing swainsonine was lyophilized and the swainsonine triacetate derivative was prepared. This derivative was chromatographed on a 3-ft column packed with either OV-1 or SP 2100 (both are methyl silicones; neither the stationary-phase support nor percentage loading was given). Swainsonine eluted at between 4 and 5 min on the OV-1 column, and between 5 and 6 min on the SP-2100 column when the oven was programmed from 120° to 220°C at 10°C/min. The method was used to obtain MS information.

F. *Alternaria* Toxins

The *Alternaria* species can occur in wheat, tobacco, corn, peanuts, and grasses. They produce a series of dibenzo[*a*]pyrone toxins. These compounds are acutely toxic and teratogenic (Cole and Cox, 1981).

Pero and co-workers (1971) examined the GC behavior of alternariol, altenuene, and alternariol monomethyl ether (AME). Standard samples of the toxins were derivatized by reaction at room temperature with a mixture of *N,O*-bis(trimethylsilyl)acetamide and trimethylchlorosilane in pyridine. Reaction was said to be complete within a few minutes, and the derivatives were stable for several weeks under refrigeration. GC was performed on 5 ft × 2 mm i.d. glass columns packed with 1, 3 and 10% OV-1 (methyl silicone) or OV-17 (50% phenyl methyl silicone) or OV-25 (75% phenyl methyl silicone) on 100 to 120 mesh Gas-Chrom Q. Detection was with an FID. Nitrogen at a flow of 25 ml/min was used as the carrier gas. All three phases were found to be unsuitable at the 1% loading level due to unacceptable peak symmetries, presumably due to active-site interactions. The most satisfactory results were reported for the 3 or 10% OV-17 phases using temperatures programmed from 100° to 300°C at 8°C/min. Using this temperature program on 10% OV-17, the derivatives eluted as follows: altenuene at

23.6 min; alternariol at 25.8 min; and AME at 26.7 min. Each toxin was detectable at the 0.1-μg injected level, and calibration curves were linear from approximately 0.1 μg injected to ~100 μg injected. The separation obtained on the OV-17 column was sufficient to allow quantitation of the toxins in a crude tetrahydrofuran extract of inoculated rice cultures.

Lucas *et al.* (1971) determined alternariol and AME in tobacco. Flue-cured tobacco leaf (50 g) was extracted with boiling acetone. Concentrated extract was applied to a silica gel G column. AME was eluted with benzene–tetrahydrofuran (95 + 5 v/v), and alternariol was eluted with benzene–tetrahydrofuran (85 + 15 v/v). Each of the fractions was evaporated, and the toxins were derivatized with a mixture of *N,O*-bis(trimethylsilyl)acetamide and trimethylchlorosilane in tetrahydrofuran. GC was carried out on a 3% OV-17 (50% phenyl methyl silicone) column programmed from 200° to 250°C at 2°C/min. The detection limit for the method was estimated to be as low as 250 ppm. No recovery experiments were carried out.

G. Aflatoxins

The aflatoxins have traditionally been analyzed by either TLC or HPLC methods. They have been considered unsuitable for GC analysis due to their high polarity, molecular weight, low volatility, and thermal instability. However, two recent reports have described methods for the GC-MS determination of aflatoxins B_1 and B_2.

Trucksess and co-workers (1984) were able to chromatograph aflatoxin B_1 on methyl silicone-coated fused-silica columns. The sample was injected on-column at 40°C, and the oven temperature was ramped to 250°C in 4 min. Aflatoxin B_1 eluted in 5 to 6 min.

Rosen *et al.* (1984) used a 15 m × 0.32 mm i.d. fused-silica capillary column coated with a 0.25-μm-thick film of DB-5 (bonded 5% phenyl 1% vinyl methyl silicone) to chromatograph successfully aflatoxin B_1 and B_2. Helium at a linear velocity of 50 cm/sec was used as the carrier gas, and samples were injected on-column. Temperatures in the range of 200° to 290°C, apparently programmed at various rates (exact temperature conditions for the various chromatograms displayed were not specified), caused the aflatoxins to elute in the range of 8 to 15 min.

REFERENCES

Bata, A., Vanyi, A., and Lasztity, R. (1983). Simultaneous detection of some fusariotoxins by gas-liquid chromatography. *J. Assoc. Off. Anal. Chem.* **66**, 577–581.

Bata, A., Vanyi, A., Lasztity, R., and Galacz, J. (1984). Determination of trichothecene toxins in foods and feed. *J. Chromatogr.* **286**, 357–362.

Bennett, G. A., Stubblefield, R. D., Shannon, G. M., and Shotwell, O. L. (1983). Gas chromatographic determination of deoxynivalenol in wheat. *J. Assoc. Off. Anal. Chem.* **66,** 1478–1480.

Bennett, G. A., Megalla, S. E., and Shotwell, O. L. (1984). Method of analysis for deoxynivalenol and zearalenone from cereal grains. *J. Assoc. Oil Chem. Soc.* **61,** 1449–1451.

Broquist, H. P., Mason, P. S., Hagler, W. M., and Harris, T. M. (1984). Identification of swainsonine as a probable contributory mycotoxin in moldy forage mycotoxicoses. *Appl. Environ. Microbiol.* **48,** 386–388.

Cohen, H., and Lapointe, M. (1982). Capillary gas-liquid chromatographic determination of vomitoxin in cereal grains. *J. Assoc. Off. Anal. Chem.* **65,** 1429–1434.

Cohen, H., and Lapointe, M, (1983). Correction to *J. Assoc. Off. Anal. Chem.* **65,** 1429–1434. *J. Assoc. Off. Anal. Chem.* **66,** 821.

Cohen, H., and Lapointe, M. (1984). Capillary gas chromatographic determination of T-2 toxin, HT-2 toxin, and diacetoxyscirpenol in cereal grains. *J. Assoc. Off. Anal. Chem.* **67,** 1105–1107.

Cole, R. J., and Cox, R. H. (1981). "Handbook of Toxic Fungal Metabolites." Academic Press, New York.

Dandeneau, R., and Zerenner, E. H. (1979). An investigation of glass for capillary chromatography. *HRC CC, J. High Resolut. Chromatogr. Chromatogr. Commun.* **2,** 351–356.

Drushel, H. V. (1983). Needs of the chromatographer-detectors. *J. Chromatogr. Sci.* **21,** 375–384.

Ettre, L. S. (1983). Open tubular columns prepared with very thick liquid phase film I. Theoretical basis. *Chromatographia* **17,** 553–559.

Ettre, L. S., McClure, G. L., and Walters, J. D. (1983). Open tubular columns with very thick liquid phase film II. Investigations on column efficiency. *Chromatographia* **17,** 560–569.

Fenimore, D. C., Whitford, J. H., Davis, C. M., and Zlatkis, A. (1977). Nickel gas chromatographic columns; an alternative to glass for biological samples. *J. Chromatogr.* **140,** 9–15.

Fujimoto, Y., Suzuki, T., and Hoshino, Y. (1975). Determination of penicillic acid and patulin by gas-liquid chromatography with an electron capture detector. *J. Chromatogr.* **105,** 99–106.

Gandara, V. M., Sanz, J., and Martinez-Castro, I. (1984). A two step method for the immobilization of stationary phases in GC capillary columns. *HRC CC, J. High Resolut. Chromatogr. Chromatogr. Commun.* **7,** 44–45.

Grob, K., and Grob, G. (1972). Techniques of capillary gas chromatography. *Chromatographia* **5,** 3–12.

Grob, K., and Grob, K., Jr. (1974). Isothermal analysis on capillary columns without stream splitting. *J. Chromatogr.* **94,** 53–64.

Hagler, W. M., and Behlow, R. F. (1981). Salivary syndrome in horses: identification of slaframine in red clover hay. *Appl. Environ. Microbiol.* **42,** 1067–1073.

Haken, J. K. (1984). Developments in polysiloxone stationary phases in gas chromatography. *J. Chromatogr.* **300,** 1–77.

Hubball, J. A., DiMauro, P. R., Barry, E. F., Lyons, E. A., and George, W. A. (1984). Developments in crosslinking of stationary phases for capillary gas chromatography by cobalt-60 gamma radiation. *J. Chromatogr. Sci.* **22,** 185–191.

Jennings, W. J. (1980). "Gas Chromatography with Glass Capillary Columns," 2nd Ed. Academic Press, New York.

Kamimura, H., Nishijima, K., Yasuda, K., Saito, K., Ibe, A., Nagayama, T., Ushiyama, H., and Naoi, Y. (1981). Simultaneous detection of several *Fusarium* mycotoxins in cereals, grains, and foodstuffs. *J. Assoc. Off. Anal. Chem.* **64,** 1067–1073.

Knapp, D. R. (1979). "Handbook of Analytical Derivatization Reactions." Wiley, New York.

Leary, J. J., Justice, J. B., Tsuge, S., Lowry, S. R., and Isenhour, T. L. (1973). Correlating gas chromatographic liquid phases by means of a nearest neighbor technique. Proposed set of twelve preferred phases. *J. Chromatogr. Sci.* **11**, 201–206.

Lee, M. L., Kong, R. C., Woolley, C. L., and Bradshaw, J. S. (1984). Fused silica capillary column technology for gas chromatography. *J. Chromatogr. Sci.* **22**, 136–142.

Leibrand, R. J., and Dunham, L. L. (1973). Preparing high efficiency packed GC columns. *Res./Dev. (Hewlett-Packard)* pp. 24–38.

Lucas, G. B., Pero, R. W., Snow, J. P., and Harvan, D. (1971). Analysis of tobacco for the *Alternaria* toxins, alternariol and alternariol monomethyl ether. *J. Agric. Food Chem.* **19**, 1274–1275.

McReynolds, W. O. (1970). Characterization of some liquid phases. *J. Chromatogr. Sci.* **8**, 685–691.

Mirocha, C. J., Schauerhamer, B., and Pathre, S. V. (1974). Isolation, detection, and quantitation of zearalenone in maize and barley. *J. Assoc. Off. Anal. Chem.* **57**, 1104–1110.

Pero, R. W., and Harvan, D. (1973). Simultaneous detection of metabolites from several toxigenic fungi. *J. Chromatogr.* **80**, 255–258.

Pero, R. W., Owens, R. G., and Harvan, D. (1971). Gas and thin-layer chromatographic methods for analysis of the mycotoxins altenuene, alternariol, and alternariol monomethyl ether. *Anal. Biochem.* **43**, 80–88.

Perry, J. A. (1981). "Introduction to Analytical Gas Chromatography." Dekker, New York.

Phillips, T. D., Ivie, G. W., Heidelbaugh, N. D., Kubena, L. F., Cysewski, S. J., Hayes, A. W., and Witzel, D. A. (1981). Confirmation of penicillic acid by high pressure liquid chromatography and gas–liquid chromoatography. *J. Assoc. Off. Anal. Chem.* **64**, 162–165.

Pohland, A. E., Sanders, K., and Thorpe, C. W. (1970). Determination of patulin in apple juice. *J. Assoc. Off. Anal. Chem.* **53**, 692–695.

Romer, T. R., Boling, T. M., and MacDonald, J. L. (1978). Gas–liquid chromatographic determination of T-2 toxin and diacetoxyscirpenol in corn and mixed feeds. *J. Assoc. Off. Anal. Chem.* **61**, 801–808.

Rosen, R. T., Rosen, J. D., and Diprossimo, V. P. (1984). Confirmation of aflatoxins B_1 and B_2 in peanuts by gas chromatography/mass spectrometry/selected ion monitoring. *J. Agric. Food Chem.* **32**, 276–278.

Scott, P. M., Lau, P.-Y., and Kanhere, S. R. (1981). Gas chromatography with electron capture and mass spectrometric detection of deoxynivalenol in wheat and other grains. *J. Assoc. Off. Anal. Chem.* **64**, 1364–1371.

Springston, S. R., Melda, K., and Novotny, M. V. (1983). Immobilization of silicone stationary phases for capillary chromatography through the action of azoisobutronitrile. *J. Chromatogr.* **267**, 395–398.

Stahr, H. M., Kraft, A. A., and Schuh, M. (1979). The determination of T-2 toxin, diacetoxyscirpenol, and deoxynivalenol in foods and feeds. *Appl. Spectrosc.* **33**, 294–297.

Suzuki, Y., Fujimoto, Y., Hoshino, Y., and Tanaka, A. (1975). Trimethyl-silylation of penicillic acid and patulin, and the stability of the products. *J. Chromatogr.* **105**, 95–98.

Swanson, S. P., Terwell, L., Corley, R. A., and Buck, W. B. (1982). Gas chromatographic method for the determination of diacetoxyscirpenol in swine plasma and urine. *J. Chromatogr.* **248**, 456–460.

Swanson, S. P., Ramaswamy, V., Beasley, V. R., Buck, W. B., and Burmeister, H. H. (1983). Gas–liquid chromatographic determination of T-2 toxin in plasma. *J. Assoc. Off. Anal. Chem.* **66**, 909–912.

Terhune, S. J., Nguyen, N. V., Baxter, J. A., Pryde, D. H., and Qureshi, S. A. (1984). Improved gas chromatographic method for quantitation of deoxynivalenol in wheat, corn, and feed. *J. Assoc. Off. Anal. Chem.* **67**, 1102–1104.

Thorpe, C. W. and Johnson, R. L. (1974). Analysis of penicillic acid by gas–liquid chromatography. *J. Assoc. Off. Anal. Chem.* **57,** 861–865.

Thouvenot, D. R., and Morfin, R. F. (1979). Quantitation of zearalenone by gas–liquid chromatography on capillary glass columns. *J. Chromatogr.* **170,** 165–173.

Trenholm, H. L., Warner, R., and Farnworth, E. R. (1980). Gas chromatographic detection of the mycotoxin zearalenone in blood serum. *J. Assoc. Off. Anal. Chem.* **63,** 604–611.

Trucksess, M. W., Brumley, W. C., and Nesheim, S. (1984). Rapid quantitation and confirmation of aflatoxins in corn and peanut butter, using a disposable silica gel column, thin layer chromatography, and gas chromatography/mass spectrometry. *J. Assoc. Off. Anal. Chem.* **67,** 973–975.

Ware, G. M., Carman, A., Francis, O., and Kuan, S. (1984). Gas chromatographic determination of deoxynivalenol in wheat with electron capture detection. *J. Assoc. Off. Anal. Chem.* **67,** 731–734.

11

High-Performance Liquid Chromatography and Its Application to the Analysis of Mycotoxins

MARTIN J. SHEPHERD

Ministry of Agriculture, Fisheries and Food
Food Laboratory
Haldin House
Norwich NR2 4SX, United Kingdom

293

MODERN METHODS IN THE ANALYSIS
AND STRUCTURAL ELUCIDATION
OF MYCOTOXINS

I. INTRODUCTION

Mycotoxins are by definition those metabolites of fungi which are toxic to humans or to animals. The various classes of toxins have little else in common apart from a typically oxygenated structure with moderate or high polarity and thus low volatility. High-performance liquid chromatography (HPLC) is particularly suited to the separation of such compounds and has become the most widely used form of chromatography in the analytical laboratory. This chapter will examine the methodology of HPLC and will review its application to the analysis and preparative fractionation of mycotoxins. Although the theory of HPLC will not be covered here, an appreciation of the underlying principles of the technique offers a useful guide to the development of separation methods. Many books describing the theory and the practical details of HPLC are available (see, e.g., Simpson, 1982; Heftmann, 1983; Poole and Schuette, 1984; Bristow, 1976). The Bristow text is rather out of date in some details but gives a great deal of miscellaneous information not collected together elsewhere, such as solvent compressibility tables and physical data on materials, including sinters and meshes. Several specialized journals and various other periodicals dedicated to chromatography cover all aspects of HPLC. The technique was last reviewed in depth in 1982 by Majors *et al.*

Attention is focused here on mycotoxin HPLC methods described in the literature from 1980 onward, although earlier contributions of general interest are included. Due to space limitations it has been necessary to be very selective, and few articles are discussed in detail. More than 110 articles describing HPLC methods for mycotoxins appeared during the years 1981 to 1983. A recent review (Scott, 1981) gives comprehensive coverage of work published before 1981, while a French language textbook, *The Study and Determination of Aflatoxins in Foods* (Blanc, 1982), containing over 1000 references, describes the HPLC analysis of these toxins.

Historically, thin-layer chromatography (TLC) has been the method of choice for mycotoxin analysis, due to the then-recent development of this technique in the period 1960–1963 when the discovery of the aflatoxins and subsequent worldwide concern about their acute toxicity and carcinogenicity stimulated the growth of research into mycotoxins. Ten years later the gradual introduction of HPLC analysis for the final determination had to compete with existing thoroughly validated and widely used TLC-based methods. The trend has been toward increased use of HPLC; McKinney (1981a) reported that of 50 participants (on average) in the Smalley aflatoxin M_1 series of 1979 and 1980, no more than 4 were using HPLC, while for the 1981–1982 exercise on B_1, up to 20% of analysts employed the technique (McKinney, 1984). In the 1982 IARC/WHO aflatoxin check sample program (Friesen, 1983a,b), 21% of the analyses (including 28 of 115 for M_1 in milk) were completed by HPLC.

However, although quantitation of mycotoxin standards is significantly improved using HPLC, it is also true that interlaboratory coefficients of variation, which are typically 30–60% (obtained from collaborative studies and check sample exercises), do not differ greatly whether HPLC or TLC is employed for the determination step (Friesen, 1982). This may be an artifact of the data caused by the difficulties of statistical analysis on relatively small numbers of results when a few near-outliers can markedly influence the calculated standard deviation, but an alternative explanation lies in the fact that analysts employing HPLC have tended to use a wide range of cleanup methods compared to those utilizing TLC—the latter having predominantly opted for the Association of Official Analytical Chemists (AOAC)-tested methods. Thus extra variability at the cleanup stage may have offset the anticipated improvement in quantitation. McKinney (1984) concluded that the inadequate validation of aflatoxin standard solutions was the principal factor responsible for that part of the coefficient of variation in excess of the component intrinsically due to the analytical technique. The American Society for Testing and Materials has evaluated the quantitative precision of HPLC analysis (Anonymous, 1981) and found that standard deviations of 3 to 8% may be expected, depending on the complexity of the mixture to be separated. The corresponding figures for TLC with densitometric and visual quantitation are approximately 10% and ~25%, respectively.

The most significant advantage of TLC is that it can be a very inexpensive technique, although in its more sophisticated forms it requires a considerable capital investment in items such as spotters and densitometers. In addition, if one-dimensional TLC gives adequate resolution, a considerable number of sampes may be analyzed on one plate. Should two-dimensional TLC be found necessary, several plates can be developed simultaneously. Thus results may be obtained rapidly by TLC, even when taking into account the time required for spotting the plates. The two principal disadvantages of TLC analysis are its lack of potential for automation and the subjective nature of the quantitation step. Use of a densitometer overcomes the latter objection to the employment of TLC but at a cost equivalent to that of a simple HPLC system. Autosamplers permit unattended running of HPLC equipment and allow the sample throughput of this sequential method of analysis to be as great as for TLC, while the recent development of short, very high-efficiency columns has demonstrated the capability of HPLC to provide extremely rapid results. Because of the general growth in the use of HPLC, many laboratories possess the necessary instrumentation and could therefore perform mycotoxin analysis in this way should it appear to offer definite advantages over the more conventional TLC methods. Comparing HPLC and TLC techniques, a similar high degree of competence is necessary with either one when establishing procedures and validating methods, but it is sometimes not appreciated that although TLC may be carried out using very simple equipment, it then demands greater operator skills and attention to detail in use than does HPLC (Coker, 1984).

Separation by HPLC may be preferred for other reasons. One factor to consider is safety; liquid chromatography offers greater protection, particularly for preparative work, because toxins are maintained in solution and contaminated silica dust does not arise. Equally, moisture- or oxygen-sensitive samples, such as xanthomegnin, are more readily chromatographed on a column. One potentially important advantage of HPLC lies in its suitability for on-line cleanup of crude extracts, and it is possible that this will ultimately be seen as one of the more compelling reasons for employing HPLC rather than TLC as the analytical technique. Whatever method is used for the final measurement, the purity of the residue obtained from the sample will have a major influence on both the detection limit achievable and also the degree of confidence which may be placed on the result.

Sample cleanup is normally by far the most time-consuming stage in the determination of mycotoxins, and an improvement here would be very welcome. Aspects of HPLC methodology are discussed in Section II, including a brief outline of instrumentation of varying degrees of sophistication and some of the procedures necessary for optimizing resolution and ensuring trouble-free operation. The analysis of the aflatoxins is then used in Section III as an example to demonstrate the results to be anticipated when changing system variables such as column type, temperature, and mobile-phase composition and flow rate. Section IV follows with a literature review highlighting current trends in HPLC analysis of the aflatoxins, ochratoxin A, zearalenone, the trichothecenes, and various miscellaneous toxins. The ergot alkaloids are not considered in detail, and in addition toxic peptides such as roseotoxin and toxic antibiotics such as griseofulvin are excluded, as are metabolites of the macroscopic fungi.

II. HIGH-PERFORMANCE LIQUID CHROMATOGRAPHIC INSTRUMENTATION AND METHODOLOGY*

HPLC systems vary widely in sophistication, and it is not possible to describe the whole range of instruments currently available. One area of HPLC recently developed is microbore chromatography, but because of the severe restrictions imposed on system dead volume, specialized instrumentation is required which will not be discussed further here in any detail. Only one application of microbore HPLC to the analysis of mycotoxins has been reported to date (Danieli *et al.*, 1980), but it is probable that this form of chromatography will be of considerable significance in the future because of a number of factors, including low solvent consumption and simpler interfacing with mass spectrometers. Although there is a reduction in acceptable injection volume, the diminished axial disper-

---

*The mention of individual manufacturer's products does not imply that they are endorsed or recommended by the U.K. Ministery of Agriculture, Fisheries and Food over other similar products not named.

sion of solutes associated with the use of microbore columns suggests that the detection limits attainable with such a system when employing concentration-dependent detectors—for example, ultraviolet (UV) or fluorescence spectro-photometers—should equal or exceed those found with the current generation of typically 4- or 5-mm diameter columns (Kucera, 1984; White and Laufer, 1984), while the associated reduction in the volume of sample extract required for cleanup should facilitate the automation of this time-consuming stage of analy-sis. A company developing robot-arm sample preparation working stations (Zymark) has recently introduced an HPLC injection interface to permit the unattended sequential cleanup and analysis of extracts. Unfortunately, this sys-tem is currently not capable of dealing with the initial extraction step for the large sample masses typically required for mycotoxin analysis.

Although a wide variety of instrumentation is available, including binary, ternary, and even quaternary solvent gradient systems, with the further option of automated optimization of solvent composition, to date most mycotoxin deter-minations have been performed using relatively straightforward HPLC methods with isocratic (constant solvent strength) elution and detection by UV or fluores-cence spectrophotometry. Requirements for this type of analysis are quite mod-est; a pump, injector, column and detector (with connecting tubing), and some means of recording the signal from the detector. Chromatographs may be pur-chased as an integrated system or as the individual components. Each method has its advantages and drawbacks. The purchase of a complete chromatograph means dealing with just one manufacturer, which should ensure cheaper servicing, while training will often be available. The instrument will, however, be less flexible, and any fault will lead to the whole system becoming unavailable. Constructing a chromatograph from the component units may be the more expen-sive option and require greater knowledge of HPLC, but allows technical im-provements in instrumentation to be accommodated more readily. In the follow-ing sections we consider the various components in greater detail.

A. Pump

Various types of pumps are available. Syringe pumps were used in the early days of HPLC, but the limited barrel capacity combined with difficulties experi-enced in flushing out one solvent with another caused them to drop from favor. They are well suited to microbore HPLC, however, as they are capable of delivering precise low volumetric flow rates. Another pump infrequently used for analytical HPLC is the pneumatic amplifier, which is powered by low-pressure gas over a large surface area to provide proportionally higher pressure over the smaller cross section of a column. Gas may contact the eluent either directly or via an intervening piston; the direct method produces eluent saturated with gas, which may cause both instrumental and chromatographic problems. In addition, eluent flow rate varies with column impedance. It is, however, possible to produce very high pressures by this method, and it is often used when packing columns. The twin-piston reciprocating pump is very widely employed because of its reliability, simple operation, and capability for providing a constant volumetric flow rate. The reproducibility and constancy of flow rate is important in HPLC to permit unambiguous assignment of chromatographic peaks when employing integrators or automated method development. It should be recognized that solvent compressibility can cause apparent variations in retention times (because the backpressure across a column will increase with use), which for accurate work require compensation. Some pumps include feedback devices to accomplish this, but these are often difficult to set up.

Inexpensive pumps can be quite satisfactory, but features such as an integral pressure gauge and automatic eluent shutoff in response to system overpressure (blockages) or underpressure (leaks) are desirable. A pulse damper should also be fitted, and as these are typically of high dead volume, it should be placed between pump and injector. Complex cam shapes, multiple heads, or microprocessor control also are often employed in the more expensive models in order to reduce eluent pulsation, and in consequence this form of baseline noise is not often found to be a problem with UV or fluorescence detectors operated at less than maximum sensitivity. Where multicomponent eluents are required for isocratic chromatography, the choice lies between manual premixing of the solvents in one reservoir or alternatively carrying out the operation instrumentally, either combining and mixing the output from two such pumps or using a single-piston pump with externally controlled proportioning valves, taking frequent samples of the individual solvents with subsequent mixing. Considering the relative hardware costs, the latter option should be the cheaper, but often there is little to choose between these two systems. The instrument-based systems for producing mixed eluents will often be preferred even when only isocratic mobile phases will be used, because they permit facile alteration of mobile-phase composition. However, isocratic eluents containing a small percentage of a strongly eluting solvent can cause unacceptable solute retention time variations when prepared instrumentally unless the mixing stage is very thorough.

If gradient elution is required, then only the instrumental methods are realistic, but in order to obtain accurate and reproducible gradients at extreme solvent compositions premixed solvents are required. Thus if it is necessary, for example, to generate a gradient between 0 and 10% of methanol in water, the reservoirs might contain water and 20% methanol in water. It is usually desirable to be able to flush the column with a more powerful eluent after the injection of a number of samples, which is why in the above example 20% methanol would be preferred instead of the more immediately obvious 10% mixture. Some separations require the use of buffer solutions, and with these eluents it is most important that the mobile phase is not left in the pump for extended periods. Eluent always creeps between the piston and the cylinder, and if this film is left to dry out and form crystals, the next time the pump is used the wall of the cylinder will be scored and performance lost. Further requirements of HPLC solvents are discussed in Section II,E.

B. Injector

Injection is the most critical instrumental feature governing overall efficiency. Either valve or syringe injection is possible, but the simplicity of the former method is such that it is widely preferred. Syringe injection has the advantage that it gives higher efficiency but at the cost of significantly increased operational complexity. Injection may be either through a septum or via a PTFE needle seal assembly. Septum materials are under considerable stress in use, and it is very difficult to find satisfactory elastomers for some solvent systems, particularly chlorinated hydrocarbons and tetrahydrofuran (THF). The superior efficiency of syringe injection is due to the introduction of the sample below the column top, at the center of the packing and with very little radial disperson. The abrupt change in eluent flow pattern at the column end fittings makes a major contribution to overall band spreading.

Further, it has been calculated that a solute injected in this manner as a near point onto a 250×5 mm column and eluted under typical HPLC conditions of flow and retention undergoes radial diffusion sufficiently slowly that it will never come into contact with the column walls, the place where maximum inhomogeneity of the packed bed is to be expected (Knox *et al.*, 1976). In addition, column deterioration in use is almost invariably due to contamination or channeling confined to the upper few millimeters of the packing, and this type of problem is more readily dealt with when using syringe injection systems. Column repair is considered in greater detail in Section II,C. With valve injection the sample plug is much more widely distributed at the top of the column. Other undesirable features of valve injection are the unavoidable pressure pulsing and extremely rapid acceleration of the sample plug as a valve is actuated together with the impact of the eluent stream on the structure of the packed bed at the column inlet.

Nevertheless, the reliability and ease of operation of injection valves has made them the first choice of most analysts. Attempts to improve the efficiency of valve injection have resulted in the development of curtain-flow methods, which involve splitting the incoming eluent stream, with the lesser proportion passing through the valve and carrying the sample onto the column via a fine needle inserted in the adsorbent bed. The remainder of the eluent is introduced at a point upstream of the tip of the needle. However, many analysts find that the use of highly efficient modern columns coupled with the ease with which selectivity may be modified by variations in HPLC conditions means that, except perhaps for rare critical pair separations, there is little need to employ the less convenient injection techniques in order to obtain adequate resolution.

Valve sample loops may be loaded in one of two ways, but in either case a good technique is essential in order to obtain consistent quantitative results. The most reliable and accurate method is to fill the loop by syringe, passing at least three loop volumes of sample through in order to fill it completely, a requirement dictated by the laminar nature of capillary flow. A coefficient of variation of better than 1% is attainable. The alternative method, of part-loop fill, is less reproducible by a factor of perhaps 2 or 3. A loop of greater volume than the maximum desired injection and filled with mobile phase, is slowly loaded by syringe with a predetermined volume of sample. It is important not to attempt to fill the loop by more than about 50 or 60% of its nominal volume; otherwise— again because of the parabolic flow profile within the loop—part of the sample may be lost. When dissolving residues for analysis it is important to use a solvent strength equal to or preferably less than that of the eluent. If a substantially weaker solvent is employed, residue components will adsorb to the column top and be concentrated there in a narrow zone. Subsequent elution with the mobile phase will give improved resolution. This technique can be used to inject very large volumes of sample, but its application may be limited by solubility constraints.

The standard six-port injection valve may also be used for other applications such as column switching for improved cleanup and stopped-flow scanning of eluting peaks as an aid in the confirmation of mycotoxin identity. For the latter method a valve is placed between column and detector. While the trapped peak is being scanned, column eluent is diverted either to waste or to an alternative detector. This introduces additional dispersion into the system; however, an alternative method of trapping a peak in the flow cell is to shut off the pump at a suitable time before elution of the peak such that solvent decompression within the system ceases just as the required solute enters the cell. This may require several practice runs to get the timing right. The recently developed diode-array UV detectors permit spectra to be taken in real time during elution. See Section II,D for additional comment on these devices.

C. Columns

As HPLC was developed, the use of pellicular packings declined (these are made by the deposition of a thin layer of absorbant onto a solid glass bead and are typically of 20 to 40 μm diameter), and completely porous microparticulate materials gained widespread acceptance. There has been a continuing reduction in particle size of the latter in pursuit of improved column efficiency. Silica has achieved overwhelming dominance as a packing material, although other materials including alumina and graphitized carbon may be used for adsorption chromatography. The latter has some advantageous properties but is not as yet readily available, although some use has been made of ground and sized gas chromatographic packings. Poly(styrene-divinylbenzene) resins and silica may be employed for size exclusion chromatography in organic solvents (Shepherd, 1984) and acrylamide-based materials or modified silicas for exclusion in aqueous media. Various other types of polystyrene gels are available for hydrophobic-interaction chromatography and other specialized applications, while silicas grafted with quarternary ammonium or sulfonate groups may be used for ion-exchange chromatography. Liquid–liquid partition systems have been little used, principally because of the problems experienced in maintaining reproducible conditions, although the method has some advantages and improved systems have been described (Crombeen *et al.*, 1983), while supercritical fluid chromatography (which may be thought of as a form of HPLC and can be performed with modified HPLC equipment) could become of importance in the future. Methods and principles have been discussed by Peaden and Lee (1982). Coupled-column (two-dimensional) chromatography can afford much improved resolution and hence function as part of the cleanup (Karger *et al.*, 1983). Typically, a fraction initially isolated using an aqueous size exclusion column is transferred on-line to a reversed-phase gradient system. Despite solvent compatibility problems, the similar combination with nonaqueous size exclusion is feasible (Shepherd, 1984).

Both spherical and irregular-shaped particles are available, but the former appear to be gaining favor. In principle, spherical materials should withstand pressure pulses better, generating fewer fines and thereby improving long-term performance stability of the column, but in practice other factors are more important in determining column lifetime. Particle size distribution and mean diameter are crucial to obtaining high efficiency. Ideally, packings should be monodisperse and of the smallest diameter possible. Column permeability and hence pump pressure capabilities place some constraints on the continued reduction in particle size; currently 3-μm packings are the smallest available, normally as 10-cm columns. While analysis time varies directly with column length, resolution increases only with the square root of length. Consequently these short, ultrahigh-efficiency columns permit extremely rapid separations, although equipment design may be a limiting factor as it is with microbore HPLC (Erni, 1983). Despite the continued popularity of 10-μm packings, there is no chromatographically justifiable reason for using these in preference to their 5-μm equivalents. A separation developed on the former may be completed with identical resolution and more rapidly on a shorter 5-μm column. Unfortunately, some packings are not available with a particle size of less than 10 μm.

Silicas can have very variable surface properties (Unger, 1979), and the selectivity attainable with a given packing depends on many factors and cannot be predicted. Reversed-phase (RP) packings are based on silica, with an organic surface layer grafted on by reaction with surface silanol groups. Toxins or other solutes are eluted by solvents relatively more polar than the packing, in distinction from the customary normal-phase (NP) chromatography on silica. Attention switched from NP to RP chromatography because of a number of factors, including primarily the very wide range of solutes which may readily be separated on bonded phases. Additionally, RP solvents are easier to optimize and in general less toxic and less expensive than those required for NP chromatography.

A wide range of selectivity is possible depending on the nature of the organic layer; much use has been made of octadecyldimethylsilyl (ODS) columns, where the hydrophobic surface can be modified dynamically in many ways by the inclusion in the mobile phase of suitable organic compounds. Thus ionic species may be resolved by ion-pair formation with mobile-phase components such as quaternary amines or alkylsulfonic acids. Alternatively, the pH of the eluent can be adjusted so that dissociation of acidic or basic toxins is suppressed. Chromatography at a pH where such compounds are only partly ionized will result in poor peak shape and possibly also multiple peaks. Most analyses for ochratoxin A are performed under acidic conditions for this reason. Similar manipulations of secondary equilibria can be extended, for example, to the use of metal complex formation with salts included in the mobile phase. One limitation on eluent pH control is the exponential increase in the solubility of silica as the mobile phase is made more alkaline. Very short column life is to be expected when buffer pH exceeds 7.5, although a precolumn may be inserted between pump and injector to saturate the eluent with silica. Ion-pair separation could be employed to avoid this problem when separating bases.

Two types of bonded phases are recognized, monomeric and polymeric. Monochloro silanes are used to prepare the former, dichloro or trichloro for the latter, and the chromatographic properties of the two kinds are slightly different. Monomeric phases are more widely used. Retention can be adjusted by selection of shorter (C_1 to C_8) or longer (C_{22}) chain alkyldimethylsilyl-modified silica. Other, more polar materials such as cyano- or phenyl-bonded phases are less used but permit some otherwise difficult separations. Cyano phases in particular have interesting properties and may be employed under either NP or RP conditions. They are less retentive than silica with NP eluents. One of the more important characteristics of a modified silica is the percentage loading of carbon; the higher the loading, the greater the retention. In addition the proportion of underivatized silanol groups has a significant effect on performance. Steric considerations indicate that there will always be residual silanols, particularly with the bulkier modifying agents, and although their concentration may be reduced significantly by "capping" the silica with a trimethylsilylating reagent after completion of the initial surface modification, the potential remains for some form of polar interaction with solutes. Indeed it is not always desirable that all silanols be removed, as in some cases they may contribute toward achieving a difficult separation. An article by Genieser et al. (1983) discusses the nature of the surface modification for many commercially available packing materials.

Another important parameter which may be usefully varied in order to modify selectivity is column temperature (Gant et al., 1979), because adsorption is principally controlled by enthalpy changes. Thus a change of column temperature may effect the resolution of two previously cochromatographing solutes and separate an interference from a toxin peak. The relative retention of very similar compounds (e.g., aflatoxins) is unlikely to be significantly altered in this fashion. If columns are to be used at temperatures above about 50°C it is desirable that both column and valve are thermostated. If only the column is controlled, cold eluent entering the column will cause turbulence and a degradation in performance. See Fig. 6 for example chromatograms. As the temperature is raised, retention of solutes decreases and faster separations are possible, although eluent composition may require adjustment. Mobile-phase viscosity is much reduced at elevated temperatures, and thus for example at 90°C the pressure drop across a column may be only one-third of that at 35°C. It is, however, important not to change column temperature too abruptly, as this may damage the packing. It is also found that columns deteriorate faster at elevated temperatures, partly because of the greater solubility of silica under these conditions but also because steel and silica have dissimilar coefficients of thermal expansion. As mentioned in Section II,B, the column wall area is where most irregularity of the adsorbant bed may be expected. One approach to improving column performance has been to replace the standard stainless-steel column tube with a flexible one made from polyethylene (Waters' Radial-Pak system). The column then becomes a cartridge which is placed inside a radial compression unit which pushes the plastic wall against the packing. The 8-mm-diameter cartridge is also well suited to semipreparative scale separations to provide material for characterization. This system does appear to give improved resolution and seems to be gaining favor. The principal disadvantage is the cost of the compression unit.

Microbore chromatography has received much attention over the past few years, although it is perhaps more talked about than practiced. The basis for the interest in a reduction in column diameter is due primarily to the resulting increase in sensitivity when using concentration-dependent detectors. The minimum detectable mass (m) has been shown (Scott, 1984) to depend on the square of the column radius r:

$$m = 2\pi r^2 \, L\psi \, (1 + k') \, C \, N_t^{-1/2} \tag{1}$$

where L is column length, ψ is the fraction of the column occupied by eluent, and C is the minimum detectable concentration. Other considerations include the reduction in solvent consumption, again proportional to r^2. The experimental difficulties in applying microbore techniques result from the relative increase in the instrumental contribution to overall dispersion. Where peak profiles are Gaussian, variances are additive. Thus the final peak width is a consequence of the sum of the variances of the individual components of the system; injector, column, detector, and connecting tubing. In practice it is desirable that the instrumental variance should be no more than 10% of that arising from the column. A 250 × 4.6 mm, 10-μm column might give a variance of ~3400 μl², which permits an acceptable instrumental variance of 340 μl². A 250 × 1 mm column, with a variance of 8 μl², implies that the remainder of the system must give rise to no more than 0.8 μl². The standard 8-μl spectrophotometer flow cell will have a variance of at least 2 μl² (Scott, 1984). Consequently, specially designed equipment is required for microbore chromatography.

Column efficiency is customarily expressed in terms of theoretical plates, a usage borrowed from distillation science. The number of theoretical plates generated by a column (N_t), may be calculated from a knowledge of the peak profile and the distance from the chromatographic origin to the centroid of the peak. Where peaks have the ideal Gaussian shape, then the simple expression

$$N_t = 5.54 \ (L/w_{1/2})^2 \tag{2}$$

may be obtained. L is the distance of the peak maximum from the origin and $w_{1/2}$ the peak width at half-height. In practice, peak shapes are usually skewed, and the actual plate count of a column will be less than that found by applying Eq. (2). No simple means of calculating this true (second moment) plate count exists, and thus it is common practice to use Eq. (2) or its equivalents. In these circumstances it is important to note peak asymmetry at 10% of peak height, that is the ratio of the distances between the perpendicular dropped from the peak maximum and the tailing and leading edges of the recorder trace, at one-tenth of the peak height from the baseline. In practice (except under column overload conditions), the result will be greater than unity. Values exceeding approximately 2 indicate that the chromatography could be improved. Eluent flow rate is one of the more important parameters controlling efficiency, and its effect on separation has been well documented (Katz et al., 1983). For 5-mm-diameter columns, optimum flow rates are often in the region of 0.5 ml/min. See Fig. 7 for example chromatograms.

Measured plate counts will also take into account the system dispersion, as well as that due to the column. Thus early eluting peaks are broadened proportionately more than those which elute later. When comparing plate counts for columns it is therefore important that either true plateages are quoted or that peak asymmetries for solutes of interest are also taken into account. Finally it should be remembered that column manufacturers' test solutes are chosen to show their products in a favorable light and that the only true measure of performance is a chromatogram of the toxin of interest, preferably in a typical sample matrix.

If an asymmetry factor of greater than 3 is regularly obtained and this cannot be reduced by alteration of chromatographic conditions, then disruption of the packing at the column inlet might be suspected. Split peaks, where the individual components remain sharp, may be due to channeling of the packing at the inlet or possibly to unsuitable selection of eluent pH or other secondary equilibrium. Buildup of insoluble sample debris on an in-line prefilter or on the column inlet frit may be responsible for loss of efficiency, as demonstrated in Fig. 1. Back-flushing will often be found to restore performance, but if necessary frits should be replaced. Column repairs may be carried out by removing affected packing at the inlet and replacing it either with identical adsorbent or perhaps ballotini (20–40 μm glass beads). Packing will in any case eventually compact due to pressure pulses, general wear, and dissolution of the packing. Conversion to syringe injection can improve poor columns. If all else fails, these may be returned to the manufacturer for repacking.

Should columns be prepared in the laboratory or purchased prepacked? No straightforward answer can be given. In-house packing requires some practice and remains something of an art, although apparently successful techniques have

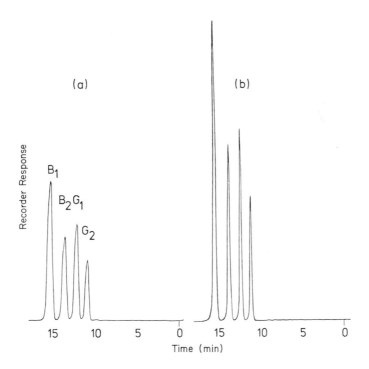

Fig. 1. Improvement in separation efficiency following backflushing of column and in-line filter. (a) Before backflush, (b) after backflush. Conditions: Varian Vista 5500 ternary liquid chromatograph. Column, 250 × 4.9 mm 5-μm Spherisorb ODS, eluted with 60:30:10 water–acetonitrile–methanol at 0.75 ml/min. Temperature 35°C. Fluorescence detection, Perkin–Elmer Model 3000 spectrofluorimeter, excitation 360 nm, emission 440 nm. Postcolumn derivatization (PCD) with iodine-saturated water. Reaction coil, 7000 × 0.3 mm; flow rate, 0.5 ml/min, temperature, 40°C. Loop volume, 10 μl. Sample: aflatoxins B_1, B_2, G_1, and G_2, retention times: 15.2, 13.5, 12.2, and 10.9 min, respectively.

been described in the literature. Generally such columns tend to be rather less efficient than similar versions obtained from a specialist manufacturer. Economy (after the initial extra expense of a packing pump and other necessary equipment) and the increased versatility may be compensating factors. Unless column consumption becomes unacceptable, which in turn depends on sample throughout and quality of cleanup, it is simpler to purchase. Preparative HPLC may be performed using large-scale versions of analytical equipment on columns $\geqslant 25$ mm in diameter. Semipreparative fractionations can be achieved by elution chromatography on standard analytical columns to provide sufficient purified material for characterization. The maximum acceptable column loading is of the order of 2 to 3 mg. Using the technique of displacement chromatography it is possible to separate perhaps hundreds of milligrams of solutes on an analytical column (Horvath and Melander, 1983).

When reporting retention data, it is preferable to quote the conditions of chromatography and give either retention times or, less usually, volumes. The alternative measure of retention, the solute capacity ratio (k'), is calculated by reference to an unretained peak:

$$k' = (t_r - t_0)/t_0 \tag{3}$$

where t_0 and t_r are the elution times (strictly, volumes) of the unretained peak and the solute under examination. The question of what exactly constitutes an unresolved peak unfortunately remains unclear. Various systems have been proposed, the most satisfactory of which refer to the elution time of an isotopically modified solvent molecule, for example, deuterium oxide. Mixed-solvent systems cause further problems, which are best avoided (as suggested above) by reporting retention times and full experimental details including column type and dimensions, packing particle size, eluent composition, and flow rate, and column temperature.

D. Detector

To date mycotoxin analysts have employed almost exclusively either UV or fluorescence detection. Fortunately, most known toxins (with the notable exception of the trichothecenes) possess moderately strong UV chromophores in suitable regions of the spectrum, while the fluorescence properties of certain compounds, particularly the aflatoxins, permit their determination in low picogram quantities. Under optimum conditions, fluorescence is about 30 to 40 times more sensitive than UV for aflatoxins. The influence of chromatographic conditions on fluorescence intensity is discussed in Section III with particular reference to the aflatoxins. Overall sensitivity is in part a function of detector design and also of cleanup efficiency, which restricts the mass equivalent of sample which can be loaded onto the column. Laser fluorescence has been used to improve detection limits for zearalenone (Diebold *et al.*, 1979a) and aflatoxins (Diebold *et al.*, 1979b), but the unavailability of a variable-wavelength UV laser has necessitated the use of exitation wavelengths significantly removed from the absorption maxima, particularly for zearalenone, with consequent reductions in attainable sensitivity. Badly designed and large-volume detector flow cells can make a substantial contribution to total extra-column dead volume and hence cause excessive loss of resolution, a problem which is particularly acute in microbore HPLC and also with the short, high-efficiency columns now widely used to achieve rapid analyses. Many currently available low-volume UV cells have been produced by the expedient of reducing cell path length with the expected consequences for sensitivity, while some fluorescence detectors have cells with volumes of up to 25 μl, thus achieving greater sensitivity at the cost of decreased resolution. A further requirement for a high-resolution detector is a rapid response or in electronic terms, a small time constant, coupled with an acceptably noise-free signal. Diode-array UV detectors are in a rapid phase of development and are relatively expensive, although their considerable potential for cost savings by eliminating the need for routine toxin confirmation tests may make them cost effective in a wide variety of applications. Several analytical methods employ wavelength ratioing techniques (see, e.g., Howell and Taylor, 1981) for confirmation of toxin identity, and the ready availability of a full spectrum provides an even more convincing extension of that principle. Other methods of obtaining spectra will normally be able to provide information on only one peak within a chromatogram. Fixed-wavelength UV detectors, though often technically superior to variable-wavelength instruments, are limited in their application, and unless required for dedication to specific methods are perhaps not suitable as a sole detector.

Nonspectroscopic detectors are typically less sensitive and less easy to use. The major exception to this generalization is the electrochemical (EC) detector, which has been utilized in the analysis of roquefortine in blue cheese (Ware *et al.*, 1980) and *cis*- and *trans*-zearalenones in cereals (Smyth and Frischkorn, 1980). EC detection is intrinsically both selective and sensitive. The oxidation mode is simpler and less restrictive than is reduction, although both are limited by eluent electrolysis. Another constraint is the requirement that the mobile phase will support current flow. An ionic strength of approximately 0.1 is usually needed in aqueous systems. Nonpolar eluents have limited application but may prove acceptable with the inclusion of compounds such as phase-transfer catalysts. Refractive index (RI) detection has been applied to the analysis of T-2 and HT-2 trichothecenes (Schmidt and Dose, 1984), with a limit of detection of approximately 1 μg. RI provides the closest HPLC approach to a universal detector. Several types are available; the interferometry principle provides the most sensitive instruments, but leaks of solvents or their vapors can cause serious damage to the plastic-based polarizers used. Furthermore, extremely stable operating conditions are required in order to attain this sensitivity, including thermostating to within 0.001°C. Pressure fluctuations, for example, caused by eluent droplets repetitively falling from the flow cell outlet can result in a very noisy baseline. All refractometers are relatively insensitive, with detection limits typically of around 0.1 to 1 μg, and in addition gradient elution is not practicable. Thus RI detection is likely to remain unused except perhaps for toxins which possess no accessible chromophore. In many cases, precolumn derivatization and the use of a spectrophotometer will be found preferable. Postcolumn reaction methods (Frei, 1982) offer many advantages.

Another mass-detection system is based on the familiar flame ionization detection (FID) principle. An early model used a thin wire to transport eluent into an evaporation zone where solvent is removed by a current of nitrogen, followed by passage of the wire through a furnace with eventual conversion of any carbon in the solute to methane, itself then detected by the FID. Problem areas included the low takeup of eluent by the wire (perhaps 1%) and the propensity of the wire to breakage. Recently a more robust and rather different instrument has appeared, where the whole of the column output is loaded into a continuous fibrous quartz belt placed around the circumference of a metal disk. After evaporation of eluent the belt passes between twin FIDs and then through a cleaning flame at high temperature. Again, sensitivity is limited with the smallest detectable mass of the trichothecene T-2 toxin being approximately 0.1 μg. The maximum flow rate of aqueous eluents in particular is restricted to under 1 ml/min. Volatile solutes are lost from the belt during the solvent evaporation step; 80–90% of dimethyl phthalate (boiling point 284°C) was removed (M. J. Shepherd, unpublished observations). Belts are expensive, although their lifetime cannot as yet be estimated. Nevertheless, this detector provides a unique perspective on samples and will be invaluable in many circumstances.

E. Tubing, Solvents, and Other Items

Several other items of equipment merit some comment. The tubing required to connect the system together should be of minimal inner diameter (i.d.); 0.25 mm (0.010 in.) is most often used, but tube is readily available down to 0.15 mm i.d. and the smaller size is to be preferred because of its reduced contribution to axial dispersion. Liquid flow through capillaries has received considerable attention, and the dispersion found is approximately proportional to the length of the tube and, more importantly, to the *fourth* power of the i.d. Experimental values of dispersion for narrow-bore tubes may be predicted from the Taylor–Golay equation as expressed in the following form (Bristow, 1976):

$$V = d^4F \, L/122D \tag{4}$$

where V is the variance in microliters squared due to the tube; d, its diameter in millimeters; F, eluent flow rate in microliters per second; L, its length in millimeters, and D, solute diffusivity in the eluent. Eq. (4) assumes a Gaussian peak profile and is subject to other potential errors, but it gives a good indication of the relative values of the variance to be expected from different tubes. It does not apply to wide-bore pipe or to packed beds. Thus connecting tubing should be kept as short and narrow-bore as possible. With standard 5-mm-diameter columns, a 40-cm length of 0.25-mm i.d. tube between column and detector will not appreciably affect resolution, although a similar length of 0.8 mm may reduce measured column efficiency by perhaps 5–10%. The actual i.d. of a capillary can differ quite widely from specification, particularly with PTFE tube. It is always advisable to check the bore before use. Magnifying lenses with an engraved scale are suitable for this purpose if care is taken.

Two problems are associated with the use of small-diameter capillaries: cutting pieces to length and blockages in operation. The latter should cause no difficulty provided reasonable care is taken when filtering solvents and (in particular) solutions of final residues, before injection. Volumes of ≥ 100 µl may be clarified by filtration from a syringe through a 0.5-µm membrane in a Swinney holder but, until very recently, lesser amounts were probably best treated by centrifugation through a similar filter to reduce liquid losses. Low-holdup filter units (retaining no more than 10–15 µl with an air purge) have just become available. Cutting stainless-steel capillary tubing of <0.25 mm i.d. is most simply done by scoring the wall with a rotary scribe and then flexing the tube until fracture occurs through metal fatigue. The tube must be gripped firmly by pliers immediately adjacent to the score mark in order to avoid "hooking" its end. Abrasive wheels should be avoided, as they are likely to plug tube of this size.

Kinked steel tubing can cause problems when connections are repeatedly to be made and broken, and PTFE tube is by virtue of its flexibility much easier to use. However, it is normally confined to low-pressure applications, partly because of its cold-flow properties and its limited burst strength (although 0.15 mm i.d. tube will withstand at least 1000 psi), and also because the negligible coefficient of friction of this material causes problems with tubing being pushed out of fittings. PTFE tubing and ferrules are commonly employed in combination, and it is important not to overtighten nuts onto these as it is possible to compact the ferrule and consequently "neck" the tube, causing a constriction and thus an excessive increase in eluent backpressure. However, the major problem with tubing of this material is that if the fitting is not tightened adequately, the tube may be pushed completely out. Impacted PTFE ferrules may readily be removed from end fittings and unions with the aid of a #1 screw extractor. One other feature of PTFE tube which should be noted is its potential for acting as a "light pipe" and raising background light levels in spectrophotometric detector flow cells, thereby increasing signal noise and reducing sensitivity. It is also important to ensure that PTFE tube is cut squarely with a sharp instrument such as a scalpel blade thus maintaining its circular cross section and also avoiding introducing extra dead volume when butting tubing into connectors. Similarly it is essential that tubing is inserted as far as possible into column end pieces and other unions. This should be checked immediately if any loss of resolution is noticed when using PTFE capillaries. The tubing is more easily gripped using small pieces of abrasive paper.

Fittings represent another potential problem. These are available from many different manufacturers and comparability of components is not always guaranteed, although experience shows that most types of ferrule, for example, are interchangeable. It is obviously preferable to standardize within a laboratory on one type of fitting but this is not always possible, particularly when a chromatograph is to be assembled using components from a number of suppliers. Connections between different types can be made using couplings and extra tubing, which is unsatisfactory, but recently specialized adapters have been introduced, allowing unions to be made with a minimum of effect on overall system efficiency. When making simple connections, it is advisable to drill the union out to permit the two tubes to be butted together before swaging. This minimizes dead volume. If this is not done, it is important to use chromatography grade unions with an i.d. of ≤ 0.25 mm.

Finally, the quality of HPLC solvents is critical. Solvents should be free of interfering impurities, both particulate and dissolved. It is desirable that solvents as well as samples should be filtered through a 0.5-μm membrane to avoid damage to pumps and injectors. Dissolved impurities also can cause problems, as for example might be encountered with reversed-phase solvent systems containing methanol. The apparent splitting of all four aflatoxin peaks in a water–acetonitrile–methanol mobile phase, an effect possibly similar to that described by Gregory and Manley (1982), is shown in Fig. 2. A reservoir of methanol which had been in use for about three weeks gave rise to the chromatogram depicted in Fig. 2. The size of the late-eluting shoulder on each toxin peak had gradually increased in successive chromatograms during this period. Experience within this laboratory has shown that methanol, though required in the eluent in order to enhance resolution, is liable in use to form some decomposition product which reacts with the toxins to give this doublet peak shape. Reaction must be nearly instantaneous, as sample comes into contact with eluent, for otherwise a less well-defined peak would be observed. The standard solution then in use was not itself degraded, as shown by the immediate restoration of normal chromatography on replacing only the methanol component of the mobile phase. Only a few samples of methanol give rise to this problem, and previously satisfactory solvent can deteriorate in use.

It had been noted both here and elsewhere (B. A. Roberts, 1984 personal communication; Beebe, 1978) that methanol is not an acceptable solvent for the

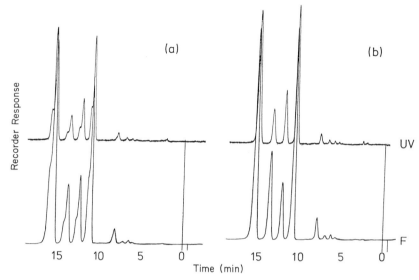

Fig. 2. Effect of mobile-phase methanol impurity on chromatography of aflatoxins B_1, B_2, G_1, and G_2. (a) Before replacing solvent in methanol reservoir; (b) after replacing solvent in reservoir. Conditions and toxin elution order: as for Fig. 1, except UV detection (before PCD) at 360 nm, 0.01AU fsd. Fluorescence: $\times 1$ (lowest sensitivity).

preparation of aflatoxin hemiacetal solutions. Whereas hemiacetal standards stored in acetonitrile–water are stable for extended periods, those prepared in methanol–water form substantial amounts of degradation products within a few hours at room temperature. Other possible problems resulting from dissolved impurities include high UV background absorptions or the quenching of analyte fluorescence. Chlorinated solvents may accumulate hydrogen chloride, which can attack stainless steel; buffers should not contain chloride counterion. One often-overlooked feature of some solvents is their stabilizer content, Chloroform will normally contain a variable amount (0.5–2%) of ethanol or methanol, which must be removed before use as a normal-phase eluent to avoid irreproducible retention times. THF may be stabilized with hydroquinone. The water content of normal-phase solvents should be rigorously controlled.

Other aspects of solvent purity include the removal of dissolved gases, particularly oxygen, both to miminize oxidation of toxins in solution and also to enhance detector sensitivity. This is most important where EC detection is to be employed, while in this case oxygen must in addition be removed from samples before injection. Degassing of solvents is critical where mixed eluents containing water are produced at the pump from reservoirs of the individual components. It is difficult to eliminate gases from water completely, and on mixing with organic modifiers bubbles are often produced. Since gases are more soluble under pressure this will cause no problems with the column, but when the pressure is released in the detector, bubbles may be formed, giving rise to a characteristic "sawtooth" baseline with a slow decrease in apparent absorption as the bubble builds up, followed by an abrupt rise as it is released.

In addition, some pumps have their performance impaired in gassy solvents due to the development of bubbles in the heads. Degassing may be carried out in several ways; unfortunately the best and simplest is also the most expensive, requiring a constant purge of oxygen-free helium through the solvents. The solubility of helium is relatively insensitive to pressure changes and bubble formation is eliminated. Other methods include refluxing and also filtering under vacuum, with or without ultrasonic agitation. The latter method is not very efficient. Whenever degassing mixed eluents, the potential preferential loss of more volatile solvent components should be considered. Bubbling can also be controlled by applying a slight backpressure at the detector outlet, and commercial devices capable of giving a controlled pressure of 25 to 100 psi. Most detector cells will withstand two or three times this level, but the manufacturer's manual should always be consulted before attempting any technique which involves restricting solvent flow at the outlet.

It is important that the dissolved organic matter content of water should be minimized, especially where it is to be used as a component of a reversed-phase gradient system. Between runs, the solvent passing through a column will under most conditions be noneluting, with a high water content. Any organics present will therefore accumulate at the top of the column, to be eluted as sharp peaks during the next gradient. Thus very good quality water is required, preferably obtained from one of the specialist purification systems available. Where consumption is too low to justify this, HPLC-grade water may be purchased in glass bottles, but as stored water is extremely susceptible to contamination, this may in the long term prove undesirable. In situ photooxidation of water in the actual HPLC reservoir could be found advantageous (Frischkorn and Schlimper, 1982). It should be remembered that trace analysis will often demand quantities of high-purity water. One alternative method for removing organics, where the pumping system permits, is to place a short column containing a 20- to 40-μm ODS packing in the solvent delivery line from the water pump. This remedy is not of course applicable where eluents are premixed. The addition of sodium azide at 0.02% will inhibit bacterial growth which seems to occur in even the best-quality water and leads to blocked filters and other problems. Some other mobile-phase components (particularly phosphate buffers) are very susceptible to bacterial contamination.

III. MANIPULATION OF HIGH-PERFORMANCE LIQUID
CHROMATOGRAPHIC CONDITIONS

It is intended in this section to use aflatoxin analysis to exemplify the effects on selectivity, retention, and sensitivity of modifying the basic HPLC system variables. For this or any other analysis based on chromatographic separation of any kind, the successful application of the technique depends on the joint optimization of these parameters and additionally on the reduction of solute dispersion to a minimum, in the case of HPLC by careful selection and connection of instrumentation and by the use of well-packed columns containing particles of ≤ 5 μm diameter, as outlined above.

It is also true that the concept of optimization should be extended to the analysis as a whole, when a balance must be achieved between complexity and length of sample cleanup and the efficiency and speed of the subsequent HPLC separation. In many cases it will be desirable to utilize the maximum resolution attainable with the chromatographic system as a means of increasing overall sample throughput. The use of selective detectors is therefore to be preferred and the standard 25- or 30-cm column will usually be employed.

The first chromatographic choice to be made is the selection of column packing material. For solutes of intermediate or high polarity such as most mycotoxins, a reversed-phase (typically, ODS) column will usually give good results. Silica columns can often be employed for the same separation, but solvent selection and preparation is normally simpler with ODS packings. For example, one commonly employed normal-phase method uses water-saturated or half-saturated chloroform as a component of the eluent and in the absence of a temperature-controlled environment toxin retention times will be quite variable. In addition, where gradient elution is essential or columns require regular flushing to remove strongly sorbed sample components, reversed-phase packings are to be preferred to silica because the former will require only a brief reequilibration with its initial solvent system, while the latter can take extended periods to stabilize.

An extra factor which must be considered in aflatoxin analysis is the effect of the chosen solvent on their fluorescence characteristics. Chlorinated hydrocarbons in NP eluents often promote intersystem crossing and hence quenching of fluorescence. In the case of aflatoxin separations, sensitivity is lost selectively for B_1 and G_1. Quenching also occurs in the aqueous mobile phases employed for RP chromatography. This problem and solutions to it are discussed in Section IV,A,1 and by Scott (1981). Some variation in emission intensity may also be expected between identical concentrations of an aflatoxin in different apparently nonquenching solvents. In addition, excitation and emission wavelengths are solvent dependent. The former may readily be optimized by obtaining a UV spectrum of the toxin in the eluent of choice. Fortunately aflatoxins exhibit broad adsorption maxima and thus excitation wavelength selection is not critical, even when several toxins are to be determined. A setting of 360 nm will often be found a satisfactory compromise. However, where sensitivity is of concern, detection wavelengths should be chosen with care (Scott, 1981).

Emission wavelength is dependent on mobile-phase polarity and for B_1 may vary between 424 and 431 nm (and for G_1, 428 and 445 nm) in chloroform solutions containing increasing amounts of methanol (Chang-Yen et al., 1984). Emission of B_1 in aqueous eluents may maximize at wavelengths of up to 450 nm (Scott, 1981), but in this laboratory it has been found that using the mobile phase of Fig. 1, the optimum excitation and emission wavelengths were respectively for B toxins (B_1, B_2, B_{2a}, and iodo-B_1), 363 and 433 nm; for the similar G toxins, 370 and 452 nm; and for M_1 and M_{2a}, 355 and 433 nm. For M_1 and M_{2a} only 1–2% sensitivity was lost at an excitation wavelength of 363 nm. Other values for these maxima have been reported (Gregory and Manley, 1982), and it

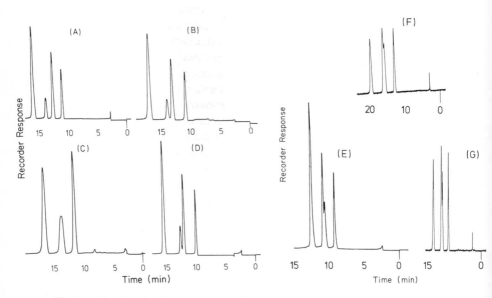

Fig. 3. Selectivity for aflatoxins B_1, B_2, G_1, and G_2 of seven reversed-phase packings each eluted with a standard mobile phase. Conditions and toxin elution order: as Fig. 1, without PCD. UV detection, 360 nm, 0.1AU fsd (columns A–E); 0.01A fsd (columns F,G). All columns, 250 × 4.9 mm: (A) Spherisorb ODS, 5 μm. (B) Zorbax ODS, 7 μm. (C) μBondapak C_{18}, 10 μm. (D) Spherisorb ODS2, 5 μm. (E) Spherisorb C_6, 5 μm. (F) Partisil ODS3, 5 μm. (G) Hypersil ODS, 5 μm. Elution order of G_1 and B_2 inverted with column E. Standards containing differing relative concentrations of the four toxins were employed when generating the various chromatograms.

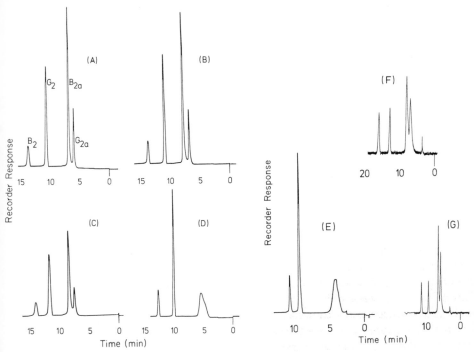

Fig. 4. Selectivity for aflatoxins B_{2a}, B_2, G_{2a}, and G_2 of seven reversed-phase packings each eluted with a standard mobile phase. Conditions: as for Fig. 3. Retention order: $B_2 > G_2 > B_{2a} > G_{2a}$.

is therefore desirable that both excitation and emission are determined for each detector in the eluent to be employed.

The following discussion of the effect on retention and selectivity of varying the chromatographic conditions will be confined to RP systems. However, the principles are exactly the same for either mode of HPLC. Whether a silica or ODS column is selected, many seemingly equivalent packings are available. Often the choice will be restricted to those currently used within the laboratory, but it is interesting to compare the selectivity of seven RP materials (including several which are frequently used for aflatoxin analysis) for B_1, B_2, G_1, and G_2 when eluted with a standard solvent mixture. These chromatograms are shown in Fig. 3. With RP analysis, the typical retention observed is $G_{2a} < B_{2a} < G_2 < G_1 < B_2 < B_1$. Inversion of retention for G_1 and B_2 was observed for the Spherisorb C_6 packing. When NP chromatography is employed, the six toxins listed above can be expected to elute in the reverse order. The separation of B_{2a} and G_{2a} on the same columns (Fig. 4) is of some interest. Nonequilibrium effects occurred during the elution of the hemiacetals from several of these columns, resulting in peak fronting and loss of resolution. Similar variations of selectivity to those depicted in Figs. 3 and 4 are to be expected with a selection of NP packings.

Two comments must be made about these chromatograms. Firstly, the Zorbax ODS and μBondapak C_{18} sorbents have particle sizes of 7 and 10 μm, respectively, while the others are 5-μm packings. Second, Figs. 3 and 4 cannot be taken in isolation as a complete indication of the suitability for aflatoxin analysis for materials other than Spherisorb ODS, for which the solvent system employed was optimized. It should also be noted that these chromatograms were obtained using relatively new columns, and it is to be expected that the selectivities will alter as the columns are used, particularly when only moderately clean extracts are analyzed. In such circumstances it may be found desirable to employ a precolumn, which can be repacked as necessary, to prevent the irreversible adsorption of sample components to the top of the analytical column. However, a gradual loss of efficiency of the column during use is inevitable.

Examination of Fig. 5 demonstrates the effect on selectivity of altering the nature of the organic component of the mobile phase. Two changes can be seen to occur as the modifier becomes progressively less polar. The proportion of modifier required to give approximately equal retention for G_2 (the earliest eluting toxin) decreases and, in addition, the retentions of B_1 and G_1 become greater relative to their dihydro analogs. As a consequence of the latter effect, the elution order of G_1 and B_2 inverts with mobile phases containing the least polar modifiers, as may be seen in Fig. 5. Further points of interest include the tailing peaks observed with methanol and the poor resolution of the G_1–B_2 pair with acetonitrile and of G_2–G_1 and B_2–B_1 with methanol. Ternary mixtures of water,

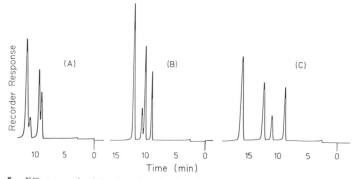

Fig. 5. Effect on selectivity for aflatoxins B_1, B_2, G_1, and G_2 of varying the nature of the organic modifier in the mobile phase. Conditions and toxin elution order: as Fig. 1, without PCD. UV detection at 360 nm, 0.1AU fsd. All flow rates, 1.5 ml/min. Mobile phases: (A) methanol–water, 60:40; (B) acetonitrile–water, 40:60; (C) THF–water, 20:80. Elution order of G_1 and B_2 is inverted with the THF-containing mobile phase.

acetonitrile, and methanol yield results qualitatively predictable from the binary eluent chromatograms. Compare, for example, Fig. 1b with Fig. 5a and b. Thus for the Spherisorb ODS and to a lesser extent, the Zorbax ODS and Spherisorb ODS2 columns, a 60:30:10 (v/v/v) mobile phase allows a near-optimum balance between the resolution of all three toxin pairs.

Although water–THF (80:20 v/v) gives perhaps rather better separation than this mixture, THF is an unsatisfactory solvent for routine use because of its poor stability, which results in the formation of peroxides and polymers on storage. Water–acetonitrile–methanol has become the preferred eluent of many aflatoxin analysts using RP HPCL (see Section IV,A,l). Gradient elution increases resolution further, but at the cost of additional complexity in operation. Solvent selectivity has been studied in detail for both RP and NP chromatography by Snyder and colleagues (see, e.g., Glajch et al., 1982; Glajch and Kirkland, 1983). Recently introduced chromatographs incorporating data-processing facilities may be programmed using decision-making algorithms capable of the automated optimization of mobile-phase composition for the resolution of a defined solute mixture. These techniques and some of their limitations have been discussed by Rafel (1983) and by Berridge (1984).

Eluent temperature also may affect selectivity. Because of the very similar structures of the B and G aflatoxins it is unlikely that the relative retentions of

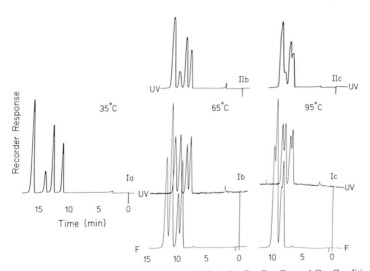

Fig. 6. The effect of temperature on retention of aflatoxins B_1, B_2, G_1, and G_2. Conditions and toxin elution order: as for Fig. 1, except either UV detection at 360 nm, 0.1AU fsd or sequential UV (360 nm, 0.01AU fsd) and PCD and fluorescence detection (setting ×1, least sensitive). Series I: Column thermostated but not valve. Series II: Both column and value thermostated; eluent preheated before entering injection value. a, 35°C; b, 65°C; and c, 95°C.

these four toxins will change significantly, but it is possible that coextracted compounds which interfere at one temperature may be resolved at another. Figure 6 illustrates the separation of a standard mixture at 35°, 65°, and 95°C. Two points should be noted; the general decrease in retention times at higher temperatures and the importance of thermostating both column *and* injector. It is essential that turbulence at the column inlet and also large-scale temperature gradients along the column be avoided. Thus eluent must be preheated before entering the injector. Some temperature variation will always occur because of frictional heating of the mobile phase within the column.

The last system variable considered here is eluent flow rate. It may be predicted from the Van Deemter equation that an optimum flow rate exists for any separation (Katz *et al.*, 1983). Often this is less than a realistic working value, but fortunately the efficiency versus flow rate curve exhibits a broad minimum for many low molecular weight solutes. The chromatography of a standard toxin mixture under various flow conditions is demonstrated in Fig. 7. For each analysis a compromise must be attained between separation efficiency and sample throughput.

In this section the resolution of the aflatoxins has been considered in detail because of their central importance to mycotoxin analysts. Similar considerations will apply to all HPLC separations, and it is almost always possible to obtain

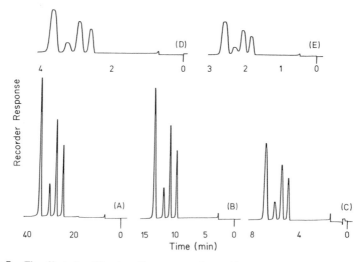

Fig. 7. The effect of mobile-phase flow rate on column efficiency. Conditions and toxin elution order: as Fig. 1, without PCD. UV detection at 360 nm, 0.1AU fsd. Column temperature 50°C. (A) Chart speed, 120 mm/hr, flow rate 0.3 ml/min. (B) 300 mm/hr, 0.75 ml/min. (C) 600 mm/hr, 1.5 ml/min. (D) 30 mm/min, 3.0 ml/min. (E) 30 mm/min, 4.5 ml/min. Note that apart from (D), all chromatograms are reproduced at the same linear flow rate in terms of column volumes of eluent per centimeter of trace.

good chromatography of toxin standards. It is usually less easy to achieve this with sample extracts, and here it is important to balance cleanup time, analysis time, and column consumption with the required degree of confidence in the results.

IV. HIGH-PERFORMANCE LIQUID CHROMATOGRAPHY: A LITERATURE REVIEW

This literature survey will concentrate on articles published since the last review of the subject (Scott, 1981), but earlier articles describing widely used methods or points of interest are also included. No attempt has been made to be comprehensive. Many publications describe details of cleanup or chromatography which differ in only minor ways from existing methods. Often the objective has been to develop more rapid cleanup techniques, although it has usually been found that cleanup schemes adequate for TLC determination also produce extracts satisfactory for HPLC analysis.

A. Aflatoxins and Related Compounds

1. Aflatoxins B_1, B_2, G_1, and G_2

The Pons NP method (Pons, 1976, 1979) remains in frequent use, although a silica gel-packed flow cell (Zimmerli, 1977; Panalaks and Scott, 1977) is required to enhance the fluorescence of B_1 and G_1 in the chlorinated mobile phase. One company (Varian) can supply prepacked flow cells, but for most instruments this may be done in the laboratory with little difficulty (Scott, 1981). It appears acceptable to replace chloroform in the eluent with dichloromethane, although this highly volatile solvent may cause cavitation problems with some pumps. The toluene-based mobile phase of Manabe et al. (1978), which does not suppress toxin fluorescence, has not been used extensively, possibly because of the corrosive properties of the formic acid incorporated.

RP systems have been gaining favor. However, a recent collaborative study (Campbell et al., 1984) of a typical RP method (Beebe, 1978) and a modified Pons method (Francis et al., 1982), applied to the determination of aflatoxins in peanut butter, showed that although both gave an improved limit of detection and better repeatability over current AOAC methods, there was some evidence for interferences in the RP system, particularly for G_2. Neither method was submitted for adoption as official first action. The problems with interferences are not characteristic of RP chromatography, and it is probable that a slight alteration in conditions would have been sufficient to eliminate them. A reversed-phase method for aflatoxins in naturally contaminated peanut butter has been shown to provide a coefficient of variation of 15% (Tarter et al., 1984). Many minor variations in the composition of the ternary water–acetonitrile–methanol mobile phase have been employed in published RP methods. It may be found necessary to adjust the proportions when changing from one column to another, apparently identical (Haghighi *et al.*, 1981). It has been suggested (Hurst and Toomey, 1978) that the water component of the mobile phase should be made 0.001 M in sodium chloride to reduce irreversible adsorption of peanut product extract components onto the column. Degradation of aflatoxins was observed at higher salt concentrations. Alternative RP eluents appear to offer no advantages. Water–acetonitrile–acetic acid was employed by DeVries and Chang (1982) but required constant stirring.

Mass sensitivity for B_1 under optimum conditions is about 1 to 10 pg for both NP and RP chromatography, while many methods give detection limits of 0.1 to 2 ng/g. Obviously the minimum measurable concentration depends on the quality of cleanup, and here there is usually a balance to be achieved between speed and efficiency of purification. The AOAC CB method, though lengthy, is generally recognized to yield clean residues, and detection limits at the lower end of the above range are possible. A number of "rapid" procedures have been developed for matrices such as feeding stuffs (Tuinstra and Haasnoot, 1983), heavily contaminated corn (Hutchins and Hagler, 1983), corn and peanuts (DeVries and Chang, 1982), cottonseed products (McKinney, 1981b), and corn and dairy feeds (Cohen and Lapointe, 1981). As might be expected, several of these procedures give detection limits rather higher than those mentioned above. The method devised by DeVries and Chang (1982) offers an interesting exception. These workers found hexane–acetonitrile partition preferable to the use of a silica column for cleanup and obtained a limit of approximately 0.1 ng/g. Prepacked silica disposable cartridges are increasingly replacing laboratory-prepared cleanup columns (Cohen and Lapointe, 1981; McKinney, 1981b; Hutchins and Hagler, 1983; Hurst *et al.*, 1984; Qian and Yang, 1984).

Hemiacetal formation using tetrafluoroacetic acid (TFA) is standard practice in RP chromatography in order to obtain adequate sensitivity with B_1 and G_1, although the occasional trace-level natural occurrence of B_{2a} caused Miller *et al.* (1982) to prefer the NP system. When using the hemiacetals, Gregory and Manley (1981) considered it necessary to precondition the column with several injections of a TFA blank before analyzing sample residues. Reaction with iodine-saturated water at 100°C for 15 sec as an alternative to the TFA-catalyzed addition of water was developed by Davis and Diener (1980). The iodine derivatives are eluted later than the corresponding hemiacetals and this is desirable in order to improve the resolution of toxins from potential interferences, which are often relatively weakly retained. Both iodine and TFA treatment of residues introduce an extra stage into sample preparation and in addition, although the water adducts are formed very cleanly (avoiding methanol-containing injection solvents), several minor products are seen after iodine derivatization.

To circumvent this problem, other investigators (Thorpe *et al.*, 1982; Tuinstra and Haasnoot, 1983) have employed postcolumn derivatization (PCD) with the iodine reagent. The optimum conditions for this method have been evaluated (Shepherd and Gilbert, 1984). Apart from the requirement for an extra pump (not necessarily of HPLC quality), the additional equipment required is minimal and the method appears to give excellent results, even with feeds containing citrus pulp (Tuinstra and Haasnoot, 1983), However, the cleanup used with these samples seems undesirably complex, involving a preliminary TLC separation. One less obvious advantage of the PCD iodination technique is that the retention time of B_1 is considerably longer than that of B_{2a} and thus the toxin peak is more widely separated from the typical early-running cluster of interferences. This is particularly apparent with confectionary samples (Shepherd and Gilbert, 1984). The published protocols suggest maintaining the derivatizing reagent over solid iodine in order to ensure saturation. Should the use of the filtered supernatant be preferred, in order to avoid the possibility of iodine particles entering the pump, it is desirable that the solution be replaced daily as the attainable sensitivity is moderately dependent on reagent concentration (Shepherd and Gilbert, 1984). The vapor pressure of iodine is sufficiently high to cause significant losses over longer periods, even from nearly enclosed reservoirs, and photolytically induced reaction of iodine and water is also significant. Chromatograms obtained with and without iodine PCD are presented in Fig. 8a. The only likely problem with

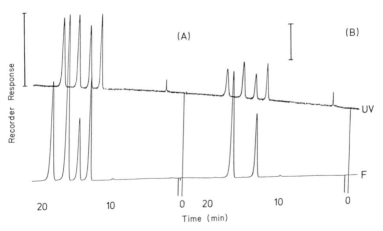

Fig. 8a. The effect of PCD with iodine-saturated water on the fluorescence of aflatoxins B_1, B_2, G_1, and G_2 under reversed-phase HPLC conditions. (A) PCD reagent at 0.5 ml/min; (B) PCD reagent pump off. Conditions and toxin elution order: as for Fig. 1, except sequential detection with, first, UV (360 nm; 0.01AU fsd, PCD on; 0.02AU fsd, PCD off; each bar represents 0.0025AU) and then fluorescence (setting ×1, least sensitive). Approximately 1 ng of each toxin injected.

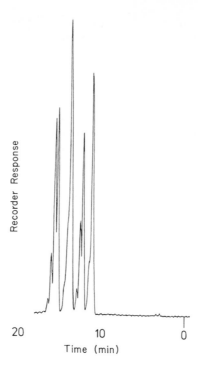

Fig. 8b. Chromatogram showing effect of malfunctioning PCD reagent pump. Conditions and toxin elution order: as for Fig. 1. A dual-piston PCD reagent pump was employed, and for the above chromatogram this was operating on one piston only. Note that splitting is observed primarily in the peaks due to the derivatizable toxins, B_1 and G_1. With their dihydro analogs, the effect is to produce a stepped peak arising from the alternate (0.75 or 1.25 ml/min) flow rates at the detector. It is important to be aware of this possible explanation for irregularities apparently originating in the chromatography, because peak splitting of such sharpness is usually caused by problems at the inlet end of the column.

this system is a malfunctioning reagent pump, and the consequences of one failure typical of twin-piston pumps are shown in Fig. 8b.

Analysis of sample extracts both unmodified and after TFA or iodine treatment gives good confirmation of B_1 and G_1. Another technique, applicable to all four toxins, is wavelength ratioing of the emission spectra. This was the method of choice of Howell and Taylor (1981) in a "multi" mycotoxin method for aflatoxins, ochratoxin A, and zearalenone in mixed feeds, where one sample cleanup scheme provides three extracts. These were analyzed using three HPLC systems, two of which are very similar, involving only a change in the proportions of a binary mobile phase. The Manabe toluene-based NP eluent was used for aflatoxins with some modifications to allow for potential interferences from grassmeal, and excellent detection limits were achieved for a wide variety of feeds. Another similar scheme (Bohm *et al.*, 1982) incorporates in addition to the above three toxins deoxynivalenol, with a claimed detection limit of 25 ng/g. Matrices said to be tractible include maize, corn silage, other cereals, mixed feed, peanut butter, and animal tissues. The sample preparation stage appears rather protracted. Tonsager *et al.* (1980) used a simplified cleanup involving gel permeation chromatography for all four of these toxins and also T-2. Aflatoxins were determined using the Pons method, with a detection limit of 2 ng/g. The principles and methodology of exclusion and gel chromatographic cleanup for trace-level contaminants in foods have been discussed in detail by Shepherd (1984).

A very different form of sample preparation was devised by a Chinese group (Wu *et al.*, 1983) who found that antibodies raised against B_{2a} precipitated B_1, B_2, G_2, and M_1 from biological fluids. After liberation from the immunoglobulin complex with methanol or acetone, the toxins were determined by RP chromatography. Detection limits in the picograms per milliliter range were claimed for urine. Other methods for human tissues were devised by Siraj *et al.* (1981), Lamplugh (1983), and Tsuboi *et al.* (1984).

2. Aflatoxin Metabolites and Reaction Products

a. Aflatoxin M_1. Current detection limits are approximately 0.01 ng/ml in liquid dairy products, limited principally by the quality of sample cleanup and the cumbersome equipment required for handling large volumes of milk. Early HPLC methods tended to employ NP chromatography with a silica gel-packed flow cell, but recently RP systems have been favored. Tuinstra and Haasnoot (1982) reported that adsorption of fluorescing compounds gave rise to high detector noise levels which precluded their use of a packed flow cell. In common with other mycotoxin analyses, the replacement of laboratory-prepared silica cleanup columns by prepacked disposable cartridges (available from several manufacturers and typified by Waters' Sep-Pak range) has simplified sample purification.

Both silica (Fremy and Boursier, 1981) and RP cartridges (Takeda, 1984) have found application. A rapid method with a detection limit of 0.08 ng/ml was developed by Cohen *et al.* (1984) involving sequential use of both ODS and silica cartridges. The ODS cartridges in particular have proved popular, perhaps partly because the equivalent bulk adsorbent has not been readily available until recently. Unfortunately, batch-to-batch variations may occur in cartridge activity (principally those containing silica), and hence each lot should be tested before routine use. Takeda (1984) found that up to 50 ml of liquid cow's milk (type not specified) could be injected directly onto an ODS Sep-Pak without overload. Cartridges could be regenerated and reused to reduce costs. The similar method of Winterlin *et al.* (1979) was considered to give inadequate resolution of M_1

from interferences at low toxin concentrations. The Takeda (1984) method was preferred in a comparison of six published protocols for M_1 (Shepherd et al., 1986). Exclusion chromatography was found useful by Miller et al. (1980) for preparing sample extracts. These authors employed a complex RP gradient which was claimed to provide excellent mass spectrometry backgrounds from M_1 fractions collected after elution for confirmation.

Hydrophilic sample absorption cartridges, onto which an aqueous solution (including milk) may be loaded and thus partitioned against an organic solvent subsequently poured through the packing, have been applied to M_1 analysis (Gauch et al., 1979), and the high-capacity versions now available might with advantage be employed for the cleanup of milk, using larger sample volumes to achieve improved detection limits. Other extraction methods include that of Chambon et al. (1983), who incorporated a zinc gel precipitation step prior to partition against chloroform, with subsequent RP analysis. Chang and DeVries (1983) emphasized the difficulties often encountered with emulsion formation at this stage and recommended rolling the mixture as an alternative to shaking. A method for cheese, with detection limit 0.01 ng/g, has been published (Hisada et al., 1984).

Typical RP eluents are similar to those employed for the B and G aflatoxins (see Section III), with perhaps a slightly greater water content. It has been reported that formation of the TFA reaction product increases sensitivity by a factor of 3 (Beebe and Takahashi, 1980; Chang and DeVries, 1983). There is some uncertainty concerning the nature of the M_1 derivative and whereas B_{2a} and G_{2a} are formed virttually instantaneously, the M_1 reaction requires about 15 min to go to completion (Beebe and Takahashi, 1980).

b. Other Metabolites. Trucksess and Stoloff (1984) have reported the determination in eggs of aflatoxicol, B_1, and M_1, with an aflatoxicol recovery of 78% at a concentration of 0.025 ng/g. RP chromatography was employed. Previously these workers had devised a method for aflatoxicol in milk, blood, and liver (Trucksess and Stoloff, 1981). The RP chromatography of B^1-dihydrodiol and the formation of complexes with amines by both this compound and B_{2a} were described by Neal et al (1981). Control of pH was found to be important. Perhaps an amino column would be found useful for the separation of these toxins and also G_{2a}, while the behavior of the product obtained from M_1 with TFA might yield further clues to its identity. B_1-DNA and sterigmatocystin-DNA hydrolysates were analyzed using an RP system by Essigmann et al. (1980). Both B_1-dihydrodiol and B_1-N^7-guanosine were resolved. The separation of other metabolites is discussed by Scott (1981).

c. Reaction Products. Chromatography of the adduct formed from B_1 and sodium bisulfite (aflatoxin B_{1S}) has been reported both using an unmodified RP system and also when incorporating an ion-pair reagent in the mobile phase (Hagler *et al.*, 1983). The decontamination of [^{14}C]aflatoxin-contaminated peanut meal with monomethylamine and calcium hydroxide gave rise to a number of unidentified products which were separated using the Pons NP system (Park *et al.*, 1981).

d. Sterigmatocystin and Biosynthetic Intermediates. The fluorescence of sterigmatocystin is weak and UV detection is preferred. Consequently, mass sensitivity for this compound is relatively poor, approximately 25 ng. RP chromatography was employed by Novotny *et al.* (1983) for sterigmatocystin in cheese, and Schmidt *et al.* (1981) similarly separated sterigmatocystin from extracts of moldy rice. Sterigmatocystin, norsolorinic acid, averufin, versicolorin A, and versiconal hemiacetal acetate were resolved by gradient RP elution (Berry *et al.*, 1984).

B. Ochratoxins

Fluorescence detection limits for ochratoxin are typically approximately 1 ng. Reversed-phase systems are invariably employed with pH control of the eluent (often with approximately 0.1% phosphoric acid) to suppress ionization of the carboxylic acid function. PCD with ammonia has been used to improve sensitivity to 0.25 ng of toxin (Hunt *et al.*, 1979). It has been suggested that addition of 0.01% sodium chloride to the eluent helps prevent irreversible adsorption of sample components to the column (Cantafora *et al.*, 1983). Coffee beans were being analyzed. Ochratoxin B was separated from an extract of moldy bread on an ODS column (Visconti and Bottalico, 1983a) and the chromatography of various hydroxylated metabolites described by Stormer *et al.* (1983). Ochratoxin A has been included in various multimycotoxin methods (see Section IV,A,1).

C. *Fusarium* Toxins

1. Zearalenone

Although the quantum efficiency of zearalenone is not as high as for the aflatoxins, fluorescence spectrophotometry remains the most sensitive means of detection with a mass limit of approximately 1 ng. Laser excitation has the potential to improve on this, but suitable wave length UV lasers are not available (Diebold *et al.*, 1979a). Again, reversed-phase chromatography is preferred. Zearalenone is frequently found in corn, and many methods exist for its determination in this matrix (Scott, 1981). See also the discussion of multimycotoxin methods in Section IV,A,1. The toxin may enter animal tissues via corn in the feed or alternatively as a metabolite of zearalenol, which has been used as a growth enhancer for livestock.

A method with a detection limit of 1 ng/g has been devised for zearalenone in chicken tissues (Turner *et al.*, 1983). Because of metabolic conversions, other closely related compounds are likely to be present and methods have been developed for α- and β-zearalenols (James *et al.*, 1982; Trenholm *et al.*, 1981) and zearalanone and zearalanols (Thouvenot *et al.*, 1981). The use of zearalenone oxime has been recommended as an internal standard (Trenholm *et al.*, 1984). Light-induced cis–trans isomerization also may occur, and the separation and EC detection of the two forms has been reported by Smyth and Frischkorn (1980), with a detection limit in corn of 5 ng/g. A rapid method for zearalenone and α-zearalenol is available with detection limits around 2 ng/g (Chang and DeVries, 1984).

2. Trichothecenes

Although at least 100 trichothecenes have been characterized, few have been found in naturally contaminated samples and fewer analyzed by HPLC. The toxins fall into several categories and in terms of liquid chromatography, the most salient distinguishing feature is the possession or otherwise of a useable UV chromophore. The 8-keto toxin group, which includes deoxynivalenol (DON), have a UV spectrum with a maximum at about 220 nm, giving a mass sensitivity for standards of 1 to 2 ng. The macrocyclics such as the roridins and verrucarins have sufficient conjugation to shift the UV maximum to the region of 260 nm with even better sensitivity. Unfortunately many other trichothecenes, including the exceptionally toxic T-2 and diacetoxyscirpenol (DAS), cannot be detected by UV. DON, the toxin of current major concern, gives excellent chromatography in RP systems containing a high percentage of water and may readily be separated from other members of the 8-keto group. Gradient elution on an ODS column—from ∼90:10 to 50:50 water–acetonitrile (v/v) linearly over 10 min—will be found to give excellent resolution of a mixture of DON, nivalenol, and ther monoacrylated derivatives 3-acetyldeoxynivalenol (3-ADON) and fusarenone X with detection limits for each of 1 to 2 ng. Under isocratic conditions 3-ADON is relatively strongly retained compared to nivalenol, and sensitivity is correspondingly reduced. Unfortunately the short wavelength absorption monitored leads to considerable difficulties with interferences from other components of sample extracts unless these have been highly purified (Visconti and Bottalico, 1983b; Ehrlich *et al.*, 1983; Chang *et al.*, 1984). DON was determined in a similar manner in a multitoxin analysis scheme (Ehrlich and Lee, 1984) and has also been preparatively purified by HPLC (Bennett *et al.*, 1981; Scott *et al.*, 1984).

Similar RP gradient methods have been developed for macrocyclics (verrucarin J, roridin E, and three satratoxins) in straw (Harrach *et al.*, 1983) and from cultures on rice (Mirocha *et al.*, 1981). Little HPLC work has been done on the other trichothecenes. T-2 and HT-2 (Schmidt and Dose, 1984), isolated from cereal cultures and separated in a RP system monitored by RI could be detected with a limit of 1 μg toxin. FID can bring this down to 0.1 μg for standards (M. J. Shepherd, unpublished observations). Supercritical fluid chromatography with mass spectrometry detection shows interesting possibilities for these toxins (Smith and Udseth, 1983).

3. Moniliformin

Highly purified extracts from corn were separated by ion-pair chromatography and visualized at 227 nm. Recoveries were low and erratic. Simple aqueous extracts produced chromatograms exhibiting severe interferences (Thiel *et al.*, 1982). Greater selectivity might have been achieved at 260 nm. A recent method employs ion-pairing for both extraction and chromatography (Shepherd and Gilbert, 1986).

D. Patulin

The toxin patulin may be chromatographed on both RP (Geipel *et al.*, 1981) and NP (Wisniewska and Piskorska-Pliszczynska, 1982) systems. Very many methods have been reported (Scott, 1981). Detection limits are generally good when columsn are monitored at 276 nm, typically 1–10 ng. Resolution from 5-hydroxymethylfurfural is well known to be critical. Confirmation techniques include subtractive reaction with cysteine (Wilson, 1981) and peak scanning.

E. Ergot Alkaloids

Ergot alkaloids and their derivatives lie in the undefined area between mycotoxins and pharmaceuticals. The articles reviewed here form only a small part of the literature, but Scott's review (1981) provides an excellent introduction to the complexities of ergot analysis. Reversed-phase systems have been favored, although other column types may be employed, and Eckers *et al.* (1982) in a study employing directly coupled HPLC-mass spectrometry found that an amino packing gave satisfactory results, although best resolution was obtained using a silica column eluted with a mobile phase containing trace amounts of ammonium hydroxide. Most RP systems have ammonium hydroxide or carbonate incorporated into the eluent (Wolff *et al.*, 1983; Young *et al.*, 1983), or bases such as di- or triethylamine. Column lifetimes have therefore often been rather limited. For separations of a more restricted number of toxins, neutral eluents may be preferred (Herenyi and Gorog, 1982).

F. Miscellaneous Toxins

Penicillic acid was determined in chicken tissues by Hanna *et al.* (1981) in a simple RP system with a detection limit of 2 ng/g. Confirmation was obtained by derivatization with diazomethane to produce a later-eluting pyrazoline compound. Conjugated metabolites of this toxin isolated from mice were separated with a highly aqueous mobile phase acidified with acetic acid (Chan *et al.*, 1982). Secalonic acid D extracted from biological fluids was chromatographed on an ODS column using a quarternary mobile phase, the proportions of which were varied to provide either improved resolution from interferences or enhanced sensitivity (Reddy *et al.*, 1981). Detection at 339 nm enabled visualization of 0.6 ng toxin. Secalonic acid D in grain dust was determined as part of a multi-mycotoxin assay (Ehrlich and Lee, 1984). The detection limit was 10 ng/g. Naphthoquinone toxins have been isolated by preparative NP chromatography (Peterson and Grove, 1983) and determined analytically at 15 ng/g in extracts from cultures using a similar system at 405 nm (Ciegler *et al.*, 1981). Reversed-phase chromatography with a C_8 column has also been found suitable for xanthomegnin. Binding of toxins to the packing was observed, and this was eliminated by incorporating sodium dodecyl sulfate in the mobile phase (Wall and Lillehoj, 1983). It has been recommended that acidic conditions should be avoided when isolating these compounds (Carman *et al.*, 1983), which are readily oxidized. EC detection may usefully be employed for these toxins, providing a detection limit of 0.5 ng xanthomegnin, equivalent to better than 15 ng/g in grain and mixed feeds (Carman *et al.*, 1984).

Interestingly, Danieli *et al.* (1980) employed microbore RP HPLC with detection at 250 nm for the determination of PR toxin with a mass limit of less than 0.2 ng. Retention time and peak shape were found to be critically dependent on the water–acetonitrile ratio of the mobile phase. The 75×0.5 mm PTFE column was eluted at 8 μl/min from a 250-μl syringe with its plunger actuated by a motor-driven screw. Engstrom and Richard (1981) showed that rubratoxin B is heat labile, at least during cleanup, and that 50% loss occurred in 90 min at room temperature. The toxin was detected (5 ng; 20 ng/g) in mixed-feed extracts using an acidified NP mobile phase to maintain it in the nonionized form. A cyano-bonded phase was preferred (Lauren and Fairclough, 1980) for the separation and purification of metabolites of sporidesmin. Various mixed mobile phases were chosen to provide different combinations of selectivity and column-loading capacity. A similar system was not so favorable for chromatography of the janthitrems (Lauren and Gallagher, 1982) and a C_8 RP column gave superior results, although resolution of the B and C toxins required gradient elution. A combination of fluorescence and UV detection provided good differentiation from interferences together with mass limits of 1 to 5 ng for the individual toxins. Penitrem A and other metabolites were also separated from the janthitrems with this system. Penitrems were readily chromatographed using water–methanol on a RP C_8 column (Maes *et al.*, 1982). Detection at the 296-nm maximum affords only one-third of the sensitivity attainable at 233 nm (7 ng) but with much improved selectivity. Dihydrocitrinone (a urinary metabolite of citrinin) and the parent toxin were separated using a RP eluent incorporating ethyl acetate (Dunn *et al.*, 1983).

Alternaria toxins including alternariol (AOH), alternatiol methyl ester (AME), and altenuene (ALT) have been preparatively isolated by a method including NP chromatography (Chu and Bennett, 1981), although these workers preferred a RP system for analysis. Simpler HPLC conditions were employed by Heisler *et al.* (1980) with detection at 324 nm for AOH, ALT, and AME. In addition tenuazonic acid was separated on the same RP column using an alternative mobile phase, although this toxin may well have been chromatographed in the ionic form as eluent pH was not controlled and it was found that retention and mobile-phase organic modifier content increased together (Heisler *et al.*, 1980). Detection limits were in the low nanogram range for all four toxins. Another acidic metabolite, altertoxin I (ATX-1), was purified by NP HPLC with an eluent incorporating formic acid. The system was monitored at 340 nm and several decomposition products were isolated (Stinson *et al.*, 1982). AOH, AME, ALT, and ATX-1 were determined in fruits with detection limits in the range 30–500 ng/g (Wittkowski *et al.*, 1983). Tenuazonic acid was resolved from its equilibrium isomeric form using acetonitrile but not methanol as organic modifier (Scott and Kanhere, 1980).

V. CONCLUSIONS

It will be apparent that in reality there is no such thing as an HPLC method for any mycotoxin or indeed any other analyte. What do exist are cleanup techniques which provide sample residues of sufficient purity to permit the final measurement to be carried out to acceptable standards of sensitivity and certainty. It is very useful if it can be shown that a given chromatographic system can be used to carry out this measurement, but inevitably most of the work entailed in devising an overall method is taken up in proving that toxins may be recovered from sample matrices reproducibly, in high yield, and with freedom from interferences.

The relative merits of HPLC and TLC have already been discussed. In practice, they each have their place. TLC provides a low-cost, fast analysis but is labor intensive. HPLC requires a considerable investment in equipment but offers advantages in such areas as on-line sample cleanup and automated analysis and data handling. HPLC is also more suitable when toxins are sensitive to environmental factors such as oxygen or light. As in other sectors of organic analysis, there has been a marked preference for the reversed-phase mode of HPLC because, though fundamentally a complex phenomenon, it is simple to apply and separations may often be developed with little difficulty.

HPLC has become indispensable in many areas of analytical chemistry. It will continue to find increasing application in all laboratories where mycotoxin analyses are performed.

REFERENCES

Anonymous (1981). *J. Chromatogr. Sci.* **19,** 338–348.

Beebe, R. M. (1978). *J. Assoc. Off. Anal. Chem.* **61,** 1347–1352.

Beebe, R. M., and Takahashi, D. M. (1980). *J. Agric. Food Chem.* **28,** 481–482.

Bennett, G. A., Peterson, R. E., Plattner, R. D., and Shotwell, O. L. (1981). *J. Am. Oil Chem. Soc.* **58,** 1002A–1005A.

Berridge, J. C. (1984). *Trends Anal. Chem.* **3,** 5–10.

Berry, R. K., Dutton, M. F., and Jeenah, M. S. (1984). *J. Chromatogr.* **283,** 421–424.

Blanc, M. (1982). "Study and Determination of Aflatoxins in Foods," Monogr. No. 29. Actual. Sci. Tech. Ind. Agroaliment., Cent. Doc. Int. Ind. Util. Prod. Agric., France.

Bohm, J., Schuh, M., and Leibetseder, J. (1982). *Proc. Int. IUPAC Symp. Mycotoxins Phycotoxins, 5th, Vienna* (W. Pfannhauser and P. B. Czedik-Eysenberg, eds.), pp. 16–19. Tech. Univ., Vienna.

Bristow, P. A. (1976). "LC in Practice." HETP, Wilmslow, England.

Campbell, A. D., Francis, O. J., Beebe, R. A., and Stoloff, L. (1984). *J. Assoc. Off. Anal. Chem.* **67,** 312–316.

Cantafora, A., Grossi, M., Miraglia, M., and Benelli, L. (1983). *Riv. Soc. Ital. Sci. Aliment.* **12,** 103–108.

Carman, A. S., Kuan, S. S., Francis, O. J., Ware, G. M., Gaul, J. A., and Thorpe, C. W. (1983). *J. Assoc. Off. Anal. Chem.* **66,** 587–591.

Carman, A. S., Kuan, S. S., Francis, O. J., Ware, G. M., and Luedtke, A. E. (1984). *J. Assoc. Off. Anal. Chem.* **67,** 1095–1098.

Chambon, P., Dano, S. D., Chambon, R., and Geahchan, A. (1983). *J. Chromatogr.* **259,** 372–374.

Chan, P. K., Hayes, A. W., and Siraj, M. Y. (1982). *Toxicol. Appl. Pharmacol.* **66,** 259–268.

Chang, H. L., and DeVries, J. W. (1983). *J. Assoc. Off. Anal. Chem.* **66,** 913–917.

Chang, H. L., and DeVries, J. W. (1984). *J. Assoc. Off. Anal. Chem.* **67,** 741–744.

Chang, H. L., DeVries, J. W., Larson, P. A., and Patel, H. H. (1984). *J. Assoc. Off. Anal. Chem.* **67,** 52–54.

Chang-Yen, I., Stoute, V. A., and Felmine, J. B. (1984). *J. Assoc. Off. Anal. Chem.* **67,** 306–308.

Chu, F. S., and Bennett, S. C. (1981). *J. Assoc. Off. Anal. Chem.* **64,** 950–954.

Ciegler, A., Lee, L. S., and Dunn, J. J. (1981). *Appl. Environ. Microbiol.* **42,** 446–449.

Cohen, H., and Lapointe, M. (1981). *J. Assoc. Off. Anal. Chem.* **64,** 1372–1376.

Cohen, H., Lapointe, M., and Fremy, J. M. (1984). *J. Assoc. Off. Anal. Chem.* **67,** 49–51.

Coker, R. D. (1984). *In* "Analysis of Food Contaminants" (J. Gilbert, ed.), pp. 207–258. Appl. Sci., London.

Crombeen, J. P., Heemstra, S., and Kraak, J. C. (1983). *J. Chromatogr.* **282,** 95–106.

Danieli, B., Bianchi-Salvadori, B., and Zambrini, A. V. (1980). *Milchwissenschaft* **35,** 423–426.

Davis, N. D., and Diener, U. L. (1980). *J. Assoc. Off. Anal. Chem.* **63,** 107–109.

DeVries, J. W., and Chang, H. L. (1982). *J. Assoc. Off. Anal. Chem.* **65,** 206–209.

Diebold, G. J., Karny, N., and Zare, R. N. (1979a). *Anal. Chem.* **51,** 67–69.

Diebold, G. J., Karny, N., Zare, R. N., and Seitz, L. M. (1979b). *J. Assoc. Off. Anal. Chem.* **62,** 564–569.

Dunn, B. B., Stack, M. E., Park, D. L., Joshi, A., Friedman, L., and King, R. L. (1983). *J. Toxicol. Environ. Health* **12,** 283–289.

Eckers, C., Games, D. E., Mallen, D. N. B., and Swann, B. P. (1982). *Biomed. Mass Spectrom.* **9,** 162–173.

Ehrlich, K. C., and Lee, L. S. (1984). *J. Assoc. Off. Anal. Chem.* **67,** 963–967.

Ehrlich, K. C., Lee, L. S., and Ciegler, A. (1983). *J. Liq. Chromatogr.* **6,** 833–843.

Engstrom, G. W., and Richard, J. L. (1981). *J. Agric. Food Chem.* **29**, 1164–1167.

Erni, F. (1983). *J. Chromatogr.* **282**, 371–383.

Essigmann, J. M., Donahue, P. R., Story, D. L., Wogan, G. N., and Brunengraber, H. (1980). *Cancer Res.* **40**, 4085–4091.

Francis, O. J., Lipinski, L. J., Gaul, J. A., and Campbell, A. D. (1982). *J. Assoc. Off. Anal. Chem.* **65**, 672–676.

Frei, R. W, (1982). *Chromatographia* **15**, 161–166.

Fremy, J.-M., and Boursier, B. (1981). *J. Chromatogr.* **219**, 156–161.

Friesen, M. (1982). *In* "Environmental Carcinogens Selected Methods of Analysis. Vol. 5: Some Mycotoxins" (H. Egan, ed.), pp. 85–106. IARC, Lyon.

Friesen, M. (1983a). "Preliminary Report on the Statistical Analysis of Results Obtained for the Analysis of Aflatoxin M1 in Lyophilised Milk." IARC, Lyon.

Friesen, M. (1983b). "Preliminary Report on the Statistical Analysis of Results Obtained for the Analysis of Aflatoxins B1, B2, G1, and G2 in Yellow Corn Meal and Finished Peanut Butter." IARC, Lyon.

Frischkorn, C. G. B., and Schlimper, H. (1982). *Fresenius' Z. Anal. Chem.* **312**, 541–542.

Gant, J. R., Dolan, J. W., and Snyder, L. R. (1979). *J. Chromatogr.* **185**, 153–177.

Gauch, R., Leuenberger, U., and Baumgartner, E. (1979). *J. Chromatogr.* **178**, 543–549.

Geipel, M., Baltes, W., Kroenert, W., and Weber, R. (1981). *Chem., Mikrobiol., Technol. Lebensm.* **7**, 93–96.

Genieser, H.-G., Gabel, D., and Jastoff, B. (1983). *J. Chromatogr.* **269**, 127–152.

Glajch, J. L., and Kirkland, J. J. (1983). *Anal. Chem.* **55**, 319A, 320A, 322A, 326A, 328A, 332A, 336A.

Glajch, J. L., Kirkland, J. J., and Snyder, L. R. (1982). *J. Chromatogr.* **238**, 269–280.

Gregory, J. F., and Manley, D. B. (1981). *J. Assoc. Off. Anal. Chem.* **64**, 144–151.

Gregory, J. F., and Manley, D. B. (1982). *J. Assoc. Off. Anal. Chem.* **65**, 869–875.

Haghighi, B., Thorpe, C. W., Pohland, A. E., and Barnett, R. (1981). *J. Chromatogr.* **206**, 101–108.

Hagler, W. M., Hutchins, J. E., and Hamilton, P. B. (1983). *J. Food Prot.* **46**, 295–300, 304.

Hanna, G. D., Phillips, T. D., Kubena, L. F., Cysewski, S. J., Ivie, G. W., Heidelbaugh, N. D., Witzel, D. A., and Hayes, A. W. (1981). *Poult. Sci.* **60**, 2246–2252.

Harrach, B., Bata, A., Bajmocy, E., and Benko, M. (1983). *Appl. Environ. Microbiol.* **45**, 1419–1422.

Heftmann, E., ed. (1983). "Chromatography. Part A: Fundamentals and Techniques." Elsevier, Amsterdam.

Heisler, E. G., Siciliano, J., Stinson, E. E., Osman, S. F., and Bills, D. D. (1980). *J. Chromatogr.* **194**, 89–94.

Herenyi, B., and Gorog, S. (1982). *J. Chromatogr.* **238**, 250–252.

Hisada, K., Tereda, H., Yamamoto, K., Tsubouchi, H., and Sakabe, Y. (1984). *J. Assoc. Off. Anal. Chem.* **67**, 601–606.

Horvath, C., and Melander, W. R. (1983). *In* "Chromatography. Part A: Fundamentals and Techniques" (E. Heftmann, ed.), pp. A27–A135. Elsevier, Amsterdam.

Howell, M. V., and Taylor, P. W. (1981). *J. Assoc. Off. Anal. Chem.* **64**, 1356–1363.

Hunt, D. C., Philp, L. A., and Crosby, N. T. (1979). *Analyst (London)* **104**, 1171–1175.

Hurst, W. J., and Toomey, P. B. (1978). *J. Chromatogr. Sci.* **16**, 372–376.

Hurst, W. J., Snyder, K. P., and Martin, R. A. (1984). *Peanut Sci.* **11**, 21–23.

Hutchins, J. E., and Hagler, W. W. (1983). *J. Assoc. Off. Anal. Chem.* **66**, 1458–1465.

James, L. T., McGirr, L. G., and Smith, T. K. (1982). *J. Assoc. Off. Anal. Chem.* **65**, 8–13.

Karger, B. L., Giese, R. W., and Snyder, L. R. (1983). *Trends Anal. Chem.* **2**, 106–109.

Katz, E., Ogan, K. L., and Scott, R. P. W. (1983). *J. Chromatogr.* **270**, 51–75.

Knox, J. H., Laird, G. R., and Raven, P. A. (1976). *J. Chromatogr.* **122**, 129–145.

Kucera, P., ed. (1984). "Microcolumn High-Performance Liquid Chromatography." Elsevier, Amsterdam.

Lamplugh, S. M. (1983). *J. Chromatogr.* **273**, 442–448.

Lauren, D. R., and Fairclough, R. J. (1980). *J. Chromatogr.* **200**, 288–292.

Lauren, D. R., and Gallagher, R. T. (1982). *J. Chromatogr.* **248**, 150–154.

McKinney, J. D. (1981a). *J. Assoc. Off. Anal. Chem.* **64**, 939–949.

McKinney, J. D. (1981b). *J. Am. Oil Chem. Soc.* **58**, 935A–937A.

McKinney, J. D. (1984). *J. Assoc. Off. Anal. Chem.* **67**, 25–32

Maes, C. M., Steyn, P. S., and Van Heerden, F. R. (1982). *J. Chromatogr.* **234**, 489–493.

Majors, R. E., Barth, H. G., and Lochmuller, C. H. (1982). *Anal. Chem.* **54**, 323R–363R.

Manabe, M., Goto, T., and Matsuura, S. (1978). *Agric. Biol. Chem.* **42**, 2003–2007.

Miller, D. M., Wilson, D. M., Wyatt, R. D., McKinney, J. K., Crowell, W. A., and Stuart, B. P. (1982). *J. Assoc. Off. Anal. Chem.* **65**, 1–4.

Miller, M., Kiermeier, F., Weiss, G., and Klostermeyer, H. (1980). *Z. Lebensm.-Unters. Forsch.* **171**, 20–23.

Mirocha, C. J., Harrach, B., Pathre, S. V., and Palyusik, M. (1981). *Appl. Environ. Microbiol.* **41**, 1428–1432.

Neal, G. E., Judah, D. J., Stirpe, F., and Patterson, D. S. P. (1981). *Toxicol. Appl. Pharmacol.* **58**, 431–437.

Nowotny, P., Baltes, W., Kroenert, W., and Weber, R. (1983). *Lebensmittelchem. Gerichtl. Chem.* **37**, 71–72.

Panalaks, T., and Scott, P. M. (1977). *J. Assoc. Off. Anal. Chem.* **60**, 583–589.

Park, D. L., Jemmali, M., Frayssinet, C., LaFarge-Frayssinet, C., and Yvon, M. (1981). *J. Am. Oil Chem. Soc.* **58**, 995A–1002A.

Peaden, P. A., and Lee, M. L. (1982). *J. Liq. Chromatogr.* **5**, Suppl. 2, 179–221.

Peterson, R. E., and Grove, M. D. (1983). *Appl. Environ. Microbiol.* **45**, 1937–1938.

Pons, W. A. (1976). *J. Assoc. Off. Anal. Chem.* **59**, 101–105.

Pons, W. A. (1979). *J. Assoc. Off. Anal. Chem.* **61**, 793–800.

Poole, C. F., and Schuette, S. A. (1984). "Contempory Practice of Chromatography." Elsevier, Amsterdam.

Qian, G.-S., and Yang, G. C. (1984). *J. Agric. Food Chem.* **32**, 1071–1073.

Rafel, J. (1983). *J. Chromatogr.* **282**, 287–295.

Reddy, C. S., Reddy, R. V., and Hayes, A. W. (1981). *J. Chromatogr.* **208**, 17–26.

Schmidt, R., and Dose, K. (1984). *J. Anal. Toxicol.* **8**, 43–45.

Schmidt, R., Mondani, J., Ziegenhagen, E., and Dose, K. (1981). *J. Chromatogr.* **207**, 435–438.

Scott, P. M. (1981). *In* "Trace Analysis" (J. F. Lawrence, ed.), Vol. 1, pp. 193–266. Academic Press, New York.

Scott, P. M., and Kanhere, S. R. (1980). *J. Assoc. Off. Anal. Chem.* **63**, 612–621.

Scott, P. M., Lawrence, G. A., Telli, A., and Iyengar, J. R. (1984). *J. Assoc. Off. Anal. Chem.* **67**, 32–34.

Scott, R. P. W., ed. (1984). "Small Bore Liquid Chromatography Columns." Wiley, Chichester, England.

Shepherd, M. J. (1984). *In* "Analysis of Food Contaminants" (J. Gilbert, ed.), pp. 1–72. Appl. Sci., London.

Shepherd, M. J., and Gilbert, J. (1984). *Food Addit. Contam.* **1**, 325–335.

Shepherd, M. J., and Gilbert, J. (1986). *J. Chromatogr.* **358**, 415–422.

Shepherd, M. J., Holmes, M., and Gilbert, J. (1986). *J. Chromatogr.* **354**, 305–315.

Simpson, C. F., ed. (1982). "Techniques in Liquid Chromatography." Wiley, Chichester, England.

Siraj, M. Y., Hayes, A. W., Unger, P. D., Hogan, G. R., Ryan, N. J., and Wray, B. B. (1981). *Toxicol. Appl. Pharmacol.* **58,** 422–430.

Smith, R. D., and Udseth, H. R. (1983). *Anal. Chem.* **55,** 2266–2272.

Smyth, M. R., and Frischkorn, C. G. B. (1980). *Anal. Chim. Acta* **115,** 293–300.

Stinson, E. E., Osman, S. F., and Pfeffer, P. E. (1982). *J. Org. Chem.* **47,** 4110–4113.

Stormer, F. C., Storen, O., Hansen, C. E., Pederson, J. I., and Aasen, A. J. (1983). *Appl. Environ. Microbiol.* **45,** 1183–1187.

Takeda, N. (1984). *J. Chromatogr.* **288,** 484–488.

Tarter, E. J., Hanchay, J.-P., and Scott, P. M. (1984). *J. Assoc. Off. Anal. Chem.* **67,** 597–600.

Thiel, P. G., Meyer, C. J., and Marasas, W. F. O. (1982). *J. Agric. Food Chem.* **30,** 308–312.

Thorpe, C. W., Ware, G. M., and Pohland, A. E. (1982). *Proc. Int. IUPAC Symp. Mycotoxins Phycotoxins, 5th, Vienna* (W. Pfannhauser and P. B. Czedik-Eysenberg, eds.), pp. 52–55. Tech. Univ., Vienna.

Thouvenot, D., Morfin, R., Di Stefano, S., and Picart, D. (1981). *Eur. J. Biochem.* **121,** 139–145.

Tonsager, S. R., Maltby, D. A., Schock, R. J., and Braselton, W. E. (1980). *Adv. Thin Layer Chromatogr., Proc. Bien. Conf., 2nd* (J. C. Touchstone, ed.), pp. 389–402. Wiley, Chichester. England.

Trenholm, H. L., Warner, R. M., and Farnworth, E. R. (1981). *J. Assoc. Off. Anal. Chem.* **64,** 302–310.

Trenholm, H. L., Warner, R. M., and Fitzpatrick, D. W. (1984). *J. Assoc. Off. Anal. Chem.* **67,** 968–972.

Trucksess, M., and Stoloff, L. (1981). *J. Assoc. Off. Anal. Chem.* **64,** 1083–1087.

Trucksess, M., and Stoloff, L. (1984). *J. Assoc. Off. Anal. Chem.* **67,** 317–320.

Tsuboi, S., Nakagawa, T., Tomita, M., Eeo, T., Ono, H., Kawamura, K., and Iwamura, N. (1984). *Cancer Res.* **44,** 1231–1234.

Tuinstra, L. G. M. T., and Haasnoot, W. (1982). *Fresenius' Z. Anal. Chem.* **312,** 622–623.

Tuinstra, L. G. M. T., and Haasnoot, W. (1983). *J. Chromatogr.* **282,** 457–462.

Turner, G. V., Phillips, T. D., Heidelbaugh, N. D., and Russell, L. H. (1983). *J. Assoc. Off. Anal. Chem.* **66,** 102–104.

Unger, K. K. (1979). "Porous Silica." Elsevier, Amsterdam.

Visconti, A., and Bottalico, A. (1983a). *J. Agric. Food Chem.* **31,** 1122–1123.

Visconti, A., and Bottalico, A. (1983b). *Chromatographia* **17,** 97–100.

Wall, J. H., and Lillehoj, E. B. (1983). *J. Chromatogr.* **268,** 461–468.

Ware, G. M., Thorpe, C. W., and Pohland, A. E. (1980). *J. Assoc. Off. Anal. Chem.* **63,** 637–641.

White, E. R., and Laufer, D. N. (1984). *J. Chromatogr.* **290,** 187–196.

Wilson, R. D. (1981). *Food Technol. N.Z.* **27,** 27, 29, 31.

Winterlin, W., Hall, G., and Hsieh, D. P. H. (1979). *Anal. Chem.* **51,** 1873–1874.

Wisniewska, H., and Piskorska-Pliszczynska, J. (1982). *Bull. Vet. Inst. Pulawy* **25,** 38–42.

Wittowski, M., Baltes, W., Kronert, W., and Weber, R. (1983). *Z. Lebensm.-Unters. Forsch.* **177,** 447–453.

Wolff, J., Ocker, H.-D., and Zwingelberg, H. (1983). *Getreide, Mehl Brot* **37,** 331–335.

Wu, S., Yang, J., and Sun, Z. (1983). *Zhanghua Zhongliu Zazhi* **5,** 81–84 [*C. A.* **100,** 46441].

Young, J. C., Chen, Z.-J., and Marquardt, R. R. (1983). *J. Agric. Food Chem.* **31,** 413–415.

Zimmerli, B. (1977). *J. Chromatogr.* **131,** 458–463.

12

Gas Chromatographic–Mass Spectrometric Analysis of Mycotoxins

RONALD F. VESONDER AND WILLIAM K. ROHWEDDER

U.S. Deparmtent of Agriculture
Agricultural Research Service
Northern Regional Research Center
Peoria, Illinois 61604

I. INTRODUCTION*

The instrumental technique of gas chromatography–mass spectrometry (GC-MS) provides an excellent method for the analysis of mycotoxins from contami-

---

*The mention of firm names or trade products does not imply that they are endorsed or recommended by the U.S. Department of Agriculture over other firms or similar products not mentioned.

MODERN METHODS IN THE ANALYSIS
AND STRUCTURAL ELUCIDATION
OF MYCOTOXINS

nated cereal grains, animal feeds, and foodstuffs. The technique has a high sensitivity and specificity for all mycotoxins that can be derivatized to a compound which is sufficiently volatile to be gas chromatographed. GC-MS has the power of gas chromatography to separate the complex mixtures resulting from the extraction of natural materials and the power of the mass spectrometer to identify the wide variety of structures associated with mycotoxins.

II. HISTORY AND INSTRUMENTATION

A. Mass Spectrometry

Mass spectrometry was invented by J. J. Thomson in the early part of this century but did not become a useful tool for chemists until the development of commercially available instruments after World War II. Mass spectra are produced by passing sample vapor through a 70-V electron beam in the mass spectrometer source. The electrons knock an electron off an atom in the molecule, leaving the molecule with a positive charge and part of the energy of the ionizing electron as vibrational energy. The mass spectrometer separates the charged molecules according to mass and records their intensities. Mass spectra are plots of these intensities, percentized versus mass. Mass spectra show the molecular ion and a series of fragment ions from the breakup of the original positively charged ion. The relative intensities of the molecular ion and the fragment ions are a function of the ability of the molecule to stabilize the positive charge and dissipate the vibrational energy.

Because of the rearrangement of the atoms in the molecule occurring during the ionization process, early mass spectroscopists were unable to interpret the fragment peaks, and correlate the spectra with compound structure. As they ran more oxygenated compounds, mass spectroscopists found that many of the fragmentations could be explained by ordinary organic mechanisms if they included the presence of the positive charge on the molecule. One of the pioneers in the interpretation of mass spectra was Fred W. McLafferty (1966), whose book is an excellent text on interpretation of mass spectra for beginners.

B. Gas Chromatography

Chromatography has been available in one form or another for generations. The development of gas chromatography in the mid-1950s provided a new convenient method of analysis of almost anything that could be volatilized. The gas chromatograph provides an excellent tool for separating small amounts of compounds into very pure fractions. As an analytical method the gas chromatograph is deficient in that only one number, the elution time, is available for identification of compound.

C. Combined Gas Chromatography–Mass Spectrometry

Analysis of the eluate of the gas chromatograph by a mass spectrometer would combine the excellent separation power of the gas chromatograph with the identification ability of the mass spectrometer. But, the two instruments are basically incompatible. The packed-column GC produces ≥40 ml of gas while the pumping systems of most spectrometers is ≤1 ml. At the time GC first became available, the commercial mass spectrometers were magnetic instruments which required 10–15 min to scan a spectrum and record the data on a strip chart recorder. Early instruments were connected with a high-quality valve which allowed 1–5% of the eluate into the mass spectrometer and vented the rest to the atmosphere. A separator based on the rapid diffusion of helium through a glass frit was developed by Watson and Biemann and provided considerable enrichment of the sample while greatly reducing the helium flow into the mass spectrometer. Ryhage developed the jet separator, which also enhances the amount of sample entering the mass spectrometer and reduces the amount of helium. The jet separator is available in many commercial instruments. Techniques of interfacing gas chromatographs to mass spectrometers are thoroughly discussed in a book by McFadden (1973).

Scan speeds of commercial magnetic mass spectrometers have been increased to full scans in 6 to 12 sec, but most GC-MS work is done with quadrupole mass spectrometers, which can do a full scan in 0.5 to 2 sec. Two seconds for 1 min equals 30 spectra, times 60 min equals 1800 spectra/hr. A sample of extracted feed corn can easily contain 180 identifiable peaks. This much data can only be handled by computer. The computer controls the scanning of the mass spectrometer, and stores the output on disk or tape. Scientists' enthusiasm would quickly fag if they had to search through 1800 spectra looking for 180 meaningful spectra. The computer can generate a total ionization plot, which is the sum of the intensities for all masses for each scan plotted against scan number. This plot is similar to the GC plot and can be used as an index to the stored spectra. Modern computer programs can automatically make hard copies and do a library search on all spectra with total ionization peaks above a certain threshold. The computer can be set to search all spectra for just a few mass peaks of interest such as mass 407 or mass 422 for deoxynivalenol, or mass 436 for T-2 toxin, and plot the data as mass chromatograms, which highlight the presence of these compounds. By putting elution times, mass numbers, and sensitivities into the computer program, quantitative reports can be produced automatically.

D. Capillary Columns Chromatography

Capillary columns have slowly replaced packed columns. The advent of fused-quartz capillary columns with their great mechanical flexibility have allowed the outlet end of the capillary to be fed directly into the mass spectrometer ionization source, eliminating many problems of gas holdup and loss of GC resolution. A 30-m, bonded-phase, 0.32-mm bore, thick film, fused-quartz capillary is almost a perfect inlet for the mass spectrometer, allowing excellent mixture separation, and good sensitivity without column bleed. There are several commercially available GC-MS data systems, which have eliminated the problems discussed in the first part of this section and are almost ideal analytical engines.

III. COMPARISON OF MASS SPECTROMETRY TECHNIQUES

A. Soft Ionization Methods

Most mass spectra generated since the mid-1960s have been electron impact (EI) mass spectra, with the sample gas ionization by a beam of electrons. The molecular ion and the fragment ions provide an excellent tool for interpretation of the spectra. Large libraries (70,000 spectra) have been collected, and unknown spectra can be compared with the libraries by computer for identification. Some compounds, such as alcohols and trimethylsilyl ether (TMS) derivatives, do not give molecular ions with this technique.

Chemical ionization (CI) uses a plasma of ammonia, methane, isobutane, or some other gas as a reagent to ionize the sample. CI is a soft ionization technique, which transfers little vibrational energy to the molecule during ionization and causes relatively little fragmentation of the molecule. The plasma is generated by putting 1000 times as much reagent gas with the sample into a tightly enclosed source. Most CI reagents are protonating agents and add a proton to the sample molecule to produce a M + 1, or pseudomolecular ion. CI is used extensively with GC to emphasize the molecular ion region of the compounds which have little or no molecular ion with EI ionization.

B. Multiple-Ion Monitoring

Most MS used in environmental and general analyses since the mid-1970s has been GC-MS. GC-MS provides several complete mass spectra for each GC peak, even for capillary columns. When this is recorded on a computer it provides a very complete analysis of a sample mixture which can be used immediately or archived for future use. If the scientist knows exactly what compound one is looking for and has no need for general identification of the compounds present, the sensitivity and precision of measurements can be increased 100- to 1000-fold, if the mass spectrometer is set to monitor only one mass peak. When the mass spectrometer is scanned from mass 10 to mass 650, most of the instrument time is spent on mass peaks of no interest or scanning the area between peaks. The sensitivity and precision of a mass spectrometer depends on the number of ions of interest measured. The technique of following one or a few mass peaks is called selected-ion monitoring (SIM), mass fragmentometry, multiple-ion detection, or multiple-ion monitoring (MIM). SIM is useful only when one knows the sample well and is interested in only one or a few compounds in it. The full spectrum is not recorded and the possibility of confirming the identity of a GC peak with the full spectrum is not possible. With dirty samples there is the possibility of false positive results caused by some other compound having a mass peak at the selected mass. SIM is very useful for the quantitative analysis of specific mycotoxins in relatively clean samples.

C. Mass Spectrometry–Mass Spectrometry

Another technique which has recently become important is MS-MS, which is two mass spectrometers in series, with the first acting as a separator and the second acting as an analyzer. The source is set up to generate molecular or pseudomolecular ions of the sample, and the mass filter of the first mass spectrometer is set to pass the ion of interest to the second mass spectrometer, which is scanned to produce a full spectrum for confirmation of identity. The sample can be a solid such as part of a leaf, put on a probe directly into the MS-MS source without sample extraction or preparation. It has been said that 60 samples/hr can be run this way, although it would take a superperson to keep up such a pace and a team of people to keep up with the data. Commercial instruments of this type are equipped with a gas chromatograph, making them a GC separator, followed by a mass spectrometer separator. The mass spectrometer separator can be programmed to focus on a different mass for each GC peak or even several mass peaks for each GC peak until either the computer or the operator are overwhelmed. The MS-MS has been stressed as an analytical tool for the rapid analysis of samples with essentially no sample preparation. Its limitations are that no complete mass spectral record of the sample is available as with GC-MS, and the researcher must know what compound is being sought so as to set the correct mass for the first mass spectrometer. MS-MS is very useful in the study of the mechanism of the formation of mass peaks and fragmentation pathways.

IV. QUANTITATIVE ANALYSIS

There are many factors which limit the accuracy of the tandem gas chromatograph–mass spectrometer. The injection of sample into the gas chromatograph is always difficult to reproduce; changes can occur in capillary column injector split ratios, and fractionation of the sample can occur at the splitter. All mass spectrometers are subject to mass discrimination, with the sensitivity varying with mass number. The mass sensitivity of quadrupoles is particularly affected by instrument tuning Ionization source efficiency varies with source cleaniness, and the fragmentation patterns vary with source temperature. The intensity of mass peaks is a direct function of the mass spectrometer multiplier gain, which is difficult to control with precision. With the use of external standards and great care to control the above-mentioned problems, accuracies of a few percentage points can be obtained. For analyzing casually sampled contaminated feedstuffs, this accuracy is probably considerably better than the sampling error involved.

The use of internal standards eliminates almost all of the above problems. A good internal standard would be the mycotoxin of interest labeled with six or more deuterium atoms. A known amount of the labeled mycotoxin is then added to the unknown sample at as early a stage in the procedure as possible. The internal standard suffers all of the same losses and enjoys all of the same gains as the unknown. Since the weight of the internal standard is known, the weight of the unknown is simply this known weight times the ratio of the height of the unknown peak to the height of the labeled internal standard peak. Relative standard deviations of 0.01 can be expected at the microgram level with the use of internal standards and SIM. Trichothecenes are relatively difficult compounds to label.

V. MYCOTOXINS

Toxins produced by *Fusarium,* especially the trichothecenes and zearalenone, have been studied intensely. These mycotoxins are important because they occur naturally in cereal grains throughout the world and have the capability to bring about pharmacological effects such as estrogenic and anabolic activity for zearalenone (Pathre and Mirocha 1976), and refusal and vomiting for trichothecenes (Vesonder and Hesseltine, 1981). The *Fusarium* toxins have been the target of GC-MS programs aimed at the development of methodology for analyses of these substances, both underivatized and as derivatives, such as TMS, heptafluorobutyrate (HFB), and trifluoroacetate (TFA). More than 40 trichothecenes have been reported to be produced by the *Fusarium* species, further complicating the selection of a method to detect the trichothecenes likely to occur. GC-MS can also be applied to the detection and identification of metabolites of trichothecenes and zearalenone, which are apparently metabolized quickly and are contained in excreta, blood, and tissues from poultry, small laboratory animals, and farm animals fed grains containing these toxins (Yoshizawa *et al.,* 1980; Pathre *et al.,* 1980).

Self (1979) reviewed 120 articles appearing in the food literature for mycotoxins, nitrosamines, polyhalogenated hydrocarbons, and other industrial contaminants in which MS has been used to measure low levels of toxic components. It has been pointed out that even though many mycotoxins are lactones or substituted coumarins, the lack of strict chemical classification causes difficulties in designing methods for assaying this group. For instance, the analyses of aflatoxins B_1, B_2, G_1, G_2, and M_1 by GC-MS has been impossible because these toxins either could not be chromatographed on packed and tubular capillary columns, or easily derivatized to make them suitable for GC. Therefore, aflatoxins were analyzed by direct-insertion probe (DIP) where the sample is volatilized directly into the high vacuum of the mass spectrometer (Haddon *et al.*, 1971, 1977). However, Friedli (1981) showed aflatoxin B_1 standard was assayable by a fused-silica capillary GC-MS direct-coupling technique. This technique allows transfer of sample from the point of injection to the ion source of a mass spectrometer. The utility of the technique is for GC-MS analyses of labile compounds. Other MS methods that have been used, include CI and field desorption techniques, which are soft ionization techniques emphasizing the molecular ion and minimizing fragmentation of the molecule (Stahr *et al.*, 1979). Capillary columns have been used to increase the separation of *Fusarium* toxins from one another and other contaminants with improved sensitivity and reproducibility (Szathmary *et al.*, 1980). Brumley *et al.* (1982) has since described negative-ion chemical ionization (NICI)-MS of trichothecenes with the use of OH⁻ as the reagent ion.

C. J. Mirocha *et al.* describe the identification of trichothecenes on the basis of their TMS, TFA, and HFB derivatives in the following chapter. Many of these are metabolic products of T-2 toxin isolated from chickens, rats, and dairy cattle. These products of metabolism found in the excreta are formed quickly and may have value in diagnosis of trichothecene intoxication. Extensive literature is available on *Fusarium* metabolites, but this review will list only literature that contains information on the use of GC-MS in the determination of mycotoxins.

A. Trichothecenes

Trichothecenes are a group of sesquiterpenoids, characterized as 12,13-epoxytrichothec-9-enes. They comprise more than 65 biologically active secondary metabolites. More than half of these are produced by the *Fusarium* sp. (Vesonder and Hesseltine, 1981), and their occurrence in North America has been summarized (Vesonder, 1983). Trichothecenes and their analyses have been reviewed by Pathre and Mirocha (1977), Scott (1982), and Eppley (1979).

Due to the limited availability of trichothecenes, and because only deoxynivalenol, diacetoxyscirpenol, T-2 toxin, and nivalenol occur naturally in cereal grains, only these four have been studied in detail. The trichothecenes are divided into four groups: Group A has a functional group other than a ketone at carbon-8 (Fig. 1); group B contains a ketone group at carbon-8 (Fig. 2); group C contains those trichothecenes having an epoxide between carbon-7 and carbon-8; group D has a macrocyclic ring between carbon-4 and carbon-15. Group A is further divided into group A_1, which are those trichothecenes with an oxygen

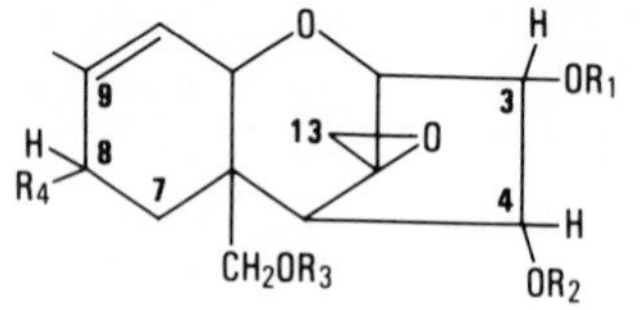

A_1 R_4- Oxygenated(hydroxyl group; ester)

Compound	R_1	R_2	R_3	R_4
T-2 Toxin	H	Ac	Ac	$O-C(=O)-CH_2CH(CH_3)_2$
HT-2 Toxin	H	H	Ac	$O-C(=O)-CH_2CH(CH_3)_2$
Acetyl-T-2 Toxin	Ac	Ac	Ac	$O-C(=O)-CH_2CH(CH_3)_2$
T-2-Tetraol	H	H	H	OH
TC-1	Ac	Ac	Ac	$O-C(=O)-CH_2-CH(OH)(CH_3)_2$
TC-3	H	Ac	Ac	$O-C(=O)-CH_2-CH(OH)(CH_3)_2$
Neosolaniol	H	Ac	Ac	OH
4,8-Diacetoxytrichothecene-3,15-diol	H	Ac	H	OAc
Neosolaniol monoacetate	H	Ac	Ac	OAc
Trichothecene-4,8-diol	a	H	a	OH

A_2 R_4- (nonoxygenated)

Compound	R_1	R_2	R_3	R_4
Diacetoxyscirpenol	H	Ac	Ac	H
Monoacetoxyscirpenol	H	H	Ac	H
Scirpentriol	H	H	H	H
4-Acetoxyscirpen-3,15-diol	H	Ac	H	H
Roridin-C	a	H	H	H
Trichodermin	a	Ac	H	H
Diacetyiverrucarol	a	Ac	Ac	H
Calonectrin	Ac	a	Ac	H
Deacetylcalonectrin	Ac	a	H	H

[a] No Oxygen; H_2

Ac = $CH_3C{=}O$

Fig. 1. Group A trichothecenes.

function at carbon-8 (e.g., hydroxyl group), and A_2 trichothecenes, which do not have an oxygen functional group at carbon-8.

GC-MS used in the mode where the full mass spectrum is recorded is sufficiently sensitive for surveying feedstocks for trichothecenes. As discussed in Section III,B, SIM of mass peaks diagnostic for the compound of interest increases the sensitivity as well as the precision. However, the samples analyzed by this method must be sufficiently purified, since many substances may have the same mass peak as that of the monitored mass ion. Mirocha *et al.* (1976) found that 1-glycerol monolinoleate and 1-glyceryl monooleate cochromatographed with T-2 toxin, and Collins and Rosen (1979) reported that mixed diglycerides interfered in the determination of T-2 toxin in milk. Visconti and Palmisano (1982) reported on a deoxynivalenol-like substance produced by *Fusarium solani*, in which its TMS derivative showed peaks in the mass spectrum of m/z 512, 422, and 235 characteristic for deoxynivalenol. Vesonder *et al.* (1981) also reported a similar observation with a compound from *Fusarium* that gave a 512 molecular ion for its TMS derivative. The trichothecenes are converted to specific derivatives for GC-MS analyses (TMS, HFB, and TFA), and each has its own diagnostic characteristic fragmentation ions and retention time. Pathre and Mirocha (1977) indicate sensitivity of 0.03 μg for T-2 toxin as its TMS derivative in normal scanning mode and 0.002 μg for SIM; 0.03–0.05 μg for neosolaniol and 0.002 μg for SIM; 0.03–0.05 μg for diacetoxyscirpenol and 0.015 μg for SIM; 0.03–0.05 μg for monoacetoxyscirpenol and 0.015 μg for SIM; 0.02 μg for deoxynivalenol and 0.007 μg for SIM. Yoshizawa *et al.* (1980) have documented the presence of 11 metabolites of T-2 toxin in broiler chickens by the use of GC-MS. Specific products of the metabolism of T-2 and HT-2 toxin were identified as T-2 tetraol and neosolaniol, and four other unidentified metabolic products were isolated from excreta of chickens. The unidentified products may have undergone chemical transformations other than deacetylation reactions. Yoshizawa *et al.* (1982) identified metabolites, TC-1 as 3'-

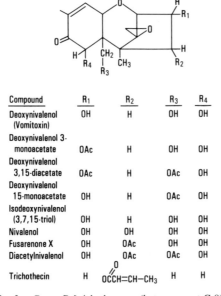

Compound	R_1	R_2	R_3	R_4
Deoxynivalenol (Vomitoxin)	OH	H	OH	OH
Deoxynivalenol 3-monoacetate	OAc	H	OH	OH
Deoxynivalenol 3,15-diacetate	OAc	H	OAc	OH
Deoxynivalenol 15-monoacetate	OH	H	OAc	OH
Isodeoxynivalenol (3,7,15-triol)	OH	H	OH	OH
Nivalenol	OH	OH	OH	OH
Fusarenone X	OH	OAc	OH	OH
Diacetylnivalenol	OH	OAc	OAc	OH
Trichothecin	H	OCCH=CH–CH₃	H	H

Fig. 2. Group B [trichothecenes (keto group at C-8)].

hydroxy-T-2 and TC-3 as 3'-hydroxy-HT-2 toxin (Fig. 1) by GC-MS as their TMS ether derivatives. Diagnostic fragmentation ions are listed in Table I.

Of the trichothecenes, it appears that deoxynivalenol (vomitoxin) has the distinction of being the most widespread in cereal grains in North America and throughout the world. Many investigators have used GC-MS in the EI mode for unambiguous detection and identification of deoxynivalenol in grains. Trenholm *et al.* (1983) reported on GC-MS analysis of TMS derivatives for deoxynivalenol, diacetoxyscirpenol, neosolaniol, and T-2 and HT-2 toxin, and found deoxynivalenol in 1980 Ontario white winter wheat. Scott *et al.* (1981) used single-ion monitoring at m/z 884 (Table III) for confirmation of deoxynivalenol as its HFB derivative.

Chaytor and Saxby (1982) developed a GC-MS method for the analysis of T-2 toxin in corn as its TMS derivative by monitoring the ions at m/z 350 and 346, with a detection limit of 5 ppb. Cohen and Lapointe (1982) determined vomitoxin in cereal grains using capillary GC with a limit of detection of 50 ppb. The method has been applied to wheat, oats, barley, and corn. Confirmation of vomitoxin was by GC-MS as its HFB ester using EI ionization and single-ion monitoring at m/z 884.05. Collins and Rosen (1979) determined T-2 toxin in milk as its TMS derivative by EI and CI, using selected-ion monitoring. Using GC-MS, Rosen and Rosen (1982) analyzed samples of yellow rain, an alleged biological warfare agent, used to expose populations to mycotoxins in Southeast

TABLE I

Selected-Ion Monitoring of Type A_1 Trichothecenes as Trimethylsilyl Ether (TMS) Derivatives and as Deuterated TMS Derivatives, and Bisheptafluorobutyrate

Compound	M^+	Ion (m/z)	Ion species	Reference
T-2 toxin TMS		436	M-Isovaleric	Rosen and Rosen (1982)
		350	$[\text{M-CH}_2 = \text{CH} - \text{O}_2$	
		122	$\text{C-C}_4\text{H}_9 - \text{HOAc}]^+$	
		85		
			Indicates $[\text{CH}_3]_2\ \text{CH} - \text{CH}_2 - \text{CHO}^+$	
T-2 toxin ($^2\text{H}_9$) TMS		445	M-Isovaleric	Rosen and Rosen (1982)
		359	$[\text{C-C}_4\text{H}_9 - \text{HOAc}]^+$	
HT-2 toxin TMS		466	M-Isovaleric	Rosen and Rosen (1982)
		347	466-HOAc$-$OAc$^+$	
		85	$[\text{CH}_3]_2\ \text{CH}-\text{CH}_2-\text{CHO}^+$	
T-2 toxin HFB	884			Scott *et al.* (1981)
TC-1 TMS	626	436	Indicates 7-9 diene	Yoshizawa *et al.* (1982)
		466		
TC-3 TMS	656	83	Indicates $[\text{CH}_2 = \text{C(CH}_3) - \text{CH}_2\text{CO}]$	

TABLE II

Selected-Ion Monitoring of Type A_2 Trichothecenes as
Trimethylsilyl Ether (TMS) Derivative and as Deuterated
TMS Derivative

Compound[a]	M^+	Ion (m/z)	Ion species	Reference
DAS-TMS		378	$[M-HOAc]^+$	Rosen and Rosen (1982)
		350	$[387-C_2H_4]^+$	
		290	$[350-HOAc]$	
DAS ($^2H_9$ TMS)		387	$[M-HOAc]^+$	Rosen and Rosen (1982)
		359	$[387-C_2H_4]^+$	
NEO TMS		466		
		436		
		350		
		193		
		117		
DAS HFB	502			Scott et al. (1981)

[a]DAS, Diacetoxyscirpenol; NEO, neosolaniol.

Asia, for the presence of T-2 toxin, HT-2, diacetoxyscirpenol, nivalenol, deoxynivalenol, and zearalenone as their TMS derivatives. Selected ions for monitoring of type A_1 trichothecenes (Fig. 1) are listed in Table I, type A_2 trichothecenes in Table II, and type B trichothecenes (Fig. 2) in Table III.

Gilbert et al. (1983) analyzed extracts from United Kingdom-grown barley and imported corn by GC-MS for deoxynivalenol as its TMS derivative (selected-ion monitoring) as m/z 512. In this same study, deoxynivalenol was added to flour and baked. About 50% of the deoxynivalenol was recovered; the remainder was involved in decomposition products generated by heat and was not identified. Stahr et al. (1979) used GC-MS by EI ionization for the identification of underivatized trichothecenes.

TABLE III

Selected-Ion Monitoring of Type B Trichothecenes as Trimethylsilyl
Ether (TMS) Derivative and Deuterated TMS Derivative

Compound[a]	M^+ (molecular ion[+])	Ion (m/z)	Ion species	Reference
DON-TMS	512	513	$[M + 1]^+$	Rosen and Rosen (1982)
DON-($^2H_{27}$)TMS	539	521	$[M-C^2H_3]^+$	Rosen and Rosen (1982)
DON-trisheptafluorobutyrate	884			Scott et al. (1981)

[a] DON, Deoxynivalenol.

B. Zearalenone

Zearalenone is an estrogenic substance elaborated mainly by *Fusarium* members of the discolor section, especially *Fusarium graminearum* (perfect stage *Gibberella zeae*). It is also produced by *F. sporotrichiodies* and *F. oxysporum* (Christensen, 1979; Hesseltine, 1974), and has been reported to be produced in small amounts by some isolates of *F. moniliforme* (Mirocha *et al.*, 1969).

Zearalenone, a resorcylic acid lactone [6-(10-hydroxy-6-oxo-*trans*-1-undecenyl)-β-resorcylic acid lactone], is also referred to as F-2 and RAL. It generally is not lethal to the consuming animal but it does induce hyperestrogenism exhibited by swelling and reddening of the vulva, increase in the size of the uterus, and growth and lactation in the mammary glands. This mycotoxin is found to occur naturally in corn and in a wide variety of agricultural commodities. Information on the natural occurrence of zearalenone and assay methods for zearalenone can be found in several review articles (Shotwell, 1977; Bennett and Shotwell, 1979; Mirocha *et al.*, 1977; Yoshizawa, 1983). Several zearalenone derivatives have been elaborated by *F. graminearum* in cultures in the laboratory (Fig. 3), and recently Richardson *et al.* (1985) reported cis-zearalenone, cis-α-zearalenol, cis- and trans-β-zearalenol, and α- and β-zearalenol were produced by *Fusarium* spp. in rice cultures. However, only zearalenone and trans-α-zearalenol have been found to occur naturally in cereal grains (Christensen, 1979) and the α- and β-isomers of zearalenol in corn stems from southern Italy (Bottalico *et al.*, 1985).

Mirocha *et al.* (1974) described methods for quantitation of zearalenone in a number of grains and feedstuffs that employed either thin-layer chromatography

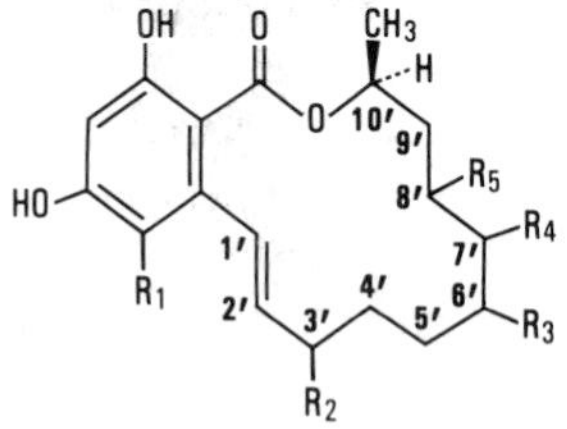

	R_1	R_2	R_3	R_4	R_5
Zearalenone	H	H_2	=O	H_2	H_2
Zearalenol	H	H_2	OH	H_2	H_2
8'-Hydroxyzearalenone	H	H_2	=O	H_2	OH
3'-Hydroxyzearalenone	H	OH	=O	H_2	H_2
7'-Dehydrozearalenone	H	H_2	=O	H	H
6',8'-Dihydroxyzearalene	H	H_2	OH	H_2	OH
5-Formylzearalenone	CHO	H_2	=O	H_2	H_2

Fig. 3. Zearalenone and its derivatives.

(TLC), GC, ultraviolet (UV) spectroscopy, or GC-MS, or a combination of these. In the GC-MS analysis, zearalenone was converted to its di-TMS ether derivative, and the diagnostic ions at m/z 462 (M$^+$), 477, 444, 429, 333, 305, 260, and 151 (base) were monitored. Shotwell *et al.* (1976) confirmed the presence of zearalenone as its TMS derivative in extracts from wheat, monitoring the ions as Mirocha. Scott *et al.* (1978) detected low levels of zearalenone in cornflakes and products made from corn for human consumption by GC-MS through monitoring the parent ion m/z 462. In this same study, 5 ppb of zearalenone was determined by use of a fluorescence detector with high-pressure liquid chromatography (HPLC). In other studies, HPLC was used to quantitate zearalenone in corn and feedstuffs at levels of 10 ppb (Ware and Thorpe, 1978; Holder *et al.*, 1977). Eppley (1968) reported on a multiple screening for detection of zearalenone, aflatoxin, and ochratoxin using TLC. This method has been evaluated in a collaborative study by Shotwell *et al.* (1976), and has been accepted in official first action by the Association of Official Analytical Chemists (AOAC) and the American Association of Cereal Chemists (AACC). The level of detection in corn was 0.5 ppm.

Mirocha *et al.* (1982) determined the metabolic products of zearalenone by GC-MS and radioimmunoassay in excreta and edible tissues of broiler chickens fed feed amended with 100 ppm zearalenone for 8 days prior to intubation with [^{3}H]zearalenone. Zearalenone and α- and β-zearalenols were detected mainly in the excreta at 39 to 43.5 ppb at 48 hr, and in the liver at concentrations ranging from 17.3 to 254.3 ppb when measured by GC-MS. The concentration of zearalenone in the muscle ranged from 23 to 25 ppb at 0.5 hr to 4 ppb at 48 hr. It was suggested that analysis of chicken broiler feces for zearalenone could be an indicator of that type of intoxication, since feed sample analyses may be an unreliable indicator for mycotoxin intoxication because of the inherent errors in sampling.

C. Other Mycotoxins

Aflatoxins are often found as natural contaminants on agricultural commodities. In the United States, since aflatoxins are carcinogenic, they are routinely analyzed for on corn, peanuts, and cottonseeds. Regulatory guidelines have been set by the Food and Drug Administration. Methods for analysis have been reviewed by Dr. O. Shotwell in this book. A reliable confirmatory method is highly desirable for the aflatoxins found in agricultural products. The advent of fused-silica capillary GC-MS direct-coupling technology (Friedli, 1981) permits the analysis of labile substances such as aflatoxin B$_1$. Rosen *et al.* (1984) have applied this technology with medium-resolution selected-ion monitoring for the conformation of aflatoxin B$_1$ and B$_2$ in peanuts. The extract from the peanut sample used for TLC quantitation is eluted through a silica gel Sep-Pak cartridge and analyzed by GC-MS selected-ion monitoring at m/z 312.063 and 314.079 at 30,000 resolution for aflatoxin B_1 and B_2 respectively. Limits of detection for these aflatoxins were 0.1 ppm. Trucksess *et al.* (1984) used this same GC-MS technique for the conformation of aflatoxin B_1 contained in the extract from 3-g samples (corn, peanut butter: 20 ng/g). The column temperature of the gas chromatograph was raised to 250°C after injection and the extract analyzed by NICI-MS. The NICI mass spectrum of aflatoxin B_1 showed major ions of m/z 312.34 and 297.

Patulin, a lactone metabolite of several species of *Penicillium* and *Aspergillus*, is found to occur naturally on apples, bananas, pineapples, grapes, peaches, and apricots, and also is found in apple cider and apple juice (Ciegler *et al.*, 1983). The AOAC procedure for analyses of patulin in apple juice consists of ethyl acetate extraction, a silica gel column chromatography with solvent system benzene–ethyl acetate (75:25), followed by TLC (toluene–ethyl acetate–90% formic acid, 5:4:1). The limit of detection is 20 µg patulin per liter of apple juice.

Scott and Kennedy (1973) detected patulin in sweet apple cider by GC and confirmed its presence by MS. Rosen and Pareles (1974) used low-resolution SIM to estimate patulin at the 1-ppb level as its TMS derivative. Price (1979) made a comparison of quantitative MS methods for the analysis of patulin in apple juice, using high-resolution ion monitoring and low-resolution SIM. The former was more sensitive and reliably measured 0.2 ppb patulin in apple juice, whereas 5 ppb could be detected by SIM and 10 ppb was reliably measured. The perdeuterated analog of silylated patulin as an internal standard was found to be unsuitable, because the deuterated silyl group exchanged with the excess BSA reagent [N,O-bis(trimethylsilyl)acetamide].

Sterigmatocystin is a carcinogenic metabolite produced by various *Aspergillus* species. Salhab *et al.* (1976) developed a direct GC-MS technique for analyses of underivatized sterigmatocystin. The GC-MS system was utilized in the mass fragmentography mode, with SIM of the m/e 295, 306, and 324 peaks. Limit of detection for sterigmatocystin was 20 ng, but accurate quantitation was not feasible. However, 1 ppb of sterigmatocystin was detected in grains. The labile characteristics of trace amounts of sterigmatocystin (levels <5 ppb) under GC conditions restrict the detection limits.

VI. CONCLUSIONS

Gas chromatography–mass spectrometry provides the analyst with an excellent tool to study mycotoxins in research or in contaminated feedstuffs. Fused-quartz capillary column, full-scan GC-MS provides a complete recorded analysis of all significant components of very compex mixtures of contaminated feed-stuffs. If greater precision or sensitivity is desired, and the samples can be cleaned up, selected-ion monitoring with electron impact ionization or with chemical ionization can be used. If a great many samples must be analyzed for a list of specific compounds and it is desired to do no sample preparation, then MS-MS may be the best technique. Mycotoxins have been studied for a good many years by GC-MS, and many excellent sample-handling and analytical techniques have been developed for them. Since the GC-MS system is very specialized and costly, its availability may be limited as a routine method for analyses.

REFERENCES

Bennett, G. A., and Shotwell, O. L. (1979). Zearalenone in cereal grains. *J. Am. Oil Chem. Soc.* **56,** 812–819.

Bottalico, A., Visconti, A., Logerico, A., Solfrizzo, M., and Mirocha, C. J. (1985). Occurrence of zearalenols (diastereoisomeric mixture) in corn stalk rot and their production by associated *Fusarium* species. *Appl. Environ. Microbiol.* **49,** 547–551.

Brumley, W. C., Andrezjewski, D., Trucksess, M. W., Dreifuss, P. A., Roach, J. H., Eppley, R. M., Thomas, F. S., Thorpe, C. W., and Spohn, J. A. (1982). Negative ion chemical ionization mass spectrometry of trichothecenes: Novel fragmentation under OH⁻ conditions. *Biomed. Mass Spectrom.* **9,** 451–458.

Chaytor, J. P., and Saxby, M. J. (1982). Development of a method for the analysis of T-2 toxin in maize by gas chromatography-mass spectrometery. *J. Chromatogr.* **237,** 103–113.

Christensen, C. M. (1979). Conference on Mycotoxins in Animal Feeds and Grains related to Animal Health. *Food Drug Adm., Bur. Vet. Med. [Tech. Rep.] FDA/BVM (U.S.)* **FDA/BVM-79/139,** pp. 1–80.

Ciegler, A. C., Burmeister, H. R., and Vesonder, R. F. (1983). Poisonous fungi: mycotoxins and mycotoxicoses. *In* "Fungi Pathogenic for Humans and Animals" (D. H. Howard, ed.), p. 429. Marcel Dekker, Inc. New York, and Basel.

Cohen, H., and Lapointe, M. (1982). Capillary gas–liquid chromatographic determination of vomitoxin in cereal grains. *J. Assoc. Off. Anal. Chem.* **65,** 1429–1434.

Collins, G. J., and Rosen, J. D. (1979). Gas–liquid chromatographic/mass spectrometric screening method for T-2 toxin in milk. *J. Assoc. Off. Anal. Chem.* **62,** 1274–1280.

Eppley, R. M. (1968). Screening method for zearalenone, aflatoxin and ochratoxin. *J. Assoc. Off. Anal. Chem.* **51,** 74–78.

Eppley, R. M. (1979). Trichothecenes and their analysis. *J. Am. Oil Chem. Soc.* **56,** 824–829.

Friedli, F. (1981). Fused silica capillary GC/MS coupling: A new innovative approach. *HRC CC, J. High Resolut. Chromatogr. Chromatogr. Commun.* **4,** 459–499.

Gilbert, J., Shepherd, M. J., and Startin, J. R. (1983). A survey of the occurrence of the tri-chothecene mycotoxin deoxynivalenol (vomitoxin) in UK grown barley and in imported maize by combined gas chromatography-mass spectrometry. *J. Sci. Food Agric.* **34,** 86–92.

Haddon, W. F., Wiley, M., and Waiss, A. C., Jr. (1971). Aflatoxin detection by thin-layer chro-matography-mass spectrometry. *Anal. Chem.* **43,** 268–270.

Haddon, W. F., Masri, S. M., Randall, R. H., Elsken, R. H., and Meneghell, B. J. (1977). Mass spectral confirmation of aflatoxins. *J. Assoc. Off. Anal. Chem.* **60,** 107–113.

Hesseltine, C. W. (1974). Natural occurrence of mycotoxins in cereals. *Mycopathol. Mycol. Appl.* **53,** 141–153.

Holder, C. L., Nony, C. R., and Bowman, M. C. (1977). Trace analysis of zearalenone and/or zearalanol in animal chow by high pressure liquid chromatography and gas–liquid chromatography. *J. Assoc. Off. Anal. Chem.* **60,** 272–278.

McFadden, W. (1973). "Techniques of Combined Gas Chromatography/Mass Spectrometry." Wiley, New York.

McLafferty, F. W. (1966). "Interpretation of Mass Spectra." Benjamin, New York.

Mirocha, C. J., Christensen, C. M., and Nelson, G. H. (1969). Biosynthesis of the fungal estrogen F-2 and a naturally occurring derivative (F-3) by *Fusarium moniliforme. Appl. Microbiol.* **17,** 482–483.

Mirocha, C. J., Schauerhamer, B., and Pathre, S. V. (1974). Isolation, detection and quantitation of zearalenone. *J. Assoc. Off. Anal. Chem.* **57,** 1104–1110.

Mirocha, C. J., Pathre, S. V., and Behrens, J. (1976). Substances interfering with the gas–liquid chromatographic determination of T-2 mycotoxin. *J. Assoc. Off. Anal. Chem.* **59,** 221–225.

Mirocha, C. J., Pathre, S. V., and Christensen, C. M. (1977). Zearalenone. *In* "Mycotoxins in Human and Animal Health" (J. V. Rodricks, C. W. Hesseltine, and M. A. Mehlman, eds.), pp. 345–364. Pathotox, Park Forest South, Illinois.

Mirocha, C. J. Robison, T. S., Pawlosky, R. J., and Allen, N. K. (1982). Distribution and residue determination of [^{3}H] zearalenone in broilers. *Toxicol. Appl. Pharmacol.* **66,** 77–87.

Pathre, S. V., and Mirocha, C. J. (1976). Zearalenone and related compounds. *In* "Mycotoxins and Other Fungal Related Food Problems" (J. V. Rodrick, ed.), Advances in Chemistry Series, No. 149, pp. 178–227. Am. Chem. Soc., Washington, D.C.

Pathre, S. V., and Mirocha, C. J. (1977). Assay methods for trichothecenes and review of natural occurrence. *In* "Mycotoxins in Human and Animal Health" (J. V. Rodricks, C. W. Hesseltine, and M. A. Mehlman, eds.), pp. 229–253. Pathotox, Park Forest South, Illinois.

Pathre, S. V., Fenton, S. V., and Mirocha, C. J. (1980). 3′-Hydroxyzearalenones, two new metabolites produced by *Fusarium roseum. J. Agric. Food Chem.* **28,** 421–424.

Price, K. R. (1979). A comparison of two quantitative mass spectrometric methods for the analysis of patulin in apple juice. *Biomed. Mass Spectrom.* **6,** 573–574.

Richardson, K. E., Hagler, Jr., W. M., and Mirocha, C. J. (1985). Production of zearalenone, α- and β-zearalenol and α- and β-zearalenol by *Fusarium* spp. in rice culture. *J. Agric. Food Chem.* **33,** 862–866.

Rosen, J. D., and Pareles, S. R. (1974). Quantitative analysis of patulin in apple juice. *J. Agric. Food Chem.* **22,** 1024–1026.

Rosen, R. T., and Rosen, J. D. (1982). Presence of four Fusarium mycotoxins and synthetic material in 'yellow rain,' *Biomed. Mass Spectrom.* **9,** 443–450.

Rosen, R. T., Rosen, J. D., and Di Prossimo, V. P. (1984). Confirmation of aflatoxin B_1 and B_2 in peanuts by gas, chromatography/mass spectrometry/selected ions monitoring. *J. Agric. Food Chem.* **32,** 276–278.

Salhab, A. S., Russell, G. F., Coughlin, J. R., and Hsieh, D. P. H. (1976). Gas–liquid chromatography and mass spectrometric ion selective detection of sterigmatocystin in grains. *J. Assoc. Off. Anal. Chem.* **59,** 1037–1044.

Scott, P. M. (1982). Assessment of quantitative methods for the determination of trichothecenes in grains and grains products. *J. Assoc. Off. Anal. Chem.* **65,** 876–883.

Scott, P. M., and Kennedy, B. P. C. (1973). Improved method for the thin layer chromatographic determination of patulin in apple juice. *J. Assoc. Off. Anal. Chem.* **56,** 813–816.

Scott, P. M., Panalaks, T., Kanhere, S., and Miles, W. F. (1978). Determination of zearalenone in corn flakes and other corn-based foods by thin layer chromatography, high pressure liquid chromatography, and gas–liquid chromatography/high resolution mass spectrometry. *J. Assoc. Off. Anal. Chem.* **61,** 593–600.

Scott, P. M., Lau, P. Y., and Kanhere, S. R. (1981). Gas chromatography with electron capture and mass spectrometric detection of deoxynevalinol in wheat and other grains. *J. Assoc. Off. Anal. Chem.* **64,** 1364–1371.

Self, R. (1979). Recent developments in the estimation of trace toxic substances in food. *Biomed. Mass Spectrom.* **6,** 361–373.

Shotwell, O. L. (1977). Assay methods for zearalenone and its natural occurrence. *In* "Mycotoxins in Human and Animal Health" (J. V. Rodricks, C. W. Hesseltine, and M. A. Mehlman, eds.), pp. 403–416. Pathotox, Park Forest South, Illinois.

Shotwell, O. L., Goulden, M. L., and Bennett, G. A. (1976). Determination of zearalenone in corn: Collaborative study. *J. Assoc. Off. Anal. Chem.* **59,** 666–670.

Stahr, H. M., Kraft, A. A., and Schuh, M. (1979). The determination of T-2 toxin, diacetoxyscirpenol and deoxynevalenol in foods and feeds. *Appl. Spectrosc.* **33,** 294–297.

Szathmary, C., Galacz, J., Vida, L., and Alexander, G. (1980). Capillary gas chromatographic–mass spectrometric determination of some mycotoxins causing fusariotoxicosis in animals. *J. Chromatogr.* **191,** 327–331.

Trenholm, H. L., Cochrane, W. P., Cohen, H., Elliot, J. I., Farnworth, E. R., Friend, D. W., Hamilton, R. M. G., Standish, J. F., and Thompson, B. K. (1983). Survey of vomitoxin contamination of 1980 Ontario white winter wheat crop: results of survey and feed trials. *J. Assoc. Off. Anal. Chem.* **66,** 92–97.

Trucksess, M. W., Brumley, W. C., and Nesheim, S. (1984). Rapid quantitation and confirmation, thin layer chromatography, and gas chromatography/mass spectrometry. *J. Assoc. Off. Anal. Chem.* **67,** 973–975.

Vesonder, R. F. (1983). Natural occurrence in North America. *In* "Trichothecenes: Chemical, Biological, and Toxicological Aspects," (Y. Ueno, ed.), pp. 210–217. Kodansha, Tokyo.

Vesonder, R. F., and Hesseltine, C. W. (1981). Vomitoxin: natural occurrence on cereal grain and significance as a refusal and emetic factor to swine. *Process Biochem.* **16**(1), 12–15, 44.

Vesonder, R. F., Ellis, J. J., and Rohwedder, W. K. (1981). Swine refusal factors elaborated by Fusarium strains and identified as trichothecenes. *Appl. Environ. Microbiol.* **41,** 323–324.

Visconti, A., and Palmisano, F. (1982). Interference in the gas chromatographic determination of deoxynivalenol in cultures of *Fussarium solani* on corn. *J. Chromatogr.* **252,** 305–309.

Ware, G. M., and Thorpe, C. W. (1978). Determination of zearalenone in corn by high pressure liquid chromatography and fluorescence detection. *J. Assoc. Off. Anal. Chem.* **61,** 1058–1062.

Yoshizawa, T. (1983). Fusarium metabolites other than trichothecenes. *In* "Trichothecenes: Chemical, Biological, and Toxicological Aspects" (Y. Ueno, ed.), pp. 60–71. Kodansha, Tokyo.

Yoshizawa, T., Swanson, S. P., and Mirocha, C. J. (1980). T-2 metabolites in the excreta of broiler chickens administered ^{3}H-labeled T-2 toxin. *Appl. Environ. Microbiol.* **39,** 1172–1177.

Yoshizawa, T., Sakamoto, T., Ayano, Y., and Mirocha, C. J. (1982). 3'-Hydroxy T-2 and 3'-hydroxy HT-2 toxins: New metabolites of T-2 toxin, a trichothecene mycotoxin, in animals. *Agric. Biol. Chem.* **46,** 2613–2615.

13

Mass Spectra of Selected Trichothecenes

C. J. MIROCHA, S. V. PATHRE, R. J. PAWLOSKY, AND D. W. HEWETSON

Department of Plant Pathology
University of Minnesota
St. Paul, Minnesota 55108

The number of trichothecenes reported in the literature must number, at the time of this writing, ~100, and most have been isolated from cultures of *Fusarium* grown in the laboratory. The greatest application of gas chromatography–mass spectrometry (GC-MS) in the area of mycotoxin analysis has been in

353

MODERN METHODS IN THE ANALYSIS
AND STRUCTURAL ELUCIDATION
OF MYCOTOXINS

the analysis of the trichothecenes. It is our intention to list the mass spectra of a selected number of these metabolites based on the usefulness of the selected spectra to the analyst, that is, those metabolites most likely to be encountered in the majority of the analyses of environmental samples.

It is also our intention to assist the analyst by publishing the spectra of derivatives of trichothecenes usually encountered during thin-layer chromatography (TLC) or gas–liquid chromatography (GLC); these are the trimethylsilyl (TMS) ether, trifluoroacetate (TFA) esters, and heptafluorobutyrate (HFB) acetates. The mass spectra of the underivatized trichothecenes are already published in the *Handbook of Toxic Fungal Metabolites* (Cole and Cox, 1981) and the *Handbook of Mycotoxin Mass Spectra* (Dusold *et al.*, 1978) and Bamburg (1969). We have included the mass spectra in the electron impact (EI) and positive and negative chemical ionization (PCI, NCI) modes. Emphasis was placed on the PCI spectra (in methane) because they are more useful in qualitative identification of the trichothecenes than the EI mass spectra. The latter have numerous fragments (molecule undergoes extensive skeletal fragmentation) but contain few unique fragments in the higher mass region useful for identification. In contrast, the CI spectra have fewer fragments but they are more distinct for the purpose of identification; moreover, the CI spectra yield the molecular ion, whereas in EI the molecular ion is infrequently found and, if found, it is very weak. The next logical progression of mass spectra are those derived through tandem mass spectrometry–mass spectrometry (MS-MS) systems and made up of the daughter ions of the main diagnostic fragments. The art presently is not sufficiently developed to allow this type of spectral library system, but we are certain that in the near future, extensive libraries of such ions will be a reality.

I. INTRODUCTION

The 12,13-epoxytrichothecenes are a group of chemically related and biologically active secondary metabolites produced by various *Fusarium* species such as *F. tricinctum, F. sporotrichiodies, F. roseum, F. oxysporum, F. solani, F. nivale, F. lateritium, F. rigidiusculum,* and *F. episphaeria,* named according to the system of Snyder and Hanson. Not all of the trichothecenes are produced by *Fusarium,* as we have examples of *Cephalosporium crotocinigenum,* which makes crotocin, *Trichoderma viride* (trichodermin), and *Trichothecium roseum* (trichothecin). The macrocyclic trichothecenes (e.g., roridins and verrucarins) are produced by *Myrothecium verrucaria* and *Stachybotrys atra.* The latter are not the subject of this chapter.

The trichothecenes have been implicated in the toxicity to farm animals and also to humans. They are unique insofar as they are the most potent small molecules among natural products that inhibit protein synthesis in eukaryote cells. There have been some claims of their potential carcinogenicity, but no definitive experimentation has demonstrated this.

II. APPLICATION TO SELECTED TRICHOTHECENE MYCOTOXINS

Trichothecenes are sesquiterpenes and possess a tetracyclic 12,13-epoxytrichothec-9-ene skeleton. Their ring structure was first established in 1964 by X-ray crystallography of the *p*-bromobenzoate derivative of trichodermol (Godtfredsen and Vangedal, 1965; Abrahamsson and Nilsson, 1964). All naturally occurring trichothecenes have at least one hydroxyl group at C-3, C-4, C-7, and C-15, and can be conveniently divided into four major groups: (A) trichothecenes having a functional group other than a ketone at C-8 (T-2 toxin); (B) those with a ketone at C-8 (deoxynivalenol, DON); (C) those characterized by an additional epoxide (oxirane) in the A ring; and (D) those with a macrocyclic bridge linking C-4 and C-15 with an ester bond. Only groups A and B will be treated here.

All the spectra have been run on a Hewlett–Packard 5787A combination gas chromatograph–mass spectrometer. The conditions when run in the EI mode are as follows: The compounds elute off a DB-5 column directly inserted into the ionization source that has been tuned with perfluorotributylamine (PFTBA) at 70 eV. The source lenses and repeller voltages are adjusted so that a ratio of 100 : (50–80) : (2–4) are achieved for ions 69 : 219 : 502 of the tuning reagent. Source temperature is set to 200°C.

PCI conditions are as follows: The ionization source is manually tuned with PFTBA at 200 eV using methane as the reagent gas. The repeller and lenses are adjusted to produce the ratio of (20–40) : 100 : (10–20) for ions 219 : 414 : 614 for the tuning reagent. The methane source pressure is kept at 0.5 torr measured at the source. Source temperature is kept at 150°C.

NCI conditions are as follows: The source is manually tuned with PFTBA at 200 eV using methane. The lenses and repeller voltages are adjusted to produce a ratio of (10–20) : (60–80) : 100 for ions 557 : 595 : 633 of the PFTBA. The source pressure is kept at 0.3 torr of methane and the temperature at 150°C.

A. T-2 Toxin

T-2 toxin (3-hydroxy-4,15-diacetoxy-8-(3-methylbutyryloxy)-12,13-epoxytrichothec-9-ene, MW 466; $C_{24}H_{34}O_9$) is normally resolved as the TMS ether or the TFA ester on either packed or capillary columns. The derivative of choice because of ease of reactivity and diagnostic value of fragment ions is the TFA derivative. Moreover, the use of TFA-generating reagents, in comparison with TMS, help keep the MS source cleaner; however, the mass spectra of the TMS ethers of T-2 are included for comparison.

The TMS ether of T-2 in EI does not show the molecular ion at 538 and, although the derivative is easily made, the spectrum is not as useful as that of the TFA derivative. The most intense masses are found in the low-mass region where 350, 290, 185, 157, and 122 are the most intense. The sensitivity of the TMS ether is low (Fig. 1).

Analysis of the TMS ether in PCI yields a molecular ion of 539 (M + 1) with an intensity of approximately 15%. The base peak of 377, and m/z^+ of 523, 479, 437, and 347 are excellent fragment ions for identification and for quantitation. PCI has potential for qualitative analysis of T-2 toxin (TMS) which should be investigated in greater detail. Unlike most CI spectra, the TMS ether of T-2 has multiple high masses which can be very useful in detection.

Analysis of T-2 (TFA) by EI does not yield the molecular ion. The most diagnostic ions in the higher mass region are the M^+ (562) when present, 401, 327, and 205 (Fig. 2). Mass : charge ratios of 85, 121 (base peak), 138, and 180 are useful in the lower mass range. Iso-T-2 TFA toxin in EI has a fragmentation pattern similar to T-2 (Fig. 3).

T-2 (TFA) in PCI (methane) has a base peak of 401 and the M + 1 of 563 (~8%). In contrast, the 563 mass of iso-T-2 (TFA) is intense (~35%). The fragment m/z^+ 287 is absent in T-2 (TFA) but present in iso-T-2 (TFA), and is very intense (~60%). In addition, these isomers can easily be distinguished from each other by the difference in retention time on a capillary column.

An important metabolic derivative of T-2 is 3′-hydroxy-T-2 toxin (TC-1). Its mass spectrum in EI as the TFA derivative is similar to T-2 toxin (Fig. 4). Resolution of TC-1 (TFA) by GLC yields two resolvable isomers formed by differential dehydration of the hydroxyl on the isovaleroxy group on C-8. Analysis by PCI yields the M + 1 at 561 (in contrast to 563 for T-2) and a base peak at 401. TC-1 can be found in the urine and feces of animals metabolizing T-2 (Yoshizawa et al., 1981) and in some cultures of some Fusarium isolates that produce T-2.

The fragment ions of T-2 (TFA) in EI are shown in Fig. 2 found below the mass spectrum. The significant masses are formed by loss of the isovaleroxy group to yield 478, loss of water from the protonated oxirane oxygen to yield 460, and formation of the 401 fragment with the loss of an acetate group. In CI, m/z^+ 401 is the base peak and can be formed by two pathways of fragmentation as shown. The preferred scheme is loss of acetate from the M^+ followed by loss of the C-8 isovaleroxy group. The resulting 401 fragment is diagnostic for the T-2 series in CI. (See Addendum, p. 392.)

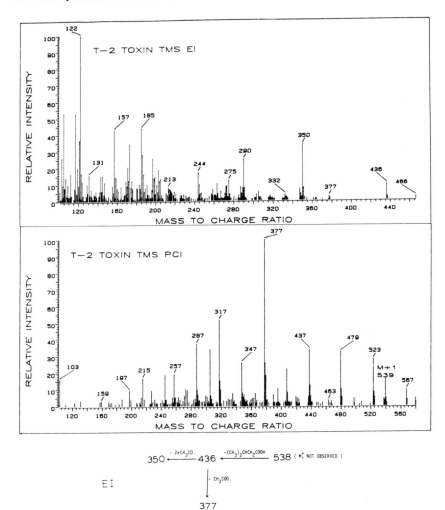

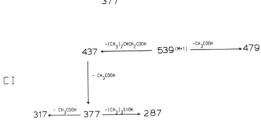

Fig. 1. EI and PCI mass spectra of the TMS ether of T-2 toxin. Fragmentation of T-2 in EI and CI is shown below.

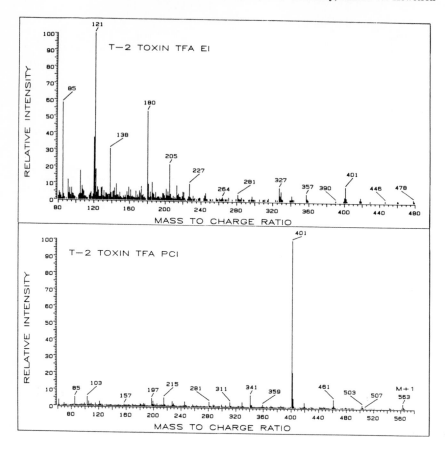

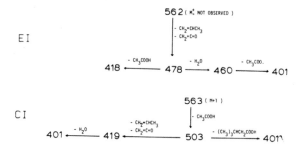

Fig. 2. EI and PCI mass spectra of the TFA ester of T-2 toxin. Origin of the major fragments is shown below.

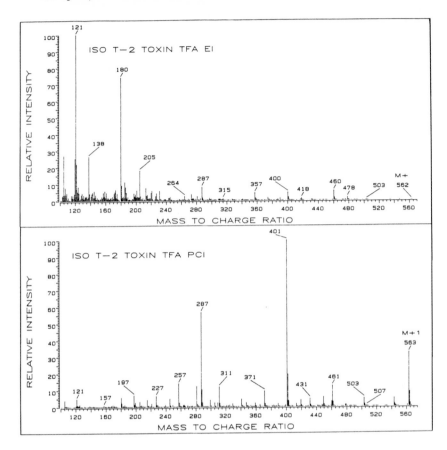

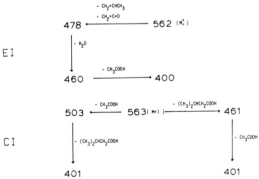

Fig. 3. EI and PCI mass spectra of the ester of iso-T-2 toxin. The major fragments are shown below.

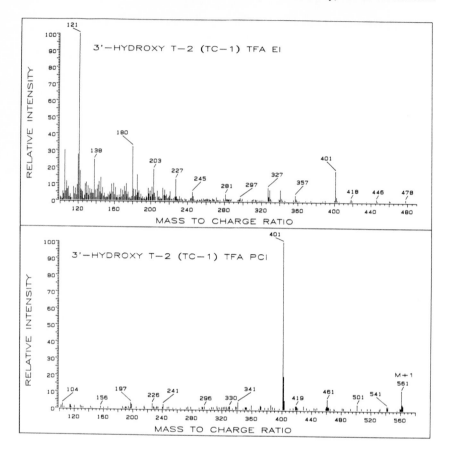

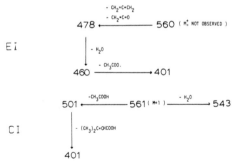

Fig. 4. EI and PCI mass spectra of the TFA ester of 3′-hydroxy-T-2 toxin (TC-1). The major fragments are shown below.

B. HT-2 Toxin

HT-2 (TFA, MW 424; $C_{22}H_{32}O_8$), like T-2, in EI does not normally produce a detectable molecular ion at 616 (Fig. 5). It is similar to T-2 in that it has m/z^+ 121 as the base peak and fairly intense fragments at m/z^+ 138, 180, and 205. It differs from T-2 in a diagnostic fragment at 455. The PCI of HT-2 (TFA) is most helpful in recognition because of a base peak of 455 and M + 1 of 617 (15–20%).

HT-2 (TFA) in EI has an important fragment (455) which is also present in PCI. It is formed by loss of the C-8 substituent followed by losses of water and an acetate group. In PCI, this fragment is postulated to be formed by a different mechanism, that is, loss of the isovaleric acid moiety followed by loss of an acetate group.

TC-3 (3'-hydroxy-HT-2 toxin) is comparable to the TC-1 derivative of T-2 toxin in that it yields two isomers when reacted with TFA anhydride and can be resolved by capillary GLC (Fig. 6). The base peak of TC-3 (TFA) like in HT-2 is 455, but in addition it has relatively strong fragments at 615 (M$^+$) and 515 which are definitive in identification. This metabolite can be found as a metabolic product in the urine and feces of animals (Yoshizawa *et al.*, 1981).

C. Neosolaniol

Neosolaniol (4,15-diacetoxy-3,8-dihydroxy-12,13-epoxytrichothec-9-ene; MW 382; $C_{19}H_{26}O_8$) is a metabolic product of T-2 toxin found in cultures of *Fusarium* as well as in excreta of chickens fed T-2 (Yoshizawa *et al.*, 1980). Its TFA derivative in EI normally does not yield a M$^+$, but it does have many diagnostic fragments in both the low- and high-mass regions. In PCI, neosolaniol (TFA) has a M + 1 of 575 (base peak) and a large fragment at 515. It is unusual for a trichothecene to display a M$^+$ as the base peak (Fig. 7).

The fragmentation scheme of neosolaniol in EI shows m/z^+ 401 (present in T-2). This ion is formed after a loss of a TFA from the parent molecule followed by a deacetylation. The mass 327 can also be accounted for by combination of losses of TFA and acetate groups. In PCI, the prominent mass 575 (M + 1) loses an acetate group to form the fragment 515 (~50%). Fragment 401 is formed by loss of a TFA group from 515. All of the other masses are minor in intensity and not useful in quantitation.

D. T-2 Tetraol

T-2 tetraol (MW 298; $C_{15}H_{22}O_6$) is one of the final reduction products of T-2 metabolism. Its mass spectrum in EI shows a base peak at 217, minor fragments at 330 and 455, and a major mass at 568 (~30%); usually the M$^+$ ion is not found (Fig. 8).

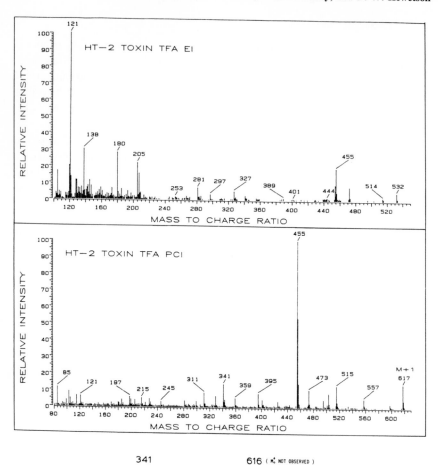

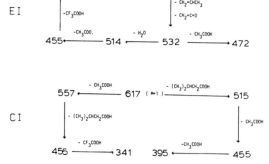

Fig. 5. EI and PCI mass spectra of the trimethylfluoroacetate ester of HT-2 toxin. The major fragments are shown below.

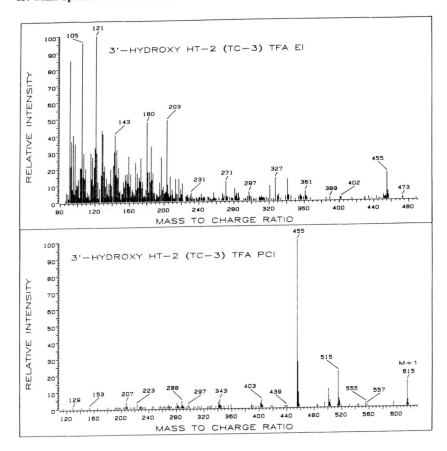

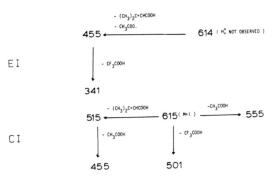

Fig. 6. EI and PCI of the TFA ester of 3'-hydroxy-HT-2 toxin. Major fragments are shown below.

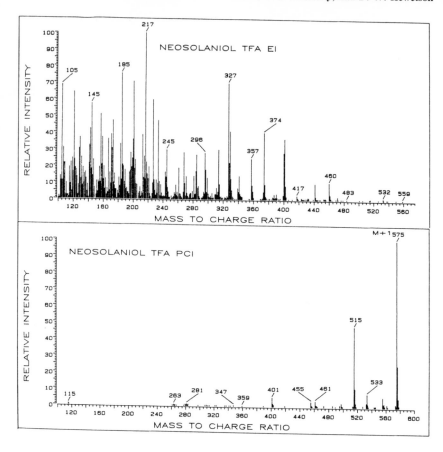

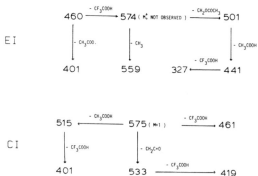

Fig. 7. EI and PCI mass spectra of the TFA ester of neosolaniol. The major EI and CI fragments are shown below.

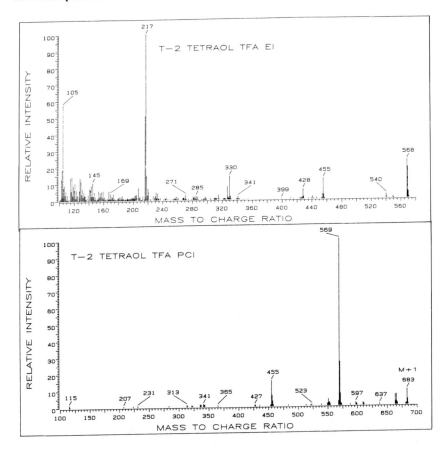

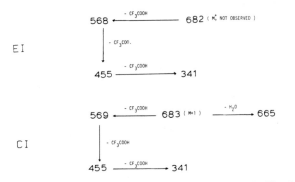

Fig. 8. EI and PCI mass spectra of the TFA ester of T-2 tetraol. The major EI and CI fragments are shown below.

In PCI, the base peak is found at 569 and the M + 1 at 683. The fragment 455 is present in both EI and CI of T-2 tetraol and is also an important diagnostic ion for HT-2 in PCI.

Fragment 568 in EI is formed by loss of TFA from the M^+, followed by an additional loss of TFA to yield 455. In CI, the M + 1 (683) loses a TFA group to form the base peak (569).

E. Deoxynivalenol

Deoxynivalenol (3,7,15-trihydroxy-2-deoxy-8-one-12,13-epoxytrichothec-9-ene; MW 296; $C_{15}H_{20}O_6$; DON), also known as vomitoxin, is one of the most common and abundant trichothecenes found naturally in the environment. It is found on cereal grains throughout the world, particularly on wheat barley and maize. This derivative occurs together with zearalenone on maize and at times on wheat. It is not as toxic as T-2 toxin, but it is partially responsible for causing farm animals to refuse their feed.

DON is closely related to nivalenol and fusarenone X; all three give a yellow-colored product when reacted with p-anisaldehyde on TLC plates. They belong to group B of trithothecenes (i.e., ketone substitution on C-8 and a hydroxyl group on C-7). Two derivatives of DON found both in culture and in the natural environment are 3-acetoxy-DON and 15-acetoxy-DON.

The derivatives of choice for resolution of DON on capillary columns are the TMS ethers and HFB esters. The TFA anhydride is not a useful reagent because it does not react with all of the hydroxyl groups, leaving a mixture of products.

The EI spectrum of DON (TMS) shows a fairly intense molecular ion at 512 and a (M − 15) at 497 (Fig. 9). There are many useful fragments (422, 393, 295, and 235) available for identification and quantitation of the molecule. The fragment 422 is formed by loss of 90 units (TMS group).

The PCI spectrum of DON (TMS) has 497 (M − 15) as its base peak and a strong molecular ion at M + 1. Fragmentation takes place similar to that in the EI spectrum (Fig. 10).

The NCI spectrum is entirely different from all others. Its base peak (298) is not due to demethylation as in the PCI spectrum but rather to some other mechanism. The M^- (512) is relatively weak. Analysis by NCI is the most sensitive (~20 pg) detection method. Unfortunately, only one reliable mass is available for quantitation.

The HFB derivative of DON (HFB-DON) is frequently used in electron capture detection and for this reason is included here. In PCI, the HFB-DON has a M + 1 at 885 (~25%). Fragments 215 and 671 lend themselves for quantitation and confirmation. They are especially useful in the selected-ion monitoring (SIM) mode, where quantitation can be evaluated at each mass value and compared with one another for agreement; however, m/z^+ 885 is preferred (Fig. 11).

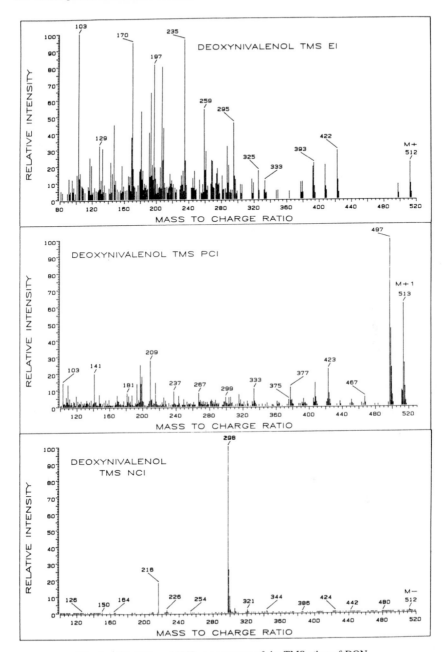

Fig. 9. EI, PCI, and NCI mass spectra of the TMS ether of DON.

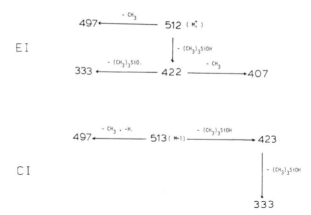

Fig. 10. Major fragments obtained in EI and CI analysis of DON.

The HFB derivative of DON in NCI has 214 as its base peak and the M − 1 of 884. The latter is an excellent fragment for monitoring because it is far enough upfield in the mass range so that very few substances will interfere with its analysis. Mass 884 is excellent for quantitation and is used for analysis by many laboratories for DON detection; it has limitations when used as a single fragment for identification.

F. 3-Acetoxy- and 15-Acetoxydeoxynivalenol

Important derivatives of DON found in cultures and in the environment are 3-acetoxy-DON and 15-acetoxy-DON. The latter, along with nivalenol, have been found in barley naturally infected with the *Fusarium* head blight organism in Japan. It is important to distinguish between these isomers, and hence their spectra are included here.

The TMS ether of 3-acetoxy-DON in EI has several fragments in the high-mass range useful for analyses (Fig. 12). They are 482 (M$^+$), 467 (M − 15), 392, 377, and 287. The fragment 377 is formed by loss of a CH_3 group from the parent ion followed by loss of a TMS group. It can be distinguished from the 15-acetoxy-DON (TMS) by the presence of m/z^+ 377 in the latter and its absence in the former, as well as by its retention time on a capillary column.

The TMS ether of 3-acetoxy-DON in PCI has 483 as its base peak and also its molecular ion (M + 1). The M − 15 (467) is very intense, and in this respect it is similar to the M − 15 in the PCI spectrum of DON. m/z^+ 393 is formed by loss of the TMS group (Fig. 13). The 3-acetoxy-DON isomer can be distinguished from 15-acetoxy-DON by masses 209, 237, 407, and 467 present in the latter but either missing in the former or being very low in abundance.

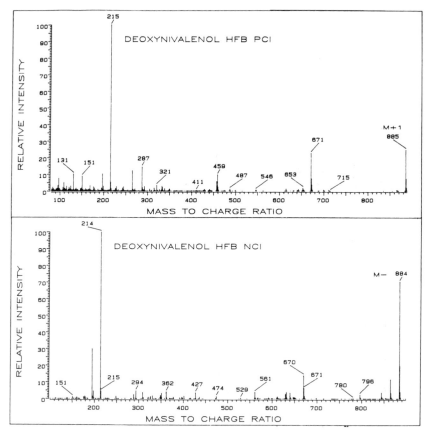

Fig. 11. NCI and PCI mass spectra of the HFB ester of DON.

The NCI spectrum of 3-acetoxy-DON (TMS) has only one useful fragment for quantitation; mass 481 (M + 1) serves as the base peak and molecular ion.

The EI spectrum of 15-acetoxy-DON (TMS) has a molecular ion at 482 and a base peak at 193 (Fig. 14). The M − 15 (CH$_3$) at 467 is useful in defining the molecular ion. Masses 392 and 350 are helpful in qualitative and quantitative analysis.

The PCI spectrum (Fig. 14) has a very intense M + 1 (483) and a M − 15 (467), which is also the base peak. In this spectrum, the molecular ion can be used for quantitation because of its intensity (70%). Refer to Fig. 15 for fragmentation scheme.

The NCI spectrum of the TMS ether of 15-acetoxy-DON displays fragment 481 (M − 1) as the molecular ion and base peak. Although NCI does not offer

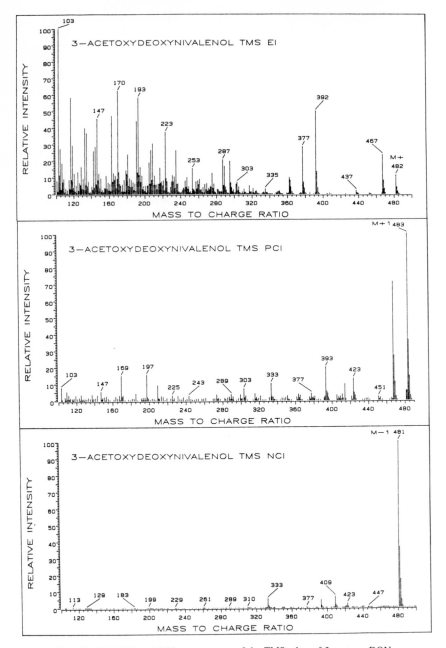

Fig. 12.   EI, NCI, and PCI mass spectra of the TMS ether of 3-acetoxy-DON.

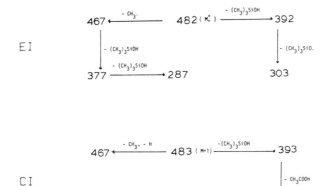

Fig. 13. Major fragments obtained from the EI and CI of the TMS ether of 3-acetoxy-DON.

much qualitative information, it is excellent for use in quantitation because of its sensitivity (Fig. 14).

The HFB derivative of 3-acetoxy-DON (TMS) in PCI (Fig. 16) has only one useful fragment, and that is M + 1 (731). Its NCI spectrum shows two fragments, 517 and M⁻ (730). The NCI spectrum is preferred.

The HFB derivative of 15-acetoxy-DON in PCI (Fig. 17) has a strong M + 1 at 731 (identical to 3-acetoxy-DON) and m/z^+ 517, which are useful for analysis. The NCI spectrum, like the PCI spectrum, has a very intense molecular ion and lends itself to both quantitative and qualitative analysis. In addition, masses 691, 519, and 459 are excellent for identification purposes. The NCI spectrum of the HFB derivatives is preferred for analysis.

G. Nivalenol

The TMS ether of nivalenol in EI is the derivative of choice, because all the hydroxyl groups react readily whereas reactivity with the TFA reagent is incomplete. The M⁺ at m/z^+ 600 is very weak but is substantiated by its M − 15 at 585 (Fig. 18). The most intense ions appear in the low-mass region, leaving the high-mass region relatively uninformative. Fragment 510 is formed after a loss of a dimethylhydroxysilyl group (75 mass units) from the 585 fragment.

The TMS ether of nivalenol in PCI has 585 as its base peak and a molecular ion at 601 (M + 1) formed by loss of a methyl group. The fragment 511 (25%) is helpful in recognition of the molecule and is formed by loss of 24 mass units from the base peak.

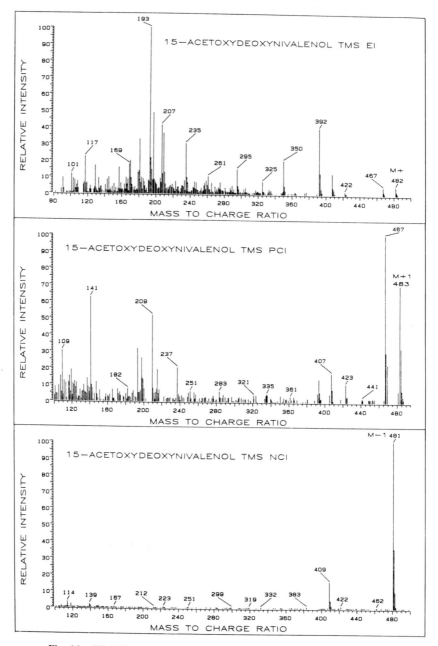

Fig. 14. EI, NCI, and PCI mass spectra of the TMS ether of 15-acetoxy-DON.

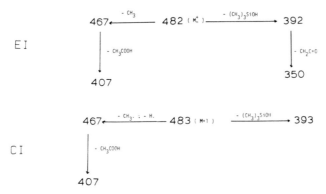

Fig. 15. Major fragments of the TMS derivative of 15-acetoxy-DON in EI and CI.

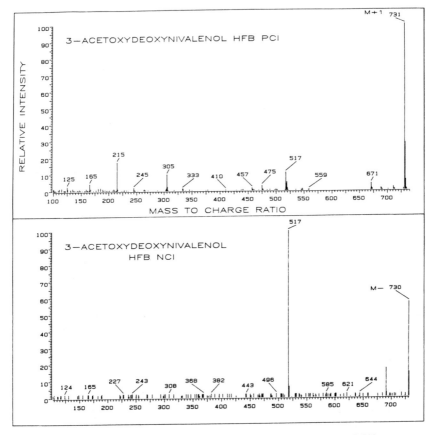

Fig. 16. PCI and NCI mass spectra of the HFB ester of 3-acetoxy-DON.

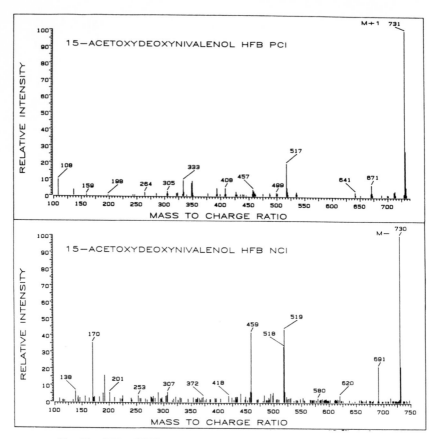

Fig. 17. PCI and NCI mass spectra of the HFB ester of 15-acetoxy-DON.

H. Fusarenone X

Fusarenone X reacts completely with the TMS reagent and in the EI mode of analysis possesses most of the intense fragments in the low-mass region (Fig. 19). Its molecular ion (570) is visible (6%), and its base peak is 170. Fragments 450, 480, 555, and the molecular ion at 570 are most useful for identification and quantitation. m/z^+ 480 (M − 90) is formed by a loss of a TMS group from the molecular ion.

The PCI spectrum in methane has 555 as its base peak, which is formed by loss of a methyl group from the molecular ion (571). These two masses plus 511 are excellent for both qualitative and quantitative analyses.

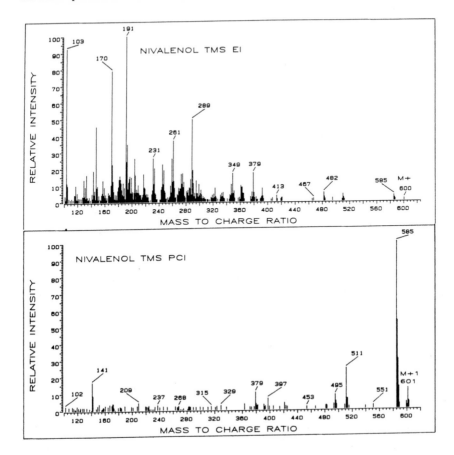

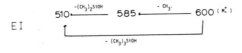

Fig. 18. EI and PCI, mass spectra of the TMS ether of nivalenol. The EI and CI major fragments are shown below.

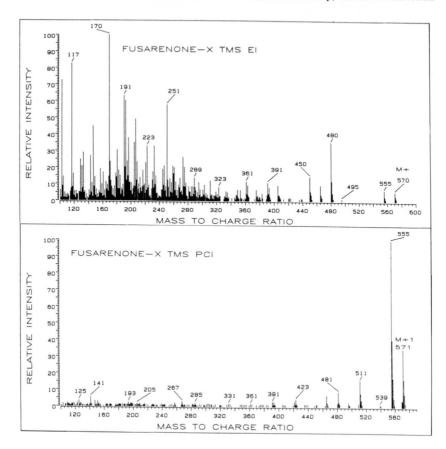

Fig. 19. EI and PCI mass spectra of the TMS ether of fusarenone X. The major fragments obtained in EI and CI are shown below.

I. Triacetoxyscirpene

Triacetoxyscirpene is produced by certain isolates of *Fusarium roseum* Graminearum, along with diacetoxyscirpenol and monoacetoxyscirpenol. It can be resolved on a capillary column (DB-5) by GLC without having to make a derivative.

Analysis by EI-MS does not reveal the molecular ion (408) but does show masses 348 and 320 in the high-mass range. Mass 348 is formed by loss of an acetate group. In the low-mass range, m/z^+ 105 (base peak) and 124 are prominent but not too useful (Fig. 20).

Analysis in PCI shows intense peaks at 349 (base peak) and 409 (molecular ion). Mass 349 is formed by loss of an acetate group. Other prominent ions in the high-mass region are 391, 331, 307, and 289. m/z^+ 391 is formed by loss of water from the M^+, and mass 289 is formed by loss of an acetate group from m/z^+ 349. Analysis of triacetoxyscirpene by PCI in methane is preferable to the use of the EI spectrum.

J. Diacetoxyscirpenol

Diactoxyscirpenol (MW 366.1671; $C_{19} H_{26} O_7$; DAS) is produced in cultures of certain isolates of *Fusarium roseum* Graminearum as three distinct isomers. The mass spectra of all three isomers are presented here in EI and PCI for comparison. All three isomers can be resolved by capillary GLC.

The TFA ester of 3,4-DAS in EI does not show the molecular ion (462) but does show a prominent ion at 402 formed by loss of an acetate group (Fig. 22). m/z^+ 374 is formed by loss of a methyl group from 462 followed by loss of a methyl acetate group. In PCI, 463 (M + 1) is intense (35%), followed by masses 403 (base peak), 385, 289, and 229 (Fig. 23). Mass 403 is formed by loss of an acetate group from 463. The PCI spectrum of 3,4-DAS is rich in diagnostic high masses and can be distinguished from both 3,15- and 4,15-DAS.

The isomer 3,15-DAS (TFA ester) in EI (Fig. 21) shows a weak molecular ion at 462 and a base peak at 91. Fragment 402 is present but much less intense than the 3,4-DAS isomer. Mass 447 is formed by loss of a methyl group from 462. The PCI spectrum (Fig. 23) of this isomer is very interesting. It has a very intense molecular ion (463) which is also its base peak. Masses 445, 403, and 343 are major and can be used for identification.

The 4,15-DAS (TFA) isomer in EI (Fig. 21) does not show the molecular ion at 462 and has very few fragments that are diagnostic. In contrast, its PCI spectrum (Fig. 23) has a base peak of 403 and a molecular ion (463) of ~10% intensity. Mass 463 is formed by loss of an acetate group followed by another acetate to yield 343 (Fig. 22).

C. J. Mirocha, S. V. Pathre, R. J. Pawlosky, and D. W. Hewetson

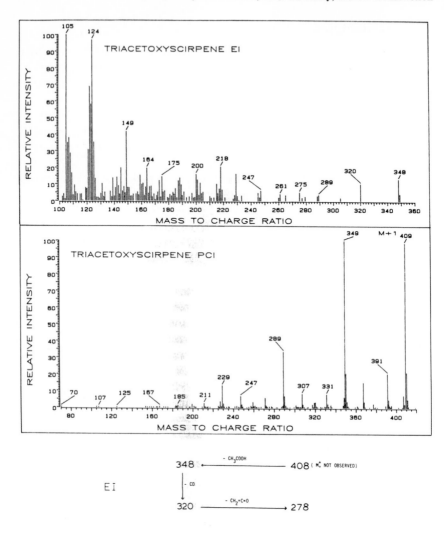

Fig. 20.   EI and PCI mass spectra of TAS. The major fragments obtained in EI and CI are shown below.

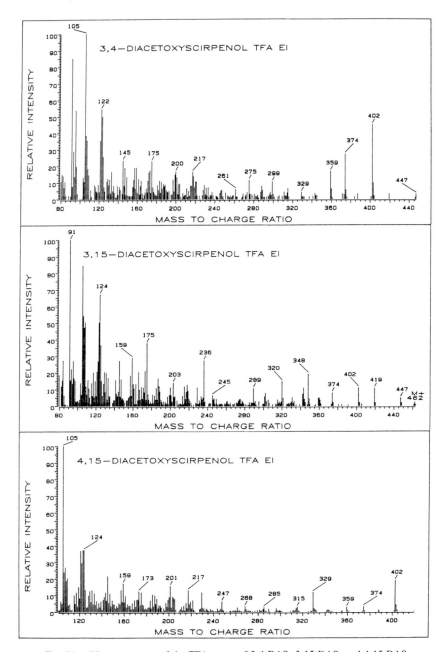

Fig. 21. EI mass spectra of the TFA esters of 3,4-DAS, 3,15-DAS, and 4,15-DAS.

3,4—DIACETOXYSCIRPENOL TFA

EI

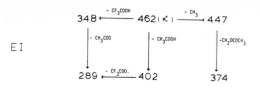

CI

3,15—DIACETOXYSCIRPENOL TFA

EI

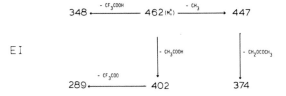

CI

4,15—DIACETOXYSCIRPENOL TFA

EI

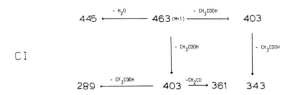

CI

Fig. 22. Major fragments obtained in EI and CI of 3,4-DAS, 3,15-DAS, and 4,15-DAS.

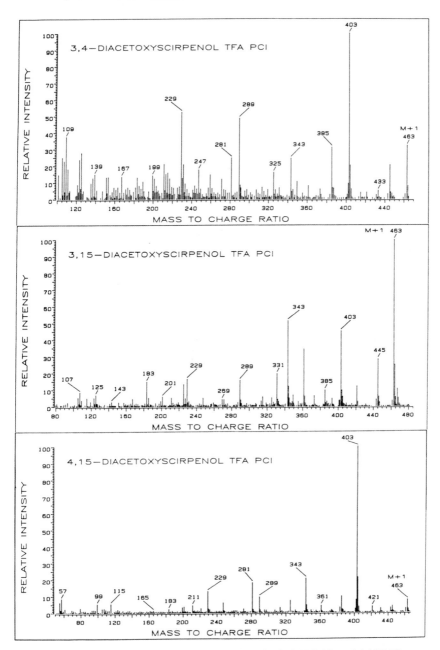

Fig. 23. PCI mass spectra of the TFA esters of 3,4-DAS, 3,15-DAS, and 4,15-DAS.

K. 7-Hydroxydiacetoxyscirpenol

This metabolite is synthesized by certain isolates of *Fusarium roseum* Graminearum when grown in laboratory culture on rice. It is usually a minor component of the culture medium and is accompanied by one or more of the isomers of DAS and monoacetoxyscirpenol (MAS).

The TFA derivative of 7-hydroxy-DAS when analyzed in EI (Fig. 24), gives very few useful masses in the high-mass region, whereas the lower region displays masses (105, 121, 185, and 207) characteristic of most trichothecenes. Characterization is best accomplished in PCI (methane), where the base peak is 401 and the molecular ion of 575 (M + 1) can be seen. Other prominent fragments in the high-mass region are 341, 461, and 515. Fragment 401 is characteristic of T-2 (TFA) in PCI.

L. Monoacetoxyscirpenol

Certain isolates of *Fusarium roseum* Graminearum produce isomers of MAS in addition to the three isomers of DAS. All of them can be resolved by capillary GLC and identified by MS. A comparison is made, where possible, between the EI and PCI spectra of the TFA derivatives.

One of the three isomers of this series is 3-MAS. When analyzed as its TFA derivative by EI, it shows a M^+ at 516 (very weak) and a prominent 473 fragment (Fig. 25). In PCI, this isomer shows a very strong M + 1 (517) with a base peak of 343 (Fig. 26). The PCI spectrum is most useful because it has numerous and intense fragments throughout the entire mass range. Mass 403 in PCI is formed by loss of a TFA group; loss of an additional acetate yields the base peak m/z^+ 343 (Fig. 27).

The TFA derivative of 4-MAS in EI (Fig. 25) is entirely different from 3-MAS because it has a base peak at 402 and strong fragment ions at 343, 374, and a M^+ at 516. The 4-MAS isomer in PCI (Fig. 26) also has strong fragment ions starting with the intense M + 1 (517) and the base peak at m/z^+ 403. Fragment 343 is almost as intense as 403 and at times may be the base peak, depending on the MS conditions and proper tuning of the source. In EI, mass 402 is formed by loss of an acetate group followed by a loss of another acetate to form m/z^+ 343. In PCI, mass 403 (base peak) and 343 is formed by successive losses of TFA groups (Fig. 27).

The isomer 15-MAS (Fig. 25), the most abundant of the three, does not show the molecular ion but displays prominent masses at 456, 428, 343, and 315. The PCI spectrum has two fragments but is diagnostic because of an intense M + 1 (517) and base peak of 457 (Fig. 26). Fragment 403 is weak compared to those found in the 3-MAS and 4-MAS isomers.

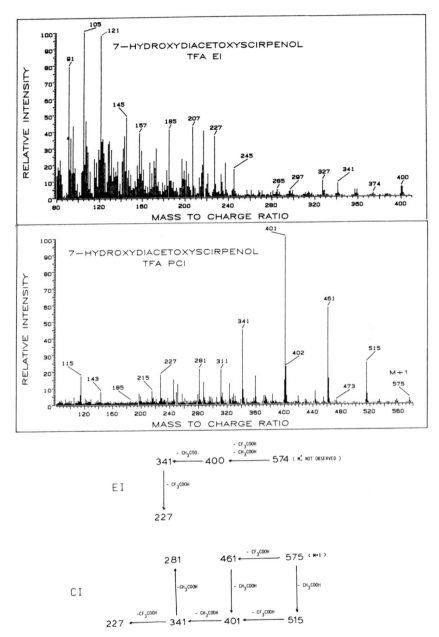

Fig. 24. EI and PCI mass spectra of 7-hydroxy-DAS. The major EI and CI mass fragments are shown below.

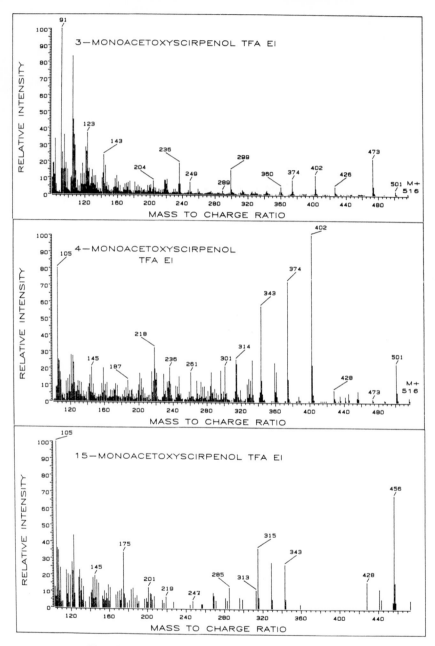

Fig. 25. EI mass spectra of 3-MAS, 4-MAS, and 15-MAS.

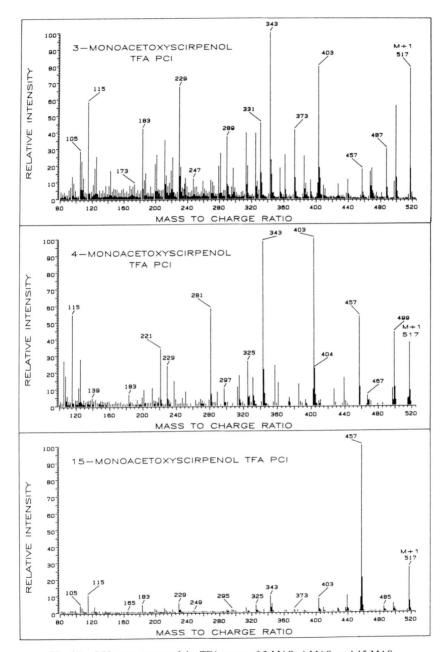

Fig. 26. PCI mass spectra of the TFA esters of 3-MAS, 4-MAS, and 15-MAS.

3−MONOACETOXYSCIRPENOL TFA

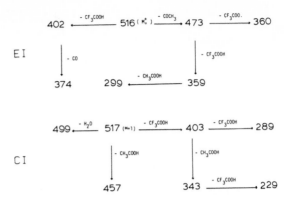

EI

CI

4−MONOACETOXYSCIRPENOL TFA

EI

CI

15−MONOACETOXYSCIRPENOL TFA

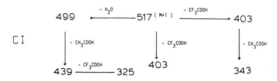

EI

CI

Fig. 27. Major fragments obtained in EI and C of 3-MAS, 4-MAS, and 15-MAS.

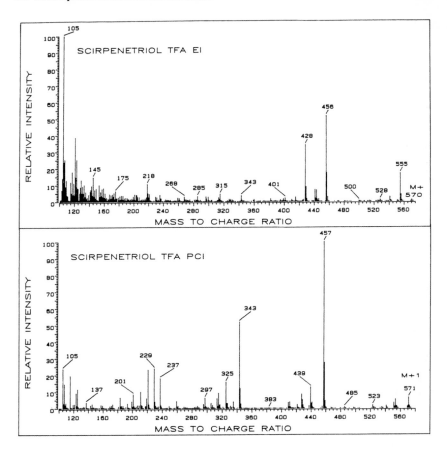

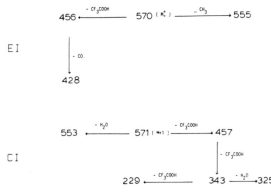

Fig. 28. EI and PCI mass spectra of the TFA ester of scirpenetriol.

M. Scirpenetriol

Scirpenetriol (3,4,15-trihydroxy-12,13-epoxytrichothec-9-ene; MW 282; $C_{15}H_{22}O_5$) is the ultimate reduction product of the metabolism of triacetoxyscirpene (TAS) by animals. It is much less toxic than its parent compounds TAS, DAS, or MAS.

In EI, the TFA derivative of scirpenetriol displays a very weak molecular ion at 570 but has a prominent M − 15, m/z^+ 456, and 428. The fragment ion 456 is formed by loss of a TFA group and mass 428 by an additional loss of a CO moiety (Fig. 28).

The PCI spectrum displays the M + 1 of 571, a base peak of 457, and a large fragment at 343. The latter two masses are essentially the same large fragments found in 4-MAS and are formed in the same manner, that is, successive loss of TFA units. The PCI spectrum is more diagnostic and sensitive in analytical procedures than the EI spectrum.

III. ANALYSIS OF A MIXTURE OF TRICHOTHECENES BY GAS CHROMATOGRAPHY AND IDENTIFICATION OF SOME OF THE COMPONENTS BY SEARCHING A COMPUTER LIBRARY OF POSITIVE CHEMICAL IONIZATION SPECTRA

The ultimate test of the utility of the mass spectra presented here is to use them in the identification of unknown components resolved by GLC. To this end, a library of the mass spectra of numerous derivatives of the trichothecenes has been assembled into a library data base designed for searching unknown components in a mixture.

An example of a computer search is given below, in which a crude extract of plant material is extracted and prepared for analysis by combination GC-MS. The derivative of choice in this example is TFA anhydride, and mass spectral analysis is done in PCI using methane as the reagent gas. The total ion chromatogram is shown in Fig. 29a (see scan 106). Each unknown mass spectrum of interest is scanned by superimposing the library mass spectra upon the unknown mass spectrum (a reverse search). If the computer determines that a given library mass spectrum could be found within the unknown mass spectrum, a match is reported. Any matches that are found in the search are printed out as shown in Fig. 29b. The results of all matches appear on the cathode ray tube (CRT) screen and can be read immediately, copied to a printer, or stored in memory for use later.

The PBM search algorithm used with the library generates a probability value (see Fig. 29b) which indicates the probability of the unknown mass spectrum and the matched library mass spectrum being the same compound. This value repre-

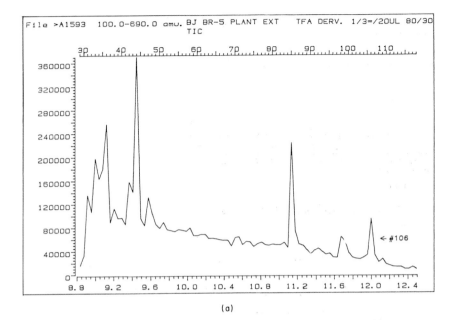

File >A1593   100.0-690.0 amu. BJ BR-5 PLANT EXT   TFA DERV. 1/3=/20UL 80/30
TIC

(a)

PBM

1. (T-2) 3-HYDROXY-4,15-DIACETOXY-8-[3-METHYL-BUTYRYLOX   563 C26H34F3O10
Y]-12,13-EPOXYTRICHOTHEC-9-ENE TRIFLUOROACETATE 200E
V CH4 PCI OP=RJP RUN 31MAR82 ORIGINALLY FROM FRN 753
7 NOW RCL,CMS103 MW=M+1

Sample file: >A1593   Spectrum #:   106
Search speed: 3   Tilting option: F   No. of ion ranges searched: 126

	Prob.	CAS #	CON #	ROOT	K	DK	#FLG	TILT	%	CON	C_I	R_IV
1.	98*	103	83	TOXINS	141	32	1	0	95	17	60	99

(b)

Fig. 29. (a) Total ion chromatogram (full scan) of a plant extract containing T-2 toxin. The analysis was done by PCI-MS in methane using TFA anhydride as the derivatizing reagent. Each peak of the entire chromatogram was searched by (b) the trichothecene library of mass spectra in order to identify the trichothecene components. Scan 106 was identified as T-2-TFA.

sents a combination of factors: the presence or absence of the molecular ion, the number of mass spectral peaks missing from the unknown mass spectrum but present in the library mass spectrum, the percentage of the library mass spectrum that would fit into the unknown mass spectrum, and the total K value. The total K value represents a summation of all the individual K values calculated for each matching mass between the two spectra in terms of uniqueness and abundance.

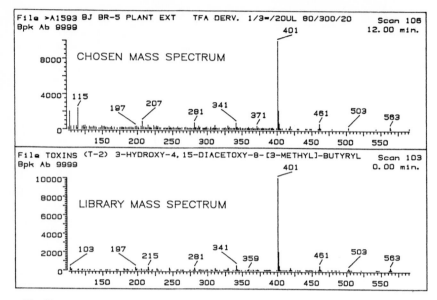

Fig. 30. PCI mass spectra of scan 106 of Fig. 29a and the corresponding match in the library. The match in the high-mass region of the spectrum of scan 106 is identical to that of authentic T-2 toxin found in the library.

Uniqueness values come from the examination of a large number of mass spectra, judging how likely it is that an ion of a given mass will appear at random (in general, masses of higher molecular weight generate higher uniqueness values). Abundance values are assigned on the basis of what percentage of the base peak a given mass represents. The search report indicates the degree of probability and suggests that the investigator use additional information to verify the identification. Figure 29b shows a probability value of 98% and a K value of 141. The component was identified as the trifluoroacetoxy derivative of T-2 toxin. The spectrum of the unknown (scan 106) is compared with that of the library spectrum and is shown in Fig. 30.

IV. CONCLUSIONS

Identification of unknown mycotoxins, as well as other chemical entities, by MS has traditionally been done by mass spectral libraries which contained spectra derived from EI ionization. Initially we depended on huge, general data bases for comparison of the unknowns, but recently the trend has been toward more specialized libraries restricted to a certain group of compounds. To this end we developed the Trichothecene Mass Spectral Library and found it very useful for identifying the many derivatives encountered in the study of the trichothecenes produced by *Fusarium*. Moreover, this library has assisted us in identifying many products of trichothecene metabolism by animals.

The spectra shown in this chapter are those of TMS ethers, TFA esters, and HFB acetates because they represent those derivatives which the analyst most often encounters when resolving the trichothecenes by GLC.

This chapter contains an attempt to update the Trichothecene Mass Spectral Library with the use of mass spectra obtained by chemical ionization (CI). In contrast to EI ionization, CI produces molecular ions, a base peak closely associated with the molecular ion, and in general a greater number of fragment ions found in the higher mass range. We have compared EI with PCI and NCI, and have evaluated the various advantages of one over the other. We conclude that CI mass spectra are superior to EI in identifying the vast number of trichothecenes and their metabolites. PCI is recommended for TFA derivatives of T-2, HT-2, DAS, and MAS, whereas both PCI and NCI have advantages in the analysis of the TMS derivatives of DON.

We also recommend the use of TFA derivatives for resolution of the majority of the trichothecenes by GLC chromatography with subsequent analysis by MS. The exception to this is DON and its derivatives, where TMS ether derivatives appear to yield the best results. The greatest sensitivity of the DON group of derivatives is found with NCI and not PCI, although the latter have greater diagnostic value in qualitative evaluations.

The TFA derivatives have sensitivity equal or perhaps slightly greater than the heptafluorobutyryl and pentafluoropropionyl esters. The latter yield a high-mass fragment, which is an advantage; however, the fragment, as in the HFB ester of DON, may be a single mass of 884, leaving the identification of the molecule with less certainty than desirable. Of greater importance is the extensive manipulation (number of steps) necessary when making the HFB ester. The TFA ester is an easily made two-step procedure.

Although PCI mass spectra yield qualitatively superior spectra in the quadrupole instrument, sensitivity of the ion source can suffer when the source is not clean. Requirements for cleanliness of the source are not as critical in the NCI mode because of inherent differences in the manner in which the ions are formed. Most of the NCI reactions with the trichothecenes proceed through electron capture; they are moderated via the reagent gas and are not proton transfers as are the PCI reactions. Source contamination does not inhibit these electron transfer reactions as it does in the PCI mode. PCI reactions are gas phase acid–base reactions that are affected by the buildup of contaminants on the source housing. This is particularly true when TMS ether derivatives are used because they have a tendency to foul the source of quadrupole instruments rapidly.

ADDENDUM

Experimentation with the use of tandem mass spectrometry revealed that the parent of fragment 401 is 461 and not 460 as postulated. Moreover, the parents of m/z^+ 400 are 401 and 460.

ACKNOWLEDGMENT

This study (paper number 14,031) from the Agricultural Experiment Station, University of Minnesota (St. Paul, Minnesota 55108), was made possible in part by contract No. DAMD17-82-C-2113.

REFERENCES

Abrahamsson, S., and Nilsson, B. (1964). Direct determination of molecular structure of trichodermin. *Proc. Chem. Soc.* p. 188.

Bamburg, J. R. (1969). Mycotoxins of the trichothecane family produced by cereal molds. Ph.D. Thesis, Univ. Microfilms, Ann Arbor, Michigan.

Cole, R. J., and Cox, R. H. (1981). "Handbook of Toxic Fungal Metabolites." Academic Press, New York.

Dusold, L. R., Pohland, A. E., Dreifus, P. A., and Sphon, J. A. (1978). "Mycotoxins Mass Spectral Data Bank." Assoc. Off. Anal. Chem., Arlington, Virginia.

Godtfredsen, W. O., and Vangedal, S. (1965). Trichodermin. A new sesquiterpene antibiotic. *Acta Chem. Scand.* **19,** 1088.

Yoshizawa, T., Swanson, S. P., and Mirocha, C. J. (1980). T-2 metabolites in the excreta of broiler chickens administered ^{3}H-labeled T-2 toxin. *Appl. Environ. Microbiol.* **39,** 1172–1177.

Yoshizawa, T., Mirocha, C. J., Behrens, J. C., and Swanson, S. P. (1981). Metabolic fate of T-2 toxin in a lactating cow. *Food Cosmet. Toxicol.* **19,** 31–39.

14

Mass Spectrometry–Mass Spectrometry as a Tool for Mycotoxin Analysis

R. D. PLATTNER

U.S. Department of Agriculture
Agricultural Research Service
Northern Regional Research Center
Peoria, Illinois 61604

I. INTRODUCTION*

Since the mid-1970s, mass spectrometry (MS) and particularly MS coupled with chromatography has become an essential tool for difficult analytical prob-

*The mention of firm names or trade products does not imply that they are endorsed or recommended by the U.S. Department of Agriculture over other firms or similar products not mentioned.

lems in many areas, including mycotoxins. The chemical specificity and sensitivity offered by MS often offset the high cost of the equipment and make it the method of choice for reliable quantitation of picogram and nanogram amounts of various target molecules. The combination of "soft" ionization techniques such as positive- and negative-ion chemical ionization (PCI, NCI) and fast atom bombardment (FAB) coupled with substantial improvements in computerized control of the mass spectrometer and data acquisition have greatly reduced detection limits and expanded the range of compounds amenable to analysis by MS. These advances and the ever-increasing complexity of the demands of research problems for rapid and unambiguous identification of compounds at very low levels help spur further improvements in MS methodology, especially in the area of sample handling and introduction.

Considerable interest has been generated recently in using the technique of tandem mass spectrometry or mass spectrometry–mass spectrometry (MS-MS) to analyze crude mixtures of organic compounds. The development of the technique stems from a new concept: using mass spectrometers in sequence, one as a device for separating the individual component of interest from others and the other to identify or quantitate the separated component. The usual mass spectrometer needs four basic parts: (1) a sample introduction system, (2) an ion source that creates positive or negative ions from the sample, (3) a mass filter that can separate from all ions formed only ions of a single mass-to-charge value, and (4) a detector to detect and quantitate the ions produced. In tandem MS there are two (or more) independently controllable mass filters. The use of multiple independent mass filters makes MS-MS analogous to gas chromatography–mass spectrometry (GC-MS) or liquid chromatography–mass spectrometry (LC-MS). The selection of a specific ion by the first mass filter, typically the parent ion of the compound of interest, is analogous to the separation of a component as a GC or LC peak. This selected ion can then react (by collision or near-collision with molecules of a target gas) in a collision cell to produce collisionally activated decomposition (CAD) products. These products are separated by the second mass filter and measured by the detector. Spectra obtained in this manner can be similar in appearance to spectra obtained with a conventional mass spectrometer and can be used to identify the structure of the parent ion or to obtain quantitative data about how much of the parent was present in the sample.

MS-MS offers several potential advantages over GC-MS for the separation and analysis of mixtures. This is particularly true for the many compounds that are not amenable to separation by GC-MS but that can be analyzed by MS. Any conventional mass-spectrometric study of these compounds in mixtures requires rigorous purification prior to the analysis. Even with compounds that are amenable to GC-MS, the speed of analysis by the MS-MS technique can significantly increase sample throughput. Response of the mass spectrometer is virtually instantaneous, since ion formation and transmission require less than 1 msec.

Thus the speed of an analysis is determined by the time required for sample introduction and data collection. Relatively high specificity is achieved by the unit mass resolution of the mass filters. The ordinary mass spectrum of a complex mixture has some signal at virtually every mass-to-charge value below mass 350. These signals interfere with the detection of almost any trace component. The MS-MS approach utilizes one of the mass filters to remove this "chemical noise" from the spectrum. This often yields improvements in signal-to-noise ratio of several orders of magnitude and detection limits equal to or superior to those attainable by GC-MS. In many cases, sample workup for analysis by MS-MS is minimal. Biological fluids, whole-plant tissues, and other complex materials have been analyzed successfully by MS-MS with no sample preparation. The only requirement is that the compound of interest can be ionized in a way such that an ion representative of the neutral species can be obtained.

This chapter briefly describes the types of MS-MS instruments currently available as well as the types of MS-MS experiments that can be performed. Several excellent reviews on the development of the MS-MS technique have appeared (McLafferty, 1980, 1981; Kondrat and Cooks, 1978; Cooks and Glish, 1981; Slayback and Storey, 1981), and the reader is referred to these reviews for more detailed discussions of instrumentation and techniques than is possible here. Several examples of the application of MS-MS to the analysis of mycotoxins in this author's laboratory will be presented.

II. MASS SPECTROMETRY–MASS SPECTROMETRY INSTRUMENTATION

MS-MS experiments can be performed with a wide variety of mass analyzers. The most common types are magnetic (B), electrostatic (E), quadrupole (Q), and time-of-flight (T) analyzers. Each selects ions from the beam using a different property. The magnetic analyzer separates on momentum-to-charge ratio (mv/z), the electrostatic analyzer uses kinetic energy-to-charge ratio (mv^2/z), and the quadrupole analyzer selects only on mass-to-charge ratio (m/z). Any combination of two or more mass analyzers can be used for MS-MS; however, best performance is obtained with a quadrupole or magnetic analyzer first. Commercial instrument vendors are now offering several types of instruments specifically designed with MS-MS experiments in mind. These include multisector magnetic mass spectrometers, quadrupole mass spectrometers, and hybrid instruments combining both magnetic sectors and quadrupoles. Additionally, Fourier transform (FT) mass spectrometers are also suitable for MS-MS studies. In a FT instrument no additional hardware is needed for MS-MS, because the experiments are defined by software using the standard hardware of the FT-MS instrument. One additional new MS-MS technique developed by Dr. C. G. Enke at

Michigan State University that requires little modification of a standard magnetic mass spectrometer is noteworthy (Stults *et al.*, 1983). This technique, time-resolved magnetic dispersion MS-MS, employs a pulsed-ion source and sophisticated computer software to generate the MS-MS experiments. No commercial instrument is yet offered using this technique. The principles of each of these instrument types are briefly discussed, along with its advantages and disadvantages.

A. Reverse-Geometry Double-Focusing Mass Spectrometer

The first MS-MS demonstrations were made on standard double-focusing magnetic mass spectrometers. The double-focusing mass spectrometer uses an electrostatic analyzer (E) in addition to the magnetic analyzer (B) to improve the energy focus of the ion beam so that higher resolution can be obtained. The electrostatic analyzer normally is coupled to the accelerating potential of the source. The ions accelerated from the source have an energy distribution due to their energy before ionization. The electrostatic analyzer retards higher energy ions coming from the source while enhancing the lower energy ions to give improved energy focus in the final mass-analyzed beam. By simply unlinking the electrostatic sector and the accelerating voltage, the electrostatic analyzer can be used as a mass filter. Linking scans of it with the magnetic scans will produce MS-MS spectra. Any reverse-geometry double-focusing mass spectrometer (i.e., one with the magnetic preceding the electrostatic analyzer, or B/E geometry) is capable of MS-MS experiments. The addition of a collision cell between the magnetic and electrostatic sectors allows CAD experiments and extends the sensitivity of the MS-MS experiment attainable. The resolution achieved by the magnetic analyzer as the first mass filter (MS-I) is on the order of approximately 1000, whereas the resolution of the electrostatic sector as MS-II is very low, typically only a few hundred at best. This means that daughter ion fragments differing by only 1 dalton will not be completely resolved. When using a normal-geometry (E/B geometry) mass spectrometer the resolution is even worse, and other "artifact" problems complicate MS-MS experiments. Poor resolution is the most serious deficiency of this system for MS-MS.

B. Multiple-Sector Magnetic Mass Spectrometer

Resolution can be improved by adding additional sectors to the standard configuration. Noncommercial multiple-sector magnetic mass spectrometers with various combinations of B and E geometry have been built in the laboratories of several pioneers in the MS-MS research, including Dr. Fred McLafferty at Cornell University. One commercial manufacturer (Kratos Scientific of Westwood, New Jersey) has added an additional electrostatic analyzer to their MS 50 high-resolution mass spectrometer to produce the "Triple Analyzer," the first commercially available multiple-sector magnetic mass spectrometer designed for MS-MS. The first instrument was installed in Dr. Michael Gross' laboratory at the University of Nebraska about 1980. V. G. Micromass, Ltd. of Cheshire, England also offers an additional mass filter for their reverse-geometry ZAB-2F to produce a high-resolution triple-sector instrument (BEEB). These elegant mass spectrometers are capable of truly high performance. High-resolution data have been obtained from these instruments at resolving powers of greater than 10,000 in MS-I (100,000 resolution has been demonstrated at Nebraska). These instruments are expensive in terms of both initial cost and the sophisticated personnel necessary to maintain and operate them. Therefore, these instruments will most likely find application for problems that cannot be solved any other way.

C. Multiple-Quadrupole Mass Spectrometer

The brightest prospects thus far for the general use of MS-MS have been shown by the multiple-quadrupole instruments. The quadrupole is a simple mass filter that is made of four stainless-steel rods held in parallel lengthwise to form a path for ions. When both a radiofrequency (rf) current and a direct-current (dc) voltage are applied to the rods, ions between the rods follow a complex trajectory. At a given ratio of the rf current to the dc voltage applied to the rods, only ions of a given mass-to-charge ratio follow a stable trajectory through the filter, whereas all other ions hit the sides of the rods or instrument and are lost. The resolution and mass range achievable with quadrupole mass filters is not nearly so high as with the classical electrostatic and magnetic analyzers, but complete unit mass resolution from mass 4 up to about mass 2000 can be achieved. This is completely adequate for most MS-MS applications. More importantly, the quadrupoles are smaller, less expensive, and more simply controlled by computerized data systems than their magnetic counterparts. Several manufacturers offer commercial multiple-quadrupole mass spectrometers that were designed for MS-MS. A considerable number (~100) of these instruments are presently in operation worldwide.

D. Hybrid Magnetic and Quadrupole Mass Spectrometer

MS-MS spectrometers combining both sectors and quadrupoles, so called hybrid instruments, have been designed to provide the best features of sector and quadrupole instruments. Several manufacturers (V. G. Micromass, Kratos Instruments, and Finnigan/MAT) currently are offering these machines that provide the high-resolution features of magnetic-sector mass spectrometers along with the desirable features of quadrupoles for efficiency of daughter ion transmission and mass resolution in the second mass filter. Because these instruments have become available only recently, the extent of advantages and/or disadvantages they provide relative to the more proven tandem-quadrupole instruments have yet to be fully demonstrated. One feature of these hybrid instruments is the ability to do CAD experiments at either high (kV) or low (V) energies. This should improve on the low-energy limited tandem-quadrupole instruments, opening up experiments involving charge inversion and other high-energy processes while maintaining the resolution advantages of quadrupoles for MS-II. They also offer resolution in MS-I, of several thousand, higher than is attainable on the tandem-quadrupole instruments. The potential chief disadvantages in the hybrid instruments may lie in the level of sophistication and expertise that will be needed to operate and maintain them, but it is still too early to make those judgments.

E. Fourier Transform Mass Spectrometer

MS-MS experiments are possible with unmodified FT mass spectrometers, because the sequence of control pulses necessary to separate parent ions from daughter ions are completely defined by software. In contrast to sector instruments, high resolution is available for daughter ions, and the ability to examine a sequence of reaction products (i.e., MS-MS-MS . . .) has been demonstrated. Furthermore, because all ions are measured simultaneously, the FT-MS technique is compatible with new pulsed desorption–ionization processes and will allow experiments to collect time profiles for collision products as well as total product intensity. In spite of its many advantages, FT-MS has yet to gain widespread use as an analytical technique because of other inherent problems, including the high cost of the hardware and the sophisticated computer power needed to drive the machine and process all the data from it. The future of FT-MS-MS, therefore, hinges on future improvements that make the instrument easier to use.

III. MASS SPECTROMETRY–MASS SPECTROMETRY TECHNIQUES AND EXPERIMENTS

MS-MS experiments, like conventional MS experiments, are usually intended to provide information in one of two broad categories: (a) experiments that determine or verify the identity or structure of the compound of interest, and (b) experiments designed to quantify the amount of one (or more) component in some matrix. The techniques and experiments available in MS-MS provide a high degree of flexibility, with several tools to approach these analytical problems. Basically, MS-MS has four major types of experiments, which can be selected in combinations to comprise an analysis for a given application. These experiments are detailed below and summarized in Table I.

TABLE I

Major MS-MS Experiments

	Configuration	
Experiment	MS-I	MS-II
Daughter experiment	Select the m/z of parent	Scan daughters
Parent experiment	Scan parents	Select the m/z of daughter
Neutral loss (gain)	Linked to MS-II with off-set equal to loss (or gain)	Scan
Normal mass spectrum	Scan either MS-I or MS-II	

A. Daughter Experiments

In daughter experiments, the first mass filter is set at a chosen m/z value of interest. If a target gas is present, these mass-selected parents undergo CADs in the collision cell. If no target gas is present, some of the mass-selected parents may still undergo metastable decompositions. The resulting daughters are then mass analyzed by scanning the second mass filter and quantified by the detector. The daughters of several ions in the normal spectrum can be recorded by successively focusing their m/z values as parents by MS-I. Daughter spectra of a given ion can be compared with those obtained from ions of known structures to obtain structural information. Identification of a given component in a matrix can be made by the coincidence of the daughter spectra from authentic sample and the daughter spectrum of the unknown. Daughter spectra obtained for either the parent ion of the molecule of interest or daughters from unique fragments (or both) are useful to identify a compound. The use of multiple daughter spectra for a parent and major fragments of a given compound can greatly strengthen the certainty of identification of a known compound. Interferences from "chemical noise" generated by other compounds present in a mixture with the target compound can often be eliminated by proper choice of daughter experiments. The yield and ratios of unique daughters from the selected parents can be used to quantitate a component.

B. Parent Experiments

For parent experiments, all ions of interest are passed sequentially by scanning MS-I. The daughter ions of only a single m/z value are allowed to pass MS-II. This experiment identifies all parent ions that decomposed in the collision cell to give the daughter ion passed by MS-II. This experiment is useful in structural and fragmentation studies and in survey studies where detection of related parents (i.e., a series of structural homologs) is desired.

C. Neutral Loss (Gain) Experiments

The third type of experiment uses a fixed-mass offset between the two mass filters. For example, if the mass filters are scanned with a constant offset of -32 amu, MS-II only passes those ions selected by MS-I which have undergone that neutral loss in the collision cell. Intense neutral losses in CAD are observed for many substituent functional groups. These common functionalities can be rapidly detected in the neutral loss experiment. Conversely, if the offset between the mass filters is positive, only ions that undergo chemical addition from ion molecule reactions in the collision cell are observed. Little has yet been reported in this interesting area of research, but addition reactions have frequently been observed, particularly in CI-MS. This technique should be a valuable one with which to study such reactions.

D. Advanced Experiments

Hybrid mass spectrometers which combine quadrupole and sector mass analyzers are capable of performing additional sophisticated scanning modes. These more selective new types of scans employ reactions in more than one reaction region and may be considered to be MS-MS-MS experiments. These scans include consecutive neutral loss, reaction intermediate, and selective or sequential parent and daughter scans. Since they require two reaction cells, they are not available on the tandem quadrupole instruments, though they are possible with the fourier transform mass spectrometers.

Finally, the MS-MS may be configured simply as a standard mass spectrometer by evacuating the collision cell and scanning either mass filter while passing all ions through the other. This configuration produces normal spectra that are identical to those obtained on a standard instrument.

The ability to switch experimental conditions in such a way to gather multiple kinds of information in a given analysis allows the operator to optimize the analysis to the questions that are to be answered. By selecting the appropriate instrument modes, the interferences that might be present in normal MS can be eliminated, yielding improved detectability of a compound.

E. Ionization Techniques

Flexibility obtained by MS-MS is also greatly enhanced by the choice of ionization techniques as well as choice of instrument configuration. Normal electron impact (EI) spectra provide a high degree of structural information but often at the expense of weak signals for important fragments, with much of the total ionization going into diagnostically unimportant ions. By contrast, CI sometimes produces spectra that contain only a single ion, the protonated molecule, MH^+. This is a great improvement in sensitivity, but at the expense of selectivity. The use of MS-MS techniques with CI allows the operator to increase selectivity. By selecting the parent, MH^+, in MS-I, the CAD daughters mass analyzed by MS-II provide assurance that the parent ion was indeed the target compound, not some other component in the matrix with the same m/z value. The improved selectivity of daughter experiments can give low-resolution MS-MS selectivities that compare favorably with high-resolution MS data but can be easier to implement routinely.

Coupling new "soft" ionization techniques such as FAB with MS-MS should provide a very important improvement. FAB spectra of important molecules typically show little fragmentation. The glycerol matrix used in FAB, moreover, shows many peaks across the mass range. This sometimes makes interpretation of FAB spectra difficult. MS-MS daughter spectra should provide improvements in signal-to-noise ratios by eliminating much of this matrix background, as well as providing valuable structural information from the daughters.

IV. EXAMPLES OF MASS SPECTROMETRY–MASS SPECTROMETRY OF MYCOTOXINS

Because of the newness of the technique, relatively few references on applications of MS-MS to the analysis of mycotoxins have appeared in the literature. Many of the references listed here are abstracts of reports presented at various mass spectrometry society meetings.* The small number of references from refereed journals is in no way an indication that the technique is now useful or is difficult to apply to mycotoxins. A tandem quadrupole MS-MS instrument has been available in the author's laboratory since 1982, and investigations of the application of MS-MS techniques to the analysis of mycotoxins have been very exciting. As time passes, this technique will take its rightful place along with other analytical methods. Several examples of MS-MS analysis of mycotoxins are presented in detail to give the reader a feeling for the types of useful and sometimes unique information that can be obtained with the technique.

A. Ergot Alkaloids

The earliest recognized form of mycotoxicosis, ergotism, has been known for much of recorded history. Ergot-contaminated feeds are known to have a severe

---

*Following the Reference section, see listing of MS-MS papers concerning mycotoxins presented at the American Society for Mass Spectrometry Meetings, 1981–1985.

economic impact on both grain utilization and livestock production (Porter and Betowski, 1981). The separation and identification of ergopeptine alkaloids produced in sclerotia of the fungus *Claviceps purpurea* is a difficult chemical problem. The ergopeptines have similar structures and properties. They are not volatile enough to be analyzed by GC-MS. When they must be detected at low levels (parts per billion, ppb) as contaminants of a food or feed matrix, determination is made by thin-layer chromatography (TLC) or high-performance liquid chromatography (HPLC), using the high sensitivity of fluorescence detection (Scott and Lawrence, 1980). Although highly sensitive, fluorescence detection is not as specific a detector as MS.

Normal MS analysis would require rigorous cleanup of relatively large amounts of sample to remove enough matrix to allow positive identification. Because these alkaloids produce intense negative-ion mass spectra, separation and identification of picogram amounts of individual alkaloids by MS-MS from simple extracts are possible. Some of the ergopeptine isomers give nearly identical CI and EI spectra, making differentiation of these compounds quite difficult. We have recorded quadrupole MS-MS spectra for 12 of the 14 known ergopeptine alkaloids and reported that all 12 of them can be differentiated by MS-MS experiments (Plattner *et al.*, 1983). For example, as shown in Fig. 1, the PCI and NCI mass spectra of ergosine and its isomer β-ergosine, which differs only in the branching in the R_2 group (see Fig. 2 and Table II for structures), are virtually indistinguishable from each other.

In PCI, small signals from protonated molecules (MH$^+$) are observed. The majority of the ions detected arise from cleavage at the bond joining the tricyclic peptide moiety with the lysergic acid amide to produce the protonated amide (*m/z* 268) and the protonated tripeptide (*m/z* 281). A weaker signal at *m/z* 211 arises from loss of the R_1 group and two carbonyl groups from the protonated peptide. In (NCI), the molecular anion (M$^-$·) is observed as a small signal. The peptide anion (*m/z* 280) accounts for virtually all of the ion intensity. The MS-MS daughters of this peptide anion (*m/z* 280) are shown in Fig. 3. The base peak (*m/z* 209) arises from loss of R_1 plus two carbonyl groups. A small signal observed in the peptide anion daughter spectra of all the alkaloids is the result of the direct loss of $R_2 + R_1$—CO—CO from the parent. When R_2 is an isobutyl group, the parent loses 43 dalton (ergosine). When R_2 is *sec*-butyl, the loss of 43 dalton is not observed, rather ions are observed for the loss of 15 dalton and 29 dalton. Both of these signals are absent from the daughters of isobutyl isomers, making discrimination of the isomers possible. The ergosines are easily differentiated from their functional isomer ergonine ($R_1 = C_2H_5$ $R_2 = i$-Pr, isopropyl) because *m/z* 195 is the base peak in its daughter spectrum.

Analysis of a crude extract of sclerotia (*Claviceps purpurea*)-infested grains demonstrates that MS-MS can be used to identify ergot peptide alkaloids directly from crude extracts. On successive scans, the peptide anion fragments of all the

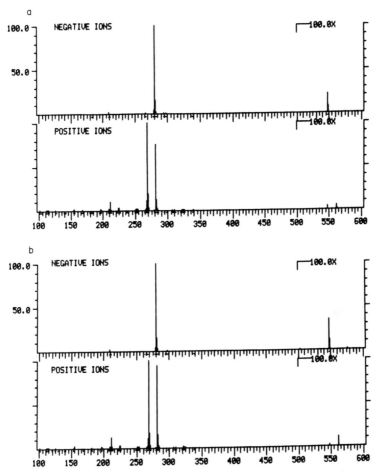

Fig. 1. PCI and NCI mass spectra of ergosine and its isomer β-ergosine.

known alkaloids are focused into the collision cell by Q_1, and the CAD daughters are mass analyzed by Q_3 and recorded while an aliquot of the extract is eluted from the direct-exposure probe. The resulting daughter spectra are compared to those generated by analysis of pure standards. Relatively little matrix interference survives this MS-MS experiment, and spectra virtually identical with standard spectra are obtained from aliquots equivalent to 1 mg of clean grain spiked at 100 ppb with standard alkaloids before extraction. The direct detection of these peptide alkaloids from simple extracts at this level could be a useful procedure to screen questionable feed samples for ergot contamination. Typical alkaloid contents from pure sclerotia were found to vary from 0.01 to 0.45%

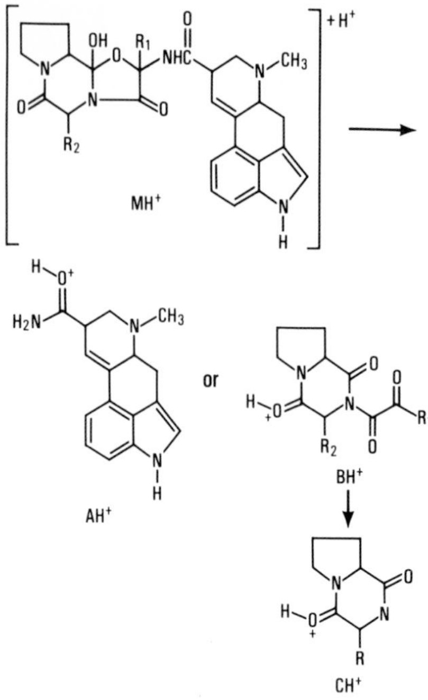

Fig. 2. Structures of ergot cyclol alkaloids.

(Young, 1981); therefore, these alkaloids should still be easily detectable at ergot contamination levels of 0.1% in grain. With some matrix cleanup, detection at even sub-part-per-billion levels should be feasible by MS-MS.

B. Fusaria Mycotoxins

Fusaria mycotoxins have generated considerable interest over the past several years. Invasion of cereal grains by *Fusarium* molds in the field or in storage occurs occasionally when field conditions are favorable or storage conditions improper. These molds can produce several mycotoxins. When the contaminated grain is fed to susceptible animals, physiological disorders of economic significance can occur.

One *Fusarium* mycotoxin, zearalenone [6-(10-hydroxy-6-oxo-*trans*-1 undecyl)-β-resorcylic acid lactone] is a synthetic estrogen and causes hyperestrogenism when fed to swine. Zearalenone has been detected by TLC and as its trimethylsilyl (TMS) derivative by GC-MS (Kamimura *et al.*, 1981). An alternative procedure is the examination of a crude extract for zearalenone by MS-MS. Figure 4 shows a portion of the CI mass spectrum of a CHCl$_3$ extract of corn

TABLE II

Structures of Ergot Cyclol Alkaloids[a]

Compound	R_2
Ergotamine group (R_1 = CH$_3$)	
Ergotamine	PhCH$_2$
Ergosine	*i*-Bu
β-Ergosine	*sec*-Bu
Ergovaline	*i*-Pr
Ergoxine group (R_1 = C$_2$H$_5$)	
Ergostine	PhCH$_2$
Ergoptine	*i*-Bu
β-Ergoptine	*sec*-Bu
Ergonine	*i*-Pr
Ergotoxine group (R_1 = *i*-Pr)	
Ergocrystine	PhCH$_2$
α-Ergokryptine	*i*-Bu
β-Ergokryptine	*sec*-Bu
Ergocornine	*i*-Pr

[a] See Fig. 2 for structures.

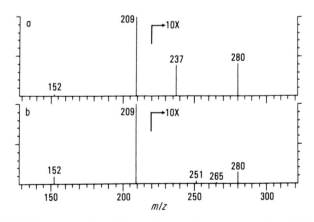

Fig. 3. MS-MS daughters of the NCI peptide anion fragment of ergosine and β-ergosine.

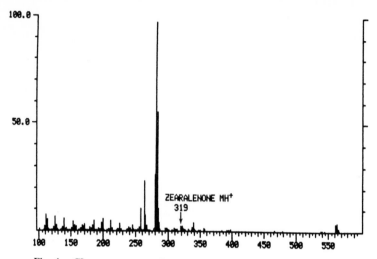

Fig. 4. CI mass spectrum of a zearalenone-contaminated corn extract.

containing 1 ppm zearalenone. The spectrum contains many signals; the signal at m/z 319, the protonated zearalenone molecule, is not particularly significant. By contrast, Fig. 5a shows the CAD daughter spectrum of m/z 319 of the same corn extract, and Fig. 5b shows the daughter spectrum of pure zearalenone's m/z 319 signal. The spectrum of the zearalenone-containing corn and the pure standard are virtually identical. Interference from the corn matrix is minimal. Zearalenone can be detected similarly in simple extracts of wheat, oats, and rice directly from a simple solvent extract without additional sample cleanup. Detection limits for this technique are approximately 100 ppb. The total analysis time is only about 3 min.

Trichothecenes are also significant mycotoxins produced by Fusaria invasion of grains. Deoxynivalenol (3,7,15-trihydroxy-12,13-epoxytrichothec-9-en-8-one; DON), also known as vomitoxin, is probably the most widespread of this class. It has been associated with feed refusal and vomiting in swine (Mirocha *et al.*, 1976). At the thirtieth annual Conference for Mass Spectrometry and Allied Topics of the American Society for Mass Spectrometry, Davidson and co-workers (1982) presented an analysis of DON and T-2 toxin by MS-MS with atmospheric-pressure chemical ionization–tandem mass spectrometry (APCI-MS-MS), using the Sciex triple-quadrupole instrument. Using this technique, the mycotoxins are thermally desorbed into the ion source, where they are ionized under APCI conditions. The parent ions (MH$^+$) are mass selected by the first quadrupole, induced to fragment by collisions with the target gas in the second quadrupole, and the fragments then are mass analyzed by the third quadrupole. The resulting daughter spectrum furnishes fragments indicative of the precursor ion's structure. Monitoring these fragments, while focusing only m/z 297 [the

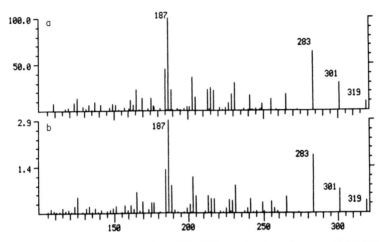

Fig. 5. The CAD daughter spectra of m/z 319 of (a) a crude extract of corn containing 1 ppm of zearalenone and (b) 10 ng of pure zearalenone.

$(MH)^+$ of DON], provides a highly selective DON detector. At the same meeting, Plattner, Bennett, and Slayback presented experience with the Finnigan TSQ triple-quadrupole MS-MS for the detection of DON in a grain matrix. CI (isobutane) was used for MS-MS experiments in both the PCI and NCI modes to identify DON. In the PCI mode, the MH^+ ion at m/z 297 was focused by Q_1. The resulting daughter spectrum was quite similar to the MH^+ ion daughter

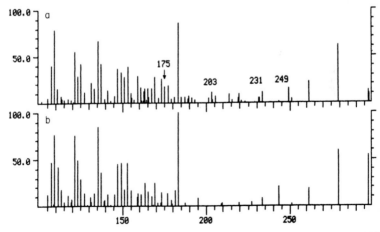

Fig. 6. CAD daughters of m/z 297 from (a) a 1-ppm DON-contaminated corn extract and (b) a DON corn extract. The ions at m/z 175, 203, 231, and 249 are the major daughters of DON.

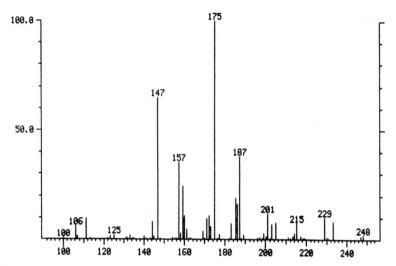

Fig. 7. CAD daughters of m/z 248 from negative ions from a 1-ppm DON-contaminated corn extract. The total spectrum is virtually identical with the spectrum of 10 ng of pure DON.

spectrum obtained by Davidson and co-workers with the atmospheric-pressure source. Monitoring specific daughters of the protonated DON parent (m/z 297) at 175, 203, and 249 afforded signal from a 1-ppm DON-containing corn extract, but none from a clean corn extract. However, full m/z 297 daughter scans of the DON-contaminated corn extract and the clean corn extract showed the presence of substantial m/z 297 ions contributed by the corn matrix. The signals represent-ing the major DON daughter ions were minor signals in the total matrix spectrum (Fig. 6). In the NCI mode, only a small signal was observed for the molecular anion of DON at m/z 296. The base peak of the spectrum was at m/Z 248, arising from the loss of 48 μm (presumably $CH_2CO + H_2O$) from the molecule. This signal was on the order of 100 times as intense as the protonated molecule (m/z 297) observed in the positive ions. The daughter spectrum of the m/z 248 ion showed intense fragments at m/z 147, 175, and 187. The spectrum of the m/z 248 daughters from a 1-ppm DON-containing extract of corn shown in Fig. 7 was virtually identical with the daughters of the authentic standard at 1-ppm DON contamination. In this cleaner background, DON can be identified at concentrations as low as 100 ppb. With some sample cleanup, lower detection limits are possible.

C. Aflatoxins

From economic and regulatory standpoints, aflatoxins are certainly the most important class of mycotoxins. Consequently, they have also been the most studied. Numerous methods have been devised to identify aflatoxins in many commodities based on both biological activity and physicochemical properties. Because aflatoxins are highly fluorescent, sensitive quantitative techniques with HPLC and TLC using fluorescence detection have detection limits in the low parts per billion range and better. By contrast, because aflatoxins are not amenable to GC-MS, MS has not been widely used for their quantitative analysis. The major limitation of MS analysis for aflatoxins is from "chemical noise" or response from residual sample matrix components not removed by sample cleanup. Molecular ions from pure aflatoxins can be routinely detected by EI-MS from 1-ng samples. However, the EI-MS of the complex matrix in which aflatoxins are found produces signal at virtually every m/z value from 50 to around 500, interfering with the identification of the aflatoxins. Even after considerable sample cleanup, chemical noise interferences still can necessitate several orders of magnitude more sample than necessary to produce a molecular ion in order to reliably identify the presence of an aflatoxin in a sample extract. Tandem MS provides a technique to filter chemical noise from the sample and thereby improve detection limits. The tandem-quadrupole MS-MS instrument, when used as a conventional mass spectrometer, is slightly more sensitive to the molecular ions of aflatoxins than it is when collecting CAD daughters. This loss in sensitivity is offset by the improvement in signal-to-noise ratio in the MS-MS mode. Setting Q_1 to pass only the molecular ion of the aflatoxin of interest and subsequently scanning for daughters in Q_3 yields less total signal, but relatively more instrument background and matrix signal is eliminated. The CAD daughters of M^+ for the aflatoxins occur at the same m/z values as the major fragments observed in the EI spectra. Functional isomers such as M_1 and G_1 which have molecular ions at the same m/z value (328) give different daughter fragments (Fig. 8). Aflatoxin G_1 (Fig. 8a) shows an intense fragment from neutral losses of CO (28) and CH_3O (31) at m/z 300 and 297, whereas M_1 (Fig. 8b) has an intense fragment at m/z 299 from neutral loss of CHO (29).

In the CI mode, differentiation of functional isomers by CAD daughter experiments is also useful. CI mass spectra typically show much less fragmentation than EI mass spectra. The decreased fragmentation minimizes noise from the matrix, while giving a strong MH^+ for the target compound. Unfortunately, the protonated molecule (MH^+) is often virtually the only significant signal in the CI spectrum of aflatoxins. Thus M_1 and G_1 cannot be easily differentiated by CI alone. However, the daughter spectra from CAD of the protonated molecule provide different daughters in both the PCI and NCI modes (Fig. 9).

In the NCI mode, aflatoxin detection limits are considerably improved for two reasons. First, many matrix interferences do not undergo resonance electron capture and become transparent in the analysis. Second, the quantity of negative ions produced for aflatoxins under CI source conditions is ~100 times as great as the positive ions. Thus, CAD daughter spectra can be obtained on picogram

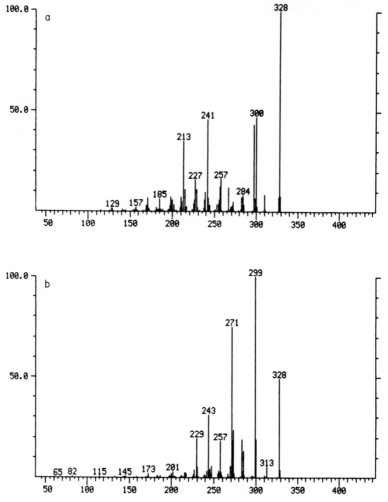

Fig. 8. CAD daughters of m/z 328 from the functional isomers (a) aflatoxin G_1 and (b) M_1.

quantities of aflatoxins in the NCI mode. To demonstrate NCI detection of aflatoxins using MS-MS, simple $CHCl_3$–water (80 : 20) extracts of aflatoxin-free and aflatoxin-contaminated corn samples containing from 8 to 40 ng/g of aflatoxin B_1 were examined. The CAD daughter at m/z 297 from the m/z 312 parent was monitored while an aliquot of the extract was heated on a sample holder of the direct-exposure probe (DEP). Figure 10a shows the daughter signal at m/z 297 from the m/z 312 parent from an aliquot equivalent to 0.2 mg of contaminated corn (8 ng/g B_1), whereas Fig. 10b and c show the response for

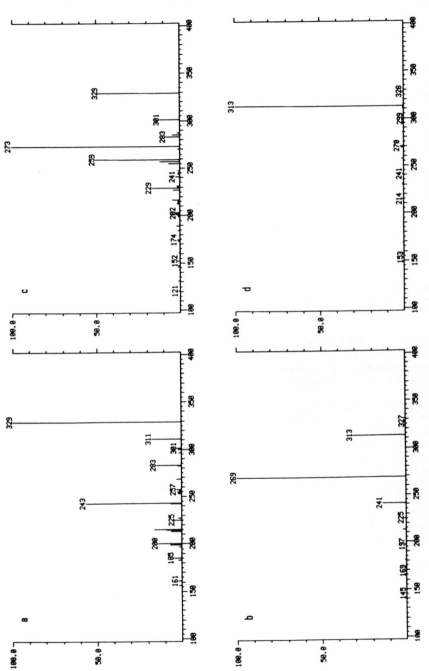

Fig. 9. CI daughter spectra of the MH$^+$ (PCI) and M$^-$ (NCI) for functional isomers aflatoxin G$_1$ (a,b) and aflatoxin M$_1$ (c,d).

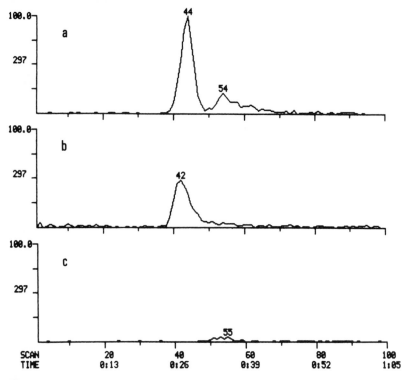

Fig. 10. Reconstructed ion current traces for the m/z 297 daughters from the m/z 312 parent from (a) an aliquot of the extract of a yellow corn contaminated with 8 ppb of aflatoxin B_1 that was equivalent to 0.2 mg, (b) 11.1 pg aflatoxin B_1, and (c) a aliquot of aflatoxin-free corn equivalent to 0.2 mg.

11.1 pg of aflatoxin B_1 and the identical aliquot (0.2 mg) of a crude extract from an aflatoxin-free corn. Thus MS-MS signals from aflatoxin B_1 can be seen in a simple solvent extract of corn without any sample cleanup. Using simple cleanup procedures, larger aliquots can be analyzed with good recoveries, and lower detection limits can be achieved. Without using any internal standard, run-to-run variation of response is considerable with large coefficients of variation (typically as high as 20 to 30%). The use of an internal standard compensates for run-to-run variations of ion yield, and better precision can be expected if a suitable standard is used.

The detection limits, speed, and the simplicity of the analytical scheme for quantitation of aflatoxin B_1 by MS-MS is comparable with other methods and with proper choice of internal standard, MS-MS could provide an attractive alternative method of analysis. The MS-MS technique, in general, so applied to mycotoxins is still very much in the development stage, but preliminary indications are that it should provide the analytical chemist with yet another useful tool, one that is flexible and has application to many problems.

REFERENCES

Cooks, R. G., and Glish, G. L. (1981). Mass spectrometry/mass spectrometry, *Chem. Eng. News* Nov. 30, pp. 40–52.

Kamimura, H., Nishyima, M., Yasuda, K., Saito, K., Ibe, A., Nagayama, T., Ushiyama, H., and Naoi, Y. (1981). Simultaneous detection of several *fusarium* mycotoxins in cereals, grains, and foodstuffs. *J. Assoc. Off. Anal. Chem.* **64**, 1067.

Kondrat, R. W., and Cooks, R. G. (1978). Direct analysis of mixtures by mass spectrometry. *Anal. Chem.* **50**, 81A.

McLafferty, F. W. (1980). Tandem mass spectrometry (MS/MS): a promising new analytical technique for specific component determination in complex mixtures. *Acc. Chem. Res.* **13**, 33–39.

McLafferty, F. W. (1981). Tandem mass spectrometry. *Science* **214**, 280–287.

Mirocha, C. J., Pathre, S. V., Schauerhaner, B., and Christensen, C. D. (1976). Natural occurrence of fusarium toxins in feedstuff. *Appl. Environ. Microbiol.* **32**, 553.

Plattner, R. D., Yates, S. G., and Porter, J. K. (1983). Quadrupole MS/MS of ergot cyclol alkaloids. *J. Agric. Food Chem.* **31**, 785–789.

Porter, J. K., and Betowski, D. J. (1981). Chemical ionization mass spectrometry of ergot cyclol alkaloids. *J. Agric. Food Chem.* **29**, 650.

Scott, P. M., and Lawrence, G. A. (1980). Analysis of ergot alkaloids in flour. *J. Agric. Food Chem.* **28**, 1258–1261.

Slayback, J. R. B., and Storey, M. S. (1981). Chemical analysis problems yield to quadrupole MS/MS. *Ind. Res. Dev. Feb.*, p. 129.

Stults, E. J. T., Enke, C. G., and Holland, J. F. (1983). Mass spectrometry/mass spectrometry of time resolved magnetic dispersion. *Anal. Chem.* **55**, 1323–1330.

Young, C. J. (1981). Variability in the content and composition of alkaloids found in Canadian ergot. *J. Environ. Sci. Health, Part B* **16**(1), 83–11.

Additional papers describing application of MS-MS to mycotoxins

Brumely, W. C., Trucksess, M. W., Adler, S. H., Cohen, C. K., White, K. D., and Sphon, J. A. (1985). Negative ion chemical ionization mass spectrometry of deoxynivalenol (DON): Application to identification of DON in grains and snack foods after quantitation/isolation by thin-layer chromatography. *J. Agric. Food Chem.* **33**, 326–330.

Davidson, W. R., Tanner, S. D., and Ngo, A. (1982). The rapid screening of food products for mycotoxins by MS/MS. *ASMS, Honolulu* Pap. FPA 12.

Games, D. E., Eckers, C., Swain, B. P., and Mallen, D. N. B. (1981). Studies of Ergot alkaloids using LC/MS and MS/MS. *ASMS, Minneapolis* Pap. RAMOA10.

Krishnamurthy, T., and Sarver, E. W. (1984). MS/MS studies on macrocyclic trichothecenes. ASMS, San Antonio Pap. RPB-18.

Krishnamurthy, T., and Sarver, E. W. (1985). Detection and quantification of roridins and baccharinols in brazilian Baccharis plant samples by DCI-MS/MS technique. ASMS, San Diego Pap. WPE-15.

Missler, S. R. (1985). Picogram detection of trichothecene mycotoxins by HPLC/MS. ASMS, San Diego Pap. WPE-2.

Pare, J. R. J., Greenhalgh, R., and Lafontaine, P. (1985). Fast atom bombardment mass spectrometry: A screening technioue for mixtures of secondary metabolites from fungal extracts of *Fusarium* species. *Anal. Chem.* **57**, 1470–1472.

Plattner, R. D., Bennett, G. A., and Slayback, J. R. B. (1982a). Detection of *Fusarium* mycotoxins in a grain matrix by quadrupole MS/MS. *ASMS, Honolulu* Pap. FOD3.

Plattner, R. D., Bennett, G. A., Stubblefield, R. D., and Slayback, J. R. B. (1982b). Identification of aflatoxins by quadrupole MS/MS. *ASMS, Honolulu* Pap. WPA 21.

Sakuma, T., and Law, B. P. (1983). Rapid screening of T-2 and diacetoxyscripenol in grain products by APCI/MS/MS. *ASMS, Boston* Pap. FOF4.

Smith, R. D., and Udseth, H. R. (1983). Mass spectrometry with direct supercritical fluid injection. *Anal. Chem.* **65**, 2266–2272.

Smith, R. D., and Udseth, H. R. (1983). Direct supercritical fluid injection mass spectrometry of trichothecenes. *Biomed. Mass Spectrom.* **10**(10), 577–580.

Smith, R. D., Udseth, H. R., and Wright, B. W. (1985). Rapid and high resolution capillary supercritical fluid chromatography (SFC) and SFC/MS of trichothecene mycotoxins. *J. Chromatog. Sci.* **22**, 192–199.

Udseth, H. R., Writht, B. W., and Smith, R. D. (1985). Supercritical Fluid extraction and chromatography-mass spectrometry of mycotoxins of the trichothecene group. ASMS, San Diego Pap. WPE-4.

15

Taxonomic Approaches to Mycotoxin Identification (Taxonomic Indication of Mycotoxin Content in Foods)

JENS C. FRISVAD

Department of Biotechnology
Food Technology
The Technical University of Denmark
DK-2800 Lyngby, Denmark

I. INTRODUCTION

Mycotoxins can only be identified by chemical and physical analytical methods, and a correlation of the amount of a particular mycotoxin and the number of fungal propagules per gram of food material (or percentage of infection of food) should not be expected (Blaser and Schmidt-Lorenz, 1981; Frisvad and Viuf,

415

MODERN METHODS IN THE ANALYSIS
AND STRUCTURAL ELUCIDATION
OF MYCOTOXINS

1986). However, knowledge of the normal fungal flora or an examination of the mycoflora of the food item may provide valuable information concerning which mycotoxins to analyze for, provided that the fungi are identified correctly to species level. Several environmental and processing factors may affect the fungal population and the toxins produced. The toxigenic fungi may be removed or killed during processing, while extracellular mycotoxins produced by these fungi remain in the food or feedstuff. In the latter cases knowledge concerning the mycoflora of raw material and fungal growth and toxin production on living plants may provide an indication of which mycotoxin occurs in the particular processed food product. In cases of a suspected mycotoxicosis, correct identification of each fungal species in the toxin-containing food or feed item is of utmost importance. In such cases, the investigator should always cite the taxonomic work used for the identifications and provide additional data such as colony diameters and mycotoxin production under defined conditions (Frisvad, 1986).

In the following sections the factors affecting mycotoxin production and the mycological examination of foods will be briefly mentioned, while the connection between known fungal species (mycelial microfungi) and profiles of mycotoxins will be emphasized, especially for the important foodborne genera *Penicillium* Link, *Aspergillus* Micheli ex Link, *Fusarium* Link, and *Alternaria* Nees.

II. FACTORS AFFECTING MYCOTOXIN PRODUCTION

Before discussing the factors affecting the "final" content of different mycotoxins in food and feedstuffs, it is very important to consider the statistical sampling problems in the determination of both mycotoxins and fungal propagules (see also Chapter 2). Mycotoxins and sites of active fungal growth on individual kernels, nuts, fruits, and so on may be heterogeneously distributed in a lot of these items. Sampling plans for small samples may be based on the negative binomial distribution or the compound Poisson γ distribution (Knutti and Schlather, 1982; Waibel, 1981), but where possible, very large samples should be used to minimize the risk of food or feed samples containing unacceptable levels of mycotoxins.

Factors which may affect the concentration of mycotoxins in food are listed in Table I. The dramatic effect of these factors on the fungal flora and the mycotoxins explains the lack of correlation between numbers of fungal propagules and concentration of different mycotoxins. For recent reviews on factors affecting mycotoxin content in foods see Bullerman *et al.* (1984), Northolt and Bullerman (1982), Ray and Bullerman (1982), Rippon (1980), Moss (1984), Lillehoj and Elling (1983), Orth (1981a), and Reiss (1981b).

TABLE I.

Factors Affecting the Connection between Mycotoxin Content and Fungal Propagules in Foods and Feedstuffs

Factor	Examples
Analytical	Methods of analysis of fungi and mycotoxins
Chemical	Preservatives, pH
Physical	Light, temperature, water activity, redox potential, atmosphere
Processing	Filtration, heat treatments, cooling and freezing, irradiation, extrusion and blending, pressure treatment
Biological	Autolysis of fungal cells, microbial competition and succession, microbial transformation and breakdown of mycotoxins, biological interactions of other types, carryover of mycotoxins

In spite of the lack of correlation between mycotoxin content and fungal propagules, toxigenic fungi present in food samples are often recoverable when the proper mycological methods are used (Frisvad and Viuf, 1986; Frisvad *et al.*, 1986a; Hesseltine *et al.*, 1976).

III. MYCOLOGICAL EXAMINATION OF FOODS

A. Fermented Foods

Fermented foods may contain mycotoxins due to (a) carryover or transformation of mycotoxins from raw materials (e.g., aflatoxin M_1 in milk and fermented cheeses originating from aflatoxin B_1-containing feedstuffs), (b) contaminated starter cultures (alien cultures producing mycotoxins), (c) mycotoxin production by the starter cultures, and (d) mycotoxin production by contaminating fungi after the fermentation process. Mycological control of starter cultures used for fermented foods is very important. The starter culture should be examined for purity, correct species, and strain. Some important fermented foods are blue-mold cheeses (*Penicillium roqueforti* Thom) and white-mold cheeses (*Penicillium camemberti* Thom). Contaminants are easily detected in the latter cheeses because of the lanose and white appearance of *P. camemberti*. Mycotoxins produced by this species are restricted to cyclopiazonic acid, but all strains examined so far produce this mycotoxin (Still *et al.*, 1978; Leistner and Eckardt, 1979). Aflatoxin M_1 has also been detected in white-mold cheeses (Kiermeier *et al.*, 1977). The prediction of mycotoxins such as aflatoxin M_1 in cheeses is based on a profound knowledge of feedstuffs used for cows, the processing and storage of these feedstuffs, and knowledge of the transformation of mycotoxins in animals.

Control of *P. roqueforti* starter cultures for purity may be based on the fact that all *P. roqueforti* strains grow well on media containing 0.5% acetic acid, while other penicillia fail to grow on such media (Engel and Teuber, 1978; Frisvad, 1981). However, *P. roqueforti* will not grow on Czapek agar containing sodium acetate instead of sucrose, in contrast to *P. verrucosum* Dierckx var. *cyclopium* (Westling) Samson, Stolk et Hadlok (Veau *et al.*, 1981), a common contaminant in cheese products. The latter contains several distinct species (Frisvad, 1986), and of these, we have found *P. crustosum* Thom and *P. camemberti* group II (wild-type isolates of *P. camemberti*) in cheese products (O. Filtenborg and J. C. Frisvad, unpublished observations). *Penicillium crustosum* produced penitrem A in cream cheese (Richard and Arp, 1979), and *P. camemberti* II produced cyclopiazonic acid and rugulovasines A and B in pure culture (Frisvad and Filtenborg, 1983). *P. roqueforti* Group I can produce PR toxin, roquefortines, and mycophenolic acid, and group II can produce roquefortines, mycophenolic acid, patulin, penicillic acid, and botryodiploidin. Isolates of group II have never been used as starter cultures (Orth, 1981b; Engel and Teuber, 1978; Engel *et al.*, 1982), but starter cultures should be continuously examined for the presence of these contaminants. In an attempt to check isolates of *P. roqueforti*, Engel and Teuber (1983) used metabolite profiles as determined by thin-layer chromatography (TLC) to categorize 78 strains into 15 groups.

Aspergillus oryzae (Ahlburg) Cohn and *A. sojae* Sakaguchi et Yamada are considered to be domesticated forms of the wild-type species *A. flavus* Link and *A. parasiticus* Speare, respectively (Wicklow, 1984; Samson, 1985). The domesticated forms have lost their ability to produce aflatoxins and aflatrem, but many isolates can produce other mycotoxins such as β-nitropropionic acid (Nakamura and Shimoda, 1954; Orth, 1977), cyclopiazonic acid (Orth, 1977), kojic acid (Birkinshaw *et al.*, 1931), and perhaps aspergillic acid (Hamsa and Ayres, 1977; Pitt *et al.*, 1983). *Aspergillus oryzae* is used in the fermentation of oriental foods and in the production of several enzymes (Rehm, 1980). Mycotoxins do not appear to cause problems in other fermented food products (Wang and Hesseltine, 1982). Mold-fermented meat products should be inoculated with known nontoxigenic starter cultures; thus *P. nalgiovense* Laxe ("Edelschimmel Kulmbach") and *P. chrysogenum* Thom ("Sp. 1947") are available for this purpose (Leistner, 1984). *Penicillium chrysogenum* strains used for food fermentation products should be examined for the production of penicillin, PR toxin, and roquefortine C (Frisvad and Filtenborg, 1983).

B. Contaminated Foods and Feedstuffs

Many different methods are available for the qualitative determination of mycoflora in foods and feedstuffs. The most important factors affecting the mycoflora spectrum of a commodity are the isolation techniques, the isolation medium, and the incubation temperature. The incubation conditions should approach the conditions of the commodity under natural conditions and possibly also worst-case situations.

1. Isolation Techniques

Obviously moldy commodities should be examined directly and different types of fungal mycelium and spores transferred to the selected growth medium. If possible, moldy kernels, fruit pieces, and so forth should be examined separately. The remaining kernels may be incubated under conditions of higher humidity and examined directly, or they may be examined by traditional food mycological analysis.

Two different methods are commonly used for the determination of the mycoflora of foods: the dilution plate count technique and the direct plating assay. The former is a quantitative procedure used to determine the number of colony-forming propagules in a sample on a weight or volume basis, while the latter is a procedure used to determine the percentage of food "pieces" that contain one or more viable propagules. Both methods may include a prior surface disinfection treatment of kernels, nuts, fruits, and the like before homogenization or direct plating. The surface disinfection treatment should remove a significant part of the superficial and casual flora of food pieces (Mislivec and Bruce, 1977a,b; Mislivec et al., 1983). Studies by Andrews (1986a) and Frisvad et al. (1986b) have shown that it is important to employ *both* the dilution plate count and the direct-plating assay (using surface-disinfected food pieces, if possible, in the latter case) in the mycological examination of foods to obtain an accurate picture of the mycoflora. The use of the direct-plating assay is especially important in the evaluation of the mycotoxin risk of a product from a mycological viewpoint (Andrews, 1986a). The surface disinfection treatment recommended is to submerge the food pieces in 1% sodium hypochlorite (0.4–0.5% free chlorine) for 1 min, followed by rinsing with sterile water (King et al., 1986; Andrews, 1986c).

2. Selective and Indicative Media

a. General-Purpose Media. The dichloran rose bengal chlortetracycline (DRBC) medium of King et al. (1979) worked well for the general isolation of toxigenic molds from foods (King et al., 1986). For intermediate- and low-moisture foods, the low-dichloran 18% glycerol (DG18) medium of Hocking and Pitt (1980) worked very well (King et al., 1986). A combination of DRBC and DG18 seems to be best (Frisvad et al., 1986c). The recovery of *Paecilomyces* Bainier and many field fungi appears to be better on media without added carbohydrate (Andrews, 1986b; Frisvad et al., 1986a). The dichloran chloramphenicol peptone agar of Andrews (1986b) can be used for field fungi and proteinophilic fungi like *Scopulariopsis* Bainier and *Paecilomyces*. The latter medium gives a good recovery of these fungi, and sporulation is much better than on DRBC agar. A combination of all three media will sometimes be necessary depending on the food or feedstuff investigated.

b. Selective and Indicative Media for Toxigenic Fungi. Two indicative media for toxigenic fungi have been developed: the *Aspergillus flavus* and *parasiticus* agar (AFPA), for the species containing aflatoxin producers (Pitt *et al.*, 1983), and the pentachloronitrobenzene rose bengal yeast extract sucrose agar (PRYES) for subgroups of *Penicillium viridicatum* Westling producing different nephrotoxins (Frisvad, 1983). The presence of *A. flavus* and *A. parasiticus* in foods and feedstuffs is often indicative of aflatoxin contamination, and chemical analysis for the toxin should be performed (Blaser and Schmidt-Lorenz, 1981). Ochratoxin-containing barley samples always contained propagules of *P. viridicatum* group II as determined by the dilution plating technique, the direct-plating assay, or both (Frisvad and Viuf, 1986).

3. Incubation Temperature

Most toxigenic fungi will grow well at 25°C, but many penicillia will compete better at temperatures of 10° to 20°C (Frisvad, 1983; Hald *et al.*, 1983). Other fungi like *Aspergillus fumigatus* Fresenius compete better (and will often only appear) on selective media at higher incubation temperatures (30°, 37° or 45°C). Knowledge of the environmental factors affecting the actual food or feed sample will indicate which incubation temperature(s) is necessary.

IV. MYCOTOXINS AND FUNGAL TAXONOMY

Classical fungal taxonomy has traditionally been based on morphology, which means that form genera have been used in the Deuteromycetes, although some genera are separated on the basis color. Melanins, produced by polyketide precursors (secondary metabolites), have gained primary importance in the protection against radiation for many fungi such as the dematiaceous fungi, *Emericella* Berk. et Br., *Penicillium,* and so on (Wheeler, 1983; Martinelli and Bainbridge, 1974; J. C. Frisvad, unpublished observations; Ha-Huy-Ke and Luckner, 1979). Secondary metabolites are also used to separate the closely related ascomyceteous genera *Nectria* (Fr.) Fr. and *Gibberella* Sacc. (Booth, 1971). However, the concept of form-species should not be used because species have an objective reality in nature (Carmichael *et al.,* 1980). Table II lists the most important species concepts applied for living organisms. The taxonomy of important genera like *Penicillium* and *Aspergillus* has been based on the omnispective species concept, with different character weighting and selection (Pitt, 1979a; Raper and Fennell, 1965; Raper and Thom, 1949; Thom, 1954; Samson, 1979), but several

TABLE II

Species Concepts Used for the Fungi

Concepts	Connection to other characters[a]	Identification of new isolates	Reference
Botanico-anatomical[b]	Fair	Simple	Gams (1984)
Biological	—	Not operational	Mayr (1969)
Omnispective	Good	Sometimes difficult (subjective criteria and character weighting)	Blackwelder (1964)
Genetic	Very good	Difficult, time-consuming	Sneath and Sokal (1973)
Phyletic	—	Not operational	Sneath and Sokal (1973)
Taxometric	Very good	Often time-consuming	Sneath and Sokal (1973)
Differentiation	Very good	Rather simple	Frisvad (1984)

[a] The predictive value of a specific name, judged on the basis of the value of these species concepts in *Penicillium, Aspergillus,* and *Fusarium.*

[b] A pure anatomical species concept is seldom used, and conidium color and other criteria are often included as important characters in species descriptions (Hawksworth *et al.,* 1983).

investigators emphasize micromorphological criteria (Ramirez, 1982; Samson *et al.,* 1976, 1977a,b; Stolk and Samson, 1983; Abe *et al.,* 1982, 1983a,b,c,d). Future species concepts may be based on numerical treatments of several characters, on differentiation characters (Frisvad, 1984), or on genetic evidence from such factors as DNA homology, DNA base sequences, or somatic hybridization (Burnett, 1983; Samson and Gams, 1984). Species based on differentiation characters (Frisvad, 1984) may prove to be both fundamental and practical taxa (Frisvad and Filtenborg, 1983) as the characters of micromorphology, macromorphology, and secondary metabolites gain acceptance (Samson and Gams, 1984; Onions *et al.,* 1984). Unfortunately, no key or monographs based on this concept have as yet been published.

Profiles of mycotoxins and other secondary metabolites seem to be closely related to known and accepted terverticillate *Penicillium* species (Frisvad, 1981; Frisvad and Filtenborg, 1983; Samson and Gams, 1984), and an agreement on the more difficult *Penicillium* species (the *Penicillium cyclopium–P. viridicatum* complex) appears to be possible. The seemingly poor relationship between some fungal species and profiles of secondary metabolites (Bu'Lock, 1980; Bennett, 1983) is caused by many misidentifications of fungi and fungal toxins (Frisvad and Filtenborg, 1983; Pitt, 1979b) and the different species concepts employed. A correct identification of a fungus is essential to obtain all information connected with a name (Samson and Gams, 1984; Frisvad, 1984; Samson *et al.,* 1986). Most mycotoxins can be isolated and detected easily by the screening methods (based on TLC) of Filtenborg and Frisvad (1980) and Filtenborg *et al.* (1983). The next section contains information on the mycotoxins produced by known homogeneous species or species subgroups (Ciegler *et al.*, 1973; Frisvad and Filtenborg, 1983) and a guide to the descriptions of these species.

V. IMPORTANT GENERA CONTAINING TOXIGENIC ISOLATES

The most important genera containing toxigenic isolates are *Penicillium, Aspergillus, Fusarium,* and *Alternaria.* Species in these genera are widespread and occur with high frequency in foods and feedstuffs and many other items. The taxonomy of these four genera is in a state of change. Most recent work has been done in *Penicillium* and *Fusarium* and some series in *Aspergillus* (Samson and Gams, 1984). The tables of toxigenic fungi given here are based on the new classification of the fungi outlined by Hawksworth *et al.* (1983), but the species of *Penicillium* and *Aspergillus* reported to produce mycotoxins have been examined by Frisvad (1983 and unpublished observations) and revised accordingly. Where possible, references from primary literature for the penicillia are given to the correctly identified fungi.

A. *Penicillium*

The close fundamental relationships between profiles of mycotoxins and *Penicillium* species have been obscured by many misidentifications and diverging taxonomic viewpoints and species concepts (Frisvad, 1984). Table III lists an array of different names given to fungi representative of *Penicillium crustosum* Thom (Raper and Thom, 1949; Pitt, 1979b) as well as a variety of isolates which were representative of other species, but were given the name *P. crustosum. Penicillium griseum* Bonorden ATCC 20068 (not the original type, which was lost) is also a *P. crustosum* isolate. The former name was used by Ramirez (1982), but it is of uncertain application. This example shows how important it is to confirm identifications with additional criteria (Frisvad, 1986) or send the isolates to major culture collections for confirmation of identity.

1. Subgenus Aspergilloides Dierckx

As pointed out by Samuels (1984) and Moss (1984), particular mycotoxins may be produced by very different fungi and even plants. The production of the same mycotoxin does not necessarily indicate a close relationship between different fungi (Frisvad and Filtenborg, 1983), but a complete profile of secondary metabolites seems to be species-specific. Section- or series-specific secondary metabolites seem to be nonexistent in the genus *Penicillium.* However, metabolites such as penicillic acid, brefeldin A, citreoviridin, patulin, and xanthocillin

TABLE III

Taxonomic Problems in *Penicillium crustosum*

A. Isolates typical of *P. crustosum sensu stricto* but given other names

Incorrect name	Culture collection number	Reference
P. casei Staub	CBS 483.75	Engel and Teuber (1978)
P. commune Thom	AUA 827	Wagener *et al.* (1980)
P. cyclopium Westling	NRRL 3476	Wilson *et al.* (1968)
	NRRL 3477	Wilson *et al.* (1968)
	NRRL 6093	Vesonder *et al.* (1980)
	ATCC 32014	Wells and Payne (1976)
P. lanoso-coeruleum Thom	ATCC 32017	Wells and Cole (1977)
	ATTC 32020	Wells and Cole (1977)
P. meleagrinum Biourge	ATCC 32021	Wells and Payne (1976)
P. palitans Westling	NRRL 3468	Ciegler (1969)
P. piceum[a] Raper et Fennell	V76/170A/28	Patterson *et al.* (1979)
P. roqueforti Thom	ATCC 32027	Wells and Payne (1976)
P. verrucosum Dierckx var. *cyclopium*	Leistner Sp 458	Leistner and Eckardt (1979)
(Westling) Samson, Stolk et Hadlok	Leistner Sp 1191	
P. verrucosum var. *melanochlorum*	Leistner Sp 2304	Leistner and Eckardt (1979)
Samson, Stolk et Hadlok		
P. viridicatum Westling	LC 212	
P. terrestre Jensen *sensu* Raper et	IMI 89384	
Thom	IMI 89385	
P. farinosum Novobranova	IMI 174717	Kozlovsky *et al.* (1981)

B. Isolates called *P. crustosum*, but representing other species

Isolate number	Correct identity[b]	Reference
ATCC 48414	*P. aurantiogriseum* Dierckx	Hald *et al.* (1983)
CCM F-389	*P. viridicatum* group IV	L. Marvanova (personal communication)
IMI 285524	*P. viridicatum* group IV	J. I. Pitt (personal communication)
LC 1628	*P. camemberti* Thom group II	J. Nicot (personal communication)
Leistner Sp 607	*P. camemberti* group II	Leistner and Eckardt (1979)
ATCC 32028 (as *P. terrestre*)	*P. camemberti* group II	Wells and Payne (1976)
IMI 285507	*P. camemberti* group II	J. I. Pitt (personal communication)
CBS 487.75	*P. mali* Gorlenko et Novobranova	Pitt (1979a)
FRR 1621	*P. echinulatum* Raper et Thom ex Fassatiova	Pitt (1979a)

[a] Received as *P. piceum*, called *P. canescens* by Patterson *et al.* (1979). The culture received was a typical *P. crustosum*.

[b] Some of these names are provisional and experimental (Frisvad and Filtenborg, 1983), but none of the fungi represent *P. crustosum sensu stricto* (see also Soderstrom and Frisvad, 1984). *P. viridicatum* IV will be described as a new taxon, *P. camemberti* II = *P. commune* Thom and *P. mali* = *P. melanochlorum* (Samson, Stolk et Hadlock) Frisvad = *P. solitum* Westling.

are only found in *Eupenicillium* Ludwig and the subgenera *Aspergilloides, Furcatum* Pitt, and *Penicillium,* while mycotoxins like rugulosin and emodin are only found in the genus *Talaromyces* C.R. Benjamin and subgenus *Biverticillium* Dierckx. The only toxins produced exclusively by isolates of subgenus *Aspergilloides* are gliotoxin and costaclavin. Gliotoxin is produced even by species in different series in the latter subgenus. Citrinin and penicillic acid are common in the series *Glabra,* but citrinin is also produced by *P. citreonigrum* in the series *Citreonigra,* and penicillic acid is produced by *P. restrictum* in the *Restricta* series (see Table IV). Several isolates representative of each species should be tested for mycotoxin production. The most common species in foods are *P. glabrum, P. spinulosum,* and *P. citreonigrum.*

2. Subgenus Furcatum *Pitt*

Mycotoxins found only in *Furcatum* are the janthitrems, β-nitropropionic acid, verruculogen, oxalin, secalonic acid D, terrein, and paxilline. Penicillic acid and griseofulvin are particularly common in this subgenus (Table V). Citrinin is produced by three species (in two different series) and roquefortine C by two species. The producers of the latter toxin may be more related to subgenus *Penicillium* than other species in *Furcatum* (*P. oxalicum* and *P. sclerotigenum*). *Furcatum* species are often found in soil, but some species such as *P. janczewskii, P. fellutanum, P. oxalicum, P. simplicissimum, P. raistrickii, P. citrinum,* and *P. miczynskii* are common in foods and feedstuffs. These species are, with the exception of *P. citrinum,* not as abundant as the terverticillate penicillia. However, tremorgen producers are found in all series in *Furcatum* except *Fellutana* Pitt and *Megaspora* Pitt (janthitrems by *P. janthinellum,* penitrem A by *P. janczewskii,* verruculogen by *P. simplicissimum,* an unknown tremorgen by *P. novae-zeelandiae,* and paxilline by *P. paxilli*). Common soil isolates from *Furcatum* are therefore important producers of griseofulvin, penicillic acid, and tremorgens, while important foodborne *Furcatum* species (*P. citrinum* and *P. miczynskii*) produce yellow toxins such as citrinin and citreoviridin.

3. Subgenus Penicillium

Subgenus *Penicillium* taxonomy is in a state of change (Onions *et al.,* 1984; Samson and Gams, 1984), but profiles of secondary metabolites seem to support the classification of these fungi based on morphology (Frisvad and Filtenborg, 1983; Samson and Gams, 1984). Subgenus *Penicillium* contains species which are rather closely related to each other, and many of these species are very important food and feed contaminants. Nearly all species (with the notable exception of *P. italicum* and *P. digitatum* which cause rot in citrus fruits) produce a number of potent mycotoxins (Table VI). *Penicillium crustosum* (penitrem A and roquefortines), *P. expansum* (patulin and roquefortine C), *P. aurantiogriseum* I and II, *P. viridicatum* I (penicillic acid and nephrotoxic naphtho-

TABLE IV

Profile of Mycotoxins in the Subgenus *Aspergilloides* Dierckx[a]

Section, series, species	Mycotoxins	References
Aspergilloides Pitt		
Glabra Pitt		
P. *glabrum*[b] Wehmer	Citromycetin	Hetherington and Raistrick (1931a)
P. *purpurescens* Sopp Biourge	Citrinin	Leistner and Pitt (1977); Lillehoj and Goransson (1980)
	Ochratoxin A	Mintzlaff *et al.* (1972)
P. *spinulosum* Thom	Fumigatin	Anslow and Raistrick (1938)
	Penitrem A	Leistner and Pitt (1977)
	Spinulosin	Anslow and Raistrick (1938)
P. *lividum*	Citrinin	Pollick (1947)
	Penicillic acid	Gorbach and Friederick (1948)
P. *thomii*	Citrinin	Moreau (1974)
	Penicillic acid	Karow *et al.* (1944)
Implicata Pitt		
P. *bilaii* Chalabuda	Gliotoxin[c]	Brian (1946)
P. *chermesinum* Biourge	Costaclavin	Agurell (1964)
Exilicaulis Pitt		
Restricta Raper and Thom ex Pitt		
P. *restrictum* Gilman et Abbott	Penicillic acid	Gorbach and Friederick (1948)
P. *roseopurpureum* Dierckx	Citromycetin	Hetherington and Raistrick (1931a)
P. *vinaceum* Gilman et Abbott	Gliotoxin	Kharchenko (1970)
Citreonigra Pitt		
P. *citreonigrum* Dierckx	Citreoviridin[c]	Sakabe *et al.* (1964)
	Citrinin	Pollock (1947)
	Viomellein[d]	Zeeck *et al.* (1979)
P. *turbatum* Westling	A 26771 (near gliotoxin)	Michel *et al.* (1974)
P. *decumbens* Thom	Brefeldin A	Singleton *et al.* (1958)
P. *cyaneum* (Bain et Sartory) Biourge	Brefeldin A	Betina *et al.* (1962)
P. *adametzii* Zaleski	Gliotoxin	Kharchenko (1970)

[a] Taxonomic references and circumscriptions: Pitt (1979a).

[b] P. *glabrum* has been reported to produce penitrem A (Mintzlaff *et al.*, 1972), but later publications (Leistner and Eckardt, 1979) show that either toxin or mold was misidentified.

[c] Confirmed for several cultures including the culture ex type (J. C. Frisvad, unpublished observations).

[d] Viomellein is produced in connection with several other biosynthetically related metabolites (semivioxanthin, 3,4-dehydroxanthomegnin a.o.).

TABLE V

Profile of Mycotoxins in the Subgenus in the *Furcatum* Pitt[a]

Section, series, species	Mycotoxins	Taxonomic reference[b]	Reference
Divaricatum Raper et Thom ex Pitt			
Janthinella Thom ex Pitt			
P. *janthinellum* Biourge	Janthitrem B	1	Gallagher *et al.* (1980)
Canescentia Raper et Thom ex Pitt			
P. *raciborskii* Zaleski	Mycophenolic acid	2–4[c]	Quintanilla (1983)
P. *damascenum* Baghdadi	Citrinin	2	J. C. Frisvad (unpublished)
P. *janczewskii* Zaleski	Griseofulvin	1[d]	Brian *et al.* (1949)
			Jeffrey *et al.* (1953)
	Penicillic acid		J. C. Frisvad (unpublished)
	Penitrem A		J. C. Frisvad (unpublished)
P. *lanosum*	Griseofulvin	5	J. C. Frisvad (unpublished)
	Kojic acid		J. C. Frisvad (unpublished)
			Birkinshaw *et al.* (1931)[e]
P. *atrovenetum* G. Smith	β-Nitropropionic acid	2	Raistrick and Stossl (1958)
Fellutana Pitt			
P. *fellutanum* Biourge	Carolic acid	1	Clutterbuck *et al.* (1934)
	Spinulosin		Bracken and Raistrick (1947)
P. *jensenii* Zaleski	Griseofulvin	2	J. C. Frisvad (unpublished)
P. *kojigenum* G. Smith	Kojic acid	2	Smith (1961)
Furcatum Pitt			
Oxalica Raper et Thom ex Pitt			
P. *oxalicum* Currie et Thom	Oxalin	1	Nagel *et al.* (1974)
	Roquefortine C		Vleggaar and Wessels (1980)
	Secalonic acid D		Steyn (1970)
P. *simplicissimum* (Oudem.)	Penicillic acid	1	Betina *et al.* (1969)
Thom	Verruculogen		Fayos *et al.* (1974)
			Pitt (1979b)
	Viridicatumtoxin		Leistner and Eckardt (1979)
P. *sclerotigenum* Yamamoto	Griseofulvin	1	Clarke and McKenzie (1967)
	Roquefortine C		J. C. Frisvad (unpublished)[a]
P. *matriti* G. Smith	Penicillic acid	1	Birkinshaw and Gowlland
			(1962)
P. *raistrickii* G. Smith	Penicillic acid	1	J. C. Frisvad (unpublished)
	Griseofulvin		Brian *et al.* (1955)
	Terrein		Grove (1954)
P. *novae-zeelandiae* van Beyma	Patulin	1	Burton (1949)
	Tremorgen		Di Menna and Mantle (1978)
P. *fennelliae* Stolk	Penicillic acid	1	van Eijk (1969)
P. *michaelis* Quintanilla	Mycochromanic acid	6	Quintanilla (1982a)
P. *smithii* Quintanilla	Citreoviridin	7	J. C. Frisvad (unpublished)[a]
Citrina Raper et Thom ex Pitt			
P. *citrinum* Thom	Citrinin	1	Hetherington and Raistrick
			(1931b)
P. *steckii* Zaleski	Citromycetin	2,3	Turner and Aldridge (1983)

TABLE V   (*Continued*)

Section, series, species	Mycotoxins	Taxonomic reference[b]	Reference
P. miczynskii Zaleski	Citreoviridin	1	Leistner and Pitt (1977) Pitt (1979a) Rebuffat *et al.* (1980)
	Citrinin		Pollock (1947)
P. paxilli Bainier	Paxilline	1	Cole *et al.* (1974)

[a] Part of the data in this table is taken from a reevaluation of the *Furcatum* species. A great number of published *Furcatum* species–mycotoxin connections are not included in this table because they were based on misidentified fungi or mycotoxins.

[b] Taxonomic references: 1. Pitt (1979a); 2. Ramírez (1982); 3. Raper and Thom (1949); 4. Quintanilla (1983); 5. Samson *et al.* (1976); 6. Quintanilla (1982a); 7. Quintanilla (1982b).

[c] Fresh isolates of *P. raciborskii* (= *P. fagi* Martínez et Ramírez = *P. caerulescens* Quintanilla) produce a blackish to brownish dark green reverse on most diagnostic media in age (Martínez and Ramírez, 1978; Quintanilla, 1983).

[d] Isolates intermediate between *P. canescens* Sopp and *P. janczewskii* (= *P. kapuscinskii* Zaleski) (see Pitt, 1979a, p. 253) are considered to be *P. janczewskii* by the author.

[e] *P. lanosum* was considered close to *Furcatum* species like *P. jensenii* (Samson *et al.*, 1976). *P. lanosum* IMI 90463 was identified as *P. daleae* by C. Thom (Birkinshaw *et al.*, 1931).

quinones), and *P. viridicatum* II and III (ochratoxins and citrinin) are particularly important and widespread toxigenic species. *Penicillium brevicompactum, P. griseofulvum, P. chrysogenum, P. roqueforti,* and *P. camemberti* II may also prove to be important toxigenic species. However, natural mycotoxicoses caused by the latter five species have seldom been reported. Cyclopiazonic acid, roquefortines (see Section V,A,2), terrestric acid, chaetoglobosin C, S toxin, brevianamides, PR toxin, and botryodiploidin are only produced by members of terverticillate asymmetric penicillia. The most important toxins produced by members of the subgenus *Penicillium* are ochratoxin A, citrinin, penitrem A, patulin, penicillic acid, roquefortines, cyclopiazonic acid, and naphthoquinones, and most of these toxins are also produced by species in the other subgenera. The identification of these species may be performed according to Pitt (1979a) combined with Frisvad and Filtenborg (1983), or according to Samson *et al.* (1976, 1977a,b) combined with Frisvad (1986). The identifications can be confirmed using known culture collection strains (recently isolated strains or cultures ex type in good condition) as standards and comparing these strains with the newly isolated strains using the simple TLC method developed by Filtenborg and Frisvad (1980) and Filtenborg *et al.* (1983).

4. *Subgenus* Biverticillium *and genus* Talaromyces

Some important mycotoxins found only in the subgenus *Biverticillium* and the genus *Talaromyces* are rubratoxins, verruculotoxin, cyclochlorotin, emodin,

TABLE VI

Profile of Mycotoxins in the Subgenus *Penicillium*

Species and group[a]	Mycotoxins	Taxonomic references[b]	Reference
P. camemberti I	Cyclopiazonic acid	1	Still *et al.* (1978)
P. camemberti II (= *P. commune* Thom)	Cyclopiazonic acid	8[c]	Leistner and Pitt (1977)
			Hermansen *et al.* (1984)
	Rugulovasines A and B		Dorner *et al.* (1980)
P. crustosum	Penitrem A	Table IV	Wilson *et al.* (1968)
	Roquefortine A		Cole *et al.* (1983)
	Roquefortine C		Kyriakidis *et al.* (1981)
	Terrestric acid		Birkinshaw and Raistrick (1936)
P. hirsutum Dierckx I (= *P. corymbiferum* Westling)	Roquefortine A	2,3	Ohmomo *et al.* (1980)
	Roquefortine C		Ohmomo *et al.* (1980)
	Terrestric acid		Frisvad and Filtenborg (1983)
P. hirsutum II	Citrinin	8	Frisvad and Filtenborg (1983)
	Penicillic acid		Frisvad and Filtenborg (1983)
	Roquefortine C		Frisvad and Filtenborg (1983)
	Terrestric acid		Frisvad and Filtenborg (1983)
P. hirsutum III (= *P. hordei* Stolk)	Roquefortine C	5	Frisvad and Filtenborg (1983)
	Terrestric acid		Frisvad and Filtenborg (1983)
P. hirsutum IV[d] (close to *P. echinulatum*)	Chaetoglobosin C	8	Springer *et al.* (1976)
P. aurantiogriseum Dierckx I (= *P. cyclopium*)	Penicillic acid	1,8	Birkinshaw *et al.* (1936)
	S-Toxin		Leistner (1984)
P. aurantiogriseum II	Penicillic acid	8	Alsberg and Black (1913)
			Wirth and Klosek (1972)
	Penitrem A		Ciegler and Pitt (1970)
	Terrestric acid		Frisvad and Filtenborg (1983)
	Xanthomegnin[e]		Stack and Mislivec (1978)
P. viridicatum I	Brevianamide A	8	Wilson *et al.* (1973)
	Penicillic acid		Mintzlaff *et al.* (1972)
	Xanthomegnin[e]		Stack *et al.* (1977)
P. chrysogenum	PR Toxin	9	Frisvad and Filtenborg (1983)
	Roquefortine C		Engel and Teuber (1978)
P. roqueforti I	Mycophenolic acid	10	Lafont *et al.* (1979)
	Roquefortine A		Scott *et al.* (1976)
	Roquefortine C		Scott *et al.* (1976)
	PR Toxin		Wei *et al.* (1973)
P. roqueforti II	Botryodiploidin	8	Moreau *et al.* (1982)
	Mycophenolic acid		Lafont *et al.* (1979)
			Engel *et al.* (1982)
	Patulin		Olivigni and Bullerman (1977)
	Penicillic acid		Olivigni and Bullerman (1977)
	Roquefortine C		Engel and Teuber (1978)
P. brevicompactum Dierckx	Brevianamide A	1	Birch and Wright (1969)
	Botryodiploidin[f]		Fujimoto *et al.* (1980)
			Tsunoda *et al.* (1977)
	Mycophenolic acid		Clutterbuck *et al.* (1932)
P. expansum Link	Citrinin	1	Haese (1963)
			Harwig *et al.* (1973)

TABLE VI (*Continued*)

Species and group[a]	Mycotoxins	Taxonomic references[b]	Reference
	Patulin		van Luijk (1983)
			Anslow *et al.* (1943)
	Roquefortine C		Ohmomo *et al.* (1980)
P. viridicatum II (= *P. verrucosum* Dierckx)	Citrinin	8	Krogh *et al.* (1970)
	Ochratoxin A		van Walbeek *et al.* (1969)
P. viridicatum III (= *P. verrucosum*)	Ochratoxin A	1,8	Ciegler *et al.* (1973)
P. viridicatum IV	Griseofulvin	8	Hutchinson *et al.* (1973)
	Viridicatumtoxin		Hutchinson *et al.* (1973)
P. griseofulvum	Cyclopiazonic acid	1	Holzapfel (1968)[g]
	Griseofulvin		Oxford *et al.* (1939)
	Patulin		Anslow *et al.* (1943)
	Roquefortine C		Ohmomo *et al.* (1980)
P. concentricum Samson, Stolk et Hadlok I [= *P. coprophilum* (Berk. et Curt.) Seifert et Samson]	Griseofulvin	5	Leistner and Eckardt (1979)
	Roquefortine C		Frisvad and Filtenborg (1983)
P. concentricum II[h]	Patulin	8	Frisvad and Filtenborg (1983)
	Roquefortine C		Frisvad and Filtenborg (1983)
P. granulatum Bainier [= *P. glandicola* (Oud.) Seifert et Samson]	Patulin	1,5	Barta and Mecir (1948)
	Penitrem A		Ciegler and Pitt (1970)
	Roquefortine C		Frisvad and Filtenborg (1983)
P. claviforme Bainier[i] [= *P. vulpinum* (Cooke et Massee) Seifert et Samson]	Patulin	1,5	Chain *et al.* (1942)
P. clavigerum Demelius	Penitrem A	5	Shreeve *et al.* (1978)

[a] Some of the groups are provisional experimental names (known species subdivided into groups) (see Frisvad and Filtenborg, 1983).

[b] 1. Pitt (1979a); 2. Ramirez (1982); 3. Raper and Thom (1949); 5. Samson *et al.* (1976); 8. Frisvad and Filtenborg (1983); 9. Samson *et al.* (1977b); 10. Samson *et al.* (1977a).

[c] *P. camemberti* II is composed of creatine-positive (see Frisvad, 1981) isolates in *P. aurantiogriseum, P. olivicolor* Pitt, *P. puberulum, P. viridicatum,* and *P. verrucosum.* Fast-growing isolates in *P. camemberti* II looks superficially like *P. crustosum,* but can be distinguished from the latter by the inability to produce a rot in pomaceous fruits, the darkening of YES agar in age, and the rather moderate production of conidia (no crust formation) (Frisvad, 1983, 1984, 1985; Hermansen *et al.,* 1984; Soderstrom and Frisvad, 1984; Frisvad and Filtenborg, 1983).

[d] *P. hirsutum* IV is more closely related to *P. echinulatum* than to *P. hirsutum.* The conidiophores are more divaricate in *P. hirsutum* IV than in any of the latter species, however.

[e] Usually only one of the biosynthetically related mycotoxins is listed. Besides xanthomegnin, viomellein, and vioxanthin, other naphthoquinones are produced, though in smaller amounts.

[f] Botryodiploidin was reported as a mycotoxin of *P. carneolutescens* G. Smith. Two strains were received from Dr. Fujimoto and both were typical of *P. brevicompactum.*

[g] The original producers of cyclopiazonic acid, *P. cyclopium* CSIR 1082 and CSIR 1389, are both representative of *P. griseofulvum.*

[h] *P. concentricum* II is very closely related to *P. granulatum.* Older strains (like NRRL 2034) have smooth stipes and have lost the ability to produce patulin and penitrem A (see Ciegler and Pitt, 1970).

[i] *P. claviforme* and *P. clavigerum* (under the name *P. duclauxii* Delacr.) are placed in *Biverticillium* Dierckx by Pitt (1979a).

erythroskyrin, simatoxin, rugulosin, and tardin (see Tables VII, VIII). The only toxins shared with other subgenera in *Penicillium* are kojic acid, ochratoxin A, and rugulovasines (Table VI). Species of the series *Islandica* seem to be closely related to species in the genus *Talaromyces* as judged by the secondary metabolites produced. The most important foodborne species are *P. purpurogenum, P. islandicum, P. variabile,* and *P. rugulosum.* Species of *Talaromyces* may be found in heat-processed foods.

5. The genus **Eupenicillium** *Ludwig*

Species of the genus *Eupenicillium* may also be found in heat-processed foods. Few of the species have been tested for mycotoxin production, but some of the species are known to produce important mycotoxins (Table IX). Cleistothecia of *Eupenicillium ochrosalmoneum* have been found in corn (Wicklow *et al.*, 1982).

TABLE VII

Profile of Mycotoxins in the Subgenus *Biverticillium* Dierckx

Section, series, species	Mycotoxins	Taxonomic references[a]	References
Simplicium (Biourge) Pitt			
Miniolutea Pitt			
P. purpurogenum	Kojic acid	1	Parrish *et al.* (1966)
	Rubratoxin		Wilson and Wilson (1962a,b)
	Rugulovasines		Dorner *et al.* (1980)
P. verruculosum	Ochratoxin A		Lillehoj and Goransson (1980)
	Verruculotoxin		Cole *et al.* (1975)
Islandica Pitt			
P. islandicum Sopp	Islanditoxin	1	Marumo and Sumiki (1955)
	Cyclochlorotin		
	Emodin		Howard and Raistrick (1954)
	Erythroskyrin		Howard and Raistrick (1954)
	Luteoskyrin		Uraguchi *et al.* (1961)
	Rugulovasine		Cole *et al.* (1976)
	Simatoxin		Ghosh *et al.* (1977)
	(−)Rugulosin		Takeda *et al.* (1973)
P. brunneum Udagawa	Emodin	1	Shibata and Udagawa (1963)
	Rugulosin		Tsunoda (1963)
P. variabile Sopp	Ochratoxin A	1	Mintzlaff *et al.* (1972)
	Rugulosin		Leistner and Pitt (1977)
P. rugulosum Thom	Rugulosin	1	Yamamoto *et al.* (1956)
	Rugulovasine		Abe *et al.* (1969)
P. tardum Thom	Emodin	3	Tatsuno *et al.* (1975)
	Tardin		Borodin *et al.* (1947)

[a] 1. Pitt (1979a); 3. Raper and Thom (1949).

TABLE VIII

Profile of Mycotoxins in the Genus *Talaromyces*
C.R. Benjamin

Section, series, species[a]	Mycotoxins	References
Talaromyces		
Lutei Pitt		
T. *luteus* (Zukal) C.R. Benjamin	Sterigmatocystin	Dean (1963)
T. *wortmannii* (Klöcker) C.R. Benjamin	Rugulosin	Breen *et al.* (1955)
Genus *Merimbla* Pitt anamorph		
T. *avellaneus* (Thom et Turesson) C.R. Benjamin	Emodin	Natori *et al.* (1965)

[a] According to Pitt (1979a).

This species is an excellent producer of citreoviridin. *Eupenicillium lapidosum*, cited as a very heat-resistant mold (Williams *et al.*, 1941), is a good producer of patulin.

B. *Aspergillus*

The taxonomy of the genus *Aspergillus* is still based on the monograph of Raper and Fennell (1965). Samson (1979) has treated new species of *Aspergillus* taxonomically, and some of the groups of the genus such as the *Aspergillus glaucus* group (Blaser, 1976), the *Aspergillus niger* group (Al-Musallam, 1980), the *Aspergillus flavus* group (Christensen, 1981), the *Aspergillus ochraceus* (and *Petromyces*) group (Christensen, 1982), and the *Aspergillus nidulans* (*Emericella*) group (Christensen and States, 1982) have been reevaluated (see also Samson and Pitt, 1985). The aspergilli are widespread in nature and occur with high frequency in foods, feeds, and many other substrates. They include effective producers of the most important carcinogenic mycotoxins such as the aflatoxins, sterigmatocystin, and ochratoxin A. Table X lists the mycotoxins produced by species in *Aspergillus*. Obvious misidentifications of molds or mycotoxins have not been included. The production of cyclopiazonic acid by *Aspergillus versicolor* (Vuill.) Tiraboschi, for example, has very often been cited, even though the original isolate producing the mycotoxin has been reidentified to *A. oryzae* (Ahlburg) Cohn (Domsch *et al.*, 1980).

Aspergillus species are especially interesting from a chemotaxonomic point of view. Some species (*A. clavatus, A. fumigatus,* and *A. terreus*) produce a large

TABLE IX

Profile of Mycotoxins in the Genus
Eupenicillium **Ludwig**

Series, species	Mycotoxins	References
Fracta Pitt		
E. *ochrosalmoneum* Scott et Stolk	Citreoviridin	Ueno and Ueno (1972)
Javanica Pitt		
E. *brefeldianum* (B. Dodge) Stolk et Scott	Brefeldin A	Härri *et al.* (1963)
E. *ehrlichii* (Klebahn) Stolk et Scott	Penicillic acid	Gorbach and Friederick (1949)
Lapidosa Pitt		
E. *lapidosum* Scott et Stolk	Patulin	Myrchink (1967)
Crustacea Pitt		
E. *baarnense* (van Beyma) Stolk et Scott	Penicillic acid	Burton (1949)
E. *egyptiacum* (van Beyma) Stolk et Scott	Xanthocillin	Vesonder (1979)

number of biosynthetically different secondary metabolites, indicating a broad species concept, while other species produce few known mycotoxins in common with other species in the same group, indicating a narrow species concept. The *Aspergillus nidulans* group (and *Emericella*) contains several taxa which could be regarded as the same species. These are E. *nidulans* var. *nidulans*, E. *nidulans* var. *lata*, E. *nidulans* var. *echinulata*, E. *nidulans* var. *dentata*, E. *acristata*, E. *quadrilineata*, E. *rugulosa*, and E. *cleistominuta*. They all produce the same profile of secondary metabolites and are alike in most morphological characters except ascospore morphology (Frisvad, 1985). These results are supported by protoplast fusion experiments (Kevei and Peberdy, 1984). Members of the *A. ochraceus* group seem to be related as judged by their known profile of secondary metabolites. It is not known whether some secondary metabolites are "group specific" and other secondary metabolites are species specific. Even though many metabolites are common to several species in the *A. flavus* group, these species can be separated on the basis of profiles of secondary metabolites including data on the relative amount of biosynthetically related toxins (Dorner *et al.*, 1984). The chemotaxonomy of *Aspergillus* species is complicated by the fact that some metabolites are morph or structure connected. For example, asperthecin is only produced in the cleistothecia of E. *nidulans* (Neclakantan *et al.*,

TABLE X

Profile of Mycotoxin in the Genus *Aspergillus*
and in Related Teleomorphic Genera[a]

Group and telomorphic state[b]	Species[b]	Mycotoxin
A. clavatus group (subgenus *Clavati*)	*A. giganteus* Wehmer	Kojic acid, Patulin
	A. clavatus Desm.	Ascladiol, clavatol, cytochalasin E, kojic acid, kotanine, patulin, tryptoquivalines, xanthocillin dimethyl ether
A. glaucus group and *Eurotium* Link ex Fries (anamorphic subgenus *Aspergillus*)	*E. amstelodami* Mangin	Sterigmatocystin
	E. chevalieri Mangin	Emodin, gliotoxin, physicon, xanthocillin, emodin
	E. cristatum (Raper et Fennel) Malloch et Cain	Emodin
	E. echinulatum Delacr.	Emodin, physicon
	E. herbariorum Link Link ex Fries	Sterigmatocystin
	E. repens De Bary	Physicon, sterigmatocystin
A. fumigatus group and *Neosartorya* Malloch et Cain (anamorphic subgenus *Fumigati*)	*A. fumigatus* Fresenius var. *fumigatus*	Fumigaclavines, fumigallin, fumigatin, fumitoxins, fumitremorgins, gliotoxin, kojic acid, spinulosin, tryptoquivalines, verruculogen
	A. fumigatus var. *ellipticus* Raper et Fennell	Fumitoxins
	A. viridi-nutans Ducker et Thrower	Viriditoxin
	A. brevipes Smith	Viriditoxin
	N. fischeri (Wehmer) Malloch et Cain var. *fischeri*	Avenaciolide, fumitremorgins, verruculogen
A. ochraceus group and *Petromyces* Malloch et Cain (anamorphic subgenus *Circumdati* section *Circumdati*)	*P. alliaceus* (Thom et Church) Malloch et Cain	Kojic acid, ochratoxins, penicillic acid
	A. sulphureus (Fres.) Thom et Church	Neoaspergillic acids, ochratoxins, penicillic acid, viomellein, xanthomegnin
	A. sclerotiorum Huber	Neoaspergillic acids, ochratoxins, penicillic acid
	A. auricomus (Gueguen) Saito	Neoaspergillic acids, penicillic acid, viomellein, xanthomegnin
	A. melleus Yukawa	Neoaspergillic acids, ochratoxins, penicillic acid, viomellein, xanthomegnin
	A. ochraceus Wilhelm (= *A. alutaceus* Berk. et Curt.)	Emodin, kojic acid, neoaspergillic acids, ochratoxins, penicillic acid,

(continued)

TABLE X *(Continued)*

Group and telomorphic state[b]	Species[b]	Mycotoxin
	A. petrakii Vörös	secalonic acid A, viomellein, xanthomegnin Emodin, ochratoxins
A. niger group (subgenus Circumdati section Niger)	A. niger van Tiegh. var. niger	Malformins, naphtho-γ-pyrones, nigragillin
	A. niger var. phoenicis (Corda) Al-Musallam	Malformins, nigragillin
	A. niger var. phoenicis. f. pulverulentus (McAlp.) Al-Musallam	Malformins
	A. niger var. awamori (Nakazawa) Al-Musallam	Kojic acid, malformins
	A. japonicus Saito var. aculeatus (Iizuka) Al-Musallam	Secalonic acid B, D, and F, emodin
A. candidus group (subgenus Circumdati section Candidi)	A. candidus Link	Candidulin, terphenyllin, xanthoascin
A. flavus group (subgenus Circumdati section Flavi)	A. flavus Link var. flavus	Aflatoxins, aflatrem, alfavinin, aspergillic acids, cyclopiazonic acid, kojic acid, neoaspergillic acids, β-nitropropionic acid, paspalinine, (sterigmatocystin)
	A. flavus var. columnaris Raper et Fennell	Aflatoxin B_2, aspergillic acids
	A. parasiticus Speare	Aflatoxins, aspergillic acids, kojic acid, neoaspergillic acids (sterigmatocystin)
	A. oryzae var. oryzae	Aspergillomarasmin, β-nitropropionic acid, oryzacidin
	A. oryzae var. effusus (Tiraboschi) Ohara	Kojic acid
	A. tamarii Kita	Cyclopiazonic acid, fumigaclavine A
	A. subolivaceus Raper et Fennell	Aspergillic acids
	A. avenaceus Smith	Avenaciolide, β-nitropropionic acid
A. wentii group (subgenus Circumdati section Wentii)	A. wentii Wehmer	Emodin, kojic acid, β-nitropropionic acid, wentilactone, physicon
	A. terricola Marchal	Emodin, kojic acid
	A. thomii Smith	Kojic acid

TABLE X *(Continued)*

Group and telomorphic state[b]	Species[b]	Mycotoxin
A. *versicolor* group (subgenus *Nidulantes* section *Versicolores*)	A. *versicolor* (Vuill.) Tiraboschi	Griseofulvin, nidulotoxin, sterigmatocystin
	A. *sydowii* (Bain. et Sart.) Thom et Church	Nidulotoxin, sterigmatocystin
	A. *caespitosus* Raper et Thom	Fumitremorgins, verruculogen
A. *nidulans* group and *Emericella* Berk. et Broome (anomorphic subgenus *Nidulantes* section *Nidulantes*)	E. *nidulans* (Eidam) Vuill. var. *nidulans*	Nidulin, nidulotoxin, sterigmatocystin
	E. *nidulans* var. *lata* (Thom et Raper) Subram.	Sterigmatocystin
	E. *nidulans* var. *dentata* (Sandhu et Sandhu) Subram.	Sterigmatocystin
	E. *nidulans* var. *echinulata* (Fennell et Raper) Godeas	Sterigmatocystin[c]
	E. *nidulans* var. *acristata* (Fennell et Raper) Subram.	Sterigmatocystin striatin
	E. *quadrilineata* (Thom et Raper) C.R. Benjamin	Sterimatocystin striatin
	E. *rugulosa* (Thom et Raper) C.R. Benjamin	Sterigmatocystin
	E. *cleistominuta* Mehrothra et Prasad	Sterigmatocystin[c]
	E. *striata* (Rai, Tewari et Mukerji) Malloch et Cain	Sterigmatocystin striatin
	E. *parvathecia* (Raper et Fennell) Malloch et Cain	Sterigmatocystin[c] striatin
	E. *aurantiobrunnea* (Atkins, Hindson et Russell) Malloch et Cain	Sterigmatocystin[c]
	E. *unguis* Malloch et Cain	Sterigmatocystin[c] nidulin
	E. *fructiculosa* (Raper et Fennell) Malloch et Cain	Sterigmatocystin[c]
	E. *foveolata* Horie	
	E. *corrugata* Udagawa et Horie	Sterigmatocystin[c]
	E. *navahoensis* Christensen et States	Sterigmatocystin[c]
	E. *heterothallica* (Kwon, Fennell et Raper) Malloch et Cain	Sterigmatocystin[c]
	E. *spectabilis* Christensen et Raper	Sterigmatocystin[c]
	E. *variecolor* Berk. et Br. var. *variecolor*	Sterigmatocystin[c], asteltoxin

(continued)

TABLE X *(Continued)*

	A. *multicolor* Sappa	Sterigmatocystin
	A. *egyptiacus* Moubasher et Moubasher	Sterigmatocystin[c]
A. *ustus* group (subgenus *Nidulantes* section *Usti*)	A. *ustus* (Bain.) Thom et Church	Austamid, austdiol, austins, austocystins, kojic acid, sterigmatocystin, xanthocillin x
A. *flavipes* group and *Fennellia* Wiley et Simmons (anamorphic subgenus *Nidulantes* section *Flavipedes*)	F. *flavipes* Wiley and Simmons	Sterigmatocystin
	F. *nivea* (Wiley et Simmons) Samson	Citrinin
	A. *carneus* (van Tiegh.) Blochwitz	Citrinin
A. *terreus* group (subgenus *Nidulantes* section *Terrei*)	A. *terreus* Thom var. *terreus*	Citroviridin, citrinin, gliotoxin, patulin, terrein, terreic acid, terretonin, territrems, cytochalasin E, terredionol
	A. *terreus* var. *aureus* Thom et Raper	Citreoviridin
	A. *terreus* var. *africanus* Fennell et Raper	Citreoviridin

[a] Primary literature references can be found in Turner (1971), Turner and Aldridge (1983), Moss (1977), Moreau (1979), Cole and Cox (1981), Wilson (1971), and Reiss (1981a,b). See also the paper of Semeniuk *et al.* (1971).

[b] For correct names according to the botanical code, see Samson and Pitt (1985).

[c] Toxin production in *Emericella* species according to Horie and Udagawa (personal communication) and/or Frisvad (1985). The sterigmatocystin production of *E. aurantiobrunnea, E. unguis, E. varicolor* var. *varicolor,* and *A. egyptiacus* could not be confirmed by Y. Horie and S. Udagawa (personal communication) or Frisvad (1985).

1957), and secondary metabolites are linked either to conidia, sclerotia, or both in *A. flavus* and *A. parasiticus* (Wicklow and Shotwell, 1983).

The most important mycotoxins, the aflatoxins, are only produced by *A. flavus* and *A. parasiticus.* Tremorgens are produced by *A. clavatus* (tryptoquivalines), *A. fumigatus* and *N. fischeri* var. *fischeri* in the *A. fumigatus* group (fumitremorgins and verruculogen), *A. caespitosus* in the *A. versicolor* group (fumitremorgins and verruculogen), *A. flavus* (aflatrem), and *A. terreus* (territrems). The carcinogen sterigmatocystin is produced by most species possessing hulle cells (*A. versicolor,* the *A. nidulans/Emericella* group, *A. ustus,* and the *A. flavipes/Fennellia* group). Production of the nephrotoxin citrinin is confined to species which are reported to form aleuriospores (Pore and Larsh, 1967): *A. terreus, A. carneus,* and *A. niveus* (*Fennellia nivea*). Temperature relations and production of secondary metabolites confirm the relationship between species in the *A. flavipes* group and the *A. terreus* group. The production of pen-

icillic acid, ochratoxins, and the nephrotoxic naphthoquinones is restricted to the
A. ochraceus/Petromyces group. Other important foodborne aspergilli are found
in the *A. niger* group and the *Aspergillus glaucus/Eurotium* group. Most toxins
in these groups are moderately toxic, and the most toxic metabolites are produced in small amounts (e.g., sterigmatocystin in *Eurotium,* Schroeder and
Kelton, 1975; Karo and Hadlok, 1982). Even though much work still has to be
done concerning chemotaxonomy of the aspergilli, a strong correlation between
fungal taxa and profiles of secondary metabolites appears to exist in this genus.

C. *Alternaria*

The taxonomy of *Alternaria* has been treated by Neergaard (1945) and Joly
(1964), and further keys and descriptions are given by Ellis (1971, 1976) and
Simmons (1981, 1982). Table XI lists species of *Alternaria* known to produce
mycotoxins. *Alternaria alternata* (Fr.) Keissler or the unresolved species group
A. tenuis Nees *auct.* Wiltshire is the most important species in foods and feeds,
but other species such as *A. longipes, A. tenuissima, A. brassicicola, A. cheiran-
ti, A. raphani, A. citri, A. brassicae,* and the *Alternaria* anamorph of *Pleospora
infectoria* (Bruce *et al.,* 1984) have been recorded also. Apparently most *Alternaria* common in foods and feeds produce mycotoxins. The use of secondary
metabolites in *Alternaria* taxonomy should probably be based on metabolites
other than the alternariols and tenuazonic acid. It is very important to determine
the circumstances under which *Alternaria* species produce toxins in foods and
feeds, since species in this genus are ubiquitous and are very potent toxin producers in pure culture. Heavy growth or internal infection with *Alternaria* species indicates that the mycotoxins alternariol, alternariol monomethyl ether,
tenuazonic acid, and altertoxin may be present in the foods or feedstuffs investigated.

D. *Fusarium*

Many important *Fusarium* species can produce mycotoxins. The producers of
the most important toxins are listed in Table XII. Many monographs and books
have been written on the Fusaria, but Nelson *et al.* (1983) have incorporated
important results from these books to produce a workable and practical taxonomic system. The connection between *Fusarium* species and profiles of mycotoxins has also been obscured by different taxonomic viewpoints and some
misidentifications. The species, which resulted in names for some of the trichothecenes, *F. nivale* and *F. solani* were reidentified to *F. sporotrichiodies* and
F. tricinctum (Ichinoe and Kurata, 1983). *Fusarium nivale* (Fr.) Ces. = *Ger-
lachia nivalis* (Ces. ex Sacc.) W. Gams et E. Muller = *Microdochium nivalis*
(Fr.) Samuels et Hallett var. *nivalis* does not appear to be a producer of known
mycotoxins. Recently, however, El-Banna *et al.* (1984) reported that *F. solani*

TABLE XI

Profile of Mycotoxins in the Genus *Alternaria*
and the Teleomorph *Pleospora* Rabenh[a]

Species	Mycotoxins
A. alternata (Fr.) Keissler	Alternariols[b], altertoxins[c], tenuazonic acid
A. brassicae (Berk.) Sacc.	Tenuazonic acid
A. brassicicola (Schw.) Wiltsh.	Alternariols, tenuazonic acid
A. cheiranti (Fr.) Bolle	Alternariols, tenuazonic acid
A. citri Ell. et Pierce emend Bliss et Fawcet	Tenuazonic acid
A. cucumerina (Ell. et Ev.) Elliott	Alternariols
A. dauci (Kuehn) Groves et Skolko[d]	Alternariols
A. japonica Yoshii	Tenuazonic acid
A. kikuchiana Tanaka	Alternariols, tenuazonic acid
A. longipes (Ell. et Ev.) Mason	Alternariols, tenuazonic acid
A. porri (Ell.) Cif.	Altersonasol A and B
A. porri f. sp. *solani* (Ell et Martin) Neerg.	Alternariols, tenuazonic acid
A. raphani Groves et Skolko	Alternariols, tenuazonic acid
A. tenuissima (Fr.) Wiltsh.	Alternariols, tenuazonic acid
Pleospora infectoria Fuckel	Alternariols, tenuazonic acid

[a] Primary literature references can be found in Turner (1971), Turner and Aldridge (1983), King and Schade (1984), and Bruce *et al.* (1984).
[b] Alternariol and alternariol monomethyl ether.
[c] Altertoxin I and II.
[d] *A. dauci* can be included in *A. porri* according to Neergaard (1945).

isolates produced trichothecenes. The four types of mycotoxins produced by *Fusarium* are apparently not sufficient to separate all species in the genus.

Some taxonomic conclusions can be drawn from Table XII, however. Zearalenone, butenolide, and moniliformin have not been detected in species from sections "belonging" to the teleomorphs *Nectria* (Fr.) Fr. or *Calonectria* de Not. (i.e., in the sections *Martiella, Eupionnotes,* and *Spicarioides*). However, the rare species *F. aqueductuum* has been reported to produce moniliformin. The trichothecenes have been detected in nearly all important foodborne species of *Fusarium,* except *F. nivale.* Butenolide and moniliformin are not produced by isolates in the important section *Discolor,* but most isolates of this section are

TABLE XII

Production of Mycotoxins by *Fusarium* Species

Section, species	Mycotoxins	References
Sporotrichiella Wollenw.		
F. *chlamydosporum* Wollenw. et Renking	Moniliformin	Turner and Aldridge (1983)
F. *poae* (Peck) Wollenw.	Trichothecenes A[a]	Turner and Aldridge (1983); Palti (1978); Ichinoe and Kurata (1983); Ueno (1980)
	Trichothecenes B[b]	Abbas *et al.* (1984)
	Butenolide	Burmeister *et al.* (1971)
F. *sporotrichiodies* Sherb.	Zearalenone	Szathmany *et al.* (1976)
	Trichothecenes A	See F. *poae*
	Trichothecenes B	Ichinoe and Kurata (1983)
	Butenolide	Burmeister *et al.* (1971)
F. *tricinctum*	Trichothecenes A	See F. *poae*
	Trichothecenes B	Ichinoe and Kurata (1983)
	Butenolide	Burmeister *et al.* (1971)
Discolor Wollenw.		
F. *culmorum* (W. G. Smith) Sacc.	Zearalenone	Ichinoe *et al.* (1978) Chelkowski and Manka (1983)
	Trichothecenes A	See F. *poae*
	Trichothecenes B	Ichinoe and Kurata (1983)
F. *graminearum* Schwabe	Zearalenone	Turner (1971) Ichinoe *et al.* (1978)
	Trichothecenes A, B	See F. *poae*
F. *heterosporum* Nees	Zearalenone	Chelkowski and Manka (1983)
	Trichothecenes A	Cole *et al.* (1981)
F. *sambucinum* Fuckel	Zearalenone	Chelkowski and Manka (1983)
	Trichothecenes A	Turner and Aldridge (1983) Ichinoe and Kurata (1983) El-Banna *et al.* (1984)
	Trichothecenes B	El-Banna *et al.* (1984)
Roseum Wollenw.		
F. *avenaceum* (Fr.) Sacc.	Zearalenone	Chelkowski and Manka (1983)
	Trichothecenes A	Palti (1978)
	Trichothecenes B	Abbas *et al.* (1984) Palti (1978)
	Moniliformin	Turner and Aldridge (1983)
Arthrosporiella Sherb.		
F. *camptoceras* Wollenw. et Reinking	Trichothecenes A, B	Abbas *et al.* (1984)
F. *semitectum* Berk. et Rav.	Zearalenone	Ichinoe *et al.* (1978)
	Trichothecenes A, B	Ichinoe and Kurata (1983) Ueno (1980)
	Butenolide	Burmeister *et al.* (1971)
	Moniliformin	Turner and Aldridge (1983)

(*continued*)

TABLE XII   *(Continued)*

Section, species	Mycotoxins	References
Gibbosum Wollenw.		
F. acuminatum Ell. et Ev.	Trichothecenes A	Ichinoe and Kurata (1983)
		Ueno (1980)
F. equiseti (Corda) Sacc.	Zearalenone	Ichinoe *et al.* (1978)
	Trichothecenes A, B	See *F. poae*
	Butenolide	Burmeister *et al.* (1971)
	Moniliformin	Turner and Aldridge (1983)
Elegans Wollenw.		
F. oxysporum Schlecht. emend Snyder	Zearalenone	Milczewski *et al.* (1981)
et Hansen	Trichothecenes A, B	See *F. poae*
	Moniliformin	Turner and Aldridge (1983)
Liseola Wollenw., Sherb., Reinking,		
Johann et Bailey		
F. moniliforme	Zearalenone	Turner (1971)
	Trichothecenes A	Ichinoe and Kurata (1983)
	Moniliformin	Palti (1978)
Martiella Wollenw.		
F. solani (Mart.) Appel et Wollenw.	Trichothecenes A, B	El-Banna *et al.* (1984)
emend Snyder et Hansen		
Lateritium Wollenw.		
F. lateritium Nees	Zearalenone	Milczewski *et al.* (1981)
	Trichothecenes A	See *F. poae*
	Trichothecenes B	Ueno (1980)
	Butenolide	Burmeister *et al.* (1971)
Eupionnotes Wollenw.		
F. aquaeductuum	Trichothecenes B	Ichinoe and Kurata (1983)
	Moniliformin	Turner and Aldridge (1983)
Spicarioides Wollenw., Sherb.,		
Reinking, Johann et Bailey		
F. decemcellulare Brick	Trichothecenes A	Palti (1978)

[a] Type A trichothecenes are diacetoxyscirpenol, T-2 toxin, HT-2 toxin, and neosolaniol.

[b] Type B trichothecenes are fusarenone X, vomitoxin (deoxynivalenol), and nivalenol.

excellent producers of zearalenone and trichothecenes, especially in the species *F. culmorum* and *F. graminearum*. Other important trichothecene producers are *F. poae* and *F. sporotrichiodies* from the section *Sporotrichiella*. The active growth of any *Fusarium* species in foods and feeds indicates possible contamination of the substrate with trichothecenes, zearalenone, moniliformin, or butenolide, but some *Fusarium* species (*F. graminearum, F. poae, F. sporotrichoides, F. culmorum, F. moniliforme,* and *F. avenaceum*) are certainly better producers of these mycotoxins than others.

TABLE XIII

Genera of Filamentous Fungi Containing Toxigenic Isolates

Zygomycotina
  Zygomycetes
    Mucorales

Mucoraceae:	*Absidia* van Tiegh.	
	Mucor Micheli ex Fries	
	Rhizopus Ehrenb	
Mortitierellaceae:	*Mortierella* Coemans	
Zoopagales		
Piptocephalidacea:	*Piptocephalis* de Bary	
Ascomycotina		
Gymnoascales:	*Arachnoitus* Schröter	
	Nannizzia Stockdale	an[a]: *Microsporum* Gruby
	Arthroderma Currey	an: *Trichophyton* Malmsten
Eurotiales:	*Byssochlamys* Westling	an: *Paecilomyces* Bainier
	Dactylomyces Sopp	
	Dichotomyces Saito ex Scott	an: *Polypaecilum* G. Smith
	Emericella	an: *Aspergillus*
	Eupenicillium	an: *Penicillium*
	Eurotium	an: *Aspergillus*
	Fennellia	an: *Aspergillus*
	Neosartorya	an: *Aspergillus*
	Penicilliopsis Solms-Laubach	
	Petromyces	an: *Aspergillus*
	Talaromyces	an: *Penicillium*
Hypocreales:	*Gibberella* Sacc.	an: *Fusarium*
	Nectria (Fr.) Fr.	an: *Fusarium*
		Acremonium Link
		Cylindrocarpon Wollenw.
	(*Hypocrea* Fr.)	*Trichoderma* Pers.
	—	an: *Gliocladium* Corda
	—	an: *Metarhizium* Sorok.
	—	an: *Verticimonosporium* Matsuch.
	(*Nectria*)	an: *Verticillium* Nees
Clavicipitales:	*Claviceps* Tul.	an: *Sphacelia* Lév
	Epichloë (Fr.) Tul.	an: *Acremonium*
Dothidiales:	(*Mycosphaerella* Johanson)	an: *Cladosporium* Link
	Cochliobolus Drechsler	an: *Bipolaris* Shoem.
		Curvularia Boedijn
	Pyrenophora Fr.	an: *Drechslera* Ito
	—	an: *Periconia* Tode
	Pleospora	an: *Alternaria*
	Didymella Sacc.	an: *Phoma* Sacc.

(*continued*)

TABLE XIII (*Continued*)

	(*Herpotrichia* Fuckel)	an: *Pyrenochaeta* de Not.
	Sporormia de Not.	
Diaporthales:	(*Diaporthe* Nitschke)	an: *Phomopsis* (Sacc.) Bukák
	Endothia Fr.	
	Nigrosabulum Malloch et Cain	
Polystigmatales:	(*Magnaporthe* Krause et R. Webster)	an: *Pyricularia* Sacc.
Sphaeriales:	(*Physalospora* Niessl)	an: *Lasiodiplodia* Ell. et Everh.
	Apiospora Sacc.	an: *Arthrinium* Kunze
	Rossellinia de Not.	an: *Dematophora* Hartig
	(*Corollospora* Verderm.)	an: *Clavariopsis* de Wild.
	Khuskia H. Hudson	an: *Nigrospora* Zimm.
Sordariales:	*Chaetomium* Kunze	
	Farrowia Hawksw.	
	Neurospora Shear et B. Dodge	an: *Chrysonilia* von Arx
	Thielavia Zopf	
	—	an: *Stachybotrys* Corda
Ophiostomatales:	*Ceratocystis* Ell. et Halst.	
Microascales:	*Microascus* Zukal	an: *Scopulariopsis* Bain.
Helotiales:	(*Amorphotheca* Parberry)	an: *Hormoconis* von. Arx et de Vries
	Gloeotinia Wilson, Noble et Gray	
	(*Botryotinia* Whetzel)	an: *Botrytis* Micheli ex Pers.
	Sclerotinia Fuckel	an: *Monilia* Bonorden
Deuteromycotina		
Agonomycetes:	*Papulaspora* Preuss	
	Rhizoctonia DC.[b]	
	Sclerotium Tode	
Hyphomyctes:	*Beauveria* Vuill.	
	Coniosporium Link	
	Dendrodochium Bonorden	
	Epicoccum Link	
	Fulvia Cif.	
	Humicola Traaen	
	Nodulisporium Preuss	
	Oedemium Link	
	Pithomyces Berk. et Broome	
	Sepedonium Link	
	Thermomyces Tsiklinsky	
	Torula Pers.	
	Trichothecium Link	
	Wallemia Johan-Olsen	
	Zygosporium Mont.	
Coelomycetes:	*Septoria* Sacc.	

[a] an, The anamorph state of the fungus.
[b] Teleomorph in the Basidiomycotina.

E. Other Genera

Many genera of filamentous fungi, other than *Penicillium, Aspergillus, Fusarium,* and *Alternaria,* contain toxigenic isolates (Table XIII). Typically, most species of these genera are not toxic, but important species which contain toxigenic isolates are listed in Table XIV. Guides to the identification of species in these genera are given by Samson *et al.* (1984, 1985), Wyllie and Morehouse (1977), Carmichael *et al.* (1980), Domsch *et al.* (1980), Samuels (1984), Onions *et al.* (1981), Hanlin (1973), Hawksworth *et al.* (1983), von Arx (1981), and Barnett and Hunter (1972). Samuels (1984) discusses the taxonomy of ascomycetous genera containing toxigenic isolates, and Wyllie and Morehouse (1977) and Moreau (1979) provide further details on species containing toxigenic isolates. There have been no systematic studies of the chemotaxonomy of genera other than *Penicillium* (Section V,A) and *Chaetomium* (Udagawa, 1984) in the filamentous fungi, but the latter studies indicate that secondary metabolites may prove to be very valuable characters in fungal taxonomy. Knowledge of internal invaders and the normal internal flora of foods and feedstuffs will be excellent indicators of which mycotoxins to analyze chemically, but additional research is needed to determine the complete mycotoxin profiles of known fungal taxa and the profiles of mycotoxins in foods and feedstuffs under different environmental conditions.

TABLE XIV

**Important Filamentous Fungal Species
Containing Toxigenic Isolates[a]**

Species	Mycotoxins
Acremonium sp.	Oosporein
A. crotocinigenum (Schol-Schwarz) W. Gams	Crotocin
Archnoitus aureus (Eidam) Schröter =	Aranotin
Amauroascus aureus Schröter	
Beauveria bassiana (Balsamo) Vuill.	Oosporein
Botrytis cinerea Pers.: Fr.	Botrydial
Byssochlamys fulva Oliver and Smith	Byssochlamic acid, patulin
Byssochlamys nivea Westling	Byssochlamic acid, patulin, malformins
Ceuthospora sp.	Citrinin
Chaetomium abuense Lodha	Cochliodinol
C. amygdalisporum Sekita	Mollicellins G and H
C. aureum Chivers	Oosporein
C. caprinum Bainier	Chaetochromin
C. coarctorum Sergejeva	Chaetoglobosins

(continued)

TABLE XIV *(Continued)*

Species	Mycotoxins
C. cochlioides Palliser	Chaetoglobosins, chetomin, cochliodinol
C. elatum Kunze: Fr.	Cochliodinol
C. funicola Cooke	Chetomin
C. globosum Kunze: Fr. var. *globosum*	Chaetoglobosins, chetomin, cochliodinol, chaetocin
C. globosum var. *rectum* (Sergejeva) Dreyfuss	Chaetoglobosins
C. gracile Udagawa	Chaetochromin
C. mollicellum Ames	Mollicellins
C. millipilium Ames	Chaetoglobosins
C. subaffine Sergejeva	Chaetoglobosins
C. subglobosum Sergejeva	Chetomin
C. tenuissimum Sergejeva	Chetomin
C. tetrasporum Hughes	Chaetochromin
C. thielavioideum Chen	Chaetocrocin, chaetochromin, eugenitin, sterigmatocystin
C. trilaterale Chivers	Oosporein
C. udagawae Sergejeva	Sterigmatocystin
C. umbonatum Brewer	Chetomin
Cladosporium herbarum (Pers.: Fr.) Link	Epi- or fagi-cladosporic acid
Clavariopsis aquatica de Wild.	Citrinin
Claviceps paspali Stev. et Hall	Ergot alkaloids, paspalicin, paspaline, paspalinine, paspalitrem A and B
C. purpurea (Fr.:Fr.) Tul.	Ergot alkaloids, secalonic acid B and C
Cochliobolus victoriae Nelson	Victorin
Curvularia lunata (Wakker) Boedjin = *Cochliobolus lunatus* Nelson et Haasis	Cytochalasin B, brefeldin A
Cylindrocarpon sp.	Isororidin E
Dendrocochium toxicum (= ? *Myrothecium roridum* Tode: Fr.)	Dendrodochin
Diplodia maydis (Berk.) Sacc.	Diplodiatoxin
Drechslera dematoidea (Bubak et Wrob.) Subram. et Jain	Cytochalasin A and F
D. catenaria (Drechsler) Ito	Emodin
D. sorokiniana (Sacc.) Subram. et Jain	Sterigmatocystin
Endothia fluens (Sow.) Shear et Stevens	Rugulosin
E. gyrosa (Schw.:Fr.) Fr.	Rugulosin
E. parasitica (Murrill) P.J. et H.W. Anderson = *Cryphonectria parasitica* (Murrill) Barr	Rugulosin
E. viridistroma Wehmeyer	Rugulosin

TABLE XIV *(Continued)*

Species	Mycotoxins
Epicoccum nigrum Link - *E. purpurascens* Ehrenb.	Flavipin
Farrowia sp.	Sterigmatocystin
Gliocladium virens Miller, Giddens et Foster	Gliotoxin, viridin
Khuskia oryzae Hudson	Griseofulvin
Metarhizium anisopliae (Metschnikoff) Sorokin.	Cytochalasin C and D
Microsporum cookei Ajello	Xanthomegnin
Myrothecium roridum	Gliotoxin, rordins
M. verrucaria (Abl. et Schw.:Fr.) Ditmar	Muconomycin, roridins, (-)-Rugulosin, verrucarins
Nectria radicicola Gerlach et Nilsson (stat. anam. *Cylindrocarpon destructans* (Zins.) Scholten)	Brefeldin A
Nigrosabulum. sp.	Cytochalasin G
Nigrospora sacchari (Speg.) Mason	Griseofulvin
Nodulisporium hinnuleum G. Smith	Demethoxyviridin
Oospora colorans van Beyma	Oosporein
Paecilomyces lilacinus (Thom) Samson	Penicillic acid
P. variotii Bainier	Variotin
Penicilliopsis clavariaeformis Solms-Laubach	Emodin, penicilliopsin
Phoma sp.	Terrein
P. exigua Desm. var. *exigua*	Cytochalasin A and B
P. exigua var. *inoxydabilis* Boerema et Vegh	Phomenone
P. herbarum Westend. var. *medicaginis* Westend.: Rabenh.	Brefeldin A
P. sorghina (Sacc.) Boerema, Dorenbosch et van Kestern	Tenuazonic acid
Pithomyces chartarum (Berk. et Curt.) M.B. Ellis	Sporidesmins
Pyricularia oryzae Cav.	Tenuazonic acid
Rhizoctonia leguminicola Gough et Elliot	Slaframine
Rhizopus oryzae Went et Prinsen Geerligs	Isofumigaclavine A
Rhizopus stolonifer (Ehrenb.: Fr.) Vuill.	Cyclic peptide
Rosellinia necathrix (Hartig) Berl.:Prill	Cytochalasin E
Sepedonium ampullosporum Damon	Rugulosin
Septoria nodorum (Berk.) Berk.	Mycophenolic acid
Sirodesmium diversum (Cooke) Hughes	Sirodesmins
Stachybotrys atra Corda	Satratoxins
Stemphylium sarciniforme (Cav.) Wiltsh.	Stemphone
Thielavia sepedonium Emmons = *Corynascus sepedonium* (Emmons) van Arx	Malformins
Torula sp.	Cytochalasin B
Trichoderma lignorum (Tode) Harz—*T. viride* Pers.:Fr.	Trichodermin, T-2 toxin
Trichophyton megninii Blanchard	Xanthomegnin
T. rubrum (Castell.) Sabour.	Xanthomegnin

(continued)

TABLE XIV *(Continued)*

Species	Mycotoxins
T. violaceum Sabour. apud Boedin	Xanthomegnin
Trichothecium roseum (Pers.:Fr.) Link	Crotocin, roseotoxin, tri-
	chothecin, trichothecene
Verticillium cinnabarinum (Corda) Reinke et	Melanacidines
Bethold	
V. psalliotae Treschow	Oosporein
Verticimonosporium diffractum Matsushima	Vertisporin
Zygosporium masonii Hughes	Zygosporins

[a] Species of *Penicillium, Aspergillus, Fusarium, Alternaria,* and their teleomorphs have not been included (see Tables IV–XII). The list of species is taken from many sources: see Moreau (1979), Cole and Cox (1981), Turner (1971), Turner and Aldridge (1983), Reiss (1981a,b), and Dean (1963); see also the culture collection catalogs from the Centraalbureau voor Schimmelcultures (1983), Commonwealth Mycological Institute (1983), and American Type Culture Collection (1982).

ACKNOWLEDGMENTS

I thank Ulf Thrane for help with the preparation of Table XII. This work was supported by the Danish Council for Scientific and Industrial Research.

REFERENCES

Abbas, H. K., Mirocha, C. J., and Shier, W. T. (1984). Mycotoxins produced from fungi isolated from foodstuffs and soil: comparison of toxicity in fibroplasts and rat feeding tests. *Appl. Environ. Microbiol.* **48,** 654–661.

Abe, M., Ohmomo, S., Ohashi, T., and Tabuchi, T. (1969). Isolation of chanoclavine-(I) and two new interconvertible alkaloids, rugulovasine A and B, from cultures of *Penicillium concavorugulosum. Agric. Biol. Chem.* **33,** 469–471.

Abe, S., Iwai, M., and Sugihara, T. (1982). Taxonomic studies of *Penicillium* I. Species of the section *Aspergilloides. Nippon Kingakkai Kaiho* **23,** 149–163.

Abe, S., Iwai, M., and Awano, M. (1983a). Taxonomic studies of *Penicillium* IV. Species in the sections *Divaricatum* (continued) and *Furcatum. Nippon Kingakkai Kaiho* **24,** 109–120.

Abe, S., Iwai, M., and Ishikawa, T. (1983b). Taxonomic studies of *Penicillium* V. Species in the section *Furcatum* (continued). *Nippon Kingakkai Kaiho* **24,** 409–418.

Abe, S., Iwai, M., and Tanaka, H. (1983c). Taxonomic studies of *Penicillium* III. Species in the section *Divaricatum. Nippon Kingakkai Kaiho* **24,** 95–108.

Abe, S., Iwai, M., and Yoshioka, H. (1983d). Taxonomic studies of *Penicillium* II. Species in the section *Exilicaulis. Nippon Kingakkai Kaiho* **24,** 39–51.

Agurell, S. L. (1964). Costaclavine from *Penicillium chermesinum. Experientia* **20,** 25–26.

Al-Musallam, A. (1980). ''Revision of the Black *Aspergillus* Species.'' Univ. of Utrecht, Utrecht.

Alsberg, C. I., and Black, O. F. (1913). Contribution to the study of maize deterioration; biochemical and toxicological investigations of *Penicillium puberulum* and *Penicillium stoloniferum. U.S. Dep. Agric. Bur. Plant Ind. Bull.* No. 270, 7–48.

Andrews, S. (1986a). Dilution plating versus direct plating of various cereal samples. *In* "Methods for the Mycological Examination of Foods" (A. D. King, I. Corry, J. I. Pitt, and L. R. Beuchat, eds.). Plenum, New York. In press.

Andrews, S. (1986b). Selective media for the isolation and characterisation of potentially mycotoxigenic fungi from cereal. *In* "Methods for the Mycological Examination of Foods" (A. D. King, I. Corry, J. I. Pitt, and L. R. Beuchat, eds.). Plenum, New York. In press.

Andrews, S. (1986c). Optimization of conditions for the surface sterilization of sorghum and sultanos using sodium hypochlorite solutions. *In* "Methods for the Mycological Examination of Foods" (A. D. King, I. Corry, J. I. Pitt, and L. R. Beuchat, eds.). Plenum, New York. In press.

Anslow, W. K., and Raistrick, H. (1938). Fumigatin and spinulosin, metabolic products respectively of *Aspergillus fumigatus* Fresenius and *Penicillium spinulosum* Thom. *Biochem. J.* **32**, 687–696.

Anslow, W. K., Raistrick, H., and Smith, G. (1943). Antifungal compounds from moulds; patulin (anhydro-3-hydroxymethylene-tetrahydro-1:4-pyrone-carboxylic acid), a metabolic product of *Penicillium patulum* Bainier and *Penicillium expansum* (Link) Thom. *J. Soc. Chem. Ind., London, Trans. Commun.* **62**, 236–238.

Barnett, H. L., and Hunter, B. B. (1972). "Illustrated Genera of Imperfect Fungi." Burgess, Minneapolis, Minnesota.

Barta, J., and Mecir, R. (1948). Antibacterial activity of *Penicillium divergens* Bainier. *Experientia* **4**, 277–278.

Bennett, J. W. (1983). Secondary metabolism as differentiation. *J. Food Safety* **5**(1), 1–11.

Betina, V., Nemec, P., Dobias, J., and Borath, Z. (1962). Cyanein, a new antibiotic from *Penicillium cyaneum. Folia Microbiol. (Prague)* **7**, 353–357.

Betina, V., Gasparikova, E., and Nemec, P. (1969). Isolation of penicillic acid from *Penicillium simplicissimum. Biologia (Bratislava)* **24**, 482–485.

Birch, A. J., and Wright, J. J. (1969). The brevianamides, a new class of fungal metabolites. *J.C.S. Chem. Commun.* pp. 644–645.

Birkinshaw, J. H., and Gowlland, A. (1962). Studies in the biochemistry of microorganisms 110. Production and biosynthesis of orsellinic acid by *Penicillium madriti* G. Smith. *Biochem. J.* **84**, 342–347.

Birkinshaw, J. H., and Raistrick, H. (1936). Studies in the biochemistry of microorganisms LII. Isolation, properties and constitution of terrestric acid (ethylcarolic acid), a metabolic product of *Penicillium terrestre* Jensen. *Biochem. J.* **30**, 2194–2200.

Birkinshaw, J. H., Charles, J. H. V., Lilly, C. H., and Raistrick, H. (1931). The biochemistry of microorganisms VII. Kojic acid (5-hydroxy-2-hydroxymethyl-pyrone). *Philos. Trans. R. Soc. London, Ser. B* **220**, 127–138.

Birkinshaw, J. H., Oxford, A. E., and Raistrick, H. (1936). Studies in the biochemistry of microorganisms XLVIII. Penicillic acid, a metabolic product of *Penicillium puberulum* Bainier and *P. cyclopium* Westling. *Biochem. J.* **30**, 394–411.

Blackwelder, R. E. (1964). Phyletic and phenetic versus omnispective classification. *In* "Phenetic and Phylogenic Classification" (V. H. Heywood and J. McNeill, eds.), Publ. No. 6, pp. 17–18. Syst. Assoc., London.

Blaser, P. (1976). Taxonomische und Physiologische Untersuchungen über die Gattung *Eurotium* Link ex Fries. *Sydowia* **28**, 1–51.

Blaser, P., and Schmidt-Lorenz, W. (1981). *Aspergillus flavus* Kontamination von Nussen, Mandeln und Mais mit bekannten Aflatoxin-Gehalten. *Lebensm.-Wiss. Technol.* **14**, 252–259.

Booth, C. (1971). "The Genus *Fusarium.*" Common. Mycol. Inst., Kew, England.

Borodin, N., Philpot, F. J., and Florey, H. W. (1947). An antibiotic from *Penicillium tardum*. *Br. J. Exp. Pathol.* **28,** 31–34.

Bracken, A., and Raistrick, H. (1947). Studies in the biochemistry of microorganisms 75. Dehydrocarolic acid, a metabolic product of *Penicillium cinerascens* Biourge. *Biochem. J.* **41,** 569–575.

Breen, J., Dacre, J. C., Raistrick, H., and Smith, G. (1955). Studies in the biochemistry of microorganisms 95. Rugulosin, a crystalline colouring matter of *Penicillium rugulosum* Thom. *Biochem. J.* **60,** 618–626.

Brian, P. W. (1946). Production of gliotoxin by *Penicillium terlikowskii* Zal. *Trans. Br. Mycol. Soc.* **29,** 211–218.

Brian, P. W., Curtis, P. J., and Heming, H. G. (1949). A substance causing abnormal development of fungal hyphae produced by *Penicillium janczewskii* Zal. *Trans. Br. Mycol. Soc.* **32,** 30–33.

Brian, P. W., Curtis, P. J., and Heming, H. G. (1955). Production of griseofulvin by *Penicillium raistrickii*. *Trans. Br. Mycol. Soc.* **38,** 305–308.

Bruce, V. R., Starck, M. E., and Mislivec, P. B. (1984). Incidence of toxic *Alternaria*. Species in small grains from the U.S.A. *J. Food Sci.* **49,** 1626–1627.

Bullerman, L. B., Schroeder, L. L., and Park, K.-Y. (1984). Formation and control of mycotoxins in food. *J. Food Prot.* **47,** 637–646.

Bu'Lock, J. D. (1980). Mycotoxins as secondary metabolites. *In* "The Biosynthesis of Mycotoxins. A Study in Secondary Metabolism" (P. S. Steyn, ed.), pp. 1–16. Academic Press, New York.

Burmeister, H. R., Ellis, J. J., and Yates, S. G. (1971). Correlation of biological to chromatographic data for two mycotoxins elaborated by *Fusarium*. *Appl. Microbiol.* **21,** 673–675.

Burnett, J. H. (1983). Speciation in fungi. *Trans. Br. Mycol. Soc.* **81,** 1–14.

Burton, H. S. (1949). Antibiotics from penicillia. *Br. J. Exp. Pathol.* **30,** 151–158.

Carmichael, J. W., Kendrick, W. B., Conners, I. L., and Sigler, L. (1980). "Genera of Hyphomycetes." Univ. of Alberta Press, Edmonton.

Chain, E., Florey, H. W., and Jennings, M. A. (1942). An antibacterial substance produced by *Penicillium claviforme*. *Br. J. Exp. Pathol.* **23,** 202–205.

Chelkowski, J., and Manka, M. (1983). The ability of *Fusaria* pathogenic to wheat, barley and corn to produce zearalenone. *Phytopathol. Z.* **106,** 354–359.

Christensen, M. (1981). An synoptic key and evaluation of species in the *Aspergillus flavus* group. *Mycologia* **73,** 1056–1084.

Christensen, M. (1982). The *Aspergillus ochraceus* group: two new species from western soil and a synoptic key. *Mycologia* **74,** 210–225.

Christensen, M., and States, J. S. (1982). *Aspergillus nidulans* group: *Aspergillus navahoensis*, and a revised synoptic key. *Mycologia* **74,** 226–235.

Ciegler, A. (1969). A tremogenic toxin from *Penicillium palitans*. *Appl. Microbiol.* **18,** 128–129.

Ciegler, A., and Pitt, J. I. (1970). Survey of the genus *Penicillium* for tremorgenic toxin production. *Mycopathol. Mycol. Appl.* **42,** 119–124.

Ciegler, A., Fennell, D. I., Sansing, G. A., Detroy, R. W., and Bennett, G. A. (1973). Mycotoxin producing strains of *Penicillium viridicatum*: classification into subgroups. *Appl. Microbiol.* **26,** 271–276.

Clarke, S. M., and McKenzie, M. (1967). *Penicillium sclerotigenum*, a new source of griseofulvin. *Nature (London)* **213,** 504–505.

Clutterbuck, P. W., Oxford, A. E., Raistrick, H., and Smith, G. (1932). Studies in the biochemistry of microorganisms XXIV. The metabolic products of the *Penicillium brevicompactum* series. *Biochem. J.* **26,** 1441–1458.

Clutterbuck, P. W., Haworth, W. N., Raistrick, H., Smith, G., and Stacay, M. (1934). Studies in the biochemistry of microorganisms XXXVI. The metabolic products of *Penicillium charlesii* G. Smith. *Biochem. J.* **28,** 94–110

Cole, R. J., and Cox, R. H. (1981). "Handbook of Toxic Fungal Metabolites." Academic Press, New York.

Cole, R. J., Kirksey, J. W., and Wells, J. M. (1974). A tremorgenic metabolite from *Penicillium paxilli. Can. J. Microbiol.* **20,** 1159–1162.

Cole, R. J., Kirksey, J. W., and Morgan-Jones, G. (1975). Verruculotoxin, a new mycotoxin from *Penicillium verruculosum. Toxicol. Appl. Pharmacol.* **31,** 465–468.

Cole, R. J., Kirksey, J. W., Clardy, J., Eichman, N., Weinreb, S. M., Singh, P., and Kim, D. (1976). Structures of rugulovasine-A and -B and 8-chlororugulovasine-A and -B. *Tetrahedron Lett.* **43,** 3849–3852.

Cole, R. J., Dorner, J. W., Cox, R. H., Cunfer, B. M., Cutler, H. G., and Stuart, B. P. (1981). The isolation and identification of several trichothecene mycotoxins from *Fusarium heterosporum. J. Nat. Prod.* **44,** 324–330.

Cole, R. J., Dorner, J. W., Cox, R. H., and Raymond, L. W. (1983). Two classes of alkaloid mycotoxins produced by *Penicillium crustosum* Thom isolated from contaminated beer. *J. Agric. Food Chem.* **31,** 655–657.

Dean, F. M. (1963). "Natural Occurring Oxygen Ring Compounds," p. 526. Butterworth, London.

Di Menna, M. E., and Mantle, P. G. (1978). The role of penicillia in ryegrass staggers. *Res. Vet. Sci.* **24,** 347–351.

Domsch, K. H., Gams, W., and Anderson, T.-M. (1980). "Compendium of Soil Fungi," Vols. 1 and 2. Academic Press, New York.

Dorner, J. W., Cole, R. J., Hill, R., Wicklow, D. T., and Cox, R. H. (1980). *Penicillium rubrum* and *Penicillium biforme,* new sources of rugulovasines A and B. *Appl. Environ. Microbiol.* **40,** 685–687.

Dorner, J. W., Cole, R. J., and Diener, U. L. (1984). The relationship of *Aspergillus flavus* and *Aspergillus parasiticus* with reference to production of aflatoxin and cyclopiazonic acid. *Mycopathologia* **87,** 13–15.

El-Banna, A. A., Scott, P. M., Law, P.-Y., Sakuma, T., Platt, H. W., and Campbell, V. (1984). Formation of trichothecenes by *Fusarium solani* var. *coeruleum* and *Fusarium sambucinum* in potatoes. *Appl. Environ. Microbiol.* **47,** 1169–1171.

Ellis, M. B. (1971). "Dematiaceous Hyphomycetes." Commonw. Mycol. Inst., Kew, England.

Ellis, M. B. (1976). "More Dematiaceous Hyphomycetes." Commonw. Mycol. Inst., Kew, England.

Engel, G., and Teuber, M. (1978). Simple aid in the identification of *Penicillium roqueforti* Thom. Growth in acetic acid. *Eur. J. Appl. Microbiol.* **6,** 107–111.

Engel, G., and Teuber, M. (1983). Differentiation of *Penicillium roqueforti* strains by thin layer chromatography of metabolites. *Milchwissenschaft* **38,** 513–516.

Engel, G., von Milczewski, K. E., Prokopek, D., and Teuber, M. (1982). Strain-specific synthesis of mycophenolic acid by *Penicillium roqueforti* in blue-veined cheese. *Appl. Environ. Microbiol.* **43,** 1034–1040.

Fayos, J., Lokensgard, D., Clardy, J., Cole, R. J., and Kirksey, J. W. (1974). Structure of verruculogen, a tremor producing peroxide from *Penicillium verruculosum. J. Am. Chem. Soc.* **96,** 6785–6787.

Filtenborg, O., and Frisvad, J. C. (1980). A simple screening method for toxigenic mould in pure cultures. *Lebensm.-Wiss. Technol.* **13,** 128–130.

Filtenborg, O., Frisvad, J. C., and Svendsen, J. A. (1983). Simple screening method for moulds producing intracellular mycotoxins in pure cultures. *Appl. Environ. Microbiol.* **45,** 581–585.

Frisvad, J. C. (1981). Physiological criteria and mycotoxin production as aids in identification of common asymmetric penicillia. *Appl. Environ. Microbiol.* **41,** 568–579.

Frisvad, J. C. (1983). A selective and indicative medium for groups of *Penicillium viridicatum* producing different mycotoxins in cereals. *J. Appl. Bacteriol.* **54,** 409–416.

Frisvad, J. C. (1984). Expressions of secondary metabolism as fundamental characters in *Penicillium* taxonomy. *In* "Toxigenic Fungi—Their Toxins and Health Hazard" (H. Kurata and Y. Ueno, eds.), Developments in Food Science, Vol. 7, pp. 99–106. Elsevier, Amsterdam.

Frisvad, J. C. (1985). Secondary metabolites as an aid in *Emericella* classification. *In* "Advances in *Penicillium* and *Aspergillus* Systematics" (R. A. Samson and J. R. Pitt, eds.), pp. 437–444. Plenum, New York.

Frisvad, J. C. (1986). Aids in the identification of important food-borne species of filamentous fungi. *In* "Methods for the Mycological Examination of Foods" (D. A. King, I. Corry, J. I. Pitt, and L. F. Beuchat, eds.). Plenum, New York. In press.

Frisvad, J. C., and Filtenborg, O. (1983). Classification of terverticillate penicillia based on profiles of mycotoxins and other secondary metabolites. *Appl. Environ. Microbiol.* **46**, 1301–1310.

Frisvad, J. C., and Viuf, B. T. (1986). Comparison of direct and dilution plating for detecting *Penicillium viridicatum* group II in barley containing different levels of ochratoxin A. *In* "Methods for the Mycological Examination of Foods" (A. D. King, I. Corry, J. I. Pitt, and L. R. Beuchat, eds.). Plenum, New York. In press.

Frisvad, J. C., Thrane, U., and Filtenborg, O. (1986a). A note on the combined use of several media and plating methods for the evaluation of the quantitative mycoflora of foods. *In* "Methods for the Mycological Examination of Foods" (A. D. King, I. Corry, J. I. Pitt, and L. R. Beuchat, eds.). Plenum, New York. In press.

Frisvad, J. C., Kristensen, A. B., and Filtenborg, O. (1986b). Comparison of methods used for surface desinfection of food and feed commodities before mycological analysis. *In* "Methods for the Mycological Examination of Foods" (A. D. King, I. Corry, J. I. Pitt, and L. R. Beuchat, eds.). Plenum, New York. In press.

Frisvad, J. C., Thrane, U., and Filtenborg, O. (1986c). Comparison of five media for the determination of the mesophilic mycoflora of foods. *In* "Methods for the Mycological Examination of Foods" (A. D. King, I. Corry, J. I. Pitt, and L. R. Beuchat, eds.). Plenum, New York. In press.

Fujimoto, Y., Kamiya, M., Tsunoda, H., Ohtsubo, K., and Tatsuno, T. (1980). Recherche toxicologique des substances metabolique de *Penicillium carneo-lutescens*. *Chem. Pharm. Bull.* **28**, 1062–1066.

Gallagher, R. T., Latsch, G. C. M., and Keogh, R. G. (1980). The janthitrems: fluorescent tremorgenic toxins produced by *Penicillium janthinellum* isolates isolated from ryegrass staggers. *Appl. Environ. Microbiol.* **39**, 272–273.

Gams, W. (1984). The importance of synanamorphs in taxonomy and nomenclature of mycotoxin-producing *Fusarium* species. *In* "Toxigenic Fungi—Their Toxins and Health Hazard" (H. Kurata and Y. Ueno, eds.), Developments in Food Science, Vol. 7, pp. 129–138. Elsevier, Amsterdam.

Ghosh, A. C., Manmade, A., and Demain, A. L. (1977). Toxins from *Penicillium islandicum* Sopp. *In* "Mycotoxins in Human and Animal Health" (J. V. Rodricks, C. W. Hesseltine, and M. A. Mehlman, eds.), pp. 621–638. Pathotox, Park Forest South, Illinois.

Gorbach, G., and Friederick, W. (1949). Beitrage zur kenntnis der Penicillinsaure. *Oesterr. Chem.-Ztg.* **50**, 93–97.

Grove, J. F. (1954). The structure of terrein. *J. Chem. Soc.* pp. 4693–4694.

Härri, E., Leffler, W., Sigg, H. P., Stahelin, H., and Tamm, C. (1963). Uber di Isolierung neuer stoffwechselprodukte aus *Penicillium brefeldianum* Dodge. *Helv. Chim. Acta* **46**, 1235–1246.

Haese, G. (1963). Uber Antimycin. *Arch. Pharm. Ber. Dtsch. Pharm. Ges.* **296**, 227–232.

Ha-Huy-Ke, and Luckner, M. (1979). Structure and function of condiospore pigments of *Penicillium cyclopium*. *Z. Allg. Mikrobiol.* **19**, 117–122.

Hald, B., Christensen, D. H., and Krogh, P. (1983). Natural occurrence of the mycotoxin viomellein in barley and the associated quinone-producing penicillia. *Appl. Environ. Microbiol.* **46**, 1311–1317.

Hamsa, T. A. P., and Ayres, J. C. (1977). A differential medium for the isolation of *Aspergillus flavus* from cotton seed. *J. Food Sci.* **42**(2), 449–453.

Hanlin, K. T. (1973). "Keys to Families, Genera and Species of the *Mucorales.*" Cramer, Vaduz. Lichtenstein.

Harwig, J., Chen, Y.-K., Kennedy, B. P. C., and Scott, P. M. (1973). Occurrence of patulin and patulin-producing strains of *Penicillium expansum* in natural rots of apple in Canada. *Can. Inst. Food Sci. Technol. J.* **6**, 22–25.

Hawksworth, D. L., Sutton, B. C., and Ainsworth, G. C. (1983). "Ainsworth and Bisby's Dictionary of the Fungi," 7th Ed. Commonw. Mycol. Inst., Kew, England.

Hermansen, K., Frisvad, J. C., Emborg, C., and Hansen, J. (1984). Cyclopiazonic acid production by submerged cultures of *Penicillium* and *Aspergillus* strains. *FEMS Microbiol. Lett.* **21**, 253–261.

Hesseltine, C. W., Shotwell, O. L., Kwolek, W. F., Lillehoj, E. B., Jackson, W. W., and Bothast, R. J. (1976). Aflatoxin occurrence in 1973 corn at harvest. II, Mycological studies. *Mycologia* **68**, 341–353.

Hetherington, A. C., and Raistrick, H. (1931a). Studies in the biochemistry of microorganisms. Part XI.—Citromycetin, a new yellow colouring matter produced from species of *Citromyces*. *Philos. Trans. R. Soc. London, Ser. B* **220**, 209–244.

Hetherington, A. C., and Raistrick, H. (1931b). Studies in the biochemistry of microorganisms Part XIV.—On the production and chemical constitution of a new yellow colouring matter, citrinin, produced from glucose by *Penicillium citrinum* Thom. *Philos. Trans. R. Soc. London, Ser. B* **220**, 269–295.

Hocking, A. D., and Pitt, J. I. (1980). Dichloran glycerol medium for enumeration of xerophilic fungi from low-moisture foods. *Appl. Environ. Microbiol.* **39**, 488–492.

Holzapfel, C. W. (1968). The isolation and structure of cyclopiazonic acid, a toxic product of *Penicillium cyclopium* Westling. *Tetrahedron* **24**, 2101–2119.

Howard, B. H., and Raistrick, M. (1954). Studies 91. The colouring matters of *Penicillium islandicum* Sopp. Part 3. Skyrin and flavoskyrin. *Biochem. J.* **56**, 56–65.

Hutchinson, R. D., Steyn, P. S., and van Rensburg, S. J. (1973). Viridicatum-toxin, a new mycotoxin from *Penicillium viridicatum* Westling. *Toxicol. Appl. Pharmacol.* **24**, 507–509.

Ichinoe, M., and Kurata, H. (1983). Trichothecene-producing fungi. *In* "Trichothecenes—Chemical, Biological and Toxicological Aspects" (Y. Ueno, ed.), Developments in Food Science, Vol. 4, pp. 73–82. Elsevier, New York.

Ichinoe, M., Jurata, H., and Suzuki, T. (1978). Zearalenone production by *Fusarium* species in Japan. *Proc. Jpn. Assoc. Mycotoxicol.* **5/6**, 1–2.

Jeffrey, E. G., Brian, P. W., Hemming, H. G., and Lowe, D. (1953). Antibiotic production by the micro fungi of acid heath soils. *J. Gen. Microbiol.* **9**, 314–341.

Joly, P. (1964). Le genere *Alternaria. Encycl. Mycol.* **33**, 1–250.

Karo, M., and Hadlok, R. M. (1982). Investigations on sterigmatocystin production by fungi of the genus *Eurotium. Proc. Int. IUPAC Symp. Mycotoxins Phycotoxins, 5th, Vienna* (P. Krogh, ed.), pp. 178–181. Tech. Univ., Vienna.

Karow, E. O., Woodruff, H. G., and Forster, J. W. (1944). Penicillic acid from *Aspergillus ochraceus, Penicillium thomi* and *Penicillium suavolens. Arch. Biochem.* **5**, 279–282.

Kevei, F., and Peberdy, J. F. (1984). Further studies on protoplast fusion and interspecific hybridization within the *Aspergillus nidulans* group. *J. Gen. Microbiol.* **130**, 2229–2236.

Kharchenko, S. M. (1970). Identification of antibiotics from penicillia active against agents of mottled leaves and bunt fungi. *Mikrobiol. Zh. (Kiev)* **32**, 115–119.

Kiermeier, F., Weiss, G., Behringe, G., and Miller, M. (1977). Uber das vorkommen und den Gehalt an Aflatoxin M, in Kasen des Handels. *Z. Lebensm.-Unters. Forsch.* **163**, 268–271.

King, A. D., and Schade, J. E. (1984). *Alternaria* toxins and their importance in food. *J. Food Prot.* **47**, 886–901.

King, A. D., Hocking, A. D., and Pitt, J. I. (1979). Dicloran-rose bengal medium for enumeration and isolation of molds from foods. *Appl. Environ. Microbiol.* **37**, 959–964.

King, A. D., Corry, I., Pitt, J. I., and Beuchat, L. R., eds. (1986). "Methods for the Mycological Examination of Foods." Plenum, New York.

Knutti, R., and Schlather, C. (1982). Distribution of aflatoxin in whole peanut kernels, sampling plans for small samples. *Z. Lebensm.-Unters. Forsch.* **174**, 122–128.

Kozlovsky, A. G., Solovieva, T. F., Reschetilova, T. A., and Skruabin, G. K. (1981). Biosynthesis of roquefortine and 3,12-dihydroxyroquefortine by the culture *Penicillium farinosum.* *Experientia* **37**, 472.

Krogh, P., Hasselager, E., and Friis, P. (1970). Studies of fungal nephrotoxicity. II. Isolation of two nephrotoxic compounds from *Penicillium viridicatum* Westling: citrinin and oxalic acid. *Acta Pathol. Microbiol. Scand., Sect. B* **78**, 401–413.

Kyriakidis, N., Waight, C., Day, J. B., and Mantle, P. G. (1981). Novel metabolites from *Penicillium crustosum,* including penitrem E, a tremorgenic mycotoxin. *Appl. Environ. Microbiol.* **42**, 61–62.

Lafont, P., Debeaupuis, J.-P., Gaillardin, M., and Payen, J. (1979). Production of mycophenolic acid by *Penicillium roqueforti* strains. *Appl. Environ. Microbiol.* **37**, 365–368.

Leistner, L. (1984). Toxigenic penicillia occurring in feeds and foods. *In* "Toxigenic Fungi—Their Toxins and Health Hazard" (H. Kurata and Y. Ueno, eds.), Developments in Food Science, Vol. 7, pp. 162–171. Elsevier, Amsterdam.

Leistner, L., and Eckardt, C. (1979). Vorkommen toxinogener Penicilien bei Fleisch Erzeugnisse. *Fleischwirtschaft* **59**, 1892–1896.

Leistner, L., and Pitt, J. I. (1977). Miscellaneous *Penicillium* toxins. *In* "Mycotoxins in Human and Animal Health" (J. V. Rodricks, C. W. Hesseltine, and M. A. Mehlman, eds.), pp. 639–653. Pathotox, Park Forest South, Illinois.

Lillhoj, E. B., and Elling, F. (1983). Environmental conditions that facilitate ochratoxin contamination of agricultural commodities. *Acta Agric. Scand.* **33**, 113–128.

Lillehoj, E. B., and Goransson, B. (1980). Occurrence of ochratoxin- and citrinin-producing fungi on developing danish barley grains. *Acta Pathol. Microbiol. Scand., Sect. B* **88**, 133–137.

Martinelli, S. D., and Bainbridge, M. W. (1974). Phenoloxidases of *Aspergillus nidulans. Trans. Br. Mycol. Soc.* **63**, 361–370.

Martínez, A. T., and Ramírez, C. (1978). *Penicillium fagi* sp. nov. isolated from beech leaves. *Mycopathologia* **63**, 57–59.

Marumo, S., and Sumiki, Y. (1955). Islanditoxin, a toxic metabolite produced by *Penicillium islandicum* Sopp. *J. Agric. Chem. Soc. Jpn.* **23**, 305.

Mayr, E. (1969). The biological meaning of species. *Biol. J. Linn. Soc.* **1**, 311–320.

Michel, K. H., Chaney, M. M., Jones, N. D., Hoehn, M. M., and Nagarajan, R. (1974). Epipolythiopiperazindione antibiotics from *Penicillium turbatum. J. Antibiot.* **27**, 57–64.

Milczewski, von K. E., Engle, G., and Teuber, M. (1981). Ubersicht über die wichtigen toxinbildenden Schimmelpilze und ihre toxine. *In* "Mykotoxine in Lebensmitteln" (J. Reiss, ed.), pp. 13–84. Fischer, Stuttgart.

Mintzlaff, H.-J., Ciegler, A., and Leistner, L. (1972). Potential mycotoxin problems in mould-fermented sausages. *Z. Lebensm.-Unters. Forsch.* **150**, 133–137.

Mislivec, P. B., and Bruce, V. R. (1977a). Incidence of toxic and other mould species and genera in soybeans. *J. Food Prot.* **40**, 309–312.

Mislivec, P. B., and Bruce, V. R. (1977b). Direct plating versus dilution plating in qualitatively determining the mould flora of dried beans and soybeans. *J. Assoc. Off. Anal. Chem.* **60**, 741–743.

Mislivec, P. B., Bruce, V. R., and Gibson, R. (1983). Incidence of toxigenic and other moulds in green coffee beans. *J. Food Prot.* **46**, 969–973.

Moreau, C. (1974). "Moissisures Toxique dans l'Alimentation." Masson, Paris.

Moreau, C. (1979). "Moulds, Toxins and Food." Wiley, New York.

Moreau, S., Lablanche-Cambier, A., Biguet, J., Foulon, C., and Delfosse, M. (1982). Botryo-diploidin a mycotoxin produced by a strain of *P. roqueforti*. *J. Org. Chem.* **47,** 2358–2359.

Moss, M. O. (1977). *Aspergillus* mycotoxins. *In* "Genetics and Physiology of *Aspergillus*" (J. E. Smith and J. A. Pateman, eds.), pp. 494–524. Academic Press, New York.

Moss, M. O. (1984). Conditions and factors influencing mycotoxin formation in the field and during the storage of food. *Chem. Ind. (London)* pp. 533–536.

Myrchink, T. G. (1967). Production of patulin by a group of fungi *Penicillium lapidosum* Raper & Fennell. *Antibiotiki (Moscow)* **12,** 762–766.

Nagel, D. W., Pachler, K. G. R., Steyn, P. S., Wessels, P. L., Gafner, G., and Kruger, G. J. (1974). X-ray structure of oxaline, a novel alkaloid from *Penicillium oxalicum*. *J.C.S. Chem. Commun.* pp. 1021–1022.

Nakamura, S., and Shimoda, C. (1954). Studies on an antibiotic substance oryzacidin, produced by *Aspergillus oryzae*. V. Existence of β-nitropropionic acid. *J. Agric. Chem. Soc. Jpn.* **28,** 909–913.

Natori, S., Sato, F., and Udagawa, S. (1965). Anthraquinone metabolites of *Talaromyces avellaneus* (Thom et Turreson) C. R. Benjamin and *Preussia multispora* (Saito et Minoura) Cain. *Chem. Pharm. Bull.* **13,** 385–389.

Neclakantan, S., Pocker, A., and Raistrick, H. (1957). Studies in the biochemistry of micro-organisms 101. The colouring matter of species in the *Aspergillus nidulans* group. Part 2. Further observations on the structure of asperthecin. *Biochem. J.* **66,** 234–237.

Neergaard, P. (1945). "Danish Species of *Alternaria* and *Stemphylium*." Oxford Univ. Press, London.

Nelson, P. E., Toussoun, T. A., and Marasas, W. F. O. (1983). "*Fusarium* Species. An Illustrated Manual for Identification." Pennsylvania State Univ. Press, University Park.

Northolt, M. D., and Bullerman, L. B. (1982). Prevention of mold growth and toxin production through control of environmental conditions. *J. Food Prot.* **45,** 519–526.

Ohmomo, S., Ohashi, T., and Abe, M. (1980). Isolation and biogenetically correlated four alkaloids from the cultures of *Penicillium corymbiferum*. *Agric. Biol. Chem.* **44,** 1929–1930.

Olivigni, F. J., and Bullerman, L. B. (1977). Production of penicillic acid and patulin by an atypical *Penicillium roqueforti* isolate. *Appl. Microbiol.* **35,** 435–438.

Onions, A. H. S., Allsopp, B., and Eggins, H. O. W. (1981). "Smith's Introduction to Industrial Mycology." Arnold, London.

Onions, A. H. S., Bridge, P. D., and Patterson, R. R. (1984). Problems and prospects for the taxonomy of *Penicillium*. *Microbiol. Sci.* **1,** 185–189.

Orth, R. (1977). Mycotoxins of *Aspergillus oryzae* strains for use in the food industry as starters and enzyme producing mould. *Ann. Nutr. Aliment.* **31,** 617–624.

Orth, R. (1981a). Einfluss physikalischer Faktoren auf die Bildung von Mykotoxinen. *In* "Mykotoxine in Lebensmitteln" (J. Reiss, ed.), pp. 85–100. Fischer, Stuttgart.

Orth, R. (1981b). Mykotoxine von Pilzen der Kasenerstellung. *In* "Mykotoxine in Lebensmitteln" (J. Reiss, ed.), pp. 273–296. Fischer, Stuttgart.

Oxford, A. E., Raistrick, H., and Simonart, P. (1939). Studies in the biochemistry of micro-organisms 60. Griseofulvin. $C_{17}H_{17}O_8Cl$ a metabolic product of *Penicillium griseofulvum* Dierckx. *Biochem. J.* **33,** 240–248.

Palti, J. (1978). Toxigenic Fusaria, their distribution and significance as causes of disease in animal and man. *Acta Phytomed.* **6,** 1–110.

Parrish, F. W., Wiley, B. J., Simmons, E. G., and Lang, L. (1966). Production of aflatoxins and kojic acid by species of *Aspergillus* and *Penicillium*. *Appl. Microbiol.* **14,** 139.

Patterson, D. S. P., Roberts, B. A., Shreeve, B. J., MacDonald, S. M., and Hayes, A. W. (1979). Tremorgenic toxins produced by soil fungi. *Appl. Environ. Microbiol.* **37**, 172–173.

Pitt, J. I. (1979a). "The Genus *Penicillium* and Its Teleomorphic States *Eupenicillium* and *Talaromyces.*" Academic Press, New York.

Pitt, J. I. (1979b). *Penicillium crustosum* and *Penicillium simplicissimum*, the correct names for two common species producing tremorgenic mycotoxins. *Mycologia* **71**, 1166–1177.

Pitt, J. I., Hocking, A. D., and Glenn, D. R. (1983). An improved medium for the detection of *Aspergillus flavus* and *A. parasiticus. J. Appl. Bacteriol.* **54**, 109–114.

Pollock, A. V. (1947). Production of citrinin by five species of *Penicillium. Nature (London)* **160**, 331–332.

Pore, R. S., and Larsh, H. W. (1967). Aleuriospore formation in four related *Aspergillus* species. *Mycologia* **59**, 318–325.

Quintanilla, J. A. (1982a). Three new species of *Penicillium* isolated from soil. *Mycopathologia* **80**, 73–82.

Quintanilla, J. A. (1982b). Cuatro neuvos especies de *Penicillium* aislados en centeno: *P. mariaecrucis*, sp. nov. *P. catellae*, sp. nov., *P. cieglerii*, sp. nov. y *P. smithii*, sp. nov. *Av. Nutr. Mejora Anim. Aliment.* **23**, 333–343.

Quintanilla, J. A. (1983). *Penicillium caerulescens* sp. nov. isolated from soil. *Mycopathologia* **82**, 101–104.

Raistrick, H., and Stossl, A. (1958). Studies in the biochemistry of microorganisms 104. Metabolites of *Penicillum atrovenetum* G. Smith: β-nitropropionic acid, a major metabolite. *Biochem. J.* **68**, 647–653.

Ramírez, C. (1982). "Manual and Atlas of the Penicillia." Elsevier, Amsterdam.

Raper, K. B., and Fennell, D. I. (1965). "The Genus *Aspergillus.*" Williams & Wilkins, Baltimore, Maryland.

Raper, K. B., and Thom, C. (1949). "A Manual of the Penicillia." Williams & Wilkins, Baltimore, Maryland.

Ray, L. L., and Bullerman, L. B. (1982). Preventing growth of potentially toxic molds using antifungal agents. *J. Food Prot.* **45**, 953–963.

Rebuffat, S., Davoust, D., Molho, L., and Molho, D. (1980). Le citremontanine, nouvewlle α-pyranpolyethylenique isolee de *Penicillium pedemontanum. Phytochemistry* **19**, 427–431.

Rehm, H.-J. (1980). "Industrielle Mikrobiologie," 2nd Ed. Springer-Verlag, Berlin and New York.

Reiss, J. (1981a). "Mykotoxine in Lebensmitteln." Fischer, Stuttgart.

Reiss, J. (1981b). Einfluss chemicher Faktoren auf die Bildung von Mykotoxinen. *In* "Mykotoxine in Lebensmitteln" (J. Reiss, ed.), pp. 101–117. Fischer, Stuttgart.

Richard, J. L., and Arp, L. H. (1979). Natural occurrence of the mycotoxin penitrem A in moldy cream cheese. *Mycopathologia* **67**, 107–109.

Rippon, L. E. (1980). Wastage of post harvest fruit and its control. *CSIRO Food Res. Q.* pp. 1–12.

Sakabe, N., Goto, T., and Hirata, Y. (1964). The structure of citreoviridin, a toxic compound produced by *Penicillium citreoviride* molded on rice. *Tetrahedron Lett.* **27**, 1825–1830.

Samson, R. A. (1979). A compilation of the aspergilli described since 1965. *Stud. Mycol. (Baarn)* **18**, 1–40.

Samson, R. A. (1985). Taxonomic considerations of harmful and beneficial moulds in food. *In* "Filamentous Microorganisms: Biomedical Aspects" (T. Kuga, K. Terao, M. Yamazaki, M. Miyaji, and T. Unemoto, eds.), pp. 157–163. Jpn. Sci. Soc. Press, Tokyo.

Samson, R. A., and Gams, W. (1984). The taxonomic situation in the hyphomycete genera *Penicillium, Aspergillus* and *Fusarium. Antonie van Leeuwenhoek* **50**, 815–824.

Samson, R. A., and Pitt, J. I. (1985). "Advances in *Aspergillus* and *Penicillium* Systematics." Plenum, New York.

Samson, R. A., Stolk, A. C., and Hadlok, R. (1976). Revision of the subsection fasciculata of *Penicillium* and some allied species. *Stud. Mycol. (Baarn)* **11**, 1–247.

Samson, R. A., Eckardt, C., and Orth, R. (1977a). The taxonomy of *Penicillium* species from fermented cheeses. *Antonie von Leeuwenhoek* **43**, 341–350.

Samson, R. A., Hadlok, R., and Stolk, A. C. (1977b). A taxonomic study of the *Penicillium chrysogenum* series. *Antonie von Leeuwenhoek* **43**, 261–274.

Samson, R. A., Hoekstra, E. S., and van Oorschot, C. A. N. (1984). "Introduction to Food-Borne Fungi." Centraalbur. Schimmelcult., Baarn.

Samson, R. A., Frisvad, J. C., and Filtenborg, O. (1986). Compilation of foodborne filamentous fungi. *In* "Methods for the Mycological Examination of Foods" (A. D. King, I. Corry, J. I. Pitt, and L. R. Beuchat, eds.), Plenum, New York. In press.

Samuels, G. J. (1984). Toxigenic fungi as ascomycetes. *In* "Toxigenic Fungi—Their Toxins and Health Hazard" (H. Kurata and Y. Ueno, eds.), Developments in Food Science, Vol. 7, pp. 119–128. Elsevier, Amsterdam.

Schroeder, H. W., and Kelton, W. H. (1975). Production of sterigmatocystin by some species of the genus *Aspergillus* and its toxicity to chicken embryos. *Appl. Environ. Microbiol.* **30**, 589–591.

Scott, P. M., Merrien, M.-A., and Polonsky, J. (1976). Roquefortine and isofumigaclavine A, metabolites from *Penicillium roqueforti*. *Experientia* **32**, 140–142.

Semeniuk, G., Harshfield, G. S., Carlson, C. W., Hesseltine, C. W., and Kwolek, W. F. (1971). Mycotoxins in *Aspergillus*. *Mycopathol. Mycol. Appl.* **43**, 137–142.

Shibata, S., and Udagawa, S. (1963). Metabolic products of fungi XIX. Isolation of rugulosin from *Penicillium bruneum* Udagawa. *Chem. Pharm. Bull.* **11**, 402–403.

Shreeve, B. J., Patterson, D. S. P., Roberts, B. A., MacDonald, S. M., and Wood, E. N. (1978). Isolation of potentially tremorgenic fungi from pasture associated with a condition resembling ryegrass staggers. *Vet. Rec.* **105**, 209–210.

Simmons, E. G. (1981). *Alternaria* themes and variations. *Mycotaxon* **13**, 16–34.

Simmons, E. G. (1982). *Alternaria* themes and variations. *Mycotaxon* **14**, 17–57.

Singleton, V. L., Bohonos, N., and Ullstrup, A. J. (1958). Decumbin, a new compound from a species of *Penicillium*. *Nature (London)* **181**, 1072–1073.

Smith, G. (1961). Some new and interesting species of microfungi II. *Trans. Br. Mycol. Soc.* **44**, 42–50.

Sneath, P. H. A., and Sokal, R. (1973). "Numerical Taxonomy." Freeman, San Francisco, California.

Soderstrom, B., and Frisvad, J. C. (1984). Separation of closely related asymmetric penicillia by pyrolysis gas chromatography and mycotoxin production. *Mycologia* **76**, 408–419.

Springer, J. P., Clardy, J., Wells, J. M., Cole, R. J., Kirksey, J. W., MacFarlane, R. D., and Torgerson, D. F. (1976). Isolation and structure determination of the mycotoxin chaetoglobosin C, a new [13] cytochalasin. *Tetrahedron Lett.* **17**, 1355–1358.

Stack, M. E., and Mislivec, P. B. (1978). Production xanthomegnin and viomellein by isolates of *Aspergillus ochraceus*, *Penicillium cyclopium* and *Penicillium viridicatum*. *Appl. Environ. Microbiol.* **36**, 552–554.

Stack, M. E., Eppley, R. M., Dreifuss, P. A., and Pohland, A. E. (1977). Isolation and identification of xanthomegnin, viomellein, rubrosulphin, and viopurpurin as metabolites of *Penicillium viridicatum*. *Appl. Environ. Microbiol.* **33**, 351–356.

Steyn, P. S. (1970). The isolation, structure and absolute configuration of secalonic acid D, the toxic metabolite of *Penicillium oxalicum*. *Tetrahedron* **26**, 51–57.

Still, P., Eckardt, R., and Leistner, L. (1978). Bildung von cyclopiazosaure durch *Penicillium cammemberti*—Isolate von Kase. *Fleischwirtschaft* **58**, 876–877.

Stolk, A. C., and Samson, R. A. (1983). The ascomycete genus *Eupenicillium* and related *Penicillium* anamorphs. *Stud. Mycol. (Baarn)* **23**, 1–149.

456   Jens C. Frisvad

Szathmany, C. I., Mirocha, C. J., Palyusik, M., and Pathre, S. V. (1976). Identification of my-cotoxins produced by species of *Fusarium* and *Stachybotrys* obtained from Eastern Europe. *Appl. Environ. Microbiol.* **32**, 579–584.

Takeda, N., Seo, S., Ogihara, Y., Sankawa, U., Iitaka, I., Kitagawa, I., and Shibata, S. (1973). Studies of fungal metabolites XXXI. Anthraquinonoid colouring matters of *Penicillium is-landicum* Sopp and some other fungi (−)luteoskyrin, (−)rubroskyrin, (+)rugulosin and their related compounds. *Tetrahedron* **29**, 3703–3713.

Tatsuno, T., Kobayashi, N., Okubo, K., and Tsunoda, H. (1975). Recherches toxicologique sur les substances toxique de *Penicillium tardum* I. Isolement et identification des substance cytotoxi-que. *Chem. Pharm. Bull.* **23**, 351–354.

Thom, C. (1954). The evolution of species concepts in *Aspergillus* and *Penicillium. Ann. N.Y. Acad. Sci.* **60**, 24–34.

Tsunoda, H. (1963). Studies in the microorganisms which deteriorate the stored cereals and grains XXIX. Studies on *Penicillium brunneum* nov. sp., parasitic to rice in storage. *Proc. Food Res. Inst.* **17**, 238–242.

Tsunoda, H., Kishi, K., Okubo, K., Tatsuno, T., and Ohtsubo, K. (1977). Morphology and culture of *P. ochraceum* and *P. carneolutescens* in *Penicillium Proc. Jap. Assoc. Mycotoxicol.* **5/6**, 11–13.

Turner, W. B. (1971). "Fungal Metabolites." Academic Press, New York.

Turner, W. B., and Aldridge, D. C. (1983). "Fungal Metabolites II." Academic Press, New York.

Udagawa, S. (1984). Taxonomy of mycotoxin-producing *Chaetomium. In* "Toxigenic Fungi—Their Toxins and Health Hazard" (H. Kurata and Y. Ueno, eds.), Developments in Food Science, Vol. 7, pp. 139–147. Elsevier, Amsterdam.

Ueno, Y. (1980). Toxicological evaluation of trichothecene mycotoxins. *In* "Natural Toxins" (D. Eaker and T. Wadstrom, eds.), pp. 663–671. Pergamon, New York.

Ueno, Y., and Ueno, I. (1972). Isolation and acute toxicity of citreoviridin, a neurotoxic mycotoxin of *Penicillium citreoviride* Biourge. *Jpn. J. Exp. Med.* **42**, 91–195.

Uraguchi, K., Tatsuno, T., Sakai, F., Tsukioka, M., Sakai, Y., Yonemitsu, O., Ito, H., Miyake, M., Saito, M., Enemoto, M., Shikata, T., and Ishiko, T. (1961). Isolation of two toxic agents, luteoskyrin and chlorine-containing peptide, from the metabolites of *Penicillium islandicum* Sopp, with some properties thereof. *Jpn. J. Exp. Med.* **31**, 19–46.

van Eijk, G. W. (1969). Isolation and identification of orsenilic acid and penicillic acid produced by *Penicillium fennelliae* Stolk. *Antonie van Leeuwenhoek* **35**, 496–504.

van Luijk, A. (1983). Antagonism between various micro-organisms and different species of the genus *Pythium,* parasitizing upon grasses and lucerne. *Med. Phytopathol. Lab. Scholten, Baarn* **14**, 43–89.

van Walbeek, W., Scott, P. M., Harwig, J., and Lawrence, J. W. (1969). *Penicillium viridicatum* Westling: a new source of ochratoxin A. *Can. J. Microbiol.* **15**, 1281–1285.

Veau, P., Samson, R. A., and Breton, A. (1981). Etude comparee de *Penicillium roqueforti* et *P. verrucosum* var. *cydopium. Lait* **61**, 370–380.

Vesonder, R. F. (1979). Xanthocillin, a metabolite of *Eupenicillium egyptiacum* NRRL 1022. *J. Nat. Prod.* **42**, 232–233.

Vesonder, R. F., Tjarks, L., Rohwedder, W., and Kieswetter, D. O. (1980). Indole metabolites from *Penicillium cyclopium* NRRL 6093. *Experientia* **36**, 1308.

Vleggaar, R., and Wessels, P. (1980). Stereochemistry of the dehydrogenation of (2S)-histidine in the biosynthesis of roquefortine and oxaline. *J.C.S. Chem. Commun.* pp. 160–162.

von Arx, J. A. (1981). "The Genera of Fungi Sporulating in Pure Culture," 3rd Ed. Cramer, Vaduz, Lichtenstein.

Wagener, R. E., Davis, N. D., and Diener, U. L. (1980). Penitrem A and roquefortine production by *Penicillium commune. Appl. Environ. Microbiol.* **39**, 882–887.

Waibel, J. (1981). Probleme de Probenahme. *In* "Mykotoxine in Lebensmitteln" (J. Reiss, ed.), pp. 511–527. Fischer, Stuttgart.

Wang, H. L., and Hesseltine, C. W. (1982). Oriental fermented foods. *In* "Prescott and Dunn's Industrial Microbiology," 4th Ed., pp. 492–538. AVI, Westport, Connecticut.

Wei, R.-D., Schnoes, H. K., Still, P. E., Smalley, E. B., and Strong, F. M. (1973). Isolation and partial characterisation of a mycotoxin from *Penicillium roqueforti. Appl. Microbiol.* **25,** 111–114.

Wells, J. M., and Cole, R. J. (1977). Production of penitrem A and of an unidentified toxin by *Penicillium lanoso-coeruleum* isolated from weevil-damaged pecans. *Phytopathology* **67,** 779–782.

Wells, J. M., and Payne, J. A. (1976). Toxigenic species of *Penicillium, Fusarium,* and *Aspergillus* from weevil-damaged pecans. *Can. J. Microbiol.* **22,** 281–285.

Wheeler, M. H. (1983). Comparison of fungal melanin biosynthesis in ascomycetous, imperfect and basidiomycetous fungi. *Trans. Br. Mycol. Soc.* **81,** 29–36.

Wicklow, D. T. (1984). Adaptation in wild and domesticated yellow-green aspergilli. *In* "Toxigenic Fungi—Their Toxins and Health Hazard" (H. Kurata and Y. Ueno, eds.), Developments in Food Science, Vol. 7, pp. 78–86. Elsevier, Amsterdam.

Wicklow, D. T., and Shotwell, O. L. (1983). Intrafungal distribution of aflatoxin among conidia and sclerotia of *Aspergillus flavus* and *Aspergillus parasiticus. Can. J. Microbiol.* **29,** 1–5.

Wicklow, D. T., Horn, B. W., and Cole, R. J. (1982). Cleistothecia of *Eupenicillium* formed naturally within corn kernels. *Can. J. Bot.* **60,** 1050–1053.

Williams, C. C., Cameron, E. J., and Williams, O. B. (1941). A facultatively anaerobic mould of unusual heat resistance. *Food Res.* **6,** 69–73.

Wilson, B. J. (1971). Miscellaneous *Aspergillus* toxins. *In* "Microbial Toxins, Fungal Toxins" (A. Ciegler, S. Kadis, and S. J. Ajl, eds.), Vol. 6, pp. 207–295. Academic Press, New York.

Wilson, B. J., and Wilson, C. H. (1962a). Hepatotoxic substance from *Penicillium rubrum. J. Bacteriol.* **83,** 693.

Wilson, B. J., and Wilson, C. H. (1962b). Extraction and preliminary characterisation of a hepatotoxic substance from cultures of *Penicillium rubrum. J. Bacteriol.* **84,** 283–290.

Wilson, B. J., Wilson, C. H., and Hayes, A. W. (1968). Tremorgenic toxin from *Penicillium cyclopium* grown on food materials. *Nature (London)* **20,** 77–78.

Wilson, B. J., Yang, D. T. C., and Harris, T. M. (1973). Production, isolation, and preliminary toxicity studies of brevianamide A from cultures of *Penicillium viridicatum. Appl. Environ. Microbiol.* **26,** 633–635.

Wirth, J., and Klosek, R. (1972). Fungi. Moniliales. Relationships in *Penicillium aurantiovirens. Phytochemistry* **11,** 2615.

Wyllie, T. D., and Morehouse, L. G. (1977). "Mycotoxic Fungi, Mycotoxins, Mycotoxicosis." Vol. 1. Dekker, New York.

Yamamoto, Y., Hamaguchi, A., Yamamoto, I., and Imai, S. (1956). Studies on the metabolic products of *Penicillium tardum* Thom. *Yakugaku Zasshi* **76,** 1428–1432.

Zeeck, A., Russ, P., Laatch, H., Loeffler, W., Wehrle, H., Zahner, H., and Holst, H. (1979). Stofwechselprodukte von Mikroorganismen 172. Isolierung des Antibioticums Semiviooxanthin aus *Penicillium citreo-viride* und synthese des Xanthomegnins. *Chem. Ber.* **112,** 957–978.

Index